Encyclopedia of Peptides

Encyclopedia of Peptides

Edited by **Matthew Langer**

R CALLISTO REFERENCE

New York

Published by Callisto Reference,
106 Park Avenue, Suite 200,
New York, NY 10016, USA
www.callistoreference.com

Encyclopedia of Peptides
Edited by Matthew Langer

International Standard Book Number: 978-1-63239-276-3 (Hardback)

Printed in the United States of America.

Contents

Preface

On the basis of the method of production, peptides can be classified into various categories like milk peptides, produced from the milk protein called casein; ribosomal peptides synthesized by translation of mRNA and also often subjected to posttranslational modifications; non-ribosomal proteins which are assembled by enzymes that are specific to each peptide. The most common form of non-ribosomal protein is glutathione; peptones, derived from animal milk or meat digested by proteolysis and lastly, peptide fragments which are used to quantify the source protein.

It is on the basis of size that proteins can be distinguished from peptides. Also, experiments reveal that they contain approximately 50 or fewer amino acids. However, the size limits which distinguish peptides from polypeptides and proteins are not absolute. Amino acids have also been incorporated into peptides called residues. During the formation of the amide bond, either a hydrogen ion from the amine end or a hydroxyl ion from the carboxyl end is released, resulting in the formation of a water molecule.

Peptides play a crucial role in molecular biology for multiple reasons like allowing the creation of peptide antibodies without the need to purify the relevant protein. Peptides also find applications in the study of protein structure and function. Inhibitory peptides are also used in clinical research to study the effects of peptides on the inhibition of cancer proteins and other diseases.

This book takes a close look at all those aspects of peptides and more. I'd like to extend my gratitude to the team at the publishing house, for their constant support at every stage of the publication process. I would like to thank my friends and family who have supported me at every step.

Editor

Diverse Effects of Glutathione and UPF Peptides on Antioxidant Defense System in Human Erythroleukemia Cells K562

Ceslava Kairane, Riina Mahlapuu, Kersti Ehrlich, Kalle Kilk, Mihkel Zilmer, and Ursel Soomets

The Centre of Excellence of Translational Medicine, Department of Biochemistry, Faculty of Medicine, University of Tartu, Ravila Street 19, 50411 Tartu, Estonia

Correspondence should be addressed to Kersti Ehrlich, kersti.ehrlich@ut.ee

Academic Editor: Katsuhiro Konno

The main goal of the present paper was to examine the influence of the replacement of γ-Glu moiety to α-Glu in glutathione and in its antioxidative tetrapeptidic analogue UPF1 (Tyr(Me)-γ-Glu-Cys-Gly), resulting in α-GSH and UPF17 (Tyr(Me)-Glu-Cys-Gly), on the antioxidative defense system in K562 cells. UPF1 and GSH increased while UPF17 and α-GSH decreased the activity of CuZnSOD in K562 cells, at peptide concentration of $10\,\mu$M by 42% and 38% or 35% and 24%, respectively. After three-hour incubation, UPF1 increased and UPF17 decreased the intracellular level of total GSH. Additionally, it was shown that UPF1 is not degraded by γ-glutamyltranspeptidase, which performs glutathione breakdown. These results indicate that effective antioxidative character of peptides does not depend only on the reactivity of the thiol group, but also of the other functional groups, and on the spatial structure of peptides.

1. Introduction

Glutathione (GSH) system is an attractive target for drug discovery because of its importance and versatility [1]. GSH (γ-L-Glu-L-Cys-Gly) is a prevalent low molecular weight thiol in eukaryotic cells and has antioxidative, detoxificative, and regulatory roles [2, 3]. Decrease of GSH level and shifted GSH redox status are related to several pathological states, including neurodegenerative, cardiovascular, pulmonary, and immune system diseases [4]. Exogenous administration of GSH to compensate the decrease of GSH levels is not reasonable because of its degradation in the plasma and poor cellular uptake [5–7]. GSH and its oxidized disulfide form (GSSG) are degraded by γ-glutamyltranspeptidase (GGT) via cleavage of the amino acid γ-glutamate from the N-terminal end of the peptide. GGT is located in the outer side of the cell membrane, and one of its functions, in cooperation with dipeptidases, is to provide cells with precursor amino acids needed for GSH *de novo* synthesis. To overcome the problems with GSH administration, several GSH analogues have been created to increase the GSH

level and support the functionality of the GSH system [8]. We have previously designed and synthesized a library of peptidic GSH analogues [9]. For this study, two of them, UPF1 (Tyr(Me)-γ-Glu-Cys-Gly) and UPF17 (Tyr(Me)-Glu-Cys-Gly), were selected. Both molecules have an O-methyl-L-tyrosine residue added to the N-terminus of GSH-like Glu-Cys-Gly sequence to increase the antioxidativity and hydrophobicity. Previously, different groups have shown that various low molecular weight antioxidants, including melatonin, carvedilol, and its metabolite SB 211475, carry a methoxy moiety in their aromatic structures [10, 11]. The only structural difference between the peptides used is that UPF17 contains α-glutamyl moiety while UPF1 has γ-glutamyl moiety similarly to GSH. This switch from γ- to α-glutamyl moiety improved hydroxyl radical scavenging ability of UPF17 by approximately 500-fold compared to UPF1 whereas UPF1 itself is about 60-fold better hydroxyl radical scavenger than GSH [9]. In addition to being an excellent *in vitro* free radical scavenger, UPF1 has shown protective properties against oxidative damage in a global brain ischemia/reperfusion model and in an ischemia/reperfusion model on an isolated heart of Wistar rats [12, 13]. UPF1 and UPF17 have been shown to be nontoxic for K562 cells up to concentration of 200 μM and UPF1 has no toxic effect on the primary culture of cerebellar granule cells at concentrations up to 100 μM [9, 13].

Superoxide dismutases (SOD, EC 1.15.1.1.) are metalloproteins and the primary enzymes that keep cellular free radical production under control [14]. Cytosolic CuZnSOD is a homodimer (151 amino acids) with a molecular weight of 32500 Da and contains two cysteines (Cys57, Cys148) bound into an intramolecular disulfide bond and two free cysteines (Cys6, Cys111) [15, 16]. SOD catalyses the dismutation of superoxide into oxygen and hydrogen peroxide. Hydrogen peroxide as a diffusible cell damaging agent is further eliminated by glutathione peroxidase or catalase. One of the essential requirements for the biological activity of the glutathione peroxidase is glutathione as a cosubstrate. Consequently, SOD works synergistically with the glutathione against free radical damage.

This study examined the influence of UPF1 and UPF17 on CuZnSOD activity and intracellular GSH level in K562 cells. The aim of studying these tetrapeptides was to get information about whether and how the replacement of γ-peptide bond with α-peptide bond in the structure affects the bioactivities of the peptides. Additionally, we measured the stability of the peptides towards GGT to clarify their status in biologicalsystems and the pK$_a$ values for thiol group dissociation.

2. Materials and Methods

2.1. *Peptide Synthesis.* UPF peptides were synthesized manually by solid phase peptide synthesis using Fmoc-chemistry and by machine using *tert*-Boc-chemistry as described previously [9, 17]. The purity of the peptides was >99% as demonstrated by HPLC on an analytical Nucleosil 120-3 C18 reversed-phase column (0.4 cm × 10 cm) and the peptides were identified by MALDI-TOF (matrix-assisted laser desorption ionization time-of-flight) mass-spectrometry (Voyager DE Pro, Applied Biosystems).

2.2. *CuZnSOD Activity in K562 Cells.* The K562 cells (human erythroleukemia cells, obtained from DSMZ, Germany) were grown in T75 cell culture flasks in RPMI 1640 supplemented with 2 mM glutamine (PAA, Austria), 7.5% fetal calf serum, streptomycin (100 μg/mL), and penicillin (100 U/mL) (all from Invitrogen, USA) at 37°C in a humidified 5% carbon dioxide atmosphere. Cells were seeded at concentration of 1.0 × 10^6 per mL. Experiments were conducted 24 h after passage. Peptides (GSH, α-GSH, UPF1 and UPF17) diluted in DPBS (PAA, Austria) were added to the flasks containing the K562 cells. The cells were incubated with DPBS as control (Co) or with the peptide solution in a concentration range from 0.5 to 10 μM for 24 h at 37°C. The peptide concentrations were chosen based on the GSH concentration in the blood plasma. After treatment, the cells were washed twice with DPBS and then lysed in water by keeping on ice for 20 min. Samples were centrifuged (12000 g)

for 10 min and supernatants were transferred for experiments. The protein concentrations in the supernatants were determined by Lowry's method [18]. CuZnSOD activity was measured with the commercially available kit (Randox Laboratories Ltd, UK). This method employs xanthine and xanthine oxidase to generate superoxide radicals, which react with 2-(4-iodophenyl)-3-(4-nitrophenol)-5-phe nyltetrazolium chloride to form a red formazan dye. The superoxide dismutase activity is then measured by the degree of inhibition of this reaction. One unit of SOD inhibited 50% of the rate of reaction.

2.3. *Measurement of Total Glutathione.* Concentrations of total glutathione (tGSH) were assessed by an enzymatic method of Tietze [19]. The homogenate was deproteinated by 10% solution of metaphosphoric acid (Sigma-Aldrich, Germany) in water and centrifuged at 12000 g for 10 min. The enzymatic reaction was initiated by the addition of NADPH, glutathione reductase, and 5,5'-dithio-*bis*-2-nitrobenzoic acid in buffer containing EDTA (Sigma-Aldrich, Germany). The change in optical density was measured after 15 min at 412 nm spectrophotometrically (Sunrise Tecan). Glutathione content was calculated on the basis of a standard curve.

2.4. *Stability towards γ-Glutamyltranspeptidase.* 1 mM UPF1 was incubated with 0.3 mg/mL equine kidney γ-glutamyltranspeptidase in 0.1 M Tris-HCl buffer pH 7.4, supplemented with 0.1% EDTA (Sigma-Aldrich, Germany) at 37°C for 1 h. 6 mM Gly-Gly was added as an acceptor for γ-Glu moiety [20]. GSH was incubated with GGT under the same conditions as the control. The samples were heat-inactivated, centrifuged at 10000 g and +4°C for 5 min, and kept on ice until analyzed. Supernatants were analyzed on a Prominence HPLC (Shimadzu, Japan) and Q-Trap 3200 (Applied Biosystems, USA) mass spectrometry tandem. Luna C18 100 × 2 mm, 3 μm column from Phenomenex was used for sample separation. Solvent A was a mixture of 99.9% water and 0.1% HCOOH, and solvent B was a mixture of 99.9% acetonitrile and 0.1% HCOOH (mass spectrometry grade, Riedel-de Haën, Germany). Samples were eluted at a flow rate of 0.1 mL/min, gradient started with 5 min at isocratic flow of solvent A, concentration of solvent B increased up to 30% in 25 min, followed by wash with 100% solvent B in 20 min. Enhanced MS scans were performed in negative mode with rate 1000 amu/s between mass range 50–1700 Da. Ionspray voltage was set to −4500 V, declustering potential to −30 V and entrance potential to −10 V.

2.5. *pK$_a$ of Thiol Groups.* The ratio of thiol and thiolate concentrations were measured spectrophotometrically at 240 nm on a PerkinElmer Lambda 25 spectrometer similarly as previously for GSH and α-GSH [21]. 1 mL of 50 μM peptide solution in phosphate buffered saline (Calbiochem, USA) was titrated with 5 μL volumes of 1 M NaOH and pH and absorbance changes were determined after each addition. The results were corrected to consider the dilution of the assay mixture during titration.

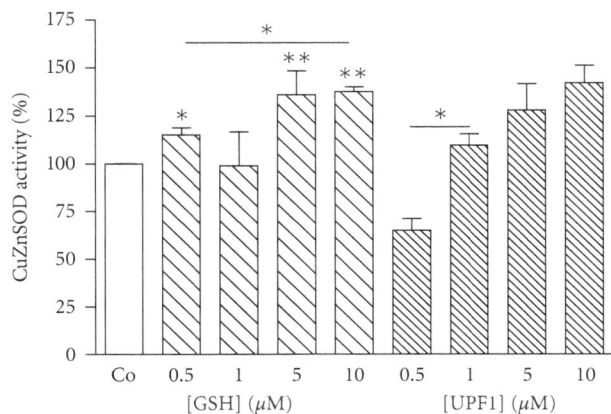

FIGURE 1: Modulation of CuZnSOD activity by GSH and UPF1 in K562 cells. The CuZnSOD activity of Co is 100%. $*P < 0.05$; $**P < 0.01$, GSH and UPF1 versus Co; $n = 4$–8.

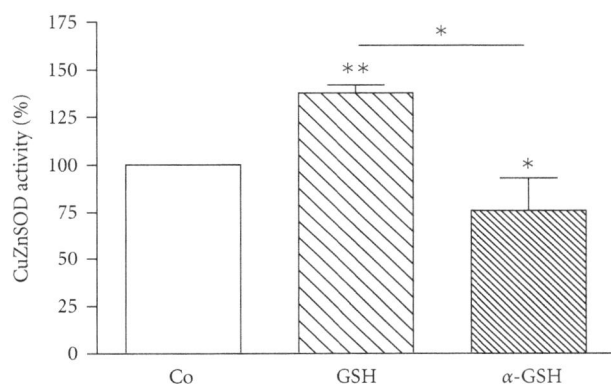

FIGURE 2: Modulation of CuZnSOD activity by GSH and α-GSH (10μM) in K562 cells. The CuZnSOD activity of Co is 100%. $*P < 0.05$; $**P < 0.01$, 10μM GSH or α-GSH versus Co; $n = 4$–8.

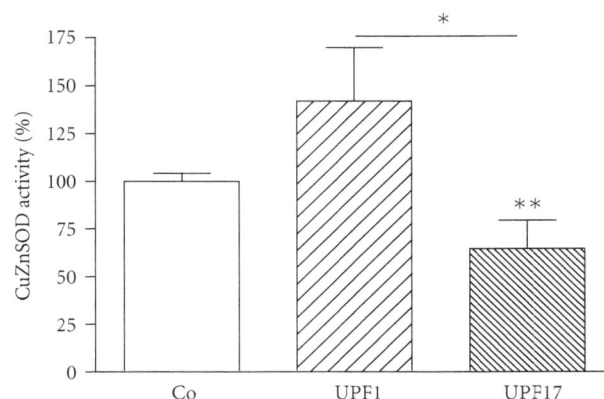

FIGURE 3: Modulation of CuZnSOD activity by UPF1 and UPF17 (10μM) in K562 cells. The CuZnSOD activity of Co is 100%. $*P < 0.05$; $**P < 0.01$, 10μM UPF1 or UPF17 versus Co; $n = 4$–8.

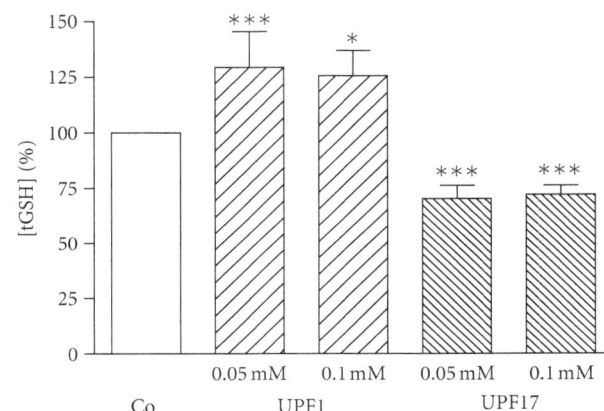

FIGURE 4: Alteration of tGSH concentration by UPF1 and UPF17 in K562 cells. The tGSH concentration of Co is 100%. $*P < 0.05$, $***P < 0.005$, UPF1 or UPF17 versus Co; $n = 6$–8.

2.6. Statistical Analysis. Data were analyzed using GraphPad Prism version 4.00 for Windows (GraphPad Software, San Diego, CA, USA). The results on the graphs are presented as the mean ± standard error of the mean (SEM).

3. Results

3.1. CuZnSOD Activity. K562 cells were incubated with investigated peptides (GSH, α-GSH, UPF1, and UPF17) for 24 h at four different concentrations: 0.5, 1.0, 5.0, and 10μM. GSH showed a concentration-dependent activating effect on CuZnSOD activity, whereas 10μM GSH increased the enzyme activity by 38% (Figure 1). α-GSH had an inhibiting effect (24%) on the enzyme activity but only at the highest concentration used (10μM) (Figure 2). UPF1 increased the activity of CuZnSOD at concentrations of 1.0, 5.0, and 10μM, but at concentration of 0.5μM showed an inhibition of the enzyme activity (Figure 1). The activation rate was concentration dependent. Contrary to UPF1, UPF17 showed an inhibitory effect and the inhibition was not concentration dependent. UPF1 increased and UPF17 decreased the activity

of CuZnSOD at peptide concentration of 10μM by 42% and 35%, respectively (Figure 3). As the peptide concentration of 10μM was the most effective, it selected for the comparison.

3.2. Intracellular GSH Level. K562 cells were incubated with UPF1 and UPF17 peptides for 3 h at concentrations of 0.05, 0.10, and 0.5 mM. Previous experiments have shown that at these concentrations UPF peptides are effective free radical scavengers and are biologically active. In addition, the 0.5 mM concentration was chosen to match with millimolar GSH concentration in number of cells. UPF1 increased and UPF17 decreased GSH concentration at concentrations of 0.05 and 0.1 mM by 29% and 26% or 26% and 28%, respectively (Figure 4). No statistical difference in tGSH concentration compared to control after incubation with 0.5 mM peptides, the highest concentration used, was detected.

3.3. Degradation by GGT. After incubating GSH with GGT, GSH was degraded and γ-Glu moiety was transferred to an acceptor Gly-Gly dipeptide, resulting in a new compound in mass spectra with MW 261.2 Da [γ-Glu-Gly-Gly]. The question arose: can the bond between γ-glutamate and cysteine be degraded by GGT in UPF1, where the access to the bond is obstructed by an additional amino acid methylated tyrosine? Results obtained from the mass spectrometry measurements demonstrated that UPF1 is not degraded by GGT as the expected peaks with or without acceptor dipeptide MW 438.4 Da [Tyr(Me)-γ-Glu-Gly-Gly] or 324.3 Da [Tyr(Me)-γ-Glu], respectively, did not appear. During the incubation, UPF1 was dimerised over disulphide bridge. GGT is also able to breakdown dimeric form of GSH, but degradation of dimerised UPF1 was not detected.

3.4. pK_a. The pK_a values of thiol groups of the peptides were measured. For GSH and α-GSH, the values were 9.0 ± 0.3 and 9.1 ± 0.1, respectively, whereas pK_a values for UPF peptides were slightly higher: 9.3 ± 0.1 for UPF1 and 9.4 ± 0.2 for UPF17.

4. Discussion

The present study focused on the effects of UPF1 and UPF17 on CuZnSOD activity and intracellular GSH level in K562 cells. For the first time we described and compared counterpoint biological activities of structural antioxidative peptide analogs differing from each other by spacial arrangement of Glu residue (γ-peptide bond in UPF1 changed to the α-peptide bond in UPF17). Previously we have shown that UPF1 and UPF17 have a tendency for MnSOD activation. However, the γ-glutamyl moiety containing UPF1 needed more time for MnSOD activation compared to UPF17, which had the effect already after 5 min incubation. UPF1 and UPF17 have also different influence on glutathione peroxidase activity (GPx): at higher concentrations than used in *in vivo* experiments, both UPF1 and UPF17 inhibited activity concentration dependently whereas the α-peptide bond containing UPF17 had stronger inhibitory effect [22]. In the present work we investigated how the replacement of γ-peptide bond with α-peptide bond on GSH and its analogue UPF1 affects CuZnSOD activity and level of GSH in K562 cells. The results showed that γ-Glu moiety containing GSH and UPF1 stimulated CuZnSOD activity and increased intracellular tGSH level, whereas α-GSH and UPF17, which have α-Glu moiety in the structure, inhibited enzymatic activity and decreased GSH level. The stability of UPF1 towards GGT activity indicated that UPF1 affects GSH level and CuZnSOD activity as intact molecule instead of being a GSH precursor. Previously, it has been shown that GSH and UPF1 are able to act as signaling molecules through G-protein activation in frontocortical membrane preparations [23]. It has been reported that plasma membranes have specific binding sites of GSH which have an interaction with the glutamate binding sites [24]. By this way GSH and UPF peptides may affect the metabolism of cells as signal molecules. The effects on the level of GSH and CuZnSOD activity may be different depending on the replacement of γ-peptide bond with α-peptide bond. GSH has been shown to bind to ionotropic glutamate receptors via gamma-glutamyl

residue in the nervous tissue [25]. Additionally, glutamate receptors have been found also in the plasma membrane of megakaryocytes and rat erythrocytes [26, 27]. By interacting with the latter receptors, GSH and UPF peptides may affect the metabolism of cells as signal molecules through the PKC pathway and affect CuZnSOD activity. The various effects of the studied molecules may be caused by structural differences between the GSH and UPF peptides (replacement of γ-peptide bond with α-peptide bond).

UPF1 and UPF17 have also shown different effects in free radical scavenging experiments. According to the classification of kinetic behavior by Sánches-Moreno et al., UPF17 is classified as fast and UPF1 as intermediate DPPH radical scavenger [9, 28]. *In silica* modeling of noncovalent complex formation by docking calculations revealed a more affine complex between DPPH radical and α-GSH compared to the complex with GSH [21]. This raised a question about pK_a values for the thiol groups of UPF peptides. Previously, the change of γ-peptide bond to α-peptide bond has also been investigated for GSH and its α-analogue: pK_a of thiol groups were similar for GSH and α-GSH (9.0 ± 0.1 and 9.1 ± 0.1) [21]. The comparison of these results with current measurements for UPF1 and UPF17 demonstrated that pK_a value is rather influenced by the addition of a methylated tyrosine moiety to the GSH backbone than by the change of the peptide bond type. Smaller pK_a values for GSH and its α-analogue showed that these molecules donate the sulfhydryl proton more easily than UPF peptides; however, UPF peptides are better radical scavengers. This indicates that reactive species elimination does not depend only of the reactivity of the thiol group.

The results of the current paper show that γ-peptide bond and α-peptide bond containing UPF peptides may influence enzyme activities in different direction, which offers a wider perspective for the usage of glutathione analogues as protective diverse regulators of the oxidative state.

Acknowledgments

This paper was financially supported by the Estonian Science Foundation Grants no. 7856 and 7494, by targeted financing from Ministry of Education and Science of Estonia (SF0180105s08) and by European Union through the European Regional Development Fund.

References

[1] M. Zilmer, U. Soomets, A. Rehema, and U. Langel, "The glutathione system as an attractive therapeutic target," *Drug Design Reviews Online*, vol. 2, no. 2, pp. 121–127, 2005.

[2] N. H. P. Cnubben, I. M. C. M. Rietjens, H. Wortelboer, J. Van Zanden, and P. J. Van Bladeren, "The interplay of glutathione-related processes in antioxidant defense," *Environmental Toxicology and Pharmacology*, vol. 10, no. 4, pp. 141–152, 2001.

[3] D. A. Dickinson, A. L. Levonen, D. R. Moellering et al., "Human glutamate cysteine ligase gene regulation through the electrophile response element," *Free Radical Biology and Medicine*, vol. 37, no. 8, pp. 1152–1159, 2004.

[4] N. Ballatori, S. M. Krance, S. Notenboom, S. Shi, K. Tieu, and C. L. Hammond, "Glutathione dysregulation and the etiology

and progression of human diseases," *Biological Chemistry*, vol. 390, no. 3, pp. 191–214, 2009.

[5] R. Franco, O. J. Schoneveld, A. Pappa, and M. I. Panayiotidis, "The central role of glutathione in the pathophysiology of human diseases," *Archives of Physiology and Biochemistry*, vol. 113, no. 4-5, pp. 234–258, 2007.

[6] O. Ortolani, A. Conti, A. R. De Gaudio, E. Moraldi, Q. Cantini, and G. Novelli, "The effect of glutathione and N-acetylcysteine on lipoperoxidative damage in patients with early septic shock," *American Journal of Respiratory and Critical Care Medicine*, vol. 161, no. 6, pp. 1907–1911, 2000.

[7] A. Wendel and P. Cikryt, "The level and half-life of glutathione in human plasma," *FEBS Letters*, vol. 120, no. 2, pp. 209–211, 1980.

[8] I. Cacciatore, C. Cornacchia, F. Pinnen, A. Mollica, and A. Di Stefano, "Prodrug approach for increasing cellular glutathione levels," *Molecules*, vol. 15, no. 3, pp. 1242–1264, 2010.

[9] K. Ehrlich, S. Viirlaid, R. Mahlapuu et al., "Design, synthesis and properties of novel powerful antioxidants, glutathione analogues," *Free Radical Research*, vol. 41, no. 7, pp. 779–787, 2007.

[10] A. Gozzo, D. Lesieur, P. Duriez, J. C. Fruchart, and E. Teissier, "Structure-activity relationships in a series of melatonin analogues with the low-density lipoprotein oxidation model," *Free Radical Biology and Medicine*, vol. 26, no. 11-12, pp. 1538–1543, 1999.

[11] T. L. Yue, P. J. Mckenna, P. G. Lysko et al., "SB 211475, a metabolite of carvedilol, a novel antihypertensive agent, is a potent antioxidant," *European Journal of Pharmacology*, vol. 251, no. 2-3, pp. 237–243, 1994.

[12] J. Kals, J. Starkopf, M. Zilmer et al., "Antioxidant UPF1 attenuates myocardial stunning in isolated rat hearts," *International Journal of Cardiology*, vol. 125, no. 1, pp. 133–135, 2008.

[13] P. Põder, M. Zilmer, J. Starkopf et al., "An antioxidant tetrapeptide UPF1 in rats has a neuroprotective effect in transient global brain ischemia," *Neuroscience Letters*, vol. 370, no. 1, pp. 45–50, 2004.

[14] I. Fridovich, "Superoxide anion radical ($O_2^{\cdot}$), superoxide dismutases, and related matters," *Journal of Biological Chemistry*, vol. 272, no. 30, pp. 18515–18517, 1997.

[15] M. D. De Beus, J. Chung, and W. Colón, "Modification of cysteine 111 in Cu/Zn superoxide dismutase results in altered spectroscopic and biophysical properties," *Protein Science*, vol. 13, no. 5, pp. 1347–1355, 2004.

[16] M. A. Hough and S. S. Hasnain, "Structure of fully reduced bovine copper zinc superoxide dismutase at 1.15 Å," *Structure*, vol. 11, no. 8, pp. 937–946, 2003.

[17] U. Soomets, M. Zilmer, and Ü. Langel, "Manual solid-phase synthesis of glutathione analogues: a laboratory-based short course," in *Peptide Synthesis and Applications*, J. Howl, Ed., pp. 241–257, Humana Press, Totowa, NJ, USA, 2006.

[18] O. H. Lowry, N. J. Rosebrough, A. L. Farr, and R. J. Randall, "Protein measurement with the Folin phenol reagent," *The Journal of biological chemistry*, vol. 193, no. 1, pp. 265–275, 1951.

[19] F. Tietze, "Enzymic method for quantitative determination of nanogram amounts of total and oxidized glutathione: applications to mammalian blood and other tissues," *Analytical Biochemistry*, vol. 27, no. 3, pp. 502–522, 1969.

[20] D. Burg, D. V. Filippov, R. Hermanns, G. A. Van der Marel, J. H. Van Boom, and G. J. Mulder, "Peptidomimetic glutathione analogues as novel γGT stable GST inhibitors," *Bioorganic and Medicinal Chemistry*, vol. 10, no. 1, pp. 195–205, 2002.

[21] S. Viirlaid, R. Mahlapuu, K. Kilk, A. Kuznetsov, U. Soomets, and J. Järv, "Mechanism and stoichiometry of 2,2-diphenyl-1-picrylhydrazyl radical scavenging by glutathione and its novel α-glutamyl derivative," *Bioorganic Chemistry*, vol. 37, no. 4, pp. 126–132, 2009.

[22] K. Ehrlich, K. Ida, R. Mahlapuu et al., "Characterization of UPF peptides, members of the glutathione analogues library, on the basis of their effects on oxidative stress-related enzymes," *Free Radical Research*, vol. 43, no. 6, pp. 572–580, 2009.

[23] E. Karelson, R. Mahlapuu, M. Zilmer, U. Soomets, N. Bogdanovic, and U. Langel, "Possible signaling by glutathione and its novel analogue through potent stimulation of frontocortical G proteins in normal aging and in Alzheimer's disease," *Annals of the New York Academy of Sciences*, vol. 973, pp. 537–540, 2002.

[24] R. Janáky, K. Ogita, B. A. Pasqualotto et al., "Glutathione and signal transduction in the mammalian CNS," *Journal of Neurochemistry*, vol. 73, no. 3, pp. 889–902, 1999.

[25] Z. Jenei, R. Janáky, V. Varga, P. Saransaari, and S. S. Oja, "Interference of S-alkyl derivatives of glutathione with brain ionotropic glutamate receptors," *Neurochemical Research*, vol. 23, no. 8, pp. 1085–1091, 1998.

[26] P. G. Genever, D. J. P. Wilkinson, A. J. Patton et al., "Expression of a functional N-methyl-D-aspartate-type glutamate receptor by bone marrow megakaryocytes," *Blood*, vol. 93, no. 9, pp. 2876–2883, 1999.

[27] A. Makhro, J. Wang, J. Vogel et al., "Functional NMDA receptors in rat erythrocytes," *American Journal of Physiology*, vol. 298, no. 6, pp. C1315–C1325, 2010.

[28] C. Sánchez-Moreno, J. A. Larrauri, and F. Saura-Calixto, "A procedure to measure the antiradical efficiency of polyphenols," *Journal of the Science of Food and Agriculture*, vol. 76, no. 2, pp. 270–276, 1998.

2

Synthesis of Hemopressin Peptides by Classical Solution Phase Fragment Condensation

P. Anantha Reddy, Sean T. Jones, Anita H. Lewin, and F. Ivy Carroll

Center for Organic and Medicinal Chemistry, Discovery Sciences Research Triangle Institute, Research Triangle Park, NC 27709-2194, USA

Correspondence should be addressed to P. Anantha Reddy, ananth@rti.org

Academic Editor: Severo Salvadori

A fragment condensation solution phase assembly of the naturally occurring CB_1 inverse agonist nonapeptides, Pro-Val-Asn-Phe-Lys-Phe/Leu-Leu-Ser-His-OH (hemopressins), and two other homologues: N-terminal 2-amino acid (dipeptide) extended undecapeptide, Val-Asp-Pro-Val-Asn-Phe-Lys-Leu-Leu-Ser-His-OH, and three-amino acid (tripeptide) extended dodecapeptide, Arg-Val-Asp-Pro-Val-Asn-Phe-Lys-Leu-Leu-Ser-His-OH, both CB_1 agonists, is reported.

1. Introduction

Naturally occurring nonapeptides, Pro-Val-Asn-Phe-Lys-Phe/Leu-Leu-Ser-His-OH (hemopressins), derived from the α_1 chain of hemoglobin of rat, human, pig, and cow are inverse agonists at the cannabinoid CB_1 receptor [1]. Sequence alignments of hemopressins from various species differ only at position 100 of the α_1-globin chain (Figure 1) where Phe (**F**) in rat is replaced by Leu (**L**) in human, pig, and cow sequences [2].

For convenience, hereafter, nonapeptide Pro-Val-Asn-Phe-Lys-Phe-Leu-Ser-His-OH, isolated from rat hemoglobin, is abbreviated as rHP and Pro-Val-Asn-Phe-Lys-Leu-Leu-Ser-His-OH, isolated from human, pig, and cow, as hHP. Interestingly, the N-terminally extended homologues of hHP: Val-Asp-Pro-Val-Asn-Phe-Lys-Leu-Leu-Ser-His-OH (VD-hHP) and Arg-Val-Asp-Pro-Val-Asn-Phe-Lys-Leu-Leu-Ser-His-OH (RVD-hHP), are in fact found to be CB_1 agonists [3, 4]. In addition, hemopressin was recently shown to self-assemble into fibrils [5] at physiological pH. Since peptide amyloid fibril formation is implicated in Alzheimer's and Parkinson's diseases, these relatively small peptides deserve a systematic investigation into their structure activity relationships (SARs).

Towards this objective, and to be able to produce the desired truncated peptides for SAR studies, a solution phase fragment condensation was adopted to synthesize these peptides and other homologues.

Though the solid phase synthesis is a fast route to synthetic peptides, the classical solution phase approach by fragment condensation has many advantages where several truncated peptides are available with minimum effort for structure activity investigations. In addition, peptide fragments from solution phase synthesis can be purified, and the pure intermediates are elaborated to the desired target peptide. Also, the smaller fragments used to make larger peptides give way for easier purifications (by size exclusion chromatography) of the target peptides by taking advantage of their size differences. Here, we report the synthetic routes to hemopressin and other related peptides using this approach. The protecting groups (Table 1) used here are orthogonal. Both Boc and Fmoc chemistries were used where necessary in the synthesis of peptide fragments. Thus, the key peptide intermediates: Boc-Pro-Val-OH (**4**), Boc-Val-Asp(OBu^t)-Pro-Val-OH (**11**), Fmoc-Arg(Pbf)-Val-Asp(OBu^t)-)-Pro-Val-OH (**15**) (Scheme 1), the N-terminal fragments and H-Asn(Trt)-Phe-Lys(Boc)-Phe/Leu-Leu-Ser(Bu^t-)His(Trt)-OBu^t-(**28**) (Scheme 1 and

Boc-Pro-OH + HCl·Val-OBn

1   **2**

TBTU/6-Cl HOBt/DIEA

Fmoc-Asp(OBu^t)-OH + TFA·Pro-Val-OBn ← TFA ← Boc-Pro-Val-OBn

6   **5**   **3**

TBTU/6-Cl HOBt/DIEA   H$_2$, Pd/C

Fmoc-Asp(OBu^t)-Pro-Val-OBn   Boc-Pro-Val-OH

7   **4**

20% Piperidine/DMF

Fmoc(or Boc)-Val-OH + H-Asp(OBu^t)-Pro-Val-OBn

9   **8**

TBTU/6-Cl HOBt/DIEA

Fmoc(or Boc)-Val-Asp(OBu^t)-Pro-Val-OBn →$\frac{H_2, Pd/C}{(Boc)}$→ Boc-Val-Asp(OBu^t)-Pro-Val-OH

10   **11**

(Fmoc) | Piperidine

Fmoc⁻Arg(Pbf)⁻OH + H-Val-Asp(OBu^t)-Pro-Val-OBn

13   **12**

TBTU/6-Cl HOBt/DIEA

Fmoc-Arg(Pbf)-Val-Asp(OBu^t)-Pro-Val-OBn →$\frac{H_2, Pd/C}{}$→ Fmoc-Arg(Pbf)-Val-Asp(OBu^t)-Pro-Val-OH

14   **15**

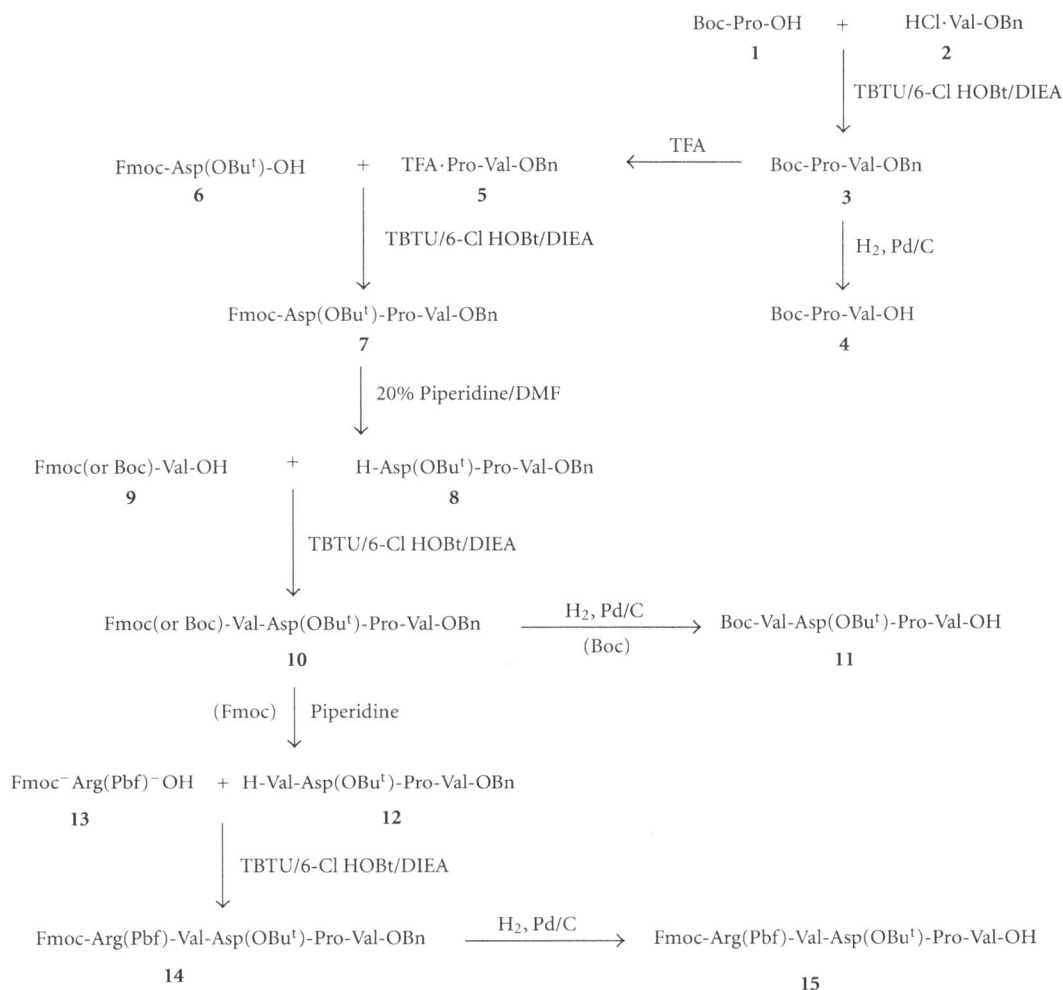

SCHEME 1: Synthesis of hemopressin (part 1).

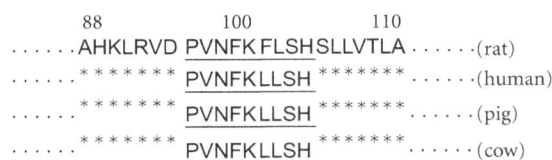

```
        88          100         110
......AHKLRVD PVNFK FLSHSLLVTLA ......(rat)
......******* PVNFKLLSH ******* ......(human)
......******* PVNFKLLSH ******* ......(pig)
......******* PVNFKLLSH ******* ......(cow)
```

FIGURE 1: Sequence alignments of the α_1-globin chain from various species. The sequence of hemopressin is underlined. Asterisks (*) indicate identical amino acids.

Scheme 2), and the C-terminal fragment common to all three peptides were prepared [6, 7] as outlined. [For the synthesis of shorter peptides, syntheses were conducted on 15 mmol to 20 mmol scale; fragment condensations were performed on 2 mmol to 5 mmol scale. The coupling agents TBTU and PyBOP were used in the preparation of shorter peptide fragments, while HATU/HOAt/DIEA (DMF) was used in the condensations of the protected peptide fragments to target peptides. In all cases, the reaction times are 3 h

to 4 h. Deprotection was carried out by exposure (40 min to 1 h) to 50% TFA/CH$_2$Cl$_2$ for Boc/Trt groups, 20% piperidine/DMF for Fmoc group and neat TFA, to deblock the Bu^t)-ether/ester protecting groups. Product yields in the preparation of shorter fragments were moderately high (80% to 95%), while yields in the condensation of fragments to get the larger peptides were in the range of 50% to 70%.]

The dipeptide fragment Boc-Pro-Val-OH (**4**) was prepared in two steps by condensation of **1** with **2** followed by hydrogenolysis of the resulting intermediate **3** (Scheme 1). Preparation of key intermediate **11** involved five easy steps (Scheme 1). Briefly, intermediate **5,** obtained by treatment of intermediate **3** with TFA, was condensed with **6** to give intermediate **7**, which after removal of the Fmoc group was coupled with **9** to provide intermediate **10**. Conversion of intermediate **10** to the desired key tetrapeptide intermediate, Boc-Val-Asp(OBu^t)-Pro-Val-OH (**11**), was accomplished by hydrogenolysis. For the synthesis of the third key N-terminal pentapeptide, Fmoc-Arg(Pbf)-Val-Asp(OBu^t)-Pro-Val-OH (**15**), intermediate **10** was converted to **12** by

Fmoc-Asn(Trt)-OH₂ + HCl·Phe-OBn   Fmoc-Lys(Boc)-OH + HCl·AA*-OBn
      16   **17**   **19**   **20**

      (1) PyBOP/HOBt/DIEA   (1) PyBOP/HOBt/DIEA
      (2) H₂, Pd/C   (2) Piperidine

      Fmoc-Asn(Trt)-Phe-OH   H-Lys(Boc)-AA*-OBn
            18   **21**

      (1) WSCI/HOBt/DIEA
      (2) H₂, Pd/C

Fmoc-Pro-Val-Asn(Trt)-Phe-Lys(Boc)-AA*-OH + p-Tos·Leu-OBn   Cbz-Ser(Buᵗ)-OH + HCl·His(Trt)-OBuᵗ
            22   **23**   **25**   **26**

            (1) WSCI/HOBt/DIEA   (1) PyBOP/HOBt/DIEA
            (2) H₂, Pd/C   (2) H₂, Pd/C

      Fmoc-Asn(Trt)-Phe-Lys(Boc)-AA*-Leu-OH   H-Ser(Buᵗ)-His(Trt)-OBuᵗ
                  24   **27**

                  (1) PyBOP/HOBt/DIEA
                  (2) Piperidine

      H-Asn(Trt)-Phe-Lys(Boc)-AA*-Leu-Ser(Buᵗ)-His(Trt)-OBuᵗ
                        28

AA* = Phe or Leu

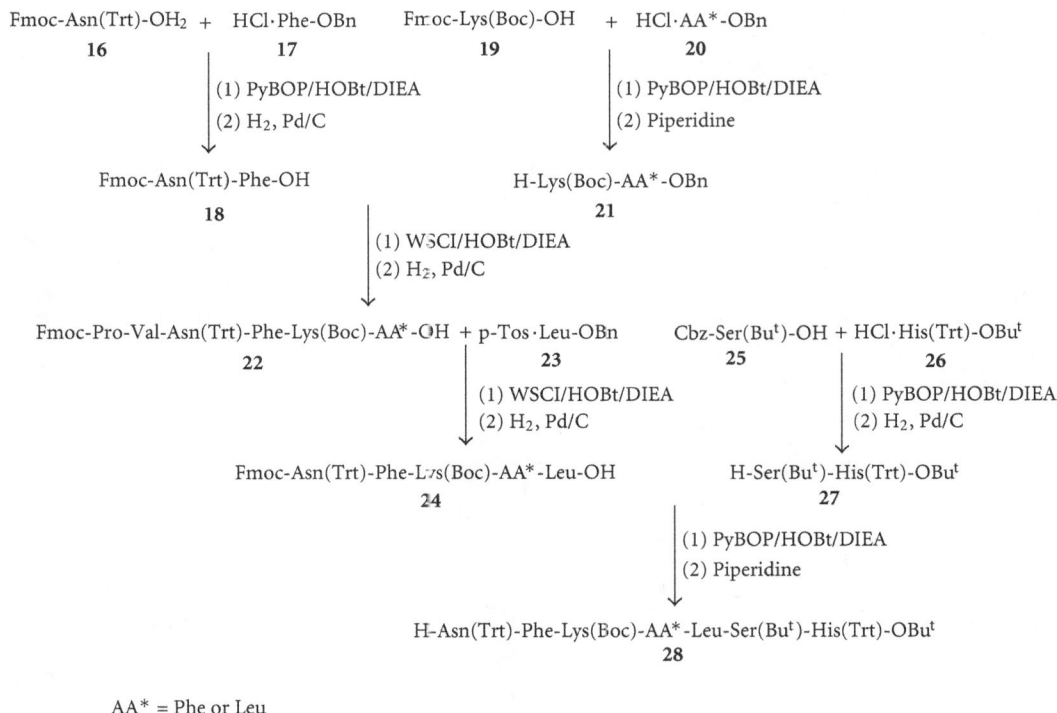

SCHEME 2: Synthesis of hemopressin (part 2).

exposure to piperidine followed by coupling with Fmoc-Arg(Pbf)-OH (**13**) to give **14**, which on hydrogenolysis afforded **15**.

Next, the final C-terminal heptapeptide fragment, H-Asn(Trt)-Phe-Lys(Boc)-Phe/Leu-Leu-Ser(Buᵗ)-)-His(Trt)-OBuᵗ)-(**28**), which is common to all target peptides was assembled as outlined in Scheme 2. The dipeptides, Fmoc-Asn(Trt)-Phe-OH (**18**) and H-Lys(Boc)-AA-OBn (**21**), where AA is either Phe or Leu, were prepared from condensations of **16** with **17** and **19** with **20**, respectively. Further, condensation of dipeptide **18** with dipeptide **21** provided tetrapeptide **22** after hydrogenolysis. Elaboration of **22** to pentapeptide **24** was accomplished by addition of C-terminal residue **23** followed by hydrogenolysis (Scheme 2). The dipeptide fragment at the C-terminus, H-Ser(Buᵗ)-)-His(Trt)-OBuᵗ)-(**27**), was assembled by coupling Cbz-Ser(Buᵗ)-)-OH (**25**) with HCl·His(Trt)-OBuᵗ)-(**26**) followed by removal of the Cbz group. The final steps in the assembly of C-terminal heptapeptide fragment **28** involve the (5 + 2) fragment condensation of pentapeptide **24** with dipeptide **27** followed by exposure to piperidine.

As shown in Scheme 3, using a (2 + 7) fragment condensation, dipeptide fragment **4** was condensed with heptapeptide fragment **28** under HATU/HOAt/DIEA coupling conditions [8] to give the fully protected non-apeptide Boc-Pro-Val-Asn(Trt)-Phe-Lys(Boc)-Phe/Leu-Leu-Ser(Buᵗ)-His(Trt)-OBuᵗ (**29**). The later individual non-apeptide(s) (**29**) on exposure to TFA furnished the target peptide(s) Pro-Val-Asn-Phe-Lys-Phe/Leu-Leu-Ser-His-OH (hemopressin) (**30a/30b**).

Similarly, VD-hemopressin (**32**) and RVD-hemopressin (**34**) were also assembled using (4 + 7) and (5 + 7) fragment condensations, respectively (Scheme 3) involving intermediates **11**, **15**, **28**, **31**, and **33**. [Representative fragment coupling procedure for the synthesis of VD-Hemopressin: to a solution of Boc-Val-Asp(OBuᵗ)-Pro-Val-OH (**11**) (1.4 g, 2 mmol), 6-Cl-HOBt (0,34 g, 2 mmol) in CH₂Cl₂ (60 mL) was added as TBTU reagent (0.65 g, 2 mmol) in CH₂Cl₂ (25 mL). To this mixture was added Asn(Trt)-Phe-Lys(Boc)-Leu-Leu-Ser(Buᵗ)-)-His(Trt)-OBuᵗ)-(**28**) (3.2 g, ~2 mmol) in CH₂Cl₂ (50 mL). The mixture was stirred overnight at room temperature. After an acid base workup, the fully protected peptide, Boc-Val-Asp(OBuᵗ)-Pro-Val-Asn(Trt)-Phe-Lys(Boc)-Leu-Leu-Ser(Buᵗ)-)-His(Trt)-OBuᵗ)-(**31**) (2.6 g), was isolated. The crude peptide was purified by gel filtration on Sephadex LH-20 using MeOH. The purified product was exposed to 50% TFA/CH₂Cl₂ to remove all the protecting groups to give VD-hemopressin, Val-Asp-Pro-Val-Asn-Phe-Lys-Leu-Leu-Ser-His-OH (**32**) (VD-hHP) (0.31 g) [MS (ESI) 1269.1 (M+H)]. The crude peptide was purified to homogeneity by preparative reversed phase HPLC. The HPLC conditions were as follows: Vydac C₁₈ column (218TP1022); flow rate: 15 mL/min; using a gradient (10% B → 65% B over 30 min) where A = 0.1% TFA/H₂O and B = 0.1% TFA/CH₃CN; UV detection 220 nm.] All target peptides: TFA·Pro-Val-Asn-Phe-Lys-Phe-Leu-Ser-His-OH (rHP) (**30a**), TFA·Pro-Val-Asn-Phe-Lys-Phe-Leu-Ser-His-OH (hHP) (**30b**), TFA·Val-Asp-Pro-Val-Asn-Phe-Lys-Phe-Leu-Ser-His-OH

TABLE 1

(VD-hHP) (**32**), and TFA·Arg-Val-Asp-Pro-Val-Asn-Phe-Lys-Phe-Leu-Ser-His-OH (RVD-hHP) (**34**), were purified to homogeneity by preparative reversed phase HPLC and characterized by TLC, HPLC, MS (ESI), and amino acid analysis. [(a) **rHP**: MS (ESI) m/z 1089.3 (M+H); $[\alpha]_D^{22} -26°$ (c 0.2, MeOH); amino acid analysis: found (*calculated*) Pro, 1.10 (*1.00*); Val 0.96 (*1.00*); Asn, 0.72 (*1.00*); Phe, 2.00 (*2.00*), Lys, 0.61 (*1.00*), Leu, 1.30 (*1.00*), Ser, 1.13 (*1.00*), His, 0.26 (*1.00*); (b) **hHP**: MS (ESI) m/z 1055 (M+H); $[\alpha]_D^{22} -5.3°$ (c 0.15, MeOH); amino acid analysis: found (*calculated*)

Pro, 1.03 (*1.00*); Val 0.96 (*1.00*); Asn, 1.00 (*1.00*); Phe, 0.84 (*1.00*), Lys, 1.03 (*1.00*), Leu, 2.03 (*2.00*), Ser, 0.94 (*1.00*), His, 0.97 (*1.00*); (c) **VD-hHP**: MS (ESI) m/z 1269.1 (M+H), m/z 635.5 (M+H)$^{+2}$; $[\alpha]_D^{22} -21°$ (c 0.1, MeOH); amino acid analysis: found (*calculated*) Pro, 1.00 (*1.00*); Val 1.90 (*2.00*); Asx, 2.13 (*2.00*); Phe, 1.00 (*1.00*), Lys, 1.06 (*1.00*), Leu, 1.96 (*2.00*), Ser, 0.89 (*1.00*), His, 0.91 (*1.00*); (d) **RVD-hHP**: MS (ESI) m/z 1424.81 (M+H), m/z 712.90 (M+H)$^{+2}$, m/z 475.60 (M+H)$^{+3}$; $[\alpha]_D^{22} -35°$ (c 0.520, MeOH); amino acid analysis: Found (*calculated*) Pro, 0.90 (*1.00*); Val 1.90 (*2.00*);

$$4 + 28 \xrightarrow{\text{HATU/HOAt/DIEA}} \text{Boc-Pro-Val-Asn(Trt)-Phe-Lys(Boc)-AA-Leu-Ser(Bu}^t\text{)-His(Trt)-OBu}^t$$

29

↓ TFA

TFA·Pro-Val-Asn-Phe-Lys-AA-Leu-Ser-His-OH

30a [AA = Phe (rHP)]

30b [AA = Leu (hHP)]

$$11 + 28 \xrightarrow{\text{HATU/HOAt/DIEA}} \text{Boc-Val-Asp(OBu}^t\text{)-Pro-Val-Asn(Trt)-Phe-Lys(Boc)-AA-Leu-Ser(Bu}^t\text{)-His(Trt)-OBu}^t$$

31

↓ TFA

TFA·Val-Asp-Pro-Val-Asn-Phe-Lys-AA-Leu-Ser-His-OH

32 [AA = Leu (VD-hHP)]

$$15 + 28 \xrightarrow{\text{HATU/HOAt/DIEA}} \text{Fmoc-Arg(Pbf)-Val-Asp(OBu}^t\text{)-Pro-Val-Asn(Trt)-Phe-Lys(Boc)-AA-Leu-Ser(Bu}^t\text{)-His(Trt)-OBu}^t$$

33

↓ (1) Piperidine
(2) TFA

TFA·Arg-Val-Asp-Pro-Val-Asn-Phe-Lys-AA-Leu-Ser-His-OH

34 [AA = Leu (RVD-hHP)]

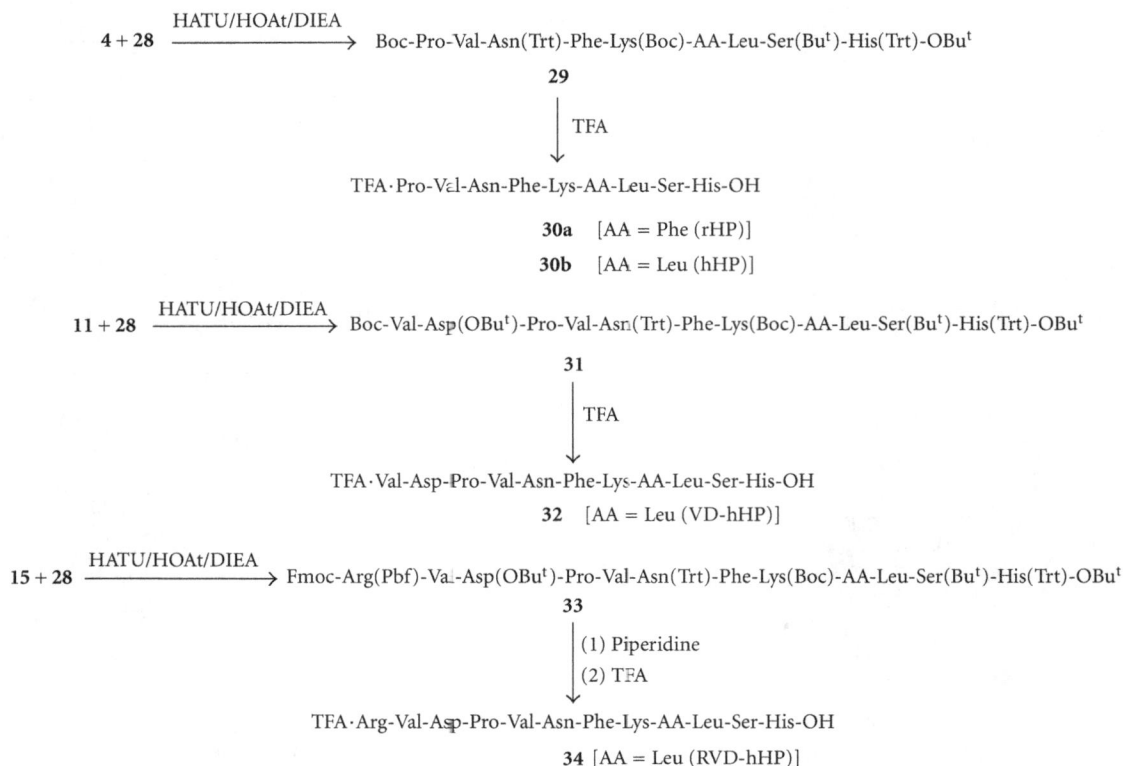

SCHEME 3: Synthesis of hemopressin (part 3).

Asx, 2.2 (*2.00*); Arg, 0.80 (*1.00*); Phe, 1.10 (*1.00*), Lys, 1.20 (*1.00*), Leu, 1.90 (*2.00*), Ser, 0.80 (*1.00*), His, 1.00 (*1.00*).]

Acknowledgments

This work was supported by the National Institute on Drug Abuse (NIDA), Contract no. NO1DA-8-7763. Amino acid analysis was done at Protein Analysis Core Lab, Wake Forest University, NC.

References

[1] A. S. Heimann, I. Gomes, C. S. Dale et al., "Hemopressin is an inverse agonist of CB1 cannabinoid receptors," *Proceedings of the National Academy of Sciences of the United States of America*, vol. 104, no. 51, pp. 20588–20593, 2007.

[2] V. T. Ivanov, A. A. Karelin, M. M. Philippova, I. V. Nazimov, and V. Z. Pletnev, "Hemoglobin as a source of endogenous bioactive peptides: the concept of tissue-specific peptide pool," *Biopolymers*, vol. 43, no. 2, pp. 171–188, 1997.

[3] C. S. Dale, R. de Lima Pagano, and V. Rioli, "Hemopressin: a novel bioactive peptide derived from the α1-chain of hemoglobin," *Memorias do Instituto Oswaldo Cruz*, vol. 100, supplement 1, pp. 105–106, 2005.

[4] I. Gomes, J. S. Grushko, U. Golebiewska et al., "Novel endogenous peptide agonists of cannabinoid receptors," *The FASEB Journal*, vol. 23, no. 9, pp. 3020–3029, 2009.

[5] M. G. Bomar, S. J. Samuelson, P. Kibler, K. Kodukula, and A. K. Galande, "Hemopressin forms self-assembled fibrillar nanostructures under physiologically relevant conditions," *Biomacromolecules*, vol. 13, no. 3, pp. 579–583, 2012.

[6] P. A. Reddy, T. McElroy, C. J. McElhinney, A. H. Lewin, and F. I. Carroll, "Synthesis of hemopressin by [(2+2+2+1)+2] segment condensation," in *Proceedings of the 21st American Peptide Symposium*, M. Lebl, Ed., pp. 42–43, Prompt Scientific, San Diego, Calif, USA, 2009.

[7] P. A. Reddy, T. McElroy, C. J. McElhinney, A. H. Lewin, and F. I. Carroll, "A [(2+2+2+1)+2] segment condensation approach to hemopressin synthesis," *Biopolymers*, vol. 92, no. 4, article 369, 2009.

[8] L. A. Carpino, "1-Hydroxy-7-azabenzotriazole. An efficient peptide coupling additive," *Journal of the American Chemical Society*, vol. 115, no. 10, pp. 4397–4398, 1993.

3

Development of the Schedule for Multiple Parallel "Difficult" Peptide Synthesis on Pins

Ekaterina F. Kolesanova, Maxim A. Sanzhakov, and Oleg N. Kharybin

Orekhovich Institute of Biomedical Chemistry, Russian Academy of Medical Sciences, 10 Pogodinskaya Ulica, Moscow 119121, Russia

Correspondence should be addressed to Ekaterina F. Kolesanova; ekaterina.kolesanova@ibmc.msk.ru

Academic Editor: John D. Wade

Unified schedule for multiple parallel solid-phase synthesis of so-called "difficult" peptides on polypropylene pins was developed. Increase in the efficiency of 9-fluorenyl(methoxycarbonyl) N-terminal amino-protecting group removal was shown to have a greater influence on the accuracy of the "difficult" peptide synthesis than the use of more efficient amino acid coupling reagents such as aminium salts. Hence the unified schedule for multiple parallel solid-phase synthesis of "difficult" peptides included the procedure for N-terminal amino group deprotection modified by applying a more efficient reagent for the deprotection and the standard procedure of amino acid coupling by carbodiimide method with an additional coupling using aminium salts, if necessary. Amino acid coupling with the help of carbodiimide allows to follow the completeness of the coupling via the bromophenol blue indication, thus providing the accuracy of the synthesis and preventing an overexpenditure of expensive reagents. About 100 biotinylated hepatitis C virus envelope protein fragments, most of which represented "difficult" peptides, were successfully obtained by synthesis on pins with the help of the developed unified schedule.

1. Introduction

Development of proteomic and interactome research linked to the mass-spectral detection and amino acid analysis of peptide fragments of proteins requires extensive development of multiple solid-phase peptide synthesis in order to prepare huge sets of peptides used as calibration standards and as affinity ligands for interactome analysis and interaction site mapping [1–5]. These peptide sets are expected to contain up to several hundreds of peptides including those with modified side-chain functional groups, since the analysis of a single tissue sample from a single organism may require the preparation of more than a hundred of the so-called characteristic peptides (unique fragments of proteins under study). The field of peptide scanning usage, which includes multiple parallel peptide syntheses as an obligatory part of the method, also expands. Besides scanning proteins for B- and T-epitope motifs [6–12], kinase phosphorylation and other posttranslational modification sites [13–17], and studies of protease cleavage specificity [16, 18], multiple parallel peptide synthesis is employed for the search of antibacterial peptides [19],

receptor peptide ligands [20], and preparation of novel biomaterials based on readily structured peptides and peptoids [21]. Though immunochemical research sometimes allowed the use of peptide preparations with 70–80% purity [22], other previously mentioned fields of peptide employment required highly purified preparations, especially the use as standards for mass spectrometry [1–4]. It necessitates a thorough elaboration of multiple parallel peptide synthesis protocols and development of unified procedures that allow obtaining peptide preparations with maximal contents of target products in the shortest time and with the least material and labor expenses.

Peptides with so-called "difficult" sequences, prone to the formation of intra- and interchain stable secondary structures, form a group that is characterized by low yields of target products [23, 24]. Hindered amino acid attachment to the growing peptide chain is typical for such peptides resulting in low yields of target products and a lot of byproducts represented by truncated peptides or peptides with gaps. These impurities are often difficult to separate from target products [23–25]. The problems of the "difficult" peptide synthesis are usually solved by (a) adding chaotropic salts or solvents [26, 27], (b) elevation of reaction mixture temperature via conventional heating or microwave irradiation [28–30], (c) use of more efficient catalysts of 9-fluorenyl(methoxycarbonyl) (Fmoc) N-terminal amino-protecting group removal and amino acid acylation [31, 32], and (d) prevention of the aggregation via introducing amido bond modifying groups [25, 33], isoacyl depsipeptide structures [34], and pseudo-proline residues [35]. However, addition of chaotropic salts reduces other reagent solubility and hence is undesirable in the case of multiple parallel peptide synthesis on pins, where Fmoc-amino acids are used in high concentrations. Temperature elevation above 60°C is also impossible in this case because of the softening of pins covered with grafted polyethylene. Introduction of amido bond modifications and isoacyl moieties requires the change of coupling conditions only for the peptides, where these modifications are used, hence disrupting the multiple parallel synthesis schedule unification; pseudoproline residues cannot be put everywhere in the peptide sequence.

Formation of intra- and intermolecular stable secondary structures hampers both the Fmoc removal from α-amino-group and Fmoc-amino acid acylation of the growing peptide chain [23, 24]. We studied the influence of the Fmoc removal and Fmoc-amino acid coupling conditions on the accuracy of multiple parallel synthesis of peptides with "difficult" sequences in order to elaborate a unified multiple parallel synthesis schedule that allows the correct synthesis of multiple sets of various peptides on pins.

2. Materials and Methods

Polyethylene pins grafted with ε-Fmoc-α-Boc-Lys-Pro moiety (DKP pins) and α-Fmoc-L-amino acids (except α-Fmoc-L-Arg(Pbf)) were from "Mimotopes" (Clayton, Australia). Side-chain functional groups of Fmoc-amino acids were protected with trityl(Trt) (Cys, Ans, and Gln), t-butoxy (OtBu) (Asp and Glu), t-butyl(tBu) (Ser and Thr), t-butyl (oxycarbonyl)(Boc)(Lys and Trp), and 2,2,4,6,7-pentamethyldihydrobenzofuran(Pbf) (Arg). Fmoc-Arg(Pbf), tri(iso-propyl) silane (TIS), and 1-[bis(dimethylamino)methylene]-1H-1,2,3-triazolo[4,5-b]pyridine-3-oxide hexafluorophosphate (HATU) were from "Sigma-Aldrich" (USA). 1-Hydroxybenzotriazole, 1-[bis (dimethylamino)methylene]-5-chloro-1H-benzotriazole-3-oxide hexafluorophosphate (HCTU) was from "Merck Chemicals/Novabiochem" (Nottingham, UK). 1-[Bis(dimethylamino)methylene]-1H-benzotriazole-3-oxide hexafluorophosphate (HBTU) was from "Applied Biosystems" (USA). 4-Methyl-piperidine (4MPIP), 1-methyl-piperidinone (NMP), diazabicyclo[5.4.0]undec-7-ene (DBU), trifluoroacetic acid (TFA), 2,4,6-collidine, and D(+)-biotin were from "Acros Organics" (Belgium). N,N'-Di(iso-propyl)carbodiimide (DIPC), anisole, and 1,2-ethanedithiol (EDT) were from "Merck" (Darmstadt, Germany).

Peptides were synthesized on DKP pins by means of multiple parallel solid-phase synthesis in polypropylene 96-well V-bottom plates, well volume 0.32 mL (Matrix, USA). The choice of hepatitis C virus envelope protein sites for scanning and the preparation of the peptide list were described

elsewhere [7, 9]. The peptide synthesis from Fmoc-amino acids using DIPC as the condensation catalyst, as described in the "Mimotopes" manual and [22], was chosen as a standard procedure, except that 4-methyl-piperidine (4MPIP) was employed instead of piperidine [36, 37] and NMP was used as the Fmoc removal and amino acid attachment solvent instead of DMF [38].

2.1. Modifications of Multiple Parallel Peptide Synthesis on Pins

(A) Fmoc-group removal was performed by 20% 4MPIP and 2% DBU in NMP (here and further v/v% are used). Fmoc amino acid attachment was performed exactly as in the standard procedure. Bromophenol blue was added at 0.05 mg/mL to the reaction mixture for controlling the attachment completeness [39]. Blue color of pins meant that the Fmoc-amino acid attachment reaction should be repeated. The second attachment was carried out using the mixture of 100 mM Fmoc-amino acid, 100 mM HATU, 100 mM HOBT, and 150 mM 2,4,6-collidine in NMP.

(B) Fmoc-group removal was carried out by 20% 4MPIP (the standard procedure). Fmoc-amino acid attachment was performed using the mixture of 100 mM Fmoc-amino acid, 100 mM HATU, 100 mM HOBT, and 150 mM 2,4,6-collidine in NMP.

(C) Fmoc-group removal was carried out as in the standard procedure. Fmoc-amino acids were attached as in (B), except that HBTU was used instead of HATU in the same concentration.

(D) Fmoc-group removal was carried out as in the standard procedure. Fmoc-amino acids were attached as in (B), except that HCTU was used instead of HATU in the same concentration. The repeated Fmoc-amino acid attachment was not carried out in modifications (B)–(D). The peptides that were used for testing the previously mentioned synthesis modifications are listed in Table 1.

Peptide biotinylation was performed on pins as described earlier [7]. Removal of side-chain protecting groups was carried out by the mixture trifluoroacetic acid- (TFA-) 1,2-ethanedithiol- (EDT-) water-tri(iso-propyl)silane- (TIS-) anisole (915 : 25 : 25 : 10 : 25, v/v) for 4 hours at room temperature. Peptides were detached from pins into 40% acetonitrile in 0.1 M ammonium bicarbonate, pH 8.4 (0.8 mL per pin) by fourfold 15 min ultrasonication at 40°C.

Matrix-assisted laser desorption-ionization time-of-flight mass spectrometric analysis (MALDI-TOF MS) of peptides was carried out on MicroFlex ("Bruker Daltonics," Germany) equipped with nitrogen laser (λ = 337 nm), in a reflectron mode with 25 kV acceleration potential. α-Cyano-4-hydroxycinnamic acid was used as a matrix. Samples were applied in triplicate onto MSP AnchorChip 600/96 plate via 1 mL drop layering on the matrix.

HPLC analysis of peptide preparations upon the modification of multiple parallel synthesis procedure development

TABLE 1: Peptides synthesized with the help of modifications (A)–(D) (see Section 2) of the standard multiple parallel solid-phase peptide synthesis schedule and MALDI-TOF MS peak lists of their nonpurified preparations.

Peptide number	Amino acid sequence and calculated molecular mass (Da) of the peptide[1]	Molecular masses of peptide products, obtained by modifications (A)–(D) of the standard multiple parallel synthesis schedule, and relative intensities of the corresponding ion peaks (HPLC with ESI-MS detection)[2]			
		(A)	(B)	(C)	(D)
1a	Biotin-*SGSG* **T**TKVIGGT-(KP) 1497,8	1497.6	1497.5	1497.4	1497.7 1396.5 (−T; 17%)
2a	Biotin-*SGSG*QTR**T**T**G**GS-(KP) 1528,7	1528.5	1528.6	1528.4	1528.6 1427.5 (−T; 20%) 1471.4 (−G; 13%) 1370.2 (−G − T ; 7%)
3a	Biotin-*SGSG*N**TK**LM**GG**T-(KP) 1542,8	1542.3	1542.7 1558.8 (Met(O))	1542.8 1558.7 (Met(O)) 1441.4 (−T; 15%)	1542.6 1558.6 (Met(O)) 1485.4 (−G; 10%) 1080.7 (Fmoc-TKLMGT-(KP); 5%)
4a	Biotin-*SGSG*NNYV**T**GGA-(KP) 1516,7	1516.3	1516.4	1516.4	1516.5 1459.7 (−G; 15%)
5a	Biotin-*SGSG*D**T**RV**VG**GQ-(KP) 1552,8	1552.6	1552.6 1495.5 (−G; 10%)	1552.6 1495.5 (−G; 17%) 1465.5 (−S; 10%)	1552.7 1451.5 (−T; 17%) 1495.5 (−G; 15%)

[1] Linker sequence between the octapeptide fragment of HCV envelope protein and biotin moiety is marked by italics. ε-(Lys-Pro)-Diketopiperazine moiety is shown in brackets.
[2] The intensity of the target product mass peak is taken as 100% in each case. Mass peaks with intensities not less than 5% of target product mass peak intensities are only listed. Lacking residues are shown as (−X) and in bold in peptide sequences.

was performed on Agilent 1200 Series HPLC system (Agilent Technologies, USA) equipped with Zorbax 300SB-C18 (3.5 μm) 1.0 mm × 150 mm column (Agilent Technologies), elution with 2–80% gradient of 0.1% HCOOH in acetonitrile in 0.1% HCOOH water solution starting 5 min following injection, elution rate 50 μL/min. Target peptide and byproduct detection and analysis were performed by ESI-MS and tandem mass-spectrometry (MS/MS) by collision-induced dissociation (CID) via Ar atom (25 eV) bombardment on Apex Qe Fourier transform ion cyclotron resonance mass spectrometer ("Bruker Daltonics," Germany). Mass spectra were analyzed with the help of FlexControl software ("Bruker Daltonics").

The list of peptides for synthesis and step-by-step schedule of multiple parallel synthesis were prepared with the help of PEPMAKER software ("Mimotopes"). Possible problems in each peptide synthesis were analyzed with the help of PINSOFT2 software ("Mimotopes").

3. Results and Discussion

Tables 1 and 2 comprise the lists of biotinylated peptides containing octapeptide fragments of HCV envelope proteins that were synthesized in this work. PINSOFT2 analysis of the peptide primary structures showed that some of the listed peptides have amino acid sequences, which are prone to the aggregation during the solid-phase synthesis: amino acid residues with β-methyl groups located in a row, amino

acid residues with hydrophobic side chains or bulky side-chain protecting groups located one through one, and a big proportion of Gly residues, that are capable to form inter- and intramolecular hydrogen bonds [23, 24].

Preliminary synthesis experiments with further MALDI-TOF MS and HPLC with MS/MS analysis (ESI followed by CIS) showed that all preparations of peptides with Gly-Gly fragment and some with Gly-X (where X = amino acid residue with Trt side-chain protecting group, mainly N and Q) just following the (Lys-Pro) diketopiperazine unit prepared by the standard procedure of multiple parallel peptide synthesis on pins (see Section 2 and [22]) contained a large portion of byproducts lacking Gly residue in this pair. Some preparations contained byproducts lacking residues inside −SGSG− linker group, and many preparations contained short truncated and also lacking certain amino acid residues peptides with nonremoved Fmoc group (mass peaks from 1280 to 1340 Da in Figure 1). Preparations of peptides nos. 86–88, 90–91 contained only trace amounts of target substances.

Hence a modification of the standard synthesis schedule was needed to achieve the correct synthesis of peptides listed in Tables 1 and 2. In particular, it was necessary to reveal whether the activation of Fmoc removal from growing peptide chain or Fmoc amino acid acylation could influence the purity of the target peptide, especially the absence of byproducts lacking 1-2 amino acid residues compared to the target peptide, more efficiently. Tertiary amine DBU was used as a more efficient catalyst of the hydroxycarbonyl dibenzofulvene detachment from the peptide α-amino group,

TABLE 2: Peptides synthesized by the modification (1) of the standard parallel solid-phase peptide synthesis schedule on pins.

Peptide number	Peptide sequence	HCV protein source of octapeptide fragment	Calculated molecular mass of peptide, D_a	Masses of molecular ions in MALDI-TOF mass spectra, D_a
1	Biotin-*SGSG*PGCVPCVR-(KP)	E1	1551.9	1551.8
2	Biotin-*SGSG*GCVPCVRE-(KP)	E1	1583.9	1583.8
3	Biotin-*SGSG*YVGDLCGS-(KP)	E1	1534.8	1556.3 (+Na$^+$)
4	Biotin-*SGSG*VGDLCGSV-(KP)	E1	1470.7	1492.7 (+Na$^+$)
5	Biotin-*SGSG*GDLCGSVF-(KP)	E1	1518.8	1540.7 (+Na$^+$)
6	Biotin-*SGSG*DLCGSVFL-(KP)	E1	1574.9	1596.8 (+Na$^+$)
7	Biotin-*SGSG*QLFTFSPR-(KP)	E1	1717.0	1716.9
8	Biotin-*SGSG*QDCNCSIY-(KP)	E1	1666.9	1688.7 (+Na$^+$)
9	Biotin-*SGSG*CNCSIYPG-(KP)	E1	1577.8	1599.6 (+Na$^+$)
10	Biotin-*SGSG*NCSIYPGH-(KP)	E1	1611.9	1611.3 1633.3 (+Na$^+$)
11	Biotin-*SGSG*AWDMMMNW-(KP)	E1	1806.2	1827.7 (+Na$^+$)
12	Biotin-*SGSG*WDMMMNWS-(KP)	E1	1822.2	1843.7 (+Na$^+$)
13	Biotin-*SGSG*DMMMNWSP-(KP)	E1	1733.1	1770.7 (+Na$^+$; +O) 1754.7 (+Na$^+$) 1786.6 (+Na$^+$; +2O)
14	Biotin-*SGSG*MMMNWSPT-KPP	E1	1719.1	1756.7 (+Na$^+$; +O) 1772.7 (+Na$^+$; +2O) 1740.7 (+Na$^+$) 1788.7 (+Na$^+$; +3O)
15	Biotin-*SGSG*AGAHWGVL-(KP)	E1	1531.8	1553.8 (+Na$^+$) 1531.8
16	Biotin-*SGSG*GAHWGVLA-(KP)	E1	1531.8	1553.8 (+Na$^+$) 1531.8
17	Biotin-*SGSG*AHWGVLAG-(KP)	E1	1531.8	1553.8 (+Na$^+$) 1531.8
18	Biotin-*SGSG*SMVGNWAK-(KP)	E1	1613.9	1635.8 (+Na$^+$) 1613.8
19	Biotin-*SGSG*MVGNWAKV-(KP)	E1	1625.9	1625.8 1647.8 (+Na$^+$) 1641.8 (+O) 1663.8 (+Na$^+$; +O)
20	Biotin-*SGSG*VGNWAKVL-(KP)	E1	1607.9	1629.9 (+Na$^+$) 1607.9
21	Biotin-*SGSG*INTNGSWH-(KP)	E2	1649.8	1671.7 (+Na$^+$) 1649.7
22	Biotin-*SGSG*NTNGSWHI-(KP)	E2	1649.8	1649.8 1671.8 (+Na$^+$)
23	Biotin-*SGSG*TNGSWHIN-(KP)	E2	1649.8	1649.8 1671.8 (+Na$^+$)
24	Biotin-*SGSG*NGSWHINR-(KP)	E2	1704.9	1704.8
25	Biotin-*SGSG*ALNCNDSL-(KP)	E2	1570.8	1592.7 (+Na$^+$)
26	Biotin-*SGSG*PVVVGTTD-(KP)	E2	1508.8	1530.8 (+Na$^+$)
27	Biotin-*SGSG*VVVGTTDR-(KP)	E2	1567.8	1567.8
28	Biotin-*SGSG*WGENETDV-(KP)	E2	1670.8	1692.8 (+Na$^+$)
29	Biotin-*SGSG*GNWFGCTW-(KP)	E2	1691.9	1709.8 (+H$_2$O)
30	Biotin-*SGSG*NWFGCTWM-(KP)	E2	1766.1	1787.8 (+Na$^+$)
31	Biotin-*SGSG*FGCTWMNS-(KP)	E2	1666.9	1688.7 (+Na$^+$)
32	Biotin-*SGSG*KCGSGPWL-(KP)	E2	1535.8	1557.7 (+Na$^+$) 1535.7
33	Biotin-*SGSG*TGFTKTCG-(KP)	E2	1568.9	1568.8
34	Biotin-*SGSG*CGSGPWLT-(KP)	E2	1541.8	1563.7 (+Na$^+$)

TABLE 2: Continued.

Peptide number	Peptide sequence	HCV protein source of octapeptide fragment	Calculated molecular mass of peptide, D_a	Masses of molecular ions in MALDI-TOF mass spectra, D_a
35	Biotin-*SGSG*GSGPWLTP-(KP)	E2	1535.8	1557.7 (+Na$^+$)
36	Biotin-*SGSG*SGPWLTPR-(KP)	E2	1634.9	1634.8
37	Biotin-*SGSG*GPWLTPRC-(KP)	E2	1651.0	1650.9
38	Biotin-*SGSG*HYPCTVNF-(KP)	E2	1702.0	1701.8 1723.8 (+Na$^+$)
39	Biotin-*SGSG*RMYVGGVE-(KP)	E2	1631.9	1631.9 1647.8 (+O)
40	Biotin-*SGSG*TGFTKTCG-(KP)	E2	1612.9	1612.8 1634.8 (+Na$^+$)
41	Biotin-*SGSG*YVGGVEHR-(KP)	E2	1637.9	1637.9 1659.8 (+Na$^+$)
42	Biotin-*SGSG*VGGVEHRL-(KP)	E2	1587.8	1587.8
43	Biotin-*SGSG*AACNWTRG-(KP)	E2	1599.8	1599.8
44	Biotin-*SGSG*ACNWTRGE-(KP)	E2	1657.9	1657.7
45	Biotin-*SGSG*CNWTRGER-(KP)	E2	1743.0	1742.8
46	Biotin-*SGSG*NWTRGERC-(KP)	E2	1743.0	1742.8
47	Biotin-*SGSG*LEDRDRSE-(KP)	E2	1740.0	1740.8 1762.8 (+Na$^+$)
48	Biotin-*SGSG*EDRDRSEL-(KP)	E2	1740.0	1740.8 1762.8 (+Na$^+$)
49	Biotin-*SGSG*DRDRSELS-(KP)	E2	1698.9	1698.7
50	Biotin-*SGSG*RDRSELSP-(KP)	E2	1680.9	1680.8
51	Biotin-*SGSG*DRSELSPL-(KP)	E2	1637.9	1637.8 1659.8 (+Na$^+$)
52	Biotin-*SGSG*RSELSPLL-(KP)	E2	1636.0	1635.8
53	Biotin-*SGSG*IHLHQNIV-(KP)	E2	1570.9	1592.7 (+Na$^+$)
54	Biotin-*SGSG*TTLPALST-(KP)	E2	1524.8	1546.7 (+Na$^+$)
55	Biotin-*SGSG*TLPALSTG-(KP)	E2	1480.7	1502.7 (+Na$^+$)
56	Biotin-*SGSG*IHLHQNIV-(KP)	E2	1695.0	1694.8 1716.8 (+Na$^+$)
57	Biotin-*SGSG*SDLPALST-(KP)	E2	1524.8	1546.7 (+Na$^+$)
58	Biotin-*SGSG*TPMPALST-(KP)	E2	1538.9	1576.6 (+Na$^+$; +O)
59	Biotin-*SGSG*DLPALSTG-(KP)	E2	1494.7	1516.6 (+Na$^+$)
60	Biotin-*SGSG*PMPALSTG-(KP)	E2	1494.8	1532.6 (+Na$^+$; +O)
61	Biotin-*SGSG*ETLSVGGS-(KP)	E2	1470.7	1492.4 (+Na$^+$)
62	Biotin-*SGSG*ETIVTGGT-(KP)	E2	1498.7	1520.6
63	Biotin-*SGSG*ETAVSGGT-(KP)	E2	1442.6	1464.4 (+Na$^+$)
64	Biotin-*SGSG*ET**R**VSGGT-(KP)	E2	1527.7	1527.3
65	Biotin-*SGSG*GTYTTGGA-(KP)	E2	1448.6	1470.6 (+Na$^+$)
66	Biotin-*SGSG*TTYTTGGS-(KP)	E2	1508.7	1530.4 (+Na$^+$)
67 (1a)	Biotin-*SGSG*TTKVIGGT-(KP)	E2	1497.8	1497.6
68	Biotin-*SGSG*GT**R**TMGGA-(KP)	E2	1471.7	1471.5
69	Biotin-***SGSG***GTHVTGGS-(KP)	E2	1436.6	1436.4
70	Biotin-*SGSG*GT**R**VSGGT-(KP)	E2	1455.7	1455.4
71	Biotin-***SGS***GSTHVTGGA-(KP)	E2	1450.6	1450.5
72	Biotin-*SGS**G***STYTTGGS-(KP)	E2	1494.7	1516.3 (+Na$^+$)
73	Biotin-*SGS**G***STTITGGS-(KP)	E2	1444.6	1466.5 (+Na$^+$)
74	Biotin-*SGS**G***ST**R**VTGGA-(KP)	E2	1469.7	1469.5
75	Biotin-***SGS**G*QTHTTGGS-(KP)	E2	1528.7	1528.8 1551.1 (+Na$^+$)

TABLE 2: Continued.

Peptide number	Peptide sequence	HCV protein source of octapeptide fragment	Calculated molecular mass of peptide, D_a	Masses of molecular ions in MALDI-TOF mass spectra, D_a
76	Biotin-*SGSG*GTRVSGGT-(KP)	E2	1509.7	1509.4 1531.3 (+Na⁺)
77	Biotin-*SGSG*QTYVTGGA-(KP)	E2	1517.7	1539.4 (+Na⁺)
78	Biotin-*SGSG*KTYTTGGA-(KP)	E2	1519.7	1519.7 1541.7 (+Na⁺)
79	Biotin-*SGSG*KTHVTGGS-(KP)	E2	1507.7	1507.5
80	Biotin-*SGSG***R**THVTGGS-(KP)	E2	1535.7	1535.7
81	Biotin-*SGSG*GGTYVTGGA-(KP)	E2	1446.6	1535.7
82	Biotin-*SGSG***R**TR**L**TGGN-(KP)	E2	1595.8	1595.6
83	Biotin-*SGSG***R**TKTIGGT-(KP)	E2	1554.8	1554.6
84	Biotin-*SGSG*ATYTTGGA-(KP)	E2	1462.7	1462.4 1484.5 (+Na⁺)
85	Biotin-*SGSG*ATHVTGGT-(KP)	E2	1464.7	1464.3
86	Biotin-*SGSG*NTYTTGGS-(KP)	E2	1521.7	1543.4 (+Na⁺)
87 (3a)	Biotin-*SGSG*NTKLMGGT-(KP)	E2	1542.8	1542.5
88	Biotin-*SGSG*NT**R**TGGT-(KP)	E2	1528.7	1528.6
89 (4a)	Biotin-*SGSG*NNYVTGGA-(KP)	E2	1516.7	1538.3 (+Na⁺)
90	Biotin-*SGSG*DTHVTGGS-(KP)	E2	1494.6	1494.4
91	Biotin-*SGSG*HT**R**TTGGA-(KP)	E2	1521.7	1521.4
92 (5a)	Biotin-*SGSG*DT**R**VVGGQ-(KP)	E2	1552.8	1552.6
93	Biotin-*SGSG*HTYTTGGT-(KP)	E2	1558.7	1558.7
94	Biotin-*SGSG*HTHTTGGV-(KP)	E2	1530.7	1530.4
95	Biotin-*SGSG*HTHVTGGV-(KP)	E2	1528.7	1528.5
96	Biotin-*SGSG*DTYTTGGS-(KP)	E2	1522.7	1544.4 (+Na⁺)

Amino acid residues, for which repeated coupling was performed while using modification (A) of the standard procedure, are shown in bold.

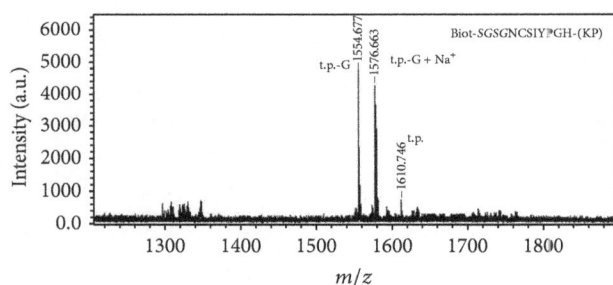

FIGURE 1: MALDI-TOF MS of the preparation of peptide Biotinyl-*SGSG*NCSIYPGH-(KP) obtained with the help of the standard multiple parallel peptide synthesis schedule on pins. t.p.: target peptide; t.p.-G: Biotinyl-*SGS*NCSIYPGH-(KP).

in addition to 4MPIP [31]. Since DBU actively catalyzes the formation of aspartimides, Asp piperidines, Asp epimerization, and Asn dehydration [23, 31, 36], it was not added to the Fmoc removal reagent after Asp or Asn introduction to the growing peptide chains. HATU, HBTU, and HCTU were employed as more efficient acylation activators, among which HATU is the most and HBTU is the least (close to DIC) efficient.

Table 1 shows HPLC and ESI MS and MS/MS analysis results of the final preparations of 5 peptides synthesized using modifications (A)–(D) of the standard multiple parallel peptide synthesis procedure on pins. Synthesis of these peptides by the standard procedure gave incorrect results (see Table 1). One can see that the application of the more efficient catalyst of Fmoc removal resulted in peptide preparation containing less byproducts lacking one or several amino acid residues compared to the target products than the application of aminium salts as acylation activators, in general. The latter also improved the results of the synthesis of peptides with "difficult" sequences; however, the use of aminium salts as acylation activators requires the addition of a tertiary amine (in our case, 2,4,6-collidine) to the reaction mixture, hence excluding the possibility to control the completeness of Fmoc-amino acid attachment to the growing peptide chain with bromophenol blue. Moreover, MS ESI and MS/MS analyses revealed noticeable (though not exceeding 5% of the target product mass peak intensity) mass peaks of Fmoc-containing truncated byproduct peptides in preparations obtained with modifications (B)–(D), but not (A) (see Figure 2). Hence the modification (A) of the standard multiple parallel solid-phase peptide synthesis schedule was used for the preparation of hepatitis C virus (HCV) envelope protein

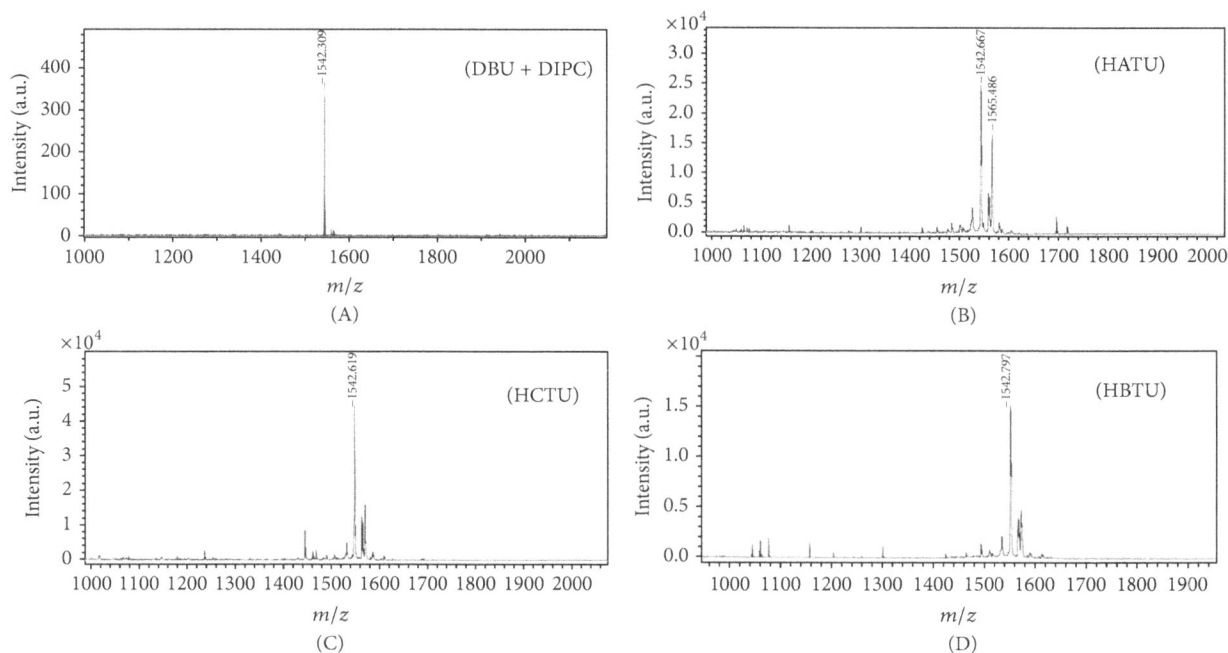

FIGURE 2: MALDI-TOF MS of preparations of peptide Biotinyl-SGSGNTKLMGGT-(KP) (mol. mass 1542.3) obtained with the help of modifications (a), (b), (c), and (d) of the standard schedule of parallel peptide synthesis on pins (see Section 2 for details).

fragments for further B-epitope mapping and characterization of the prepared anti-HCV envelope antibodies. Table 2 contains the structures of the synthesized peptides and results of MALDI-TOF MS analyses of their unpurified preparations. One can see that the enhancement of the efficiency of Fmoc removal resulted in much more correct synthesis of "difficult" peptides that we could not obtain in good purity and even could not obtain at all (peptides 86–88, 90, 91) with the help of the standard schedule.

4. Conclusion

In general, enhancement of the Fmoc removal efficiency in the parallel solid-phase peptide synthesis on pins showed a greater potential in the improvement of "difficult" peptide synthesis than the enhancement of the acylation stage efficiency. Moreover, the use of the standard DIC activation procedure better helped to control the Fmoc amino acid coupling completeness by bromophenol blue indication [39], thus excluding the overexpenditure of Fmoc amino acids, acylation reagents, and time, compared to the procedure with all coupling stages repeated in order to achieve the correct synthesis of target peptides. One should take into account that not only the aggregation of peptide chains due to interchain hydrogen bonds makes the peptide sequence difficult for synthesis. In our case the most difficult stage was the attachment of the fourth (or, maybe, fifth) residue from the C-terminus, when no peptide aggregation was supposed to occur [23]. However, the presence of rather long flexible structures (side chain of Lys, to which the first amino acid residue of HCV protein fragment was attached and Gly in

the second position from the peptide C-terminus) and Pro known to induce the turn formation could result in such turn of the growing peptide chain, augmented by the possible formation of intrachain H-bonds and hydrophobic interactions that hid the peptide N-terminal thus embarrassing its both Fmoc deblocking and further acylation. In this case, the sole enhancement of the amino acid coupling efficiency was shown to be less productive than the use of more efficient Fmoc deblocking reagent [40, 41]. Also, one could suppose that deblocking of the peptide N-terminus would disrupt the structure that hampered the further growing chain acylation, at least for short peptide chains. The same effect is frequently observed in the synthesis of peptides with Pro-Pro-, Val-Val-, -(Val-Thr(Ile))-pairs, Pro close to Val or Thr, and so forth, which are prone to β-turn formation [42]. Hence the DBU addition to 4MPIP or piperidine can be recommended as a modification of the standard procedure of multiple parallel peptide synthesis on pins, in order to obtain a unified procedure that allows the correct synthesis of "difficult" peptides. The only problem of DBU usage is its ability to catalyze aspartimide formation from Asp [36]. However, certain side-chain carboxyl protection groups greatly reduce [43–46], and Asp-X peptide bond modifications exclude this side reaction [25, 33, 46], hence permitting the use of more efficient Fmoc removal catalyst in a greater set of difficult synthesis cases.

Acknowledgments

This work was supported by RFBR Grants nos. 10-04-91054 and 13-04-00893. The authors are also grateful to the Russian

Ministry of Education and Science for a partial support of this work (Agreement no. 8274).

References

[1] V. G. Zgoda, A. T. Kopylov, O. V. Tikhonova et al., "Chromosome 18 transcriptome profiling and targeted proteome mapping in depleted plasma, liver tissue and HepG2 cells," *Proteome Research*, vol. 12, no. 1, pp. 123–134, 2013.

[2] A. Maiolica, M. A. Jünger, I. Ezkurdia, and R. Aebersold, "Targeted proteome investigation via selected reaction monitoring mass spectrometry," *Journal of Proteomics*, vol. 75, no. 12, pp. 3495–3513, 2012.

[3] H. Stephanowitz, S. Lange, D. Lang, C. Freund, and E. Krause, "Improved two-dimensional reversed phase-reversed phase LC-MS/MS approach for identification of peptide-protein interactions," *Journal of Proteome Research*, vol. 11, no. 2, pp. 1175–1183, 2012.

[4] C. Katz, L. Levy-Beladev, S. Rotem-Bamberger, T. Rito, S. G. D. Rüdiger, and A. Friedler, "Studying protein-protein interactions using peptide arrays," *Chemical Society Reviews*, vol. 40, no. 5, pp. 2131–2145, 2011.

[5] L. V. Olenina, T. I. Kuzmina, B. N. Sobolev, T. E. Kuraeva, E. F. Kolesanova, and A. I. Archakov, "Identification of glycosaminoglycan-binding sites within hepatitis C virus envelope glycoprotein E2," *Journal of Viral Hepatitis*, vol. 12, no. 6, pp. 584–593, 2005.

[6] A. M. Bray, R. M. Valerio, A. J. DiPasquale, J. Greig, and N. J. Maeji, "Multiple synthesis by the multipin method as a methodological tool," *Journal of Peptide Science*, vol. 1, no. 1, pp. 80–87, 1995.

[7] L. V. Olenina, L. I. Nikolaeva, B. N. Sobolev, N. P. Blokhina, A. I. Archakov, and E. F. Kolesanova, "Mapping and characterization of B cell linear epitopes in the conservative regions of hepatitis C virus envelope glycoproteins," *Journal of Viral Hepatitis*, vol. 9, no. 3, pp. 174–182, 2002.

[8] E. V. Kugaevskaya, E. F. Kolesanova, S. A. Kozin, A. V. Veselovsky, I. R. Dedinsky, and Y. E. Elisseeva, "Epitope mapping of the domains of human angiotensin converting enzyme," *Biochimica et Biophysica Acta*, vol. 1760, no. 6, pp. 959–965, 2006.

[9] T. I. Kuzmina, L. V. Olenina, M. A. Sanzhakov et al., "Antigenicity and B-epitope mapping of hepatitis C virus envelope protein E2," *Biochemistry*, vol. 3, no. 2, pp. 177–182, 2009.

[10] T. W. Tobery, S. Wang, X.-M. Wang et al., "A simple and efficient method for the monitoring of antigen-specific T cell responses using peptide pool arrays in a modified ELISpot assay," *Journal of Immunological Methods*, vol. 254, no. 1-2, pp. 59–66, 2001.

[11] J. Yang, E. A. James, L. Huston, N. A. Danke, A. W. Liu, and W. W. Kwok, "Multiplex mapping of CD4 T cell epitopes using class II tetramers," *Clinical Immunology*, vol. 120, no. 1, pp. 21–32, 2006.

[12] D. A. Lewinsohn, E. Winata, G. M. Swarbrick et al., "Immunodominant tuberculosis CD8 antigens preferentially restricted by HLA-B," *PLoS Pathogens*, vol. 3, no. 9, article e127, 2007.

[13] B. T. Houseman, J. H. Huh, S. J. Kron, and M. Mrksich, "Peptide chips for the quantitative evaluation of protein kinase activity," *Nature Biotechnology*, vol. 20, no. 3, pp. 270–274, 2002.

[14] F. D. Smith, B. K. Samelson, and J. D. Scott, "Discovery of cellular substrates for protein kinase A using a peptide array screening protocol," *Biochemical Journal*, vol. 438, no. 1, pp. 103–110, 2011.

[15] R. Arsenault, P. Griebel, and S. Napper, "Peptide arrays for kinome analysis: new opportunities and remaining challenges," *Proteomics*, vol. 11, no. 24, pp. 4595–4609, 2011.

[16] A. Thiele, G. I. Stangl, and M. Schutkowski, "Deciphering enzyme function using peptide arrays," *Molecular Biotechnology*, vol. 49, no. 3, pp. 283–305, 2011.

[17] S. M. Fuchs, K. Krajewski, R. W. Baker, V. L. Miller, and B. D. Strahl, "Influence of combinatorial histone modifications on antibody and effector protein recognition," *Current Biology*, vol. 21, no. 1, pp. 53–58, 2011.

[18] Y. Inoue, T. Mori, G. Yamanouchi et al., "Surface plasmon resonance imaging measurements of caspase reactions on peptide microarrays," *Analytical Biochemistry*, vol. 375, no. 1, pp. 147–149, 2008.

[19] K. Hilpert, "High-throughput screening for antimicrobial peptides using the SPOT technique," *Methods in Molecular Biology*, vol. 618, pp. 125–133, 2010.

[20] D. Koes, K. Khoury, Y. Huang et al., "Enabling large-scale design, synthesis and validation of small molecule protein-protein antagonists," *PLoS ONE*, vol. 7, no. 3, Article ID e32839, 2012.

[21] K. Kanie, R. Kato, Y. Zhao, Y. Narita, M. Okochi, and H. Honda, "Amino acid sequence preferences to control cell-specific organization of endothelial cells, smooth muscle cells, and fibroblasts," *Journal of Peptide Science*, vol. 17, no. 6, pp. 479–486, 2011.

[22] S. J. Rodda, "Synthesis of multiple peptides on plastic pins," *Current Protocols in Immunology*, Ch 9: Unit 9.7, 2001.

[23] P. Lloyd-Williams, F. Albericio F, and E. Giralt, *Chemical Approaches to the Synthesis of Peptides and Proteins*, CRC Press LLC, New York, NY, USA, 1997.

[24] J. Bedford, C. Hyde, T. Johnson et al., "Amino acid structure and "difficult sequences" in solid phase peptide synthesis," *International Journal of Peptide and Protein Research*, vol. 40, no. 3-4, pp. 300–307, 1992.

[25] V. Cardona, I. Eberle, S. Barthélémy et al., "Application of Dmb-dipeptides in the Fmoc SPPS of difficult and aspartimide-prone sequences," *International Journal of Peptide Research and Therapeutics*, vol. 14, no. 4, pp. 285–292, 2008.

[26] S. Abdel Rahman, A. El-Kafrawy, A. Hattaba, and M. F. Anwer, "Optimization of solid-phase synthesis of difficult peptide sequences via comparison between different improved approaches," *Amino Acids*, vol. 33, no. 3, pp. 531–536, 2007.

[27] S. C. F. Milton and L. R. C. De Milton, "An improved solid-phase synthesis of a difficult-sequence peptide using hexafluoro-2-propanol," *International Journal of Peptide and Protein Research*, vol. 36, no. 2, pp. 193–196, 1990.

[28] M. Erdélyi and A. Gogoll, "Rapid microwave-assisted solid phase peptide synthesis," *Synthesis*, no. 11, pp. 1592–1596, 2002.

[29] B. Bacsa, K. Horváti, S. Bõsze, F. Andreae, and C. O. Kappe, "Solid-phase synthesis of difficult peptide sequences at elevated temperatures: a critical comparison of microwave and conventional heating technologies," *Journal of Organic Chemistry*, vol. 73, no. 19, pp. 7532–7542, 2008.

[30] C. Loffredo, N. A. Assunção, J. Gerhardt, and M. T. M. Miranda, "Microwave-assisted solid-phase peptide synthesis at 60°C: alternative conditions with low enantiomerization," *Journal of Peptide Science*, vol. 15, no. 12, pp. 808–817, 2009.

[31] A. El-Faham and F. Albericio, "Peptide coupling reagents, more than a letter soup," *Chemical Reviews*, vol. 111, no. 11, pp. 6557–6602, 2011.

[32] S. A. Kates, N. A. Solé, M. Beyermann, G. Barany, and F. Albericio, "Optimized preparation of deca(L-Alanyl)-L-valinamide

by 9-fluorenylmethyloxycarbonyl (fmoc) solid-phase synthesis on polyethylene glycol-polystyrene (PEG-PS) graft supports, with 1,8-diazobicyclo[5.4.0]-undec-7-ene (DBU) deprotection," *Peptide Research*, vol. 9, no. 3, pp. 106–113, 1996.

[33] T. Johnson, M. Quibell, D. Owen, and R. C. Sheppard, "A reversible protecting group for the amide bond in peptides. Use in the synthesis of "difficult sequences"," *Journal of the Chemical Society*, no. 4, pp. 369–372, 1993.

[34] T. Haack and M. Mutter, "Serine derived oxazolidines as secondary structure disrupting, solubilizing building blocks in peptide synthesis," *Tetrahedron Letters*, vol. 33, no. 12, pp. 1589–1592, 1992.

[35] I. Coin, "The depsipeptide method for solid-phase synthesis of difficult peptides," *Journal of Peptide Science*, vol. 16, no. 5, pp. 223–230, 2010.

[36] E. Y. Aleshina, N. V. Pyndyk, A. A. Moisa et al., "Synthesis of the β-amyloid fragment 5RHDSGY10 and its isomers," *Biochemistry*, vol. 2, no. 3, pp. 288–292, 2008.

[37] J. Hachmann and M. Lebl, "Alternative to piperidine in Fmoc solid-phase synthesis," *Journal of Combinatorial Chemistry*, vol. 8, no. 2, p. 149, 2006.

[38] C. J. Bagley, K. M. Otteson, B. L. May et al., "Synthesis of insulin-like growth factor I using N-methyl pyrrolidinone as the coupling solvent and trifluoromethane sulphonic acid cleavage form the resin," *International Journal of Peptide and Protein Research*, vol. 36, no. 4, pp. 356–361, 1990.

[39] V. Krchnak, J. Vagner, P. Safar, and M. Lebl, "Noninvasive continuous monitoring of solid-phase peptide synthesis by acid-base indicator," *Collection of Czechoslovak Chemical Communications*, vol. 53, pp. 2542–2549, 1988.

[40] A. K. Tickler, C. J. Barrow, and J. D. Wade, "Improved preparation of amyloid-β peptides using DBU as Nα-Fmoc deprotection reagent," *Journal of Peptide Science*, vol. 7, no. 9, pp. 488–494, 2001.

[41] M. A. Hossain, R. A. D. Bathgate, C. K. Kong et al., "Synthesis, conformation, and activity of human insulin-like peptide 5 (INSL5)," *ChemBioChem*, vol. 9, no. 11, pp. 1816–1822, 2008.

[42] P. Y. Chou and G. D. Fasman, "Prediction of secondary structures of proteins," *Advances in Enzymology*, vol. 47, pp. 45–146, 1978.

[43] A. Karlstrim and A. Undén, "Design of protecting groups for the beta-carboxylic group of aspartic acid that minimize base-catalyzed aspartimide formation," *International Journal of Peptide and Protein Research*, vol. 48, no. 4, pp. 305–311, 1996.

[44] M. Mergler, F. Dick, B. Sax, C. Stähelin, and T. Vorherr, "The aspartimide problem in Fmoc-based SPPS—part I," *Journal of Peptide Science*, vol. 9, pp. 36–46, 2003.

[45] M. Mergler, F. Dick, B. Sax et al., "The aspartimide problem in Fmoc-based SPPS—part II," *Journal of Peptide Science*, vol. 9, no. 8, pp. 518–526, 2003.

[46] M. Mergler and F. Dick, "The aspartimide problem in Fmoc-based SPPS—part III," *Journal of Peptide Science*, vol. 11, no. 10, pp. 650–657, 2005.

4

Amyloid Beta Peptide Slows Down Sensory-Induced Hippocampal Oscillations

Fernando Peña-Ortega[1] and Ramón Bernal-Pedraza[1,2]

[1] Departamento de Neurobiología del Desarrollo y Neurofisiología, Instituto de Neurobiología,
 Universidad Nacional Autónoma de México, UNAM-Campus Juriquilla, 76230 Juriquilla, QRO, Mexico
[2] Departamento de Farmacobiología, Cinvestav-IFN, Mexico City, DF, Mexico

Correspondence should be addressed to Fernando Peña-Ortega, jfpena@unam.mx

Academic Editor: Ayman El-Faham

Alzheimer's disease (AD) progresses with a deterioration of hippocampal function that is likely induced by amyloid beta (Aβ) oligomers. Hippocampal function is strongly dependent on theta rhythm, and disruptions in this rhythm have been related to the reduction of cognitive performance in AD. Accordingly, both AD patients and AD-transgenic mice show an increase in theta rhythm at rest but a reduction in cognitive-induced theta rhythm. We have previously found that monomers of the short sequence of Aβ (peptide 25–35) reduce sensory-induced theta oscillations. However, considering on the one hand that different Aβ sequences differentially affect hippocampal oscillations and on the other hand that Aβ oligomers seem to be responsible for the cognitive decline observed in AD, here we aimed to explore the effect of Aβ oligomers on sensory-induced theta rhythm. Our results show that intracisternal injection of Aβ1–42 oligomers, which has no significant effect on spontaneous hippocampal activity, disrupts the induction of theta rhythm upon sensory stimulation. Instead of increasing the power in the theta band, the hippocampus of Aβ-treated animals responds to sensory stimulation (tail pinch) with an increase in lower frequencies. These findings demonstrate that Aβ alters induced theta rhythm, providing an *in vivo* model to test for therapeutic approaches to overcome Aβ-induced hippocampal and cognitive dysfunctions.

1. Introduction

Alzheimer's disease (AD), the most common form of dementia, is characterized by a progressive decline in cognitive function [1–5] that correlates with the extracellular accumulation of amyloid beta protein (Aβ) [1, 4, 5]. Deterioration of hippocampal function, likely induced by Aβ oligomers, contributes to the memory deficits associated with Alzheimer's disease (AD) [5–8]. Normal hippocampal function is strongly dependent on a 3 to 10 Hz oscillatory activity, namely, the theta rhythm [9–11]. Theta oscillations have been associated with various cognitive processes in several species, including humans [9–11]. Theta rhythm abnormalities are usually related to memory deficits and pathological changes in the brain [12–14]. In fact, subjects with AD show a typical "electroencephalographic slowing" that includes increased slow rhythms and decreased fast rhythms [6, 13, 15, 16]. Regarding theta rhythm, AD patients show increased theta rhythm at rest [6, 15, 16], but they also show a decrease in induced-theta rhythm; both of these changes in theta rhythm correlate with a reduced cognitive performance [17]. A similar contradictory scenario has been found in transgenic mice that overproduce Aβ and exhibit AD-like symptoms [18, 19]. The complex relationships between AD pathology and theta rhythms have been explained by the theta rhythm heterogeneity that exists both in humans and in mice [12, 20]. Experimentally, the reduction in resting hippocampal theta rhythm has been mimicked by Aβ application, both *in vitro* [21–23] and *in vivo* [24, 25]. However, just one previous study has shown that intracerebroventricular injection of monomers of a short Aβ sequence (peptide 25–35) decreases the power of the induced theta rhythm

[26]. This finding still needs to be confirmed because different $A\beta$ peptides, as well as their aggregation states, differentially affect similar hippocampal rhythms [27]. Thus, in this study we explored the effect of oligomers of the full-length $A\beta$ sequence (peptide 1–42) on induced theta rhythm *in vivo*. The use of $A\beta$1–42 oligomers has more relevance for the study of AD-related neural network disruption since early symptoms of AD are better correlated with the amount of soluble $A\beta$ than other histopathological makers [2, 3]. Our data show that intracisternal application of $A\beta$ slows down sensory-induced hippocampal oscillations, supplanting theta oscillations with a slower rhythm.

2. Materials and Methods

Experimental protocols were approved by The Local Committees of Ethics on Animal Experimentation (CICUAL-Cinvestav and INB-UNAM) and followed the regulations established in the Mexican Official Norm for the Use and Care of Laboratory Animals ("Norma Oficial Mexicana" NOM-062-ZOO-1999). For these experiments, Wistar rats (300–330 g) were briefly and lightly anesthetized with ether vapor just before receiving a single, intracisternal injection of $5\,\mu$L of either vehicle (F12 medium) or oligomerized $A\beta$1–42 (5 and 50 pmoles). The injector was connected to a Hamilton syringe mounted on dual perfusion pump (Harvard Apparatus Co., MA, USA). Animals were allowed to recover for 1 h after the intracisternal injection. Then, the animals were anesthetized with urethane (1.3 g/Kg; i.p.) and secured in a Kopf stereotaxic frame with the nose bar positioned at −3.3 mm [28, 29]. A bipolar electrode was implanted in the left dorsal hippocampus ($A = -3.6$ mm $L = 2.4$ mm and $V = 4.2$ mm from bregma, according to the atlas of Paxinos and Watson [30]) using standard stereotaxic procedures. The electrodes were attached to male connector pins, which were inserted into a connector strip. Hippocampal field recordings were amplified and filtered (highpass, 0.5 Hz; lowpass, 1.5 KHz) with a wideband AC amplifier (Grass Instruments, Quincy, MA, USA). Theta rhythm was elicited with sensory stimulation, consisting of a tail pinch produced by a plastic clamp positioned on the tail 2 cm from its base. A tail pinch, lasting 75 s, was applied each 10–20 min for at least 1 h. At the end of the hippocampal field recordings, all animals were processed for histological location of the electrode [28, 29, 31]. The recording site was visually confirmed to be located in the hippocampal fissure.

All recordings were digitized at 3–9 KHz and stored on a personal computer with an acquisition system from National Instruments (Austin, TX, USA) by using custom-made software designed in the LabView environment. The recordings obtained were analyzed offline by performing classical power spectrum analysis with a resolution of 0.61 Hz [26, 27, 32]. Segments of 30 sec were analyzed using a Rapid Fourier Transform Algorithm, with a Hamming window, in Clampfit (Molecular Devices). The power spectra during the tail pinch, at any given frequency, were also divided by their corresponding prestimulus power spectra and expressed as percentage of control (100% meaning no difference between

TABLE 1: Power and peak frequency of the hippocampal activity recorded in anaesthetized animals in control conditions and after the intracisternal injection of amyloid beta ($A\beta$). No significant differences were observed among or within groups.

Condition	Power (nV^2)	Peak Frequency (Hz)
Urethane	4.3 ± 2.5	2.5 ± 0.5
+ Tail pinch	5.1 ± 2.6	3.0 ± 0.4
Urethane + $A\beta$ 5 pmoles	1.2 ± 1.0	3.4 ± 0.6
+ Tail pinch	1.5 ± 1.3	3.4 ± 0.6
Urethane + $A\beta$ 50 pmoles	2.9 ± 1.7	3.9 ± 0.1
+ Tail pinch	3.6 ± 2.1	3.4 ± 0.4

tail-pinch and prestimulus power spectra). The mean difference spectra were then calculated by averaging the differences obtained in any given group [33–35]. For time-frequency analysis, segments of 40 s were analyzed using the Morlet wavelet basis and plotted as a time-frequency representation (TFR) [26, 32].

Data are expressed as mean $\pm$ standard error of mean (SEM). To analyze the data, the Wilcoxon signed-rank test was used to compare control versus tail-pinch spectra in the same group of animals. The Mann-Whitney U test was used to compare groups. A value of $P < 0.05$ was accepted as significant.

3. Results

Under urethane anesthesia, hippocampal local field potential showed a pattern of irregular activity (Figure 1(a); blue trace) that resembles the so-called large amplitude irregular activity (LIA) and that corresponds to the activity observed during immobility and slow-wave sleep [9, 12]. Such activity turns into more steady, oscillatory activity upon sensory stimulation (tail pinch; Figure 1(a); red trace). The spectrograms show that basal hippocampal activity under urethane anesthesia consists of a variable mixture of frequency components that vary over time (Figure 1(b)). In contrast, upon sensory stimulation, hippocampal activity exhibits a more constant oscillatory pattern (Figure 1(b)). The power spectrum shows that basal hippocampal activity under urethane anesthesia peaks at 2.5 ± 0.5 Hz, whereas theta rhythm has a frequency of 3.0 ± 0.4 Hz (Figure 1(b)). Quantification of the change in power upon tail pinch, compared with basal hippocampal activity, shows that sensory stimulation significantly increases the power in the low theta range (3.7–4.3 Hz) (Figure 1(c); inset).

When testing the effects of $A\beta$ oligomers on hippocampal activity, we did not find any significant difference in the hippocampal activity compared with control animals, due to the high variability among groups, either in power or peak frequency, due to the high variability among groups (Table 1). As illustrated in Figure 2, the hippocampal activity recorded after intracisternal injection of 5 pmoles of $A\beta$ oligomers is still characterized by a pattern of nonstationary, irregular activity under urethane anesthesia (Figure 2(a);

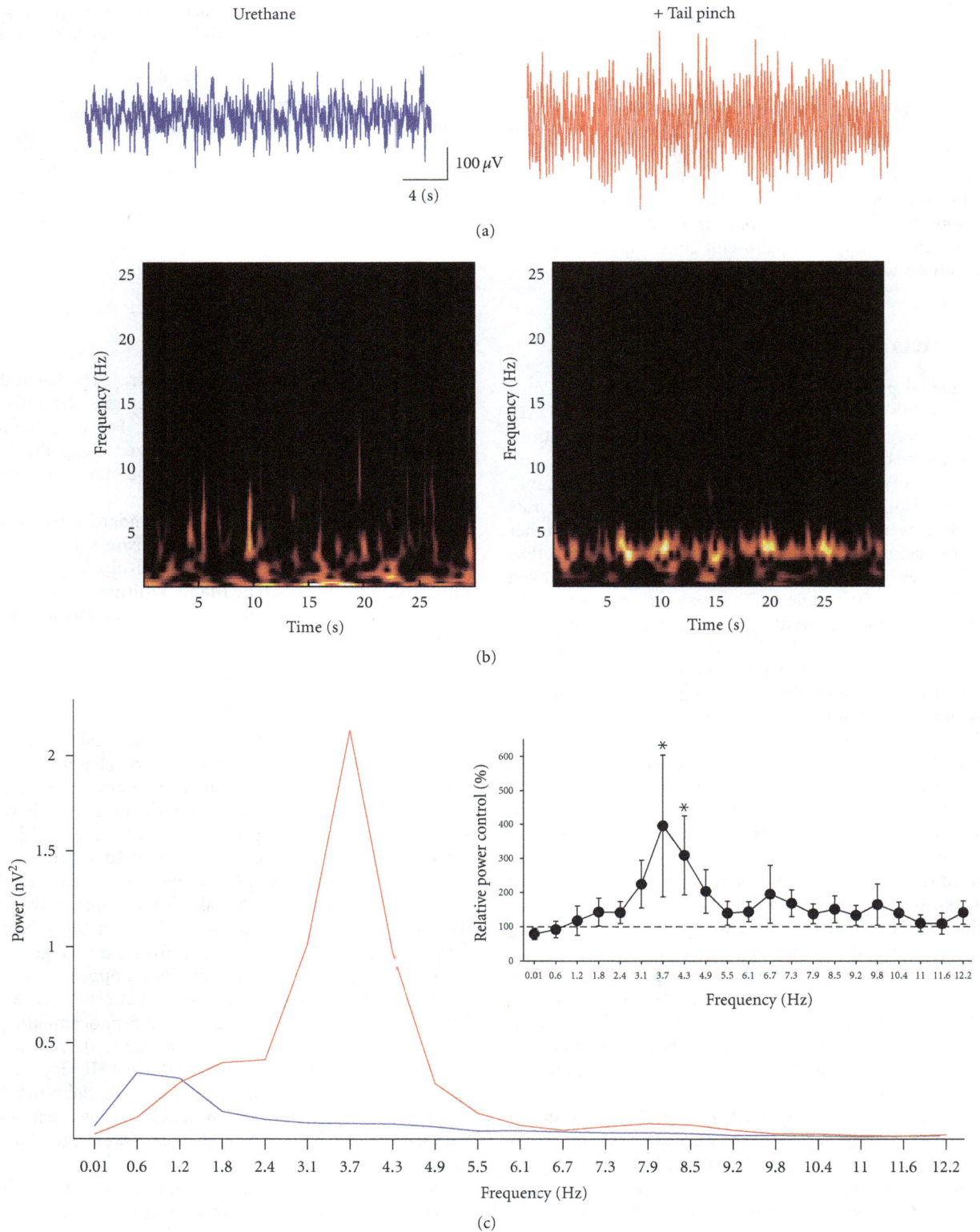

FIGURE 1: Sensory stimulation induces hippocampal theta oscillations. (a) Representative field recordings obtained from the hippocampal fissure in a urethane-anaesthetized rat at rest (blue recording) and upon sensory stimulation (red trace). (b) and (c) show the spectrograms and the power spectra, respectively, of the traces shown in (a). The blue power spectrum corresponds to the recording at rest, and the red power spectrum corresponds to the recording upon sensory stimulation. The inset in (c) shows the quantification of the change in power upon sensory stimulation, compared with basal hippocampal activity. *Indicates a significant difference compared to the control ($P < 0.05$; Wilcoxon signed-rank test).

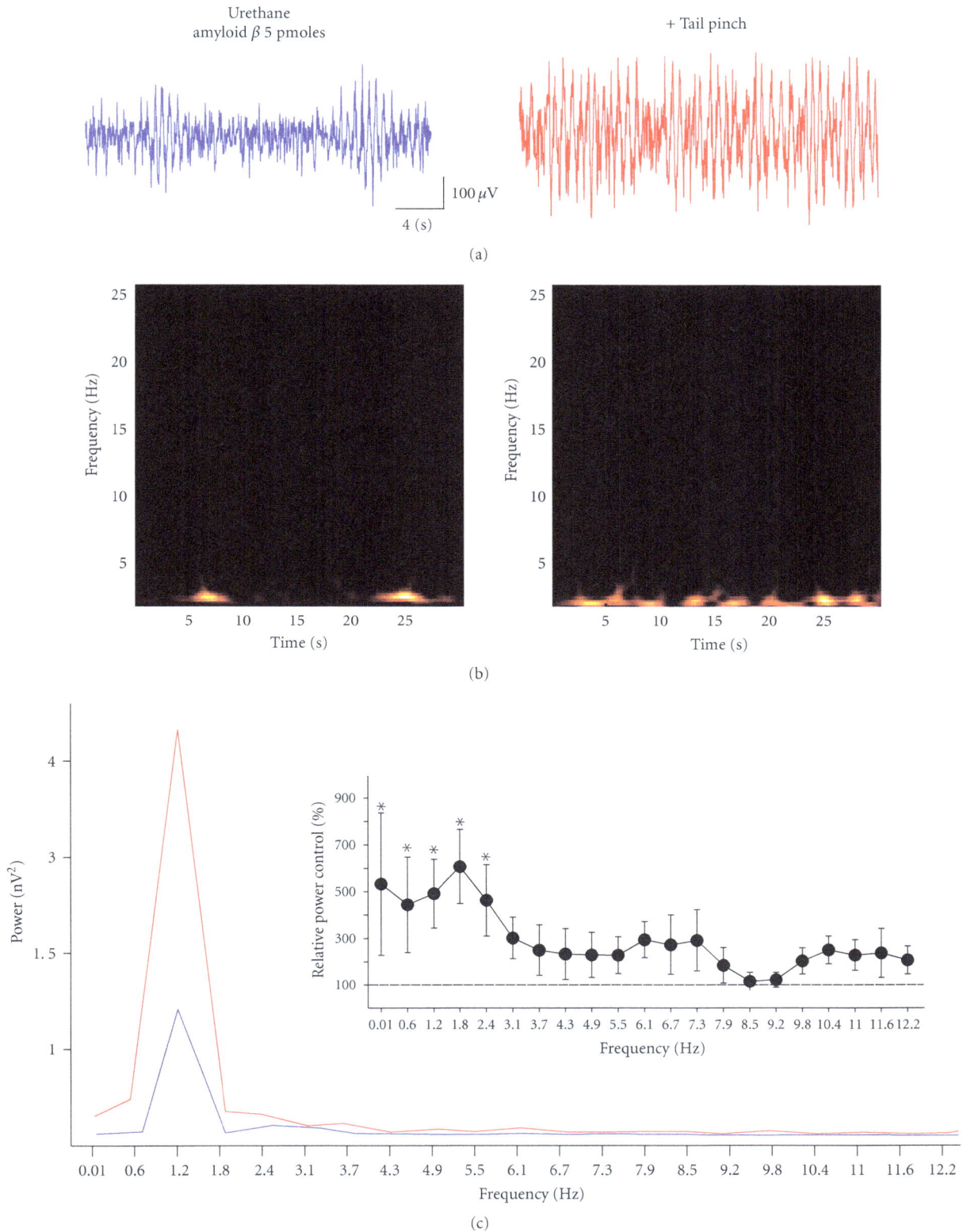

FIGURE 2: Effect of 5 pmoles amyloid beta on the sensory-induced hippocampal theta oscillations. (a) Representative field recordings obtained from the hippocampal fissure in a urethane-anaesthetized rat at rest (blue recording) and upon sensory stimulation (red recording). (b) and (c) show the spectrograms and the power spectra, respectively, of the traces shown in (a). The blue spectrum corresponds to the recording at rest, and the red power spectrum corresponds to the recording upon sensory stimulation. The inset in (c) shows the quantification of change in power upon sensory stimulation, compared with basal hippocampal activity. *Indicates a significant difference compared to control ($P < 0.05$; Wilcoxon signed-rank test).

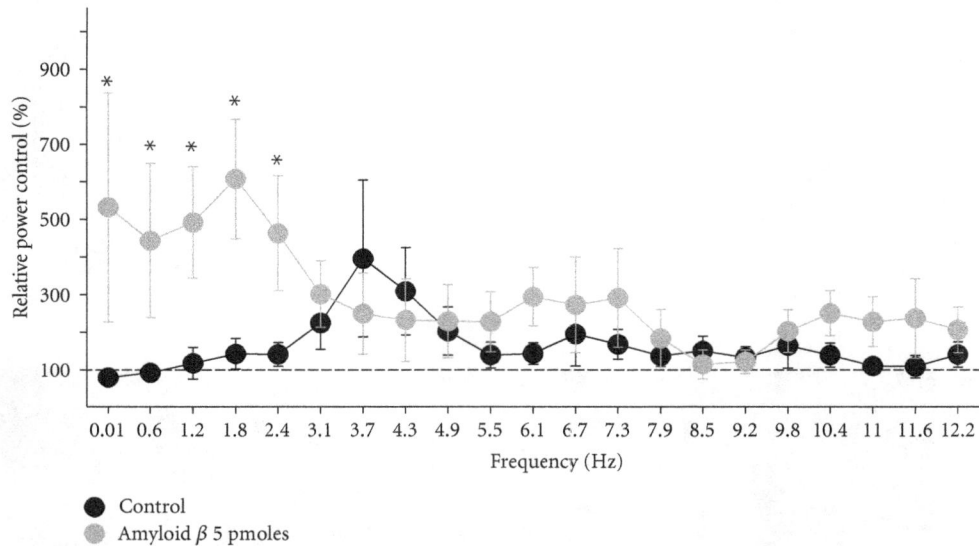

Control
Amyloid β 5 pmoles

(a)

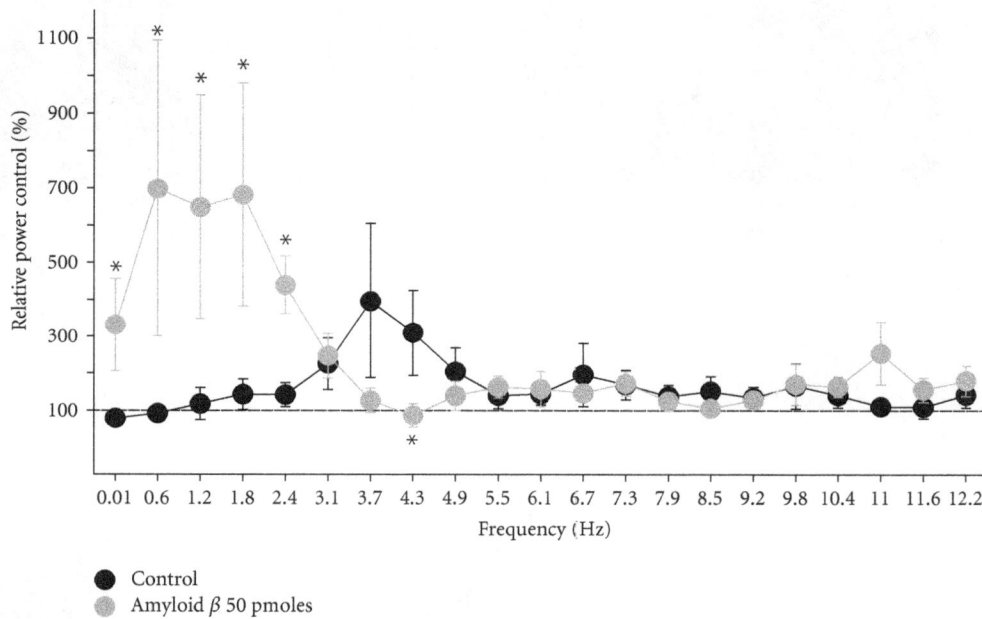

Control
Amyloid β 50 pmoles

(b)

FIGURE 3: Amyloid beta slows, in a dose-dependent manner, the oscillatory activity induced by sensory stimulation. Change in power induced by sensory stimulation in control rats (black circles; $n = 9$) compared to that in amyloid beta-injected rats (gray circles; $n = 6$). Animals were injected with two doses of amyloid beta. With 5 pmoles (a), the increase in power, upon sensory stimulation, shifts towards slow frequencies. Injection of 50 pmoles of amyloid beta (b) also shifts the increase in power, upon sensory stimulation, towards slow frequencies, and it also significantly reduces the increase in theta rhythm. *Indicates a significant difference compared to control rats ($P < 0.05$; Mann-Whitney U test).

blue trace). This activity also turns into a more homogeneous oscillatory activity upon sensory stimulation (tail pinch; Figures 2(a), 2(b); red trace). The spectrograms show that basal hippocampal activity under urethane anesthesia consists of a variable mixture of frequency components that change over time (Figure 2(b)). In contrast, upon sensory stimulation hippocampal activity turns into a more stationary, oscillatory state (Figure 2(b)). On average, in animals injected with 5 pmoles of Aβ oligomers and under urethane anesthesia, basal hippocampal activity peaks at 3.4 ± 0.6 Hz, and the tail pinch-induced rhythm has a frequency of 3.4 ± 0.6 Hz (Table 1). As mentioned, neither the power nor the peak frequency of hippocampal activity changed upon Aβ application in either basal or sensory-stimulated conditions (Table 1). However, quantification of the change in power upon tail pinch shows significant changes compared with basal hippocampal activity. Sensory stimulation in Aβ-treated animals significantly increases the power in low frequencies (0.01–2.4 Hz) (Figure 2(c); inset and Figure 3). In fact, the increase in power of those frequencies was significantly higher in Aβ-treated animals than in control (vehicle-treated) animals (Figure 3). Although sensory-induced theta rhythm was not significantly changed relative to control animals by injection of 5 pmoles of Aβ, it was significantly reduced at 4.3 Hz by a higher dose, 50 pmoles of Aβ (Figure 3).

4. Discussion

Our results show that intracisternal application of Aβ1–42 oligomers does not produce any significant effect on spontaneous hippocampal activity, but it disrupts the hippocampal activation induced by sensory stimulation. Aβ-treated animals do respond to sensory stimulation (tail pinch), but the increase occurs in lower frequencies than in control animals. These findings may correlate with the EEG slowing observed in AD patients [6, 13, 15, 16] as well as with the reduction in evoked theta rhythm [17] that was also observed in AD patients. In our previous report, we demonstrated that intracerebroventricular injection of monomers of the short Aβ sequence (25–35) reduced the power of induced theta rhythm [26]. However, in that study we did not find the change in theta frequency observed here. The simplest explanation for this difference is that oligomers of Aβ1–42 may act on different cellular targets and produce different effects than monomers of Aβ25–35 [27]. If so, without ignoring the advantages of using monomers of Aβ25–35 [23, 26], we believe that the use of Aβ1–42 oligomers may represent a more valid model to explore some of the changes related to AD pathology. A second potential explanation is that in the current study we used intracisternal application of Aβ in contrast to the intracerebroventricular injections used previously [26]. It has been found that intracerebroventricular and intracisternal administration of the same substance do not always produce the same effect, probably due to differences in the brain structures preferentially reached by the injection in those sites, as well as to the different concentrations of the injected substance reached at those structures [36–42].

Our results are in agreement with previous findings that direct application of Aβ, either in the medial septum or in the hippocampus, reduces theta-rhythm power both *in vivo* and *in vitro* [22–26, 43]. However, in our hands, intracisternal application of Aβ also shifts the frequency of sensory-evoked oscillations to the left. Several factors have been associated with the reduction in theta power. For instance, we have shown that this reduction is related to a reduction in intrahippocampal glutamatergic transmission [22, 26], but the reduction in power also has been associated with the blockade of several K+ channels [23, 44] or with Aβ-induced changes in septal neuron firing [23–25]. The shift in frequency induced by Aβ might be related to changes in the activity of interneurons in the hippocampus or elsewhere [23–25] or to the effect of Aβ on transient potassium currents

[44]. Overall, the effects of Aβ on hippocampal theta rhythm seem to involve a complex mixture of effects on several neural types within several neural networks. It is well known that hippocampal theta rhythm could be affected by a decoupling of one or several autonomous oscillators within the hippocampus [45] or in other interconnected neural networks [24, 25]. Correlative *in vitro* experiments are required to corroborate this hypothesis and to determine viable molecular targets to prevent Aβ-induced neural network disruption.

Acknowledgments

The authors would like to thank Dorothy Pless for reviewing the English version of this paper. They also thank José Rodolfo Fernandez and Arturo Franco for technical assistance. This work was sponsored by grants (to F. Peña-Ortega) from DGAPA IA201511; CONACyT 59187,151261; and from the Alzheimer's Association NIRG-11-205443.

References

[1] H. Braak and E. Braak, "Diagnostic criteria for neuropathologic assessment of Alzheimer's disease," *Neurobiology of Aging*, vol. 18, no. 4, supplement 1, pp. S85–S88, 1997.

[2] L. F. Lue, Y. M. Kuo, A. E. Roher et al., "Soluble amyloid β peptide concentration as a predictor of synaptic change in Alzheimer's disease," *American Journal of Pathology*, vol. 155, no. 3, pp. 853–862, 1999.

[3] J. Näslund, V. Haroutunian, R. Mohs et al., "Correlation between elevated levels of amyloid β-peptide in the brain and cognitive decline," *JAMA*, vol. 283, no. 12, pp. 1571–1577, 2000.

[4] F. Peña, A. I. Gutiérrez-Lerma, R. Quiroz-Baez, and C. Arias, "The role of β-amyloid protein in synaptic function: implications for Alzheimer's disease therapy," *Current Neuropharmacology*, vol. 4, no. 2, pp. 149–163, 2006.

[5] D. J. Selkoe, "Alzheimer's disease is a synaptic failure," *Science*, vol. 298, no. 5594, pp. 789–791, 2002.

[6] C. Babiloni, G. B. Frisoni, M. Pievani et al., "Hippocampal volume and cortical sources of EEG alpha rhythms in mild cognitive impairment and Alzheimer disease," *NeuroImage*, vol. 44, no. 1, pp. 123–135, 2009.

[7] W. L. Klein, G. A. Krafft, and C. E. Finch, "Targeting small A β oligomers: the solution to an Alzheimer's disease conundrum?" *Trends in Neurosciences*, vol. 24, no. 4, pp. 219–224, 2001.

[8] T. Ondrejcak, I. Klyubin, N. W. Hu, A. E. Barry, W. K. Cullen, and M. J. Rowan, "Alzheimer's disease amyloid β-protein and synaptic function," *NeuroMolecular Medicine*, vol. 12, no. 1, pp. 13–26, 2010.

[9] B. H. Bland and L. V. Colom, "Extrinsic and intrinsic properties underlying oscillation and synchrony in limbic cortex," *Progress in Neurobiology*, vol. 41, no. 2, pp. 157–208, 1993.

[10] M. J. Kahana, D. Seelig, and J. R. Madsen, "Theta returns," *Current Opinion in Neurobiology*, vol. 11, no. 6, pp. 739–744, 2001.

[11] W. Klimesch, "EEG alpha and theta oscillations reflect cognitive and memory performance: a review and analysis," *Brain Research Reviews*, vol. 29, no. 2-3, pp. 169–195, 1999.

[12] L. V. Colom, "Septal networks: relevance to theta rhythm, epilepsy and Alzheimer's disease," *Journal of Neurochemistry*, vol. 96, no. 3, pp. 609–623, 2006.

[13] C. E. Jackson and P. J. Snyder, "Electroencephalography and event-related potentials as biomarkers of mild cognitive impairment and mild Alzheimer's disease," *Alzheimer's and Dementia*, vol. 4, no. 1, supplement 1, pp. S137–S143, 2008.

[14] P. J. Uhlhaas and W. Singer, "Neural synchrony in brain disorders: relevance for cognitive dysfunctions and pathophysiology," *Neuron*, vol. 52, no. 1, pp. 155–168, 2006.

[15] C. Babiloni, E. Cassetta, G. Binetti et al., "Resting EEG sources correlate with attentional span in mild cognitive impairment and Alzheimer's disease," *European Journal of Neuroscience*, vol. 25, no. 12, pp. 3742–3757, 2007.

[16] C. Huang, L. O. Wahlund, T. Dierks, P. Julin, B. Winblad, and V. Jelic, "Discrimination of Alzheimer's disease and mild cognitive impairment by equivalent EEG sources: a cross-sectional and longitudinal study," *Clinical Neurophysiology*, vol. 111, no. 11, pp. 1961–1967, 2000.

[17] T. D. R. Cummins, M. Broughton, and S. Finnigan, "Theta oscillations are affected by amnestic mild cognitive impairment and cognitive load," *International Journal of Psychophysiology*, vol. 70, no. 1, pp. 75–81, 2008.

[18] A. Jyoti, A. Plano, G. Riedel, and B. Platt, "EEG, activity, and sleep architecture in a transgenic AβPP swe/PSEN1A246E Alzheimer's disease mouse," *Journal of Alzheimer's Disease*, vol. 22, no. 3, pp. 873–887, 2010.

[19] J. Wang, S. Ikonen, K. Gurevicius, T. Van Groen, and H. Tanila, "Alteration of cortical EEG in mice carrying mutated human APP transgene," *Brain Research*, vol. 943, no. 2, pp. 181–190, 2002.

[20] J. Shin, "Theta rhythm heterogeneity in humans," *Clinical Neurophysiology*, vol. 121, no. 3, pp. 456–457, 2010.

[21] M. Akay, K. Wang, Y. M. Akay, A. Dragomir, and J. Wu, "Nonlinear dynamical analysis of carbachol induced hippocampal oscillations in mice," *Acta Pharmacologica Sinica*, vol. 30, no. 6, pp. 859–867, 2009.

[22] H. Balleza-Tapia, A. Huanosta-Gutiérrez, A. Márquez-Ramos, N. Arias, and F. Peña, "Amyloid β oligomers decrease hippocampal spontaneous network activity in an age-dependent manner," *Current Alzheimer Research*, vol. 7, no. 5, pp. 453–462, 2010.

[23] R. N. Leão, L. V. Colom, L. Borgius, O. Kiehn, and A. Fisahn, "Medial septal dysfunction by Aβ-induced KCNQ channel-block in glutamatergic neurons," *Neurobiology of Aging*. In press.

[24] L. V. Colom, M. T. Castañeda, C. Bañuelos et al., "Medial septal β-amyloid 1–40 injections alter septo-hippocampal anatomy and function," *Neurobiology of Aging*, vol. 31, no. 1, pp. 46–57, 2010.

[25] V. Villette, F. Poindessous-Jazat, A. Simon et al., "Decreased rhythmic GABAergic septal activity and memory-associated θ oscillations after hippocampal amyloid-β pathology in the rat," *Journal of Neuroscience*, vol. 30, no. 33, pp. 10991–11003, 2010.

[26] F. Peña, B. Ordaz, H. Balleza-Tapia et al., "Beta-amyloid protein (25–35) disrupts hippocampal network activity: role of Fyn-kinase," *Hippocampus*, vol. 20, no. 1, pp. 78–96, 2010.

[27] A. Adaya-Villanueva, B. Ordaz, H. Balleza-Tapia, A. Márquez-Ramos, and F. Peña-Ortega, "Beta-like hippocampal network activity is differentially affected by amyloid beta peptides," *Peptides*, vol. 31, no. 9, pp. 1761–1766, 2010.

[28] F. Peña and R. Tapia, "Relationships among seizures, extracellular amino acid changes, and neurodegeneration induced by 4-aminopyridine in rat hippocampus: a microdialysis and electroencephalographic study," *Journal of Neurochemistry*, vol. 72, no. 5, pp. 2006–2014, 1999.

[29] F. Peña and R. Tapia, "Seizures and neurodegeneration induced by 4-aminopyridine in rat hippocampus in vivo: role of glutamate- and GABA-mediated neurotransmission and of ion channels," *Neuroscience*, vol. 101, no. 3, pp. 547–561, 2000.

[30] G. Paxinos and C. Watson, *The Rat Brain in Stereotaxic Coordinates*, Academic Press, 2005.

[31] L. Carmona-Aparicio, F. Peña, A. Borsodi, and L. Rocha, "Effects of nociceptin on the spread and seizure activity in the rat amygdala kindling model: their correlations with 3H-leucyl-nociceptin binding," *Epilepsy Research*, vol. 77, no. 2-3, pp. 75–84, 2007.

[32] R. N. Romcy-Pereira, D. B. de Araujo, J. P. Leite, and N. Garcia-Cairasco, "A semi-automated algorithm for studying neuronal oscillatory patterns: a wavelet-based time frequency and coherence analysis," *Journal of Neuroscience Methods*, vol. 167, no. 2, pp. 384–392, 2008.

[33] C. Andrew and G. Fein, "Induced theta oscillations as biomarkers for alcoholism," *Clinical Neurophysiology*, vol. 121, no. 3, pp. 350–358, 2010.

[34] J. S. Macdonald, S. Mathan, and N. Yeung, "Trial-by-trial variations in subjective attentional state are reflected in ongoing prestimulus EEG alpha oscillations," *Frontiers in Psychology*, vol. 2, article 82, 2011.

[35] J. J. Wright and M. D. Craggs, "Intracranial self-stimulation, cortical arousal, and the sensorimotor neglect syndrome," *Experimental Neurology*, vol. 65, no. 1, pp. 42–52, 1979.

[36] J. Czimmer, M. Million, and Y. Taché, "Urocortin 2 acts centrally to delay gastric emptying through sympathetic pathways while CRF and urocortin 1 inhibitory actions are vagal dependent in rats," *American Journal of Physiology*, vol. 290, no. 3, pp. G511–G518, 2006.

[37] O. Gunther, G. L. Kovacs, G. Szabo, H. Schwarzberg, and G. Telegdy, "Differential effect of vasopressin on open-field activity and passive avoidance behaviour following intracerebroventricular versus intracisternal administration in rats," *Acta Physiologica Hungarica*, vol. 71, no. 2, pp. 203–206, 1988.

[38] O. Gunther and H. Schwarzberg, "Influence of intracerebroventricularly and intracisternally administered vasopressin on the hypothalamic self-stimulation rate of the rat," *Neuropeptides*, vol. 10, no. 4, pp. 361–367, 1987.

[39] D. Harland, S. M. Gardiner, and T. Bennett, "Differential cardiovascular effects of centrally administered vasopressin in conscious Long Evans and Brattleboro rats," *Circulation Research*, vol. 65, no. 4, pp. 925–933, 1989.

[40] H. Lee, N. N. Naughton, J. H. Woods, and M. C. H. Ko, "Characterization of scratching responses in rats following centrally administered morphine or bombesin," *Behavioural Pharmacology*, vol. 14, no. 7, pp. 501–508, 2003.

[41] M. Ozawa, M. Aono, and M. Moriga, "Central effects of pituitary adenylate cyclase activating polypeptide (PACAP) on gastric motility and emptying in rats," *Digestive Diseases and Sciences*, vol. 44, no. 4, pp. 735–743, 1999.

[42] K. H. Park, J. P. Long, and J. G. Cannon, "Evaluation of the central and peripheral components for induction of postural hypotension by guanethidine, clonidine, dopamine2 receptor agonists and 5-hydroxytryptamine(1A) receptor agonists," *Journal of Pharmacology and Experimental Therapeutics*, vol. 259, no. 3, pp. 1221–1230, 1991.

[43] E. A. Mugantseva and I. Y. Podolski, "Animal model of Alzheimer's disease: characteristics of EEG and memory," *Central European Journal of Biology*, vol. 4, no. 4, pp. 507–514, 2009.

[44] X. Zou, D. Coyle, K. Wong-Lin, and L. Maguire, "Beta-amyloid induced changes in A-type K$^+$ current can alter

hippocampo-septal network dynamics," *Journal of Computational Neuroscience*. In press.

[45] R. Goutagny, J. Jackson, and S. Williams, "Self-generated theta oscillations in the hippocampus," *Nature Neuroscience*, vol. 12, no. 12, pp. 1491–1493, 2009.

5

Systemic Ghrelin Administration Alters Serum Biomarkers of Angiogenesis in Diet-Induced Obese Mice

M. Khazaei and Z. Tahergorabi

Department of Physiology, Isfahan University of Medical Sciences, Isfahan 81743638, Iran

Correspondence should be addressed to M. Khazaei; khazaei@med.mui.ac.ir

Academic Editor: Weihong Pan

Introduction. Ghrelin is a gastrointestinal endocrine peptide that was initially identified as the endogenous ligand of growth hormone secretagogue receptor; however, recently, the cardiovascular effect of this peptide has been indicated. In this study, we investigated the effect of ghrelin administration on serum biomarkers of angiogenesis including leptin, nitric oxide (NO), vascular endothelial growth factor (VEGF), and its soluble receptor (VEGF receptor 1 or sFlt-1) in control- and diet-induced obese mice. *Methods.* Male C57BL/6 mice were randomly divided into four groups, normal diet (ND) or control, ND + ghrelin, high-fat-diet (HFD) or obese and HFD + ghrelin ($n = 6$/group). Obese and control groups received either HFD or ND for 15 weeks. Then, the ghrelin was injected subcutaneously 100 μg/kg twice daily for 10 days. At the end of experiment, blood samples were collected for blood glucose, serum insulin, VEGF, sFlt-1, NO, and leptin measurements. *Results.* The obese animals had higher serum NO and leptin concentrations without changes in serum VEGF and sFlt-1 levels compared to control. Administration of ghrelin significantly increased serum VEGF and decreased serum leptin and NO concentrations in HFD group. *Conclusion.* Since ghrelin changes serum biomarkers of angiogenesis, it seems that it gets involved during states with abnormal angiogenesis.

1. Introduction

Prolonged imbalance of caloric intake and energy expenditure leads to complex metabolic disorder of obesity. It is associated with most common and chronic human diseases including type 2 diabetes, heart diseases, hypertension, and cancer [1].

Angiogenesis, the formation of new blood vessels from preexisting ones, is tightly linked with adipogenesis [2] and is considered as an essential component in development and expansion of adipose tissue [3]. Since expansion of adipose tissue (increasing cell size and number) creates adipose tissue hypoxia, it can lead to stabilization of the transcription factor hypoxia inducible factor1α (HIF-1α) [4, 5] that induces an angiogenic response [6].

Ghrelin is a gastrointestinal endocrine peptide and is identified as an endogenous ligand for the growth hormone secretagogue receptor type 1a (GHS-R Ia) [7]; however, it also regulates food intake and is associated with obesity [8]. Ghrelin and its receptors are expressed in endothelial cells and stimulate endothelial cell proliferation, migration, and angiogenesis [9]. Recently, the impact of ghrelin on cardiovascular system has been reported [10] including a decrease of peripheral vascular resistance in consequence an increase in cardiac index and stroke volume [11], improvement of ventricular remodeling [12], protection of myocytes from apoptosis [13], decrease of cardiac injury induced by ischemia/reperfusion (I/R) injury [14], and reduction of the infarct size (L). It also improves endothelial dysfunction, reduces vasoconstrictor effect of endothelin-1, and decreases blood pressure [10].

Plasma ghrelin level is associated with body mass index (BMI). It is indicated that obese patients have reduced plasma ghrelin levels [8]. The main objective of this study was to investigate the effect of ghrelin administration on serum biomarkers of angiogenesis including leptin, nitric oxide (NO), vascular endothelial growth factor (VEGF), and its soluble receptor (VEGF receptor 1 or sFlt-1) in control and obese mice.

2. Materials and Methods

2.1. Animals. Male C57BL/6 mice (5 weeks old, $n = 24$) were purchased from Pasteur Institute (Tehran, Iran), and three or four animals were housed together in one cage in controlled environment under a light-dark cycle (lights on at 19:00 and off at 07:00). The experimental procedures followed the Guiding Principles for the Care and Use of animals and were approved by the Isfahan University of Medical sciences. All mice were randomly divided into four groups: normal diet (ND) or control, ND + ghrelin, high-fat-diet (HFD) or obese and HFD + ghrelin ($n = 6$/group).

2.2. Diets and Ghrelin Administration. Mice were rendered obese by the HFD (Bio-Serv Research Diets, NJ, USA; Cat #F3282) contained with 59% from fat, 14% from protein, and 27% from carbohydrate (of total calories) starting at 5 weeks of age for 15 weeks. The ND mice were fed a standard diet (Pasteur Institute, Iran). All groups were allowed to eat food freely and had free access to water. Body weights were measured weekly. After 15 weeks, the ghrelin (Tocris Co., Bristol, UK) was administered subcutaneously 100 μg/kg twice daily for 10 days [15, 16].

2.3. Serum Measurements. Blood glucose was measured by glucometer (ACON Lab Inc San Diego, CA, USA) ELISA kits were used for determination of mice serum insulin (Mercodia, Uppsala, Sweden), VEGF and sFlt-1 (R&D systems, Minneapolis, USA), leptin (Invitrogen, Camarillo, CA 93012) and nitrite, the main metabolite of NO (Promega Corp, USA) concentrations.

2.4. Statistical Analysis. All values are expressed as mean ± SEM. The statistical software SPSS version 16 was used for data analysis. One-Way ANOVA was used to compare data between groups using LSD post-hoc test. $P < 0.05$ was considered statistically significant.

3. Results

3.1. Effect of Ghrelin on Body Weight. Figure 1 illustrates that administration of ghrelin for 10 days did not significantly change body weight in obese and control mice ($P > 0.05$).

3.2. Effect of Ghrelin on Blood Glucose and Serum Insulin Levels. As shown in Figure 2, there was a significant difference in blood glucose level between obese and control groups ($P < 0.05$). Administration of ghrelin did not significantly change blood glucose in obese and control mice ($P > 0.05$).

Serum insulin concentration in obese mice was significantly higher than that of control ($P < 0.05$). Ghrelin administration did not alter serum insulin concentration in control groups ($P > 0.05$), while significantly reduced it in obese group ($P > 0.05$) (Figure 2).

3.3. Effect of Ghrelin on Serum Biomarkers of Angiogenesis. The results indicated no significant differences in serum VEGF and sFlt-1 between obese and control animals

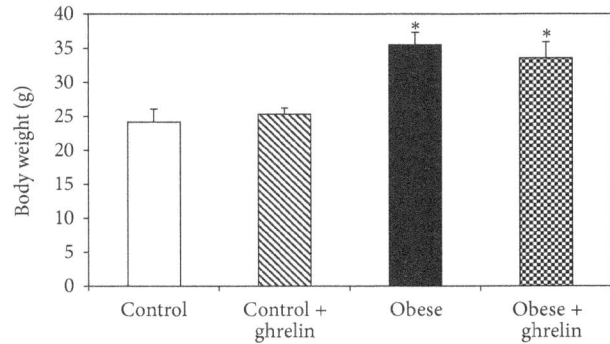

FIGURE 1: Body weight of the animals at the end of experiment. $^*P < 0.05$ compared to control groups.

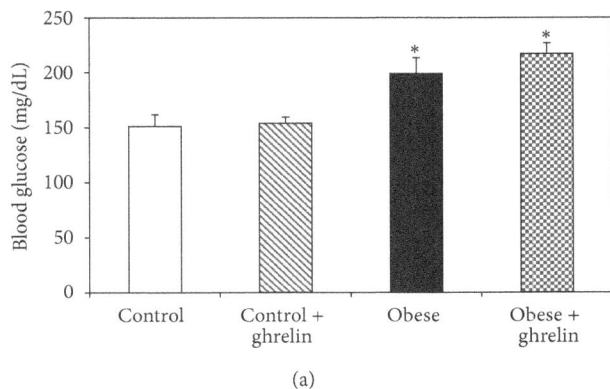

(a)

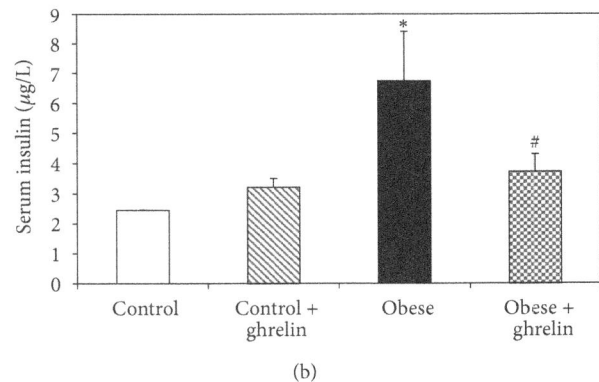

(b)

FIGURE 2: Blood glucose (a) and serum insulin (b) concentrations in experimental groups. $^*P < 0.05$ compared to control. $^\#P < 0.05$ compared to obese group.

($P < 0.05$); however, serum NO concentration in obese mice was higher than that of control ($P < 0.05$). Ghrelin administration increased serum VEGF and reduced serum NO level in obese mice and had no effect on sFlt-1 concentration (Figure 3).

3.4. Serum Leptin Measurement. Serum leptin level in obese mice was higher than that of control ($P < 0.05$), and ghrelin significantly reduced it in obese group ($P < 0.05$) (Figure 4).

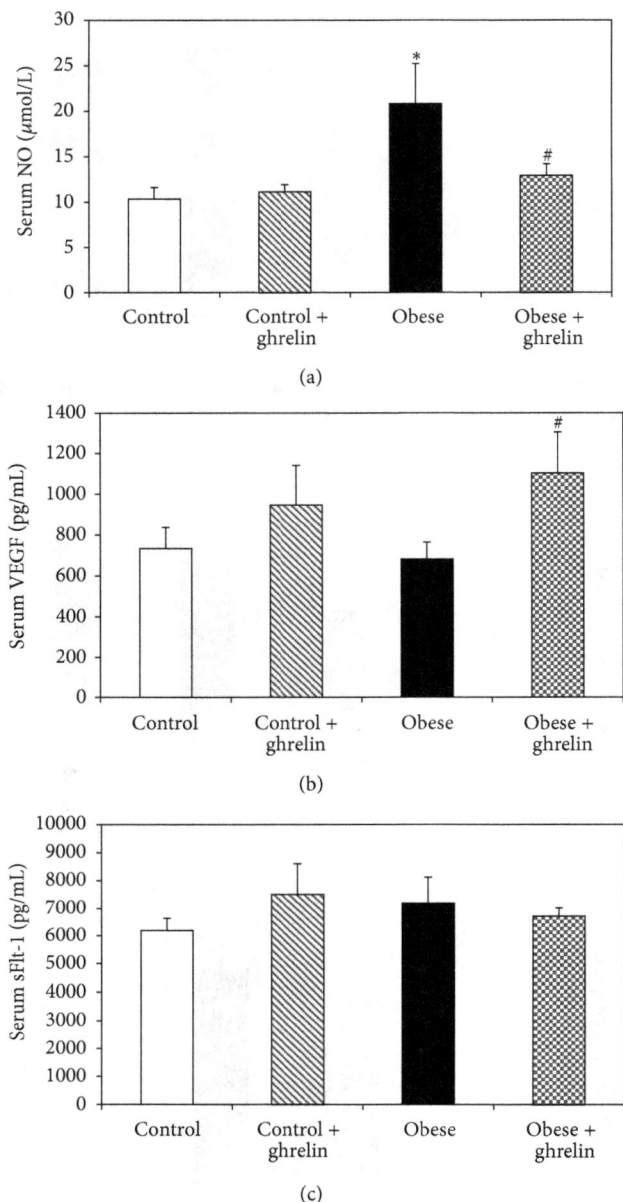

FIGURE 3: Effect of ghrelin on serum NO (a), VEGF (b), and sFlt-1 (c) concentrations. $^*P < 0.05$ compared to control. $^#P < 0.05$ compared to obese group.

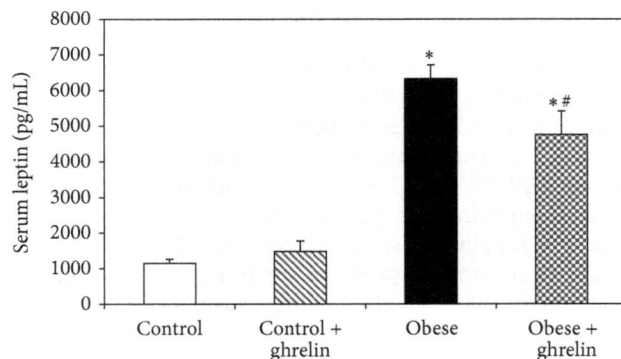

FIGURE 4: Effect of ghrelin on serum leptin level. $P < 0.05$ compared to control. $^*P < 0.05$ compared to control. $^#P < 0.05$ compared to obese group.

4. Discussion

The main finding of this study is that the obese mice had higher serum insulin, NO, and leptin concentrations compared to control without changes in serum VEGF and sFlt-1 levels. Ghrelin administration reduced serum NO, and leptin and increased serum VEGF concentrations in obese mice.

Higher blood glucose and insulin levels in HFD group indicate the insulin resistance in these animals. We demonstrated that although ghrelin treatment could not alter blood glucose level, it reduced serum insulin concentration in obese mice. Our data was in line with other studies [17, 18]. Ghrelin may also act on cellular glucose uptake [10] and may involve in control of glucose metabolism and insulin sensitivity [19]. Ghrelin stimulates insulin release; however, leptin inhibits insulin [20]. Perhaps, only ten days ghrelin treatment was the reason for unchanging of blood glucose level in the present study.

Modulation of vascular tissue and angiogenesis in adipose tissue is a strategy to affect obesity. Adipose tissue endothelial cells produce several angiogenic factors including leptin, NO, VEGF, FGF, HGF, and other growth factors [21]. NO is an endothelium-derived relaxing factor which has antiatherosclerotic effects through different mechanisms. However, it is a known angiogenic factor [22]. It is suggested that at the initial stage of obesity, a compensatory increase in NO production occurs due to upregulation of NO synthase [23]. On the other hand, adipogenesis increases upregulation of iNOS which increases NO synthesis due to chronic low-grade inflammation during obesity [2]. These data are in line with the results of the present study that we showed higher serum NO concentration in obese mice.

Leptin is an adipocyte-derived hormone that not only directly promotes angiogenesis and endothelial cell migration but also upregulates VEGF expression [24]. As we expected, in the present study, the obese animals had higher serum leptin level than that of control. These data was in agreement with the previous studies [25]. We also demonstrated that HFD did not change serum VEGF and sFlt-1 concentrations. Although some studies indicated higher serum VEGF level in obese subjects [26], a recent study showed that HFD did not affect plasma concentration of VEGF [27]. VEGF binds to two tyrosine kinase receptors of sFlt-1 and VEGFR2. sFlt-1 leads to anti- or proangiogenic signaling and inhibits angiogenic signaling through sequestration of VEGF ligands [28, 29]. In the present study, HFD did not change serum concentration of sFlt-1.

In our study, ghrelin administration reduced serum NO and leptin and increased serum VEGF concentrations in obese mice. Ghrelin is a gastrointestinal endocrine peptide which has several impacts on cardiovascular system [10]. Ghrelin and leptin circulate in the blood and have a role in regulation of body weight and energy homeostasis [13]. Study in human showed that plasma ghrelin inversely correlated to degree of obesity [30] and in this study, ghrelin reduced serum leptin level in obese mice. Thus, it seems that ghrelin has a protective mechanism including leptin resistance in setting obesity. Further studies need to clarify this. An *in vitro* studies indicated that ghrelin activates NO-dependent vasorelaxation in patients with metabolic syndrome [31]. Furthermore, there is a reciprocal regulation between VEGF and NO during angiogenesis process [32]. Thus, we expected that in the present study, ghrelin administration increased serum NO concentration. One explanation for this discrepancy is that ghrelin and leptin have mutually antagonistic effects on inflammatory cytokine expression in obesity [33] and reduced leptin after ghrelin administration may involve in reduction of serum NO level.

Recently, Yuan M.J. showed that in a rat model of myocardial infarction, chronic ghrelin treatment increased VEGF expression in peri-infarct zone and they suggested that ghrelin may induce angiogenesis after MI [34]. We also found that ghrelin altered serum biomarkers of angiogenesis and it seems that it may mediate angiogenesis through different mechanisms. Taken together, our results suggested that ghrelin administration changes the serum biomarkers of angiogenesis and can be involved during states with abnormal angiogenesis.

Conflict of Interests

There are no conflict of interests and the authors declare that they have no direct relationship with the mentioned commercial identities.

Acknowledgment

The authors thank the vice chancellor of Isfahan University of Medical Sciences for their support (Project no. 189142).

References

[1] F. Item and D. Konrad, "Visceral fat and metabolic inflammation: the portal theory revisited," *Obesity Reviews*, vol. 13, supplement 2, pp. 30–39, 2012.

[2] V. Christiaens and H. R. Lijnen, "Angiogenesis and development of adipose tissue," *Molecular and Cellular Endocrinology*, vol. 318, no. 1-2, pp. 2–9, 2010.

[3] L. Liu and M. Meydani, "Angiogenesis inhibitors may regulate adiposity," *Nutrition Reviews*, vol. 61, no. 11, pp. 384–387, 2003.

[4] J. Ye, Z. Gao, J. Yin, and Q. He, "Hypoxia is a potential risk factor for chronic inflammation and adiponectin reduction in adipose tissue of ob/ob and dietary obese mice," *American Journal of Physiology*, vol. 293, no. 4, pp. E1118–E1128, 2007.

[5] B. Wang, I. S. Wood, and P. Trayhurn, "Dysregulation of the expression and secretion of inflammation-related adipokines by hypoxia in human adipocytes," *Pflugers Archiv European Journal of Physiology*, vol. 455, no. 3, pp. 479–492, 2007.

[6] N. Halberg, T. Khan, M. E. Trujillo et al., "Hypoxia-inducible factor 1α induces fibrosis and insulin resistance in white adipose tissue," *Molecular and Cellular Biology*, vol. 29, no. 16, pp. 4467–4483, 2009.

[7] M. Kojima, H. Hosoda, Y. Date, M. Nakazato, H. Matsuo, and K. Kangawa, "Ghrelin is a growth-hormone-releasing acylated peptide from stomach," *Nature*, vol. 402, no. 6762, pp. 656–660, 1999.

[8] T. Shiiya, M. Nakazato, M. Mizuta et al., "Plasma ghrelin levels in lean and obese humans and the effect of glucose on ghrelin secretion," *Journal of Clinical Endocrinology and Metabolism*, vol. 87, no. 1, pp. 240–244, 2002.

[9] L. Wang, Q. Chen, G. Li, and D. Ke, "Ghrelin stimulates angiogenesis via GHSR1a-dependent MEK/ERK and PI3K/Akt signal pathways in rat cardiac microvascular endothelial cells," *Peptides*, vol. 33, no. 1, pp. 92–100, 2012.

[10] M. Tesauro, F. Schinzari, M. Caramanti, R. Lauro, and C. Cardillo, "Cardiovascular and metabolic effects of ghrelin," *Current Diabetes Reviews*, vol. 6, no. 4, pp. 228–235, 2010.

[11] N. Nagaya, M. Kojima, M. Uematsu et al., "Hemodynamic and hormonal effects of human ghrelin in healthy volunteers," *American Journal of Physiology*, vol. 280, no. 5, pp. R1483–R1487, 2001.

[12] T. Henriques-Coelho, J. Correia-Pinto, R. Roncon-Albuquerque et al., "Endogenous production of ghrelin and beneficial effects of its exogenous administration in monocrotaline-induced pulmonary hypertension," *American Journal of Physiology*, vol. 287, no. 6, pp. H2885–H2890, 2004.

[13] G. Baldanzi, N. Filigheddu, S. Cutrupi et al., "Ghrelin and des-acyl ghrelin inhibit cell death in cardiomyocytes and endothelial cells through ERK1/2 and PI 3-kinase/AKT," *Journal of Cell Biology*, vol. 159, no. 6, pp. 1029–1037, 2002.

[14] L. Chang, Y. Ren, X. Liu et al., "Protective effects of ghrelin on ischemia/reperfusion injury in the isolated rat heart," *Journal of Cardiovascular Pharmacology*, vol. 43, no. 2, pp. 165–170, 2004.

[15] J. P. Xu, H. X. Wang, W. Wang, L. K. Zhang, and C. S. Tang, "Ghrelin improves disturbed myocardial energy metabolism in rats with heart failure induced by isoproterenol," *Journal of Peptide Science*, vol. 16, no. 8, pp. 392–402, 2010.

[16] L. Li, L. K. Zhang, Y. Z. Pang et al., "Cardioprotective effects of ghrelin and des-octanoyl ghrelin on myocardial injury induced by isoproterenol in rats," *Acta Pharmacologica Sinica*, vol. 27, no. 5, pp. 527–535, 2006.

[17] K. Dezaki, H. Hosoda, M. Kakei et al., "Endogenous ghrelin in pancreatic islets restricts insulin release by attenuating Ca^{2+} signaling in β-cells: implication in the glycemic control in rodents," *Diabetes*, vol. 53, no. 12, pp. 3142–3151, 2004.

[18] T. Yada, K. Dezaki, H. Sone et al., "Ghrelin regulates insulin release and glycemia: physiological role and therapeutic potential," *Current Diabetes Reviews*, vol. 4, no. 1, pp. 18–23, 2008.

[19] J. Yang, M. S. Brown, G. Liang, N. V. Grishin, and J. L. Goldstein, "Identification of the acyltransferase that octanoylates ghrelin, an appetite-stimulating peptide hormone," *Cell*, vol. 132, no. 3, pp. 387–396, 2008.

[20] Y. Date, M. Nakazato, S. Hashiguchi et al., "Ghrelin is present in pancreatic α-cells of humans and rats and stimulates insulin secretion," *Diabetes*, vol. 51, no. 1, pp. 124–129, 2002.

[21] Y. Cao, "Angiogenesis modulates adipogenesis and obesity," *The Journal of Clinical Investigation*, vol. 117, no. 9, pp. 2362–2368, 2007.

[22] J. P. Cooke, "NO and angiogenesis," *Atherosclerosis Supplements*, vol. 4, no. 4, pp. 53–60, 2003.

[23] P. Codoñer-Franch, S. Tavárez-Alonso, R. Murria-Estal, J. Megías-Vericat, M. Tortajada-Girbés, and E. Alonso-Iglesias, "Nitric oxide production is increased in severely obese children and related to markers of oxidative stress and inflammation," *Atherosclerosis*, vol. 215, no. 2, pp. 475–480, 2011.

[24] E. Suganami, H. Takagi, H. Ohashi et al., "Leptin stimulates ischemia-induced retinal neovascularization: possible role of vascular endothelial growth factor expressed in retinal endothelial cells," *Diabetes*, vol. 53, no. 9, pp. 2443–2448, 2004.

[25] E. A. Hamed, M. M. Zakary, N. S. Ahmed, and R. M. Gamal, "Circulating leptin and insulin in obese patients with and without type 2 diabetes mellitus: relation to ghrelin and oxidative stress," *Diabetes Research and Clinical Practice*, vol. 94, no. 3, pp. 434–441, 2011.

[26] J. Gómez-Ambrosi, V. Catalán, A. Rodríguez et al., "Involvement of serum vascular endothelial growth factor family members in the development of obesity in mice and humans," *Journal of Nutritional Biochemistry*, vol. 21, no. 8, pp. 774–780, 2010.

[27] L. Yan, L. C. DeMars, and L. K. Johnson, "Long-term voluntary running improves diet-induced adiposity in young adult mice," *Nutrition Research*, vol. 32, no. 6, pp. 458–465, 2012.

[28] F. T. H. Wu, M. O. Stefanini, F. M. Gabhann, C. D. Kontos, B. H. Annex, and A. S. Popel, "A systems biology perspective on sVEGFR1: its biological function, pathogenic role and therapeutic use," *Journal of Cellular and Molecular Medicine*, vol. 14, no. 3, pp. 528–552, 2010.

[29] J. Tam, D. G. Duda, J. Y. Perentes, R. S. Quadri, D. Fukumura, and R. K. Jain, "Blockade of VEGFR2 and not VEGFR1 can limit diet-induced fat tissue expansion: role of local versus bone marrow-derived endothelial cells," *PLoS ONE*, vol. 4, no. 3, Article ID e4974, 2009.

[30] V. Tolle, M. Kadem, M. T. Bluet-Pajot et al., "Balance in Ghrelin and leptin plasma levels in anorexia nervosa patients and constitutionally thin women," *Journal of Clinical Endocrinology and Metabolism*, vol. 88, no. 1, pp. 109–116, 2003.

[31] M. Tesauro, F. Schinzari, M. Iantorno et al., "Ghrelin improves endothelial function in patients with metabolic syndrome," *Circulation*, vol. 112, no. 19, pp. 2986–2992, 2005.

[32] H. Kimura and H. Esumi, "Reciprocal regulation between nitric oxide and vascular endothelial growth factor in angiogenesis," *Acta Biochimica Polonica*, vol. 50, no. 1, pp. 49–59, 2003.

[33] V. D. Dixit, E. M. Schaffer, R. S. Pyle et al., "Ghrelin inhibits leptin- and activation-induced proinflammatory cytokine expression by human monocytes and T cells," *The Journal of Clinical Investigation*, vol. 114, no. 1, pp. 57–66, 2004.

[34] M. J. Yuan, H. He, H. Y. Hu, Q. Li, J. Hong, and C. X. Huang, "Myocardial angiogenesis after chronic ghrelin treatment in a rat myocardial infarction model," *Regulatory Peptides*, vol. 179, no. 1–3, pp. 39–42, 2012.

6

A Meta-Analysis of the Therapeutic Effects of Glucagon-Like Peptide-1 Agonist in Heart Failure

Mohammed Munaf, Pierpaolo Pellicori, Victoria Allgar, and Kenneth Wong

Department of Cardiovascular and Respiratory Studies, Hull and East Yorkshire Medical Research and Teaching Centre, Daisy Building, Castle Hill Hospital, Castle Road, Kingston upon Hull HU16 5JQ, UK

Correspondence should be addressed to Kenneth Wong, kenneth.wong@hey.nhs.uk

Academic Editor: Frédéric Ducancel

We conducted a meta-analysis of the existing literature of the therapeutic effects of using GLP-1 agonists to improve the metabolism of the failing heart. Animal studies showed significant improvement in markers of cardiac function, such as left ventricular ejection fraction (LVEF), with regular GLP-1 agonist infusions. In clinical trials, the potential effects of GLP-1 agonists in improving cardiac function were modest: LVEF improved by 4.4% compared to placebo (95% C.I 1.36–7.44, $P = 0.005$). However, BNP levels were not significantly altered by GLP-1 agonists in heart failure. In two trials, a modest increase in heart rate by up to 7 beats per minute was noted, but meta-analysis demonstrated this was not significant statistically. The small number of studies plus variation in the concentration and length of the regime between the trials would limit our conclusions, even though statistically, heterogeneity chi-squared tests did not reveal any significant heterogeneity in the endpoints tested. Moreover, studies in non-diabetics with heart failure yielded conflicting results. In conclusion, the use of GLP-1 agonists has at best a modest effect on ejection fraction improvement in heart failure, but there was no significant improvement in BNP levels in the meta-analysis.

1. Introduction

Heart failure (HF) is defined as "*a complex clinical syndrome that can result from any structural or functional cardiac disorder that impairs the ability of the ventricle to fill with or eject blood*" [1]. HF is a major public health issue, with a prevalence of over 5.8 million in the USA, and over 23 million (and rising) worldwide. The lifetime risk of developing HF is one in five [2]. Despite advances in treatment, the number of deaths from heart failure has increased steadily and only one quarter to one-third of people with heart failure survive 5 years after admission [3]. The cause of heart failure has shifted in the last two decades: in the late 1970s, rheumatic valvular disease was the primary cause, nowadays the leading cause is ischemic heart disease [4]. A deficit in the "pump" function as cause of signs or symptoms attributed to HF, or systolic dysfunction, is frequently well diagnosed due to widespread availability of echocardiography but, an increased left ventricular (LV)

"stiffness," or diastolic dysfunction, is often missed. To further complicate matters, the two components—systolic and diastolic dysfunction—often coexist. Some studies [5, 6] reported that isolated diastolic dysfunction could be responsible for up to 50% of heart failure admissions (often labelled as "heart failure with normal ejection fraction," HFnEF), with a major impact on patient outcome. Moreover, in patients with impaired glucose tolerance, the extent of diastolic dysfunction seems to be more severe [7] and HFnEF seems to be more common in patients with a history of hypertension and/or diabetes [8, 9].

The standard treatment of systolic heart failure is currently angiotensin-converting enzyme (ACE) inhibitors, angiotensin II receptor blockers (ARBs), beta blockers, and aldosterone antagonists. These all improve prognosis of heart failure. However, there is no specific treatment for HFnEF: diuretics are often used for symptom control; digoxin is particularly beneficial for ventricular rate control when atrial fibrillation (AF) is the predominant rhythm.

In recent years, progress in basic research has led to the identification of multiple new possible therapeutic targets for the treatment of systolic heart failure, and many promising drugs have subsequently been developed. These include novel vasodilators, such as natriuretic peptides, metabolic substrates, urocortins, guanylyl cyclase activators, and adrenomedullin. They also include drugs such as direct renin inhibitors, and aldosterone synthase inhibitors [10]. There have been numerous large randomised controlled trials (RCT) of these new drugs. They have not yet been licensed as results regarding the efficacy of these new drugs have not been entirely positive. Further evidence is needed as many of the positive results that have been observed in preclinical studies and Phase II trials have not always been confirmed in Phase III studies [10].

As mentioned above, the leading cause of systolic HF is myocardial ischaemia, whereby the myocardium is oxygen starved and thus has a decreased ability to generate ATP by oxidative metabolism. As a result, it is unable to effectively transfer the chemical energy from the metabolism of carbon fuels to contractile work. This leads the myocardium to utilise other compounds, such as free fatty acids (FFAs), for energy production. However, if the heart uses FFAs as a substrate for energy generation, there is much greater oxygen consumption per unit ATP produced than there is with glucose. This increased demand for oxygen can lead to worsening heart failure. Thus, improvement of cardiac energetics is an important therapeutic target in patients with heart failure [10].

Metabolic modulators do exactly this by altering the substrate that is oxidized by the myocardium to derive energy. They shift this substrate from FFA to glucose and thus optimize metabolic efficiency of the heart. These compounds exert their effects through several mechanisms: inhibiting carnitine O-palmitoyltransferase 1, long-chain 3-ketoacyl-CoA thiolase or malonyl-CoA decarboxylase, reducing plasma levels of FFA and myocardial uptake of FFA, and/or activating the 5′-AMP-activated protein kinase (AMPK). Thus it follows that, using metabolic manipulating agents to either promote glucose utilisation or reduce fatty acid utilisation, will improve the metabolic efficiency of the heart by decreasing oxygen demand and thus be used therapeutically in heart failure. Amongst these metabolic agents are glucagon like peptide-1 (GLP-1) agonists [10].

GLP-1 is an incretin that is released from intestinal L cells in response to glucose ingestion and is known to be a potent glucose-dependent insulinotropic hormone. It has important actions on gastric motility, on the suppression of plasma glucagon levels, and possibly on the promotion of satiety and stimulation of glucose disposal in peripheral tissues independent of the actions of insulin. It does this by increasing insulin secretion from the pancreas and myocardial glucose uptake via the translocation of glucose-transporting vesicles (glucose transporter type 1 (GLUT1) and GLUT4) to the sarcolemma. GLP-1 exerts its direct cardioprotective effects through the stimulation of G-protein-coupled receptors (i.e., GLP1Rs) that are coupled to adenylyl cyclase, and via its rapid metabolism to the GLP1 (9–36) amide [11].

Therefore, GLP-1 agonists can be used to bring about the same effects. These agents have been investigated widely as an adjunct to therapy in diabetes as they offer an obvious alternative to insulin, but their metabolic effect could also be extended to the heart as they can enable the heart to switch to the more energy-efficient glucose-dependent pathway [10]. Moreover, there are GLP-1 specific receptors in cardiac tissue so the potential for using these peptide agonists holds promise for treating heart failure [12].

However, whilst GLP-1-related compounds have proven efficacy in the treatment of hyperglycaemia associated with type 2 diabetes [13, 14], little was known about the effectiveness of GLP-1 agonist or other peptides substrates in improving cardiac function in heart failure. Because the half-life of GLP-1 in only a few minutes, several Phase III-Phase IV trials are analysing the effects of its analogues, such as exenatide, which are not degraded so quickly [15].

2. Aims and Objectives

We aimed to carry out a comprehensive review of medical literature on the therapeutic advantage of using peptide agonists to improve cardiac metabolism in heart failure. We included all papers regardless of size, whether they were preclinical or clinical trials, either randomized, blinded, or not. The results of these papers have been combined to give an overall estimate of the effectiveness of using GLP-1 agonists in heart failure. Furthermore, we conducted a meta-analysis of each primary outcome if contained in more than two papers.

3. Methods

3.1. Search Strategy of the Meta-Analysis. Highly sensitive search strategies were developed using appropriate subject headings and text word terms. Full details of the search strategies used are appended. The following electronic databases were searched: the Cochrane Library (Issue 7, 2011); MEDLINE (via OVID, from 1948 to August week 1 2011); Pubmed (via NCBI); EMBASE (via OVID, from 1996 to week 30, 2011); BMJ's Clinical Evidence; DARE (Issue 7, 2011). British and American medical journals were also hand-searched, such as The Lancet, NEJM, and BMJ. In addition, conference proceedings and reference lists of all included studies were scanned to identify additionally potentially relevant studies. There were no start year or language restrictions.

3.2. Data Extraction. One reviewer screened the titles (and abstracts if available) of all reports identified by the search strategy. Full copies of potentially relevant reports were obtained, studied, and assessed for inclusion. Data was discussed with the senior author, and disagreements were resolved by consensus.

3.3. Selection Criteria. Papers that had details of trials conducted of peptide agonists versus placebo or usual treatment alone for heart failure were included. All papers, whether they included human or animal trials were included. For humans, randomized controlled trials, regardless of whether they were blinded, were included along with pilot and observational studies.

3.4. Meta-Analysis Methodology

3.4.1. Data Synthesis. The eligible trials were entered into RevMan 5 software package, and the statistical methods were those programmed into RevMan 5.1 analysis software.

For continuous data, the mean difference and 95% confidence intervals were calculated. Where applicable, for dichotomous data, the relative risk and 95% confidence intervals would be calculated. The results from the trials were pooled using the fixed effects models. We tested for heterogeneity with the chi squared statistic, which was considered to be significant at $P < 0.10$. If significant, a random effect model would be used to allow generalisation of the results and sources of heterogeneity would be investigated. Z tests were used to test for the overall effect.

4. Results

A total of 16 papers were found in Medline and 32 in Embase. Handsearching in Pubmed yielded a further 22 papers. There were no Cochrane or DARE reviews of the use of GLP-1 agonist due to the scarcity of clinical trials on these agents and there were no additional papers found in American or British journals. The full references of the papers which contained studies are listed below in the references section.

The general finding from Medline, Embase, and Pubmed was that the papers that were found to mention GLP-1 agonists in HF, generally only detailed their pharmacology and suggested their potential for therapeutic benefit with very few containing any experimental evidence for the application of these agents [10–23]. When these papers containing studies were examined, they pertained to the use of GLP-1 agonists in diabetics with HF due to their insulinotropic effects instead of looking at their use as metabolic substrates for the ischaemic heart as has been suggested by some other papers. In the present paper, we only focused on papers that had experimental evidence for the use of GLP-1 agonists as therapeutic agents. These are discussed below.

4.1. Preclinical Experiments.
Work on rats [24, 25], rabbits [26], mice [27], and dogs [28, 29] showed favourable functional effects of GLP-1 in failing hearts with significant improvements in LV systolic and diastolic function.

Nikolaidis et al. [28] found that short-term infusion of recombinant GLP-1 over 48 hours increased myocardial insulin sensitivity and glucose uptake in a canine model of rapid pacing-induced dilated cardiomyopathy. Interestingly, GLP-1 (9–36) was found to exert similar beneficial effects to native GLP-1 in this model, supporting the growing suggestion that the metabolically inactive form of GLP-1 [GLP-1 (9–36)] may play an active role in the cardiovascular system.

Furthermore, spontaneously hypertensive heart-failure-prone rats (characterized by obesity, insulin resistance, hypertension, and dilated cardiomyopathy), treated chronically with GLP-1 from 9 months of age (when they begin to progress to advanced heart failure and death) exhibited preserved cardiac contractile function, increased myocardial glucose uptake, improved survival, and a significant reduction in cardiac myocyte apoptosis [22]. Although this study also reported GLP-1 to stimulate myocardial glucose uptake in the failing myocardium, it was unclear whether its beneficial effects on contractile function occurred due to a direct cardiac action or was secondary to its established insulinotropic effects. These promising findings led the way for clinical trials and these are discussed below.

4.2. Clinical Trials.
The beneficial effects on contractile function seen in animals treated with GLP-1 were supported by preliminary clinical studies in humans, indicating that GLP-1 may also improve LV contractile function in patients with chronic heart failure.

Thrainsdottir et al. [30], in an early nonrandomised pilot investigation conducted on 6 hospitalised type 2 diabetic hospitalised with ischaemic but stable heart failure New York Heart Association (NYHA) class II-III, with LVEF < 40%, found that short-term GLP-1 infusion for 3 days tended to improve both systolic and diastolic function, although these changes did not reach statistical significance.

However, we also found another three-day study that was conducted on 10 patients with acute myocardial infarction (AMI) or left ventricular ejection fraction (LVEF) of <40% compared with 11 controls [20]. Baseline demographics and background therapy were similar, and both groups had severe LV dysfunction at baseline (LVEF = 29 ± 2%). The study demonstrated that GLP-1 significantly improved LVEF (from 29 ± 2% to 39 ± 2%, $P \leq 0.01$), global wall motion score indexes (1.94 ± 0.11 → 1.63 ± 0.09, $P \leq 0.01$), and regional wall motion score indexes (2.53 ± 0.08 → 2.02 ± 0.11, $P \leq 0.01$) compared with control subjects. The benefits of GLP-1 were independent of AMI location or history of diabetes. Moreover, GLP-1 was well tolerated, with only transient gastrointestinal effects.

Moreover, longer-term treatment with GLP-1 has shown positive results in both diabetics and nondiabetics. Sokos and colleagues [31] compared a 5-week infusion of GLP-1 added to standard therapy in 12 patients with NYHA class III/IV heart failure and the results were compared with those of 9 patients with heart failure on standard therapy. They found that patients treated with GLP-1 infusion had significantly better LV systolic function (LVEF changed from 21 ± 3% to 27 ± 3% $P < 0.01$), exercise tolerance (VO_2 max changed from 10.8 ± .9 mL/O_2/min/kg to 13.9 ± .6 mL/O_2/min/kg; $P < 0.001$, as well as the 6-minute walk distance, from 232 ± 15 m to 286 ± 12 m; $P < 0.001$), and quality of life (Minnesota Living with Heart Failure quality of life score (MNQOL) score: from 64 ± 4 to 44 ± 5; $P < 0.01$). However, no significant changes in any of the parameters were observed in the control group on standard therapy. GLP-1 was well tolerated with minimal episodes of hypoglycaemia and gastrointestinal side effects. Like the aforementioned study [20], this study suggests a role for GLP-1 agonists beyond glycaemic control as significant improvements were seen in both diabetic and nondiabetic patients.

However, we found no further evidence for the extension of GLP-1 to nondiabetics. In a randomized, double-blind crossover trial of 20 normoglycaemic patients without diabetes and with HF with ischemic heart disease, severe left ventricular impairment, NYHA II, and III, Halbirk et al. [32] found that GLP-1 infusion over 48 h increased circulating insulin levels and reduced plasma glucose concentration but had no major cardiovascular effects in patients with chronic heart failure when compared with a placebo. The only significant cardiovascular impacts of the infusion were increases in heart rate (67 ± 2 beats/min versus 65 ± 2 beats/min; $P = 0.016$) and diastolic blood pressure (71 ± 2 mmHg versus 68 ± 2 mmHg; $P = 0.008$). GLP-1 had no effect on systolic blood pressure (113 ± 5 mmHg versus 113 ± 4 mmHg; $P = 0.95$) or on LVEF (GLP-1 treatment from $28 \pm 2\%$ to $30 \pm 2\%$ versus placebo $30 \pm 2\%$ to $30 \pm 2\%$; $P = 0.93$). Importantly, also, GLP-1 infusion did not affect exercise capacity, VO_2 max, cardiac index, stroke volume, and systemic vascular resistance during exercise. Unlike other studies, hypoglycemia was frequent with eight patients experiencing nine episodes of hypoglycaemia (capillary glucose < 3.5 mmol/L) versus none with placebo. This calls for caution in patients without diabetes but with HF and also reiterates the need for further studies with regard to the use of GLP-1 agonists in nondiabetics. Intriguingly, both GLP-1 and placebo significantly dropped BNP, although the effects of the two infusions did not differ (-112 ± 54 pg/mL versus -65 ± 54 pg/mL, $P = 0.17$). Future trials looking at changes in BNP in heart failure should bear in mind that small changes need to be interpreted with caution, as it was intriguing that placebo might have produced a significant reduction in BNP. The authors of that paper attributed this drop in natriuretic peptide to be due to patients' reduced exercise during their hospital stay, more than a direct effect of the infusion. However, a recent study conducted in healthy subjects found exenatide had significant haemodynamic effects, including natriuretic properties [33].

4.3. Meta-Analysis. Individually, some of the studies that we have discussed would suggest that GLP-1 agonist might be potentially effective for heart failure. We performed a meta-analysis on all the primary endpoints that were contained in at least two papers. The results were summarised in Table 1, and Figures 1, 2, and 3.

There was at best a modest improvement in ejection fraction (4.4%; 95% CI 1.36–7.44%). There was no significant change in BNP or heart rate in our meta-analysis. Thus, although some of the preliminary clinical studies provided some encouragement for the potential use of GLP-1 in the treatment of heart failure, it is clear that significant further research is required to confirm these initial observations, investigate the underlying mechanisms, and explore possible interactions with current heart failure therapies.

TABLE 1: Summary of all trials studying GLP-1 effects in human heart failure.

Study	Endpoints
Thrainsdottir et al., 2004 [30]	HR, BP (rest + exercise), rate pressure product, global systolic and diastolic function, LVEF, LV end-diastolic diameter
Nikolaidis et al., 2004 [20]	LVEF, ED +ESV, SV, global WMSI
Sokos et al., 2006 [31]	HR, BNP, LVEF, VO_2, 6-min walk
Halbirk et al., 2010 [32]	BNP, BP, HR, SV, CI, LVEF, SVR, 6 min hall walk test

4.4. Limitations of Meta-Analysis. As with any meta-analysis, the quality is dependent on the quality of the studies and any limitations the included studies have. Firstly, the most obvious limitation is the lack of a large number of studies available to meta-analyse. Secondly, the total sample size of patients in all four studies combined is small. A further limitation in our meta-analysis is that all four studies investigated different concentrations of GLP-1 agonist infusion: 1.0 pmol/kg/min (Halbirk); 1.5 pmol/kg/min (Nikolaidis); 2.5 pmol/kg/min (Sokos) and 4 pmol/kg/min (Thrainsdottir). Moreover, the studies measured improvements at different intervals of time, with Halbirk looking at effects after 48 hours, Thrainsdottir and Nikkolaidis at 3 days and Sokos investigating a 5-week infusion. This has definite implications for interpretation of the results. Another limitation was that not all the studies included were double blinded and randomised, for example, Thrainsdottir was an open observation study, whereas Halbirk was a double-blinded crossover placebo study. This leads to methodological heterogeneity.

4.5. Clinical Implications and Future Research. The Carvedilol Hibernating Reversible Ischaemia Trial: Marker of Success (CHRISTMAS trial) [34] found patients with more hibernation/ischaemia had greater improvement in left ventricular systolic function with beta-blocker treatment. Our Academic Cardiology Department in Hull also conducted the Heart Failure Revascularisation Trial which showed how myocardial ischaemia and hibernation could not effectively be resuscitated by revascularization in patients with chronic HF [35]. Recently, the large STITCH trial [36] did not demonstrate any survival benefit of coronary artery bypass surgery in patients with heart failure with severe coronary artery disease. Thus, to optimally treat ischaemic heart failure, we need to explore other avenues to improve myocardial metabolism, to try and optimize cardiac function.

GLP-1 is an endogenous peptide which is released from the gut following food intake. It is one of a number of factors that can augment insulin release, so as expected, its role in improving glycaemic control in diabetics is now fairly well established.

Our meta-analysis of clinical trials involving patients with heart failure demonstrated some promising evidence to suggest possible beneficial effects of the GLP-1 peptide agonist in improving cardiac function, in both diabetics and

Study or subgroup	Experimental Mean	SD	Total	Control Mean	SD	Total	Weight	Mean difference IV, fixed, 95 % Cl
Halbirk et al., 2010	2	4.9	15	0	6	15	60.2%	2[−1.92, 5.92]
Nikolaidis et al., 2004	10	6.32	10	1	6.63	11	30.1%	9[3.46, 14.54]
Sokos et al., 2006	6	10.39	12	1	12	9	9.6%	5[−4.8, 14.8]
Total (95% Cl)			**37**			**35**	**100%**	**4.4[1.36, 7.44]**

Heterogeneity: $\chi^2 = 4.1$, df = 2 ($P = 0.13$); $I^2 = 51\%$
Test for overall effect: $Z = 2.83$ ($P = 0.005$)

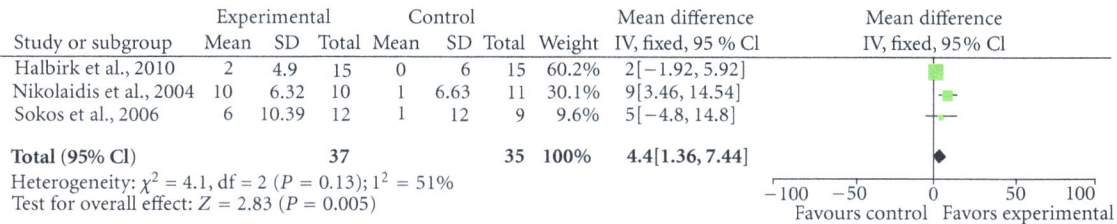

FIGURE 1: Forrest plot demonstrating GLP-1 improves ejection fraction.

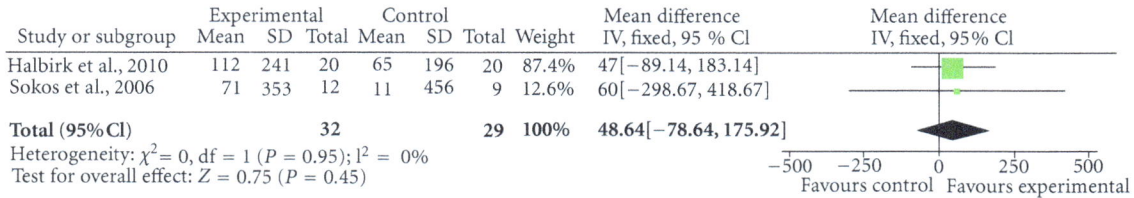

Study or subgroup	Experimental Mean	SD	Total	Control Mean	SD	Total	Weight	Mean difference IV, fixed, 95 % Cl
Halbirk et al., 2010	112	241	20	65	196	20	87.4%	47[−89.14, 183.14]
Sokos et al., 2006	71	353	12	11	456	9	12.6%	60[−298.67, 418.67]
Total (95% Cl)			**32**			**29**	**100%**	**48.64[−78.64, 175.92]**

Heterogeneity: $\chi^2 = 0$, df = 1 ($P = 0.95$); $I^2 = 0\%$
Test for overall effect: $Z = 0.75$ ($P = 0.45$)

FIGURE 2: Forrest plot demonstrating the negligible effect of GLP-1 on BNP levels.

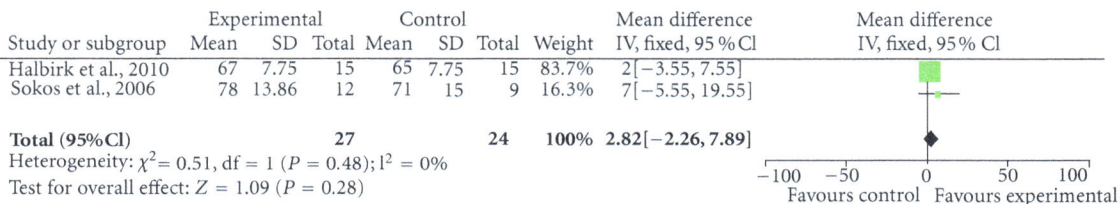

Study or subgroup	Experimental Mean	SD	Total	Control Mean	SD	Total	Weight	Mean difference IV, fixed, 95 % Cl
Halbirk et al., 2010	67	7.75	15	65	7.75	15	83.7%	2[−3.55, 7.55]
Sokos et al., 2006	78	13.86	12	71	15	9	16.3%	7[−5.55, 19.55]
Total (95% Cl)			**27**			**24**	**100%**	**2.82[−2.26, 7.89]**

Heterogeneity: $\chi^2 = 0.51$, df = 1 ($P = 0.48$); $I^2 = 0\%$
Test for overall effect: $Z = 1.09$ ($P = 0.28$)

FIGURE 3: Forrest plot demonstrating the effect of GLP-1 agonist on heart rate.

FIGURE 4: Full Medline search with MeSH terms.

nondiabetics. This was seen with the statistically significant increase in left ventricular ejection fraction, although the absolute change was very modest (4.4%). An absence of lowering effect on systolic blood pressure may be particularly appealing to clinicians who find their patients with heart failure often have relatively low blood pressure on a combination of ACE-inhibitors, beta blockers, spironolactone or eplerenone, and loop diuretics. It should be noted that the drug might drop patients' diastolic blood pressure.

Minor increase in heart rate may also turn out to be a concern as recent evidence have confirmed the hypothesis that patients with heart failure have better prognosis if their heart rate is less than 70 beats per minute [37]. However, whilst in the two individual trials (Halbirk and Sokos), there was a modest increase in heart rate by up to 7 beats

per minute, our meta-analysis demonstrated this was not significant statistically. In nondiabetics with heart failure, caution must be exercised to ensure they do not develop hypoglycaemia, which again is potentially hazardous.

Before the peptide agonist can be recommended for routine clinical use, large multicentre, double-blinded randomised controlled trials are needed, investigating the effects of GLP-1 or its analogue in patients with acute or chronic HF including hard endpoints, such as mortality, cardiovascular death, or hospitalization for heart failure. Further, as suggested previously, heart failure with normal ejection fraction (HFnEF) is often difficult to treat specifically. Future trials should study the effect of GLP-1 agonists in this challenging group of patients. Recent work suggested that advanced echocardiography techniques using speckle tracking to assess

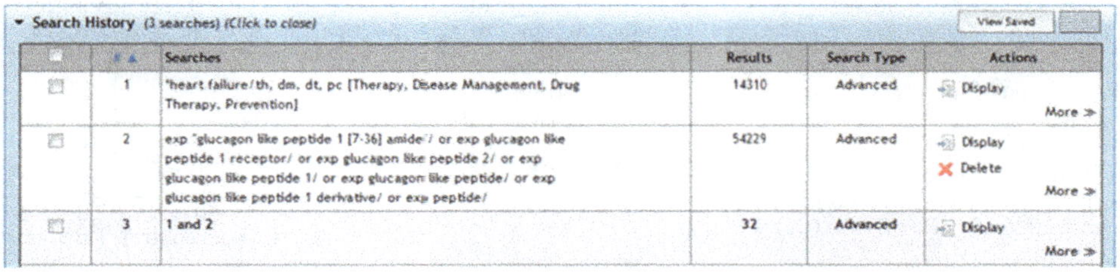

Search History (3 searches) (Click to close) View Saved

	#	Searches	Results	Search Type	Actions
	1	*heart failure/th, dm, dt, pc [Therapy, Disease Management, Drug Therapy, Prevention]	14310	Advanced	Display More »
	2	exp *glucagon like peptide 1 [7-36] amide*/ or exp glucagon like peptide 1 receptor/ or exp glucagon like peptide 2/ or exp glucagon like peptide 1/ or exp glucagon like peptide/ or exp glucagon like peptide 1 derivative/ or exp peptide/	54229	Advanced	Display Delete More »
	3	1 and 2	32	Advanced	Display More »

Figure 5: Full Embase search with MeSH terms.

the so-called global longitudinal strain (GLS) might identify patients with subtle systolic dysfunction [38] and might even be better than ejection fraction at predicting poor cardiovascular outcome in patients with chronic heart failure [39].

5. Conclusions

This meta-analysis of the potential therapeutic benefits of GLP-1 agonists in heart failure involved a thorough literature search using Embase and Medline plus hand-search strategies. The animal studies gave evidence in favour of these peptide agonists. There were only a few small clinical trials involving patients with heart failure. The use of GLP-1 agonists has at best a modest effect on ejection fraction improvement in patients with heart failure, but there was no significant improvement in BNP levels in the meta-analysis.

Appendix

Medline Search Strategy. The search focused on heart failure, mapped to subject headings, and included the following MESH terms: chemistry, drug therapy, enzymology, metabolism, physiology, and therapy.

The second search term was for peptide and this term was again mapped to include medical subject headings. From this the following mesh terms were exploded: Glucagon-like peptides; peptides and peptides, cyclic.

The third search term was oxygen, including oxygen compounds, oxygen, and oxygen consumption as these were central to our review.

The fourth search item combined the above three and produced 16 papers.

The full search is shown in Figure 4.

Embase Search Strategy. For Embase, again heart failure was the first search term, including disease management, drug therapy, prevention, and therapy.

The second search term was for peptide and all the similar MeSH including "glucagon like peptides" were selected. All these were exploded so similar terms could be included.

The third search combined the previous two searches with "AND," thus returning 32 results: see Figure 5.

Hand-Searching. Pubmed yielded a further 22 papers. Of these, only three papers contained results of studies done on humans. These, along with the papers found with Medline and Embase, were cited fully in the references section. There were no additional papers found in the medical journals that were hand-searched (BMJ, Lancet, NEJM).

References

[1] S. A. Hunt, D. W. Baker, M. H. Chin et al., "ACC/AHA guidelines for the evaluation and management of chronic heart failure in the adult: executive summary. A report of the American college of cardiology/American heart association task force on practice guidelines (committee to revise the 1995 guidelines for the evaluation and management of heart failure): developed in collaboration with the international society for heart and lung transplantation; endorsed by the heart failure society of America," *Circulation*, vol. 104, no. 24, pp. 2996–3007, 2001.

[2] A. L. Bui, T. B. Horwich, and G. C. Fonarow, "Epidemiology and risk profile of heart failure," *Nature Reviews Cardiology*, vol. 8, no. 1, pp. 30–41, 2011.

[3] J. J. V. McMurray and S. Stewart, "The burden of heart failure," *European Heart Journal*, vol. 4, supplement D, pp. D50–D58, 2002.

[4] J. G. Cleland, A. Torabi, and N. K. Khan, "Epidemiology and management of heart failure and left ventricular systolic dysfunction in the aftermath of a myocardial infarction," *Heart*, vol. 91, supplement 2, pp. ii7–ii13, ii31–ii43, 2005.

[5] K. Hogg, K. Swedberg, and J. McMurray, "Heart failure with preserved left ventricular systolic function: epidemiology, clinical characteristics, and prognosis," *Journal of the American College of Cardiology*, vol. 43, no. 3, pp. 317–327, 2004.

[6] J. G. F. Cleland, T. McDonagh, A. S. Rigby, A. Yassin, T. Whittaker, and H. J. Dargie, "The national heart failure audit for England and Wales 2008-2009," *Heart*, vol. 97, no. 11, pp. 876–886, 2011.

[7] A. M. Salmasi, P. Frost, and M. Dancy, "Left ventricular diastolic function in normotensive subjects 2 months after acute myocardial infarction is related to glucose intolerance," *American Heart Journal*, vol. 150, no. 1, pp. 168–174, 2005.

[8] T. Tsujino, D. Kawasaki, and T. Masuyama, "Left ventricular diastolic dysfunction in diabetic patients: pathophysiology and therapeutic implications," *American Journal of Cardiovascular Drugs*, vol. 6, no. 4, pp. 219–230, 2006.

[9] M. Fujita, H. Asanuma, J. Kim et al., "Impaired glucose tolerance: a possible contributor to left ventricular hypertrophy and diastolic dysfunction," *International Journal of Cardiology*, vol. 118, no. 1, pp. 76–80, 2007.

[10] J. Tamargo and J. López-Sendón, "Novel therapeutic targets for the treatment of heart failure," *Nature Reviews Drug Discovery*, vol. 10, no. 7, pp. 536–555, 2011.

[11] T. J. Kieffer and J. F. Habener, "The glucagon-like peptides," *Endocrine Reviews*, vol. 20, no. 6, pp. 876–913, 1999.

[12] C. Saraceni and T. L. Broderick, "Effects of glucagon-like peptide-1 and long-acting analogues on cardiovascular and metabolic function," *Drugs in R and D*, vol. 8, no. 3, pp. 145–153, 2007.

[13] M. B. Toft-Nielsen, S. Madsbad, and J. J. Holst, "Determinants of the effectiveness of glucagon-like peptide-1 in type 2 diabetes," *Journal of Clinical Endocrinology and Metabolism*, vol. 86, no. 8, pp. 3853–3860, 2001.

[14] J. J. Meier, D. Weyhe, M. Michaely et al., "Intravenous glucagon-like peptide 1 normalizes blood glucose after major surgery in patients with type 2 diabetes," *Critical Care Medicine*, vol. 32, no. 3, pp. 848–851, 2004.

[15] M. Monami, F. Cremasco, C. Lamanna et al., "Glucagon-like peptide-1 receptor agonists and cardiovascular events: a meta-analysis of randomized clinical trials," *Experimental Diabetes Research*, vol. 2011, Article ID 215764, 2011.

[16] W. C. Stanley, F. A. Recchia, and G. D. Lopaschuk, "Myocardial substrate metabolism in the normal and failing heart," *Physiological Reviews*, vol. 85, no. 3, pp. 1093–1129, 2005.

[17] M. F. Essop and L. H. Opie, "Metabolic therapy for heart failure," *European Heart Journal*, vol. 25, no. 20, pp. 1765–1768, 2004.

[18] H. Taegtmeyer, "Cardiac metabolism as a target for the treatment of heart failure," *Circulation*, vol. 110, no. 8, pp. 894–896, 2004.

[19] D. J. Grieve, R. S. Cassidy, and B. D. Green, "Emerging cardiovascular actions of the incretin hormone glucagon-like peptide-1: potential therapeutic benefits beyond glycaemic control?" *British Journal of Pharmacology*, vol. 157, no. 8, pp. 1340–1351, 2009.

[20] L. A. Nikolaidis, S. Mankad, G. G. Sokos et al., "Effects of glucagon-like peptide-1 in patients with acute myocardial infarction and left ventricular dysfunction after successful reperfusion," *Circulation*, vol. 109, no. 8, pp. 962–965, 2004.

[21] K. Ban, M. H. Noyan-Ashraf, J. Hoefer, S. S. Bolz, D. J. Drucker, and M. Husain, "Cardioprotective and vasodilatory actions of glucagon-like peptide 1 receptor are mediated through both glucagon-like peptide 1 receptor-dependent and -independent pathways," *Circulation*, vol. 117, no. 18, pp. 2340–2350, 2008.

[22] E. Mannucci and C. M. Rotella, "Future perspectives on glucagon-like peptide-1, diabetes and cardiovascular risk," *Nutrition, Metabolism and Cardiovascular Diseases*, vol. 18, no. 9, pp. 639–645, 2008.

[23] A. K. Bose, M. M. Mocanu, R. D. Carr, C. L. Brand, and D. M. Yellon, "Glucagon-like peptide 1 can directly protect the heart against ischemia/reperfusion injury," *Diabetes*, vol. 54, no. 1, pp. 146–151, 2005.

[24] I. Poornima, S. B. Brown, S. Bhashyam, P. Parikh, H. Bolukoglu, and R. P. Shannon, "Chronic glucagon-like peptide-1 infusion sustains left ventricular systolic function and prolongs survival in the spontaneously hypertensive, heart failure-prone rat," *Circulation*, vol. 1, no. 3, pp. 153–160, 2008.

[25] T. Zhao, P. Parikh, S. Bhashyam et al., "Direct effects of glucagon-like peptide-1 on myocardial contractility and glucose uptake in normal and postischemic isolated rat hearts," *Journal of Pharmacology and Experimental Therapeutics*, vol. 317, no. 3, pp. 1106–1113, 2006.

[26] M. Matsubara, S. Kanemoto, B. G. Leshnower et al., "Single dose GLP-1-tf ameliorates myocardial ischemia/reperfusion injury," *Journal of Surgical Research*, vol. 165, no. 1, pp. 38–45, 2011.

[27] M. H. Noyan-Ashraf, M. Abdul Momen, K. Ban et al., "GLP-1R agonist liraglutide activates cytoprotective pathways and improves outcomes after experimental myocardial infarction in mice," *Diabetes*, vol. 58, no. 4, pp. 975–983, 2009.

[28] L. A. Nikolaidis, D. Elahi, T. Hentosz et al., "Recombinant glucagon-like peptide-1 increases myocardial glucose uptake and improves left ventricular performance in conscious dogs with pacing-induced dilated cardiomyopathy," *Circulation*, vol. 110, no. 8, pp. 955–961, 2004.

[29] L. A. Nikolaidis, D. Elahi, Y. T. Shen, and R. P. Shannon, "Active metabolite of GLP-1 mediates myocardial glucose uptake and improves left ventricular performance in conscious dogs with dilated cardiomyopathy," *American Journal of Physiology-Heart and Circulatory Physiology*, vol. 289, no. 6, pp. H2401–H2408, 2005.

[30] I. Thrainsdottir, K. Malmberg, A. Olsson, M. Gutniak, and L. Rydén, "Initial experience with GLP-1 treatment on metabolic control and myocardial function in patients with type 2 diabetes mellitus and heart failure," *Diabetes & Vascular Disease Research*, vol. 1, no. 1, pp. 40–43, 2004.

[31] G. G. Sokos, L. A. Nikolaidis, S. Mankad, D. Elahi, and R. P. Shannon, "Glucagon-Like Peptide-1 Infusion Improves Left Ventricular Ejection Fraction and Functional Status in Patients With Chronic Heart Failure," *Journal of Cardiac Failure*, vol. 12, no. 9, pp. 694–699, 2006.

[32] M. Halbirk, H. Nørrelund, N. Møller et al., "Cardiovascular and metabolic effects of 48-h glucagon-like peptide-1 infusion in compensated chronic patients with heart failure," *American Journal of Physiology-Heart and Circulatory Physiology*, vol. 298, no. 3, pp. H1096–H1102, 2010.

[33] B. Mendis, E. Simpson, I. Macdonald, and P. Mansell, "Investigation of the haemodynamic effects of exenatide in healthy male subjects," *British Journal of Clinical Pharmacology*. In press.

[34] J. G. Cleland, D. J. Pennell, S. G. Ray et al., "Carvedilol hibernating reversible ischaemia trial: marker of success investigators. Myocardial viability as a determinant of the ejection fraction response to carvedilol in patients with heart failure (CHRISTMAS trial): randomised controlled trial," *The Lancet*, vol. 362, no. 9377, pp. 14–21, 2003.

[35] A. P. Coletta, J. G. F. Cleland, D. Cullington, and A. L. Clark, "Clinical trials update from heart rhythm 2008 and heart failure 2008: ATHENA, URGENT, INH study, HEART and CK-1827452," *European Journal of Heart Failure*, vol. 10, no. 9, pp. 917–920, 2008.

[36] E. J. Velazquez, K. L. Lee, M. A. Deja et al., "Coronary-artery bypass surgery in patients with left ventricular dysfunction," *The New England Journal of Medicine*, vol. 364, no. 17, pp. 1607–1616, 2011.

[37] K. Fox, I. Ford, P.G. Steg, M. Tendera, M. Robertson, and R. Ferrari, "On behalf of the BEAUTIFUL investigators/Heart rate as a prognostic risk factor in patients with coronary artery disease and left-ventricular systolic dysfunction (BEAUTIFUL): a subgroup analysis of a randomised controlled trial," *The Lancet*, vol. 372, no. 9641, pp. 817–821, 2008.

[38] M. Galderisi, V. S. Lomoriello, A. Santoro et al., "Differences of myocardial systolic deformation and correlates of diastolic function in competitive rowers and young hypertensives: a speckle-tracking echocardiography study," *Journal of the American Society of Echocardiography*, vol. 23, no. 11, pp. 1190–1198, 2010.

[39] J. Nahum, A. Bensaid, C. Dussault et al., "Impact of longitudinal myocardial deformation on the prognosis of chronic heart failure patients," *Circulation Cardiovascular Imaging*, vol. 3, no. 3, pp. 249–256, 2010.

7

Angiotensin-Converting Enzyme 2 (ACE2) Is a Key Modulator of the Renin Angiotensin System in Health and Disease

Chris Tikellis and M. C. Thomas

Division of Diabetic Complications, Baker IDI Heart and Diabetes Institute, P.O. Box 6492 Melbourne, VIC 8008, Australia

Correspondence should be addressed to Chris Tikellis, chris.tikellis@bakeridi.edu.au

Academic Editor: Suhn Hee Kim

Angiotensin-converting enzyme 2 (ACE2) shares some homology with angiotensin-converting enzyme (ACE) but is not inhibited by ACE inhibitors. The main role of ACE2 is the degradation of Ang II resulting in the formation of angiotensin 1–7 (Ang 1–7) which opposes the actions of Ang II. Increased Ang II levels are thought to upregulate ACE2 activity, and in ACE2 deficient mice Ang II levels are approximately double that of wild-type mice, whilst Ang 1–7 levels are almost undetectable. Thus, ACE2 plays a crucial role in the RAS because it opposes the actions of Ang II. Consequently, it has a beneficial role in many diseases such as hypertension, diabetes, and cardiovascular disease where its expression is decreased. Not surprisingly, current therapeutic strategies for ACE2 involve augmenting its expression using ACE2 adenoviruses, recombinant ACE2 or compounds in these diseases thereby affording some organ protection.

1. Introduction

The renin-angiotensin system (RAS) is a signalling pathway that acts as a homeostatic regulator of vascular function [1]. Its systemic actions include the regulation of blood pressure, natriuresis, and blood volume control. However, the RAS also plays an important local role, regulating regional blood flow and controlling trophic responses to a range of stimuli. The RAS is composed of a number of different regulatory components and effector peptides that facilitate the dynamic control of vascular function, in both health and disease (Figure 1). Many of these components have opposing functions to accommodate a rapid but coordinated response to specific triggers. For example, angiotensin I (Ang I) is metabolised by the dipeptide carboxypeptidase, angiotensin-converting enzyme (ACE) to form angiotensin II (Ang II) and Ang II is metabolised by the carboxypeptidase, ACE2, producing the vasodilator, angiotensin$_{(1-7)}$ (Ang 1–7) [2–4]. Historically, ACE and Ang II have been the key focus for clinical interventions targeting the RAS and its pathogenic actions. However, recent studies have also demonstrated the importance of ACE2 in maintaining the balance of the RAS. Indeed, in some settings, and the cardiovascular system

in particular, ACE2 may be more important than ACE in regulating local levels of Ang II and Ang 1–7, and therein the balance of RAS activation. For example, we have shown that acquired or genetic deficiency of ACE2 results in increased tissue and circulating levels of Ang II [5, 6] and reduced levels of Ang 1–7 [6]. By contrast, *Ace* KO mice have modestly reduced circulating Ang II, while tissue levels are not significantly modified, possibly as substantial amounts of Ang II are generated by non-ACE pathways, while degradation pathways for Ang II are more limited [7]. This paper will specifically examine the actions of ACE2 in the body and discuss their potential role in health and various disease states.

2. Angiotensin-Converting Enzyme (ACE2)

ACE2 is a type 1 integral membrane glycoprotein [8] that is expressed and active in most tissues. The highest expression of ACE2 is observed in the kidney, the endothelium, the lungs, and in the heart [2, 8]. The extracellular domain of ACE2 enzyme contains a single catalytic metallopeptidase unit that shares 42% sequence identity and 61% sequence similarity with the catalytic domain of ACE [2]. However,

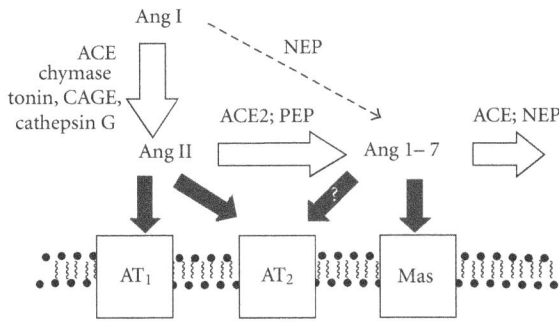

FIGURE 1: Schematic representation of the renin-angiotensin system (RAS) and the key balancing role of ACE2. Abbreviations, ACE: angiotensin-converting enzyme; ACE2: angiotensin-converting enzyme 2; NEP: neprilysin; AT1: Ang II type 1 receptor; AT2: Ang II type 2 receptor; PEP: prolyl endopeptidase; CAGE: chymostatin-sensitive angiotensin II-generating enzyme.

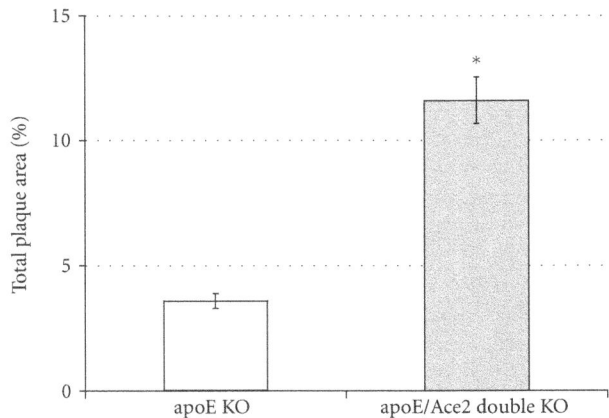

FIGURE 2: Increased plaque area accumulation in the aorta of *Apoe/Ace2* double KO mice when compared to control *Apoe* KO mice [5]. *vs control *Apoe* KO mice $P < 0.05$.

unlike ACE, it functions as a carboxypeptidase, rather than a dipeptidase, and ACE2 activity is not antagonized by conventional ACE inhibitors [4]. The major substrate for ACE2 appears to be (Ang II) [2–4], although other peptides may also be degraded by ACE2, albeit at lower affinity. For example, ACE2 is able to cleave the C-terminal amino acid from angiotensin I, vasoactive bradykinin (1–8), des-Arg-kallidin (also known as des-Arg^{10} Lys-bradykinin) [2], Apelin-13 and Apelin-36 [9] as well as other possible targets [10]. The noncatalytic C-terminal domain of ACE2 shows 48% sequence identity with collectrin [11], a protein recently shown to have an important role in neutral amino acid reabsorption from the intestine and the kidney [12]. This is highly consistent with ACE2's actions as a carboxypeptidase, as the removed amino acid then becomes available for reabsorption. The cytoplasmic tail of ACE2 also contains calmodulin-binding sites [13] which may influence shedding of its catalytic ectodomain. In addition, ACE2 has also been associated with integrin function, independent of its angiotensinase activity.

3. ACE2 and Atherosclerosis

Abnormal activation of the RAS contributes to the development and progression of atherosclerotic vascular disease [14–16]. Independent and additional to the induction of systemic hypertension and vasoconstriction, Ang II has a number of direct proatherosclerotic effects on the vascular wall [17–19], including promoting inflammation [20], endothelial dysfunction [21], oxidative stress, endothelial cell, and vascular smooth muscle cell migration, growth, proliferation [22], and thrombosis. By contrast, the major product of ACE2, Ang 1–7, has a range of anti-inflammatory and antioxidant effects [23, 24] that oppose those of Ang II in the vasculature. Indeed, an infusion of Ang 1–7 is able to attenuate vascular dysfunction and atherosclerosis in genetically susceptible *apolipoprotein E* knockout (*apoE* KO) mice [25], possibly by increased activation of the Mas receptor and the type 2 angiotensin receptor (AT$_2$). It is thought that the balance of

Ang II and Ang 1–7 represents an important driving factor for vascular disease progression. Consequently, ACE2 is also likely to play an important role in atherosclerotic plaque development. Certainly, ACE2 expression is reduced in established atherosclerotic plaques [26] and in proatherosclerotic states, such as diabetes [27]. However, direct evidence for ACE2 in the development and progression of atherosclerotic plaques has only recently become available [5].

We have shown that in *apoE* KO mice, deficiency of ACE2 is associated with increased plaque accumulation (Figure 2), comparable to that observed following angiotensin II infusion [19]. This possibly relates to an increased proinflammatory responsiveness [5], as leukocyte recruitment and adhesion to the nascent atherosclerotic lesion is generally regarded as one of the first steps toward plaque formation. While a healthy endothelium does not in general support binding of white blood cells, we show that the aortic endothelium of *apoE/Ace2* double KO mice shows increased adhesion of labeled leukocytes [5]. In addition, genetic ACE2 deficiency is associated with upregulation of putative mediators of atherogenesis, such as cytokines and adhesion molecules. The role of the RAS in these actions is further emphasized by the finding that RAS blockade is able to prevent atherogenesis in *apoE/Ace2* double KO mice. Such data emphasize the potential utility of ACE2 repletion as a strategy to reduce atherosclerosis, particularly in combination with ACE inhibition and other interventions to reduce activation of the RAS (see below).

4. ACE2 and Hypertension

Activation of the RAS is known to be a key mediator of hypertension, and interventions to block RAS activation are the most widely used of all blood pressure lowering agents. The antihypertensive efficacy of these agents is partly mediated by their ability to reduce Ang II or its signalling. However, the antihypertensive effects of conventional RAS blockade are also partly determined by the ability of both ACE inhibitors and angiotensin receptor blockers (ARBs) to increase circulating levels of Ang 1–7 [28]. Moreover, inhibiting the vascular actions of Ang 1–7 in spontaneously hypertensive rats (SHRs) receiving RAS blockade, attenuates the antihypertensive response to these agents [28, 29]. Given that the major source of Ang 1–7 is ACE2, this data suggests that ACE2, consequently influences not only the development of hypertension, but also potentially the response to its treatment. Certainly, ACE2 expression is abnormal in SHRs, in which one genetic component of this phenotype tracks to the *Ace2* locus. In addition, ACE2 deficiency is associated with modest systolic hypertension [30], although the mouse genetic background significantly alters the cardiovascular phenotype [30–33]. *Ace2* KO mice also have a heightened hypertensive response to Ang II infusion associated with exaggerated accumulation of Ang II in the kidney [30].

The RAS and ACE2 are also implicated in the pathogenesis of central hypertension. In particular, the rostral ventrolateral medulla (RVLM) is a relay point that provides supraspinal excitatory input to sympathetic preganglionic neurons in the regulation of blood pressure. In the SHRs, ACE2 expression is reduced in the RVLM [34], and persistent overexpression of ACE2 in the RVLM results in a significant attenuation of high blood pressure in this model [35, 36]. In addition, injections of the ACE2 inhibitor MLN4760 into the nucleus tractus solitarii reduce reflex bradycardia in response to the baroreceptor stimulation in rats [37], suggesting an additional role for central ACE2 in controlling baroreceptor responsiveness.

5. ACE2 in Heart Failure

In addition to effects on blood pressure, natriuresis and atherogenesis, the RAS plays a critical pathophysiological role in the maintaining and subsequently subverting cardiac function in the setting of progressive heart failure [38]. The cardiac RAS is upregulated in almost all models of cardiac injury, including volume overload [39], myocardial infarction [40], and heart failure [41]. As in the kidney, RAS upregulation appears to be a homeostatic response to restore cardiac function. For example, Ang II is an inotropic and growth factor for cardiac myocytes, stimulating compensatory hypertrophy [42]. Ang II is also important in left ventricular remodeling following myocardial infarction or with afterload-induced cardiac hypertrophy [43]. However, in the long term such actions lead to progressive functional loss and cardiac fibrosis [42], as the synthesis of extracellular matrix is increased by Ang II [44]. The key role of RAS activation in the development and progression of cardiac failure is supported by findings in a number of different models in which blockade of the RAS was able to attenuate or prevent cardiac damage, independent of blood pressure lowering [45].

In the heart, ACE2 represents the primary pathway for the metabolism of Ang II [46, 47]. ACE2 deficiency in mice results in early cardiac hypertrophy (Figure 3) [32] and accelerates adverse postmyocardial infarction ventricular remodeling [48]. Furthermore, this appears to be through the activation of the NAPDH oxidase system with the

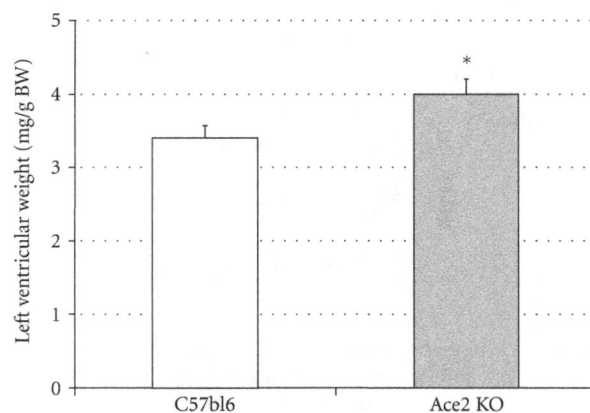

FIGURE 3: Increased LV mass in *Ace2* KO mice versus C57bl6 mice (unpublished data). *vs control C57Bl6 mice, $P < 0.05$.

p47(phox) subunit playing a critical role [49]. In some, but not all models, ACE2 deficiency also results in progressive cardiac fibrosis with aging and/or cardiac pressure overload [33, 50, 51]. Again, these changes are reversed following treatment with ACE inhibitors or AT_1 receptor blockers [33, 50, 51] suggesting that the balance of ACE and ACE2 in the heart is an important driving factor for progressive cardiac disease.

6. ACE2 and Chronic Kidney Disease (CKD)

The RAS also plays an important role in renal physiology and pathophysiology. In the adult kidney [2], ACE2 is predominantly expressed in the proximal tubule at the luminal brush border. Despite the presence of unopposed ACE activity and elevated Ang II levels, both kidney function and renal development are normal in the *Ace2* knockout mouse [33]. By comparison, *ACE, angiotensinogen,* and *AT1* receptor deficiency results in a number of alterations in kidney morphology [52]. This suggests that, at least in the healthy state, ACE2 may have a limited role in regulating renal development. However, the actions of ACE2 appear to come into its own in states of RAS activation. This is much like Ang 1–7, its major product, which shows very limited renal effects in the healthy state but profound benefits in the diabetic kidney and other states associated with renal damage and activation [10, 53]. For example, ACE2 deficient mice have been reported to show increased age-related glomerulosclerosis in susceptible mouse models [54] and enhanced renal Ang II-induced renal oxidative stress, resulting in greater renal injury [55]. Similarly, in the diabetic kidney, downregulation of tubular ACE2 (Figure 4) [27] is associated with albuminuria or tubular injury, while further inhibition of ACE2 results in augmented renal damage [56, 57]. Indeed, in most forms of CKD, including diabetes, expression of ACE2 has been reported to be reduced in tubules. However, some studies have reported that glomerular ACE2 expression may be increased in human kidney disease [58]. It is possible that this differential expression pattern of glomerular and

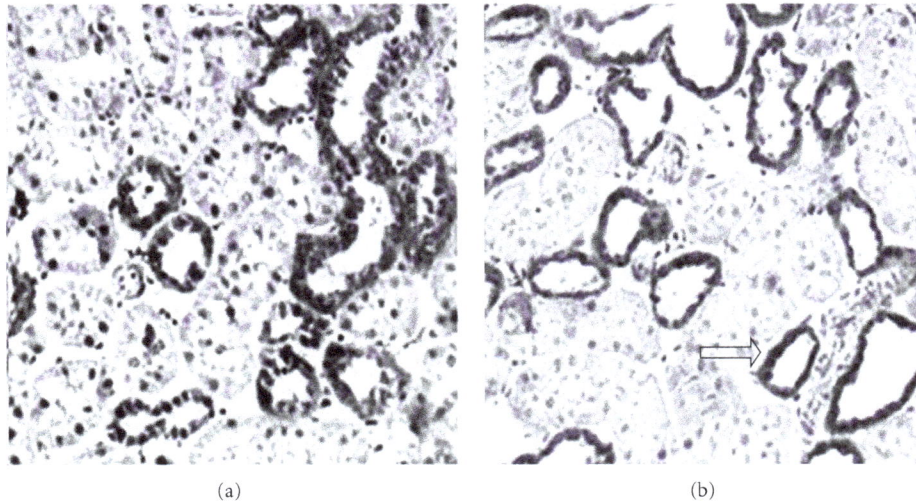

(a)   (b)

FIGURE 4: Reduced ACE2 expression (arrows) in renal cortical tubules of diabetic mice (b) when compared to control mice (a) [27].

tubular ACE2 is an important determinant for progressive renal disease.

7. ACE2 and the Lung

RAS activity is intrinsically high in the lung, which is a major source of ACE and therefore a major site of systemic Ang II synthesis. ACE2 is also highly expressed in the lung. Pulmonary ACE2 appears to have a role in regulating the balance of circulating Ang II/Ang 1–7 levels. Ang II induces pulmonary vasoconstriction in response to hypoxia, which is important in preventing shunting in patients with pneumonia or lung injury [59]. Locally increased Ang II production also triggers increasing vascular permeability facilitating pulmonary edema [60]. In Acute respiratory distress syndrome (ARDS), the RAS appears crucial in maintaining oxygenation, possibly as widespread lung injury would otherwise result in complete pulmonary shutdown. Certainly in ARDS models, ACE2 knockout mice displayed more severe symptoms of this disease compared with wild-type mice [60] while overexpression appears protective (see below). Interestingly, ACE2 protein also appears to be the entry-point receptor for the severe acute respiratory syndrome (SARS) coronavirus [61, 62].

8. Replenishing ACE2 as a Potential Therapeutic

Given the key role of ACE2, degrading Ang II and generating Ang 1–7, a number of studies have explored its potential as a treatment strategy using human recombinant ACE2 (rhACE2) or adenoviral (Ad)-ACE2 in animal disease models. For example, overexpression of ACE2 in human endothelial cells attenuates Ang II-induced oxidative stress and subsequent increase in monocyte adhesion [63]. Similarly, in rabbits, a recombinant ACE2 expressing vector stabilized atherosclerotic plaques induced by balloon injury to the abdominal aorta [64]. Treatment with a lentiviral vector containing ACE2 resulted in lower blood pressure in hypertensive mice [65, 66] or following an Ang II infusion [67]. Strategies to upregulate or replenish ACE2 are thought to be beneficial in diabetic nephropathy. For example, in diabetes the replenishment of ACE2 with rhACE2 in a mouse model of type 1 diabetes attenuated diabetic kidney injury as well as reducing in blood pressure [68]. The use of (Ad)-ACE2 has had similar beneficial effects in streptozotocin-induced diabetes, where it was shown to attenuate glomerular mesangial cell proliferation, blood pressure, oxidative stress, and fibrosis [69].

In contrast to these studies, the potential utility of ACE2 supplementation in cardiac disease remains controversial. The expression of ACE2 in the failing human heart is generally increased [70–72], consistent with the finding of elevated levels of Ang 1–7 in the same setting [73]. More importantly, overexpression of ACE2 in cardiac myocytes resulted in conduction disturbances by 2 weeks of age, ultimately leading to lethal ventricular arrhythmias and severe fibrosis [74, 75]. This may be because ACE2 is not normally expressed in high levels in myocytes, although it is present in the endocardium and other cardiac cells. However, other studies using transgenic overexpression of cardiac ACE2 have demonstrated partial protection in the heart from ischemia-induced heart failure [76]. Indeed, more recent studies using rhACE2 have shown beneficial cardiac effects [77]. However, the indication for ACE2 that appears most likely to be first tested in the clinic is the treatment of ARDS. In murine models, treatment with catalytically active recombinant ACE2 protein improved the symptoms of acute lung injury in wild-type mice as well as in ACE2 knockout mice [60]. Clinical trials in this often fatal condition are now underway.

Perhaps, the most clinically interesting, however, is the potential for rACE2 to augment the vasculoprotective effects of ACE inhibition or ARBs, in the millions of patients that take these agents, worldwide. In theory, this would be achieved by preventing feedback escape for RAS blockade or enhancing the generation of Ang 1–7, and subsequent signaling through the Mas receptor and or AT_2 receptor. Certainly, ACE2 inhibition attenuates the effects of RAS blockade, both in vitro [78] and in vivo [6]. But could rACE2 make the response to conventional RAS blockade more effective or durable? The problem is that conventional RAS blockade is highly effective in animal models of vascular and renal disease, meaning that it is difficult to explore the potential for further improvements. However, chronic intravenous infusion of ANG-(1–7), or the nonpeptide mas receptor agonist, AVE-0991, are able to improve salt-induced suppression of endothelium-dependent vasodilatation in the mesenteric arteries of male Sprague-Dawley rats, and these actions are not modified by the angiotensin receptor blocker, losartan [79], suggesting that the effects of enhancing the Ang 1–7 mas axis may be beneficial, even in the setting of conventional RAS blockade. Although it enhances the generation of Ang 1–7, whether rACE2 can also provide synergistic benefits, remains to be established.

9. ACE2 Augmenters: A New Kind of Intervention

Rather than providing exogenous ACE2, an alternative approach for augmenting ACE2 has been to increase its endogenous expression. For example, in hypertensive SHRs, all-trans retinoic acid, which increases ACE2 expression, lowers blood pressure levels, and prevents vascular damage [80]. Unfortunately retinoic acid has broader actions that make its potential utility as a therapeutic limited. However, compounds that increase activity of ACE2 could also be beneficial as a treatment in conditions where ACE2 activity is decreased. One exemplar is xanthenone (XNT). This molecule was selected following structure-based screening on compounds that would stabilize the activated form of ACE2, thereby enhancing its catalytic efficacy [81]. In experimental studies, this compound has been shown to enhance ACE2 activity in a dose-dependent manner and significantly decreased blood pressure in both SHRs rats and wild-type WKY rats [81]. Furthermore, improvements in cardiac function and reversal of myocardial, perivascular, and renal fibrosis in the SHRs were also observed [81, 82]. XNT has also shown promise in treating pulmonary hypertension (PH). For example, in a rat model of PH, treatment with XNT was shown to reduce elevated right ventricular systolic pressure, right ventricular hypertrophy, increased pulmonary vessel wall thickness, and interstitial fibrosis [83]. In a model of thrombus formation using SHRs and WKY rats, XNT has also shown antithrombotic action, reducing platelet attachment, and reducing thrombus formation [84]. This compound will not come to clinical trials because of issues of solubility that restrict its formulation. However, other drugs of the same class may prove more suitable.

10. Conclusion

ACE2 is an integral component of the RAS. It is highly expressed in the vasculature, the kidney, lungs, and heart where its actions on peptide signals balance and offset those of ACE. Its actions appear critical in a variety of disease states, including hypertension, diabetes, ageing, renal impairment, and cardiovascular disease. ACE2 deficiency leads to modest physiological changes. However, in states of RAS activation, the loss of ACE2 appears far more important in the development and progression of disease. By contrast, augmentation of ACE2 expression, either directly with recombinant ACE2 or indirectly via agonists like XNT, may have important benefits relevant in the treatment of a range of conditions.

Acknowledgments

Research in the field of ACE2 has been supported by grants from JDRF, NHF, and Australian NHMRC. Dr. C. Tikellis is supported by a JDRF CDA fellowship.

References

[1] S. D. Crowley, S. B. Gurley, M. I. Oliverio et al., "Distinct roles for the kidney and systemic tissues in blood pressure regulation by the renin-angiotensin system," *The Journal of Clinical Investigation*, vol. 115, no. 4, pp. 1092–1099, 2005.

[2] M. Donoghue, F. Hsieh, E. Baronas et al., "A novel angiotensin-converting enzyme-related carboxypeptidase (ACE2) converts angiotensin I to angiotensin 1–9," *Circulation Research*, vol. 87, no. 5, pp. E1–E9, 2000.

[3] A. J. Turner and N. M. Hooper, "The angiotensin-converting enzyme gene family: genomics and pharmacology," *Trends in Pharmacological Sciences*, vol. 23, no. 4, pp. 177–183, 2002.

[4] G. I. Rice, D. A. Thomas, P. J. Grant, A. J. Turner, and N. M. Hooper, "Evaluation of angiotensin-converting enzyme (ACE), its homologue ACE2 and neprilysin in angiotensin peptide metabolism," *Biochemical Journal*, vol. 383, part 1, pp. 45–51, 2004.

[5] M. C. Thomas, R. J. Pickering, D. Tsorotes et al., "Genetic Ace2 deficiency accentuates vascular inflammation and atherosclerosis in the ApoE knockout mouse," *Circulation Research*, vol. 107, no. 7, pp. 888–897, 2010.

[6] C. Tikellis, K. Bialkowski, J. Pete et al., "ACE2 deficiency modifies renoprotection afforded by ACE inhibition in experimental diabetes," *Diabetes*, vol. 57, no. 4, pp. 1018–1025, 2008.

[7] C. C. Wei, B. Tian, G. Perry et al., "Differential ANG II generation in plasma and tissue of mice with decreased expression of the ACE gene," *American Journal of Physiology*, vol. 282, no. 6, pp. H2254–H2258, 2002.

[8] S. R. Tipnis, N. M. Hooper, R. Hyde, E. Karran, G. Christie, and A. J. Turner, "A human homolog of angiotensin-converting enzyme: cloning and functional expression as a captopril-insensitive carboxypeptidase," *The Journal of Biological Chemistry*, vol. 275, no. 43, pp. 33238–33243, 2000.

[9] K. Kuba, L. Zhang, Y. Imai et al., "Impaired heart contractility in Apelin gene-deficient mice associated with aging and pressure overload," *Circulation Research*, vol. 101, no. 4, pp. e32–e42, 2007.

[10] C. Vickers, P. Hales, V. Kaushik et al., "Hydrolysis of biological peptides by human angiotensin-converting enzyme-related carboxypeptidase," *The Journal of Biological Chemistry*, vol. 277, no. 17, pp. 14838–14843, 2002.

[11] H. Zhang, J. Wada, K. Hida et al., "Collectrin, a collecting duct-specific transmembrane glycoprotein, is a novel homolog of ACE2 and is developmentally regulated in embryonic

kidneys," *The Journal of Biological Chemistry*, vol. 276, no. 20, pp. 17132–17139, 2001.

[12] S. Kowalczuk, A. Bröer, N. Tietze, J. M. Vanslambrouck, J. E. J. Rasko, and S. Bröer, "A protein complex in the brush-border membrane explains a Hartnup disorder allele," *The FASEB Journal*, vol. 22, no. 8, pp. 2880–2887, 2008.

[13] D. W. Lambert, N. E. Clarke, N. M. Hooper, and A. J. Turner, "Calmodulin interacts with angiotensin-converting enzyme-2 (ACE2) and inhibits shedding of its ectodomain," *FEBS Letters*, vol. 582, no. 2, pp. 385–390, 2008.

[14] C. I. Johnston, "Tissue angiotensin converting enzyme in cardiac and vascular hypertrophy, repair, and remodeling," *Hypertension*, vol. 23, no. 2, pp. 258–268, 1994.

[15] Min Ae Lee, M. Bohm, M. Paul, and D. Ganten, "Tissue renin-angiotensin systems: their role in cardiovascular disease," *Circulation*, vol. 87, no. 5, pp. IV7–IV13, 1993.

[16] D. S. Jacoby and D. J. Rader, "Renin-angiotensin system and atherothrombotic disease: from genes to treatment," *Archives of Internal Medicine*, vol. 163, no. 10, pp. 1155–1164, 2003.

[17] R. Candido, K. A. Jandeleit-Dahm, Z. Cao et al., "Prevention of accelerated atherosclerosis by angiotensin-converting enzyme inhibition in diabetic apolipoprotein E-deficient mice," *Circulation*, vol. 106, no. 2, pp. 246–253, 2002.

[18] R. Candido, T. J. Allen, M. Lassila et al., "Irbesartan but not amlodipine suppresses diabetes-associated atherosclerosis," *Circulation*, vol. 109, no. 12, pp. 1536–1542, 2004.

[19] L. A. Cassis, M. Gupte, S. Thayer et al., "ANG II infusion promotes abdominal aortic aneurysms independent of increased blood pressure in hypercholesterolemic mice," *American Journal of Physiology*, vol. 296, no. 5, pp. H1660–H1665, 2009.

[20] X. L. Chen, P. E. Tummala, M. T. Olbrych, R. W. Alexander, and R. M. Medford, "Angiotensin II induces monocyte chemoattractant protein-1 gene expression in rat vascular smooth muscle cells," *Circulation Research*, vol. 83, no. 9, pp. 952–959, 1998.

[21] A. E. Loot, J. G. Schreiber, B. Fisslthaler, and I. Fleming, "Angiotensin II impairs endothelial function via tyrosine phosphorylation of the endothelial nitric oxide synthase," *Journal of Experimental Medicine*, vol. 206, no. 13, pp. 2889–2896, 2009.

[22] M. J. A. P. Daemen, D. M. Lombardi, F. T. Bosman, and S. M. Schwartz, "Angiotensin II induces smooth muscle cell proliferation in the normal and injured rat arterial wall," *Circulation Research*, vol. 68, no. 2, pp. 450–456, 1991.

[23] J. L. Probstfield and K. D. O'Brien, "Progression of cardiovascular damage: the role of renin-angiotensin system blockade," *American Journal of Cardiology*, vol. 105, no. 1, supplement, pp. 10A–20A, 2010.

[24] C. M. Ferrario, A. J. Trask, and J. A. Jessup, "Advances in biochemical and functional roles of angiotensin-converting enzyme 2 and angiotensin-(1–7) in regulation of cardiovascular function," *American Journal of Physiology*, vol. 289, no. 6, pp. H2281–H2290, 2005.

[25] S. Tesanovic, A. Vinh, T. A. Gaspari, D. Casley, and R. E. Widdop, "Vasoprotective and atheroprotective effects of angiotensin (1–7) in apolipoprotein E-deficient mice," *Arteriosclerosis, Thrombosis, and Vascular Biology*, vol. 30, no. 8, pp. 1606–1613, 2010.

[26] J. C. Sluimer, J. M. Gasc, I. Hamming et al., "Angiotensin-converting enzyme 2 (ACE2) expression and activity in human carotid atherosclerotic lesions," *The Journal of Pathology*, vol. 215, no. 3, pp. 273–279, 2008.

[27] C. Tikellis, C. I. Johnston, J. M. Forbes et al., "Characterization of renal angiotensin—converting enzyme 2 in diabetic nephropathy," *Hypertension*, vol. 41, no. 3, pp. 392–397, 2003.

[28] L. Stanziola, L. J. Greene, and R. A. S. Santos, "Effect of chronic angiotensin converting enzyme inhibition on angiotensin I and bradykinin metabolism in rats," *American Journal of Hypertension*, vol. 12, no. 10, part 1, pp. 1021–1029, 1999.

[29] I. F. Benter, M. H. M. Yousif, F. M. Al-Saleh, R. Raghupathy, M. C. Chappell, and D. I. Diz, "Angiotensin-(1–7) blockade attenuates captopril- or hydralazine-induced cardiovascular protection in spontaneously hypertensive rats treated with NG-nitro-l-arginine methyl ester," *Journal of Cardiovascular Pharmacology*, vol. 57, no. 5, pp. 559–567, 2011.

[30] S. B. Gurley, A. Allred, T. H. Le et al., "Altered blood pressure responses and normal cardiac phenotype in ACE2-null mice," *The Journal of Clinical Investigation*, vol. 116, no. 8, pp. 2218–2225, 2006.

[31] S. B. Gurley and T. M. Coffman, "Angiotensin-converting enzyme 2 gene targeting studies in mice: mixed messages," *Experimental Physiology*, vol. 93, no. 5, pp. 538–542, 2008.

[32] G. Y. Oudit, Z. Kassiri, M. P. Patel et al., "Angiotensin II-mediated oxidative stress and inflammation mediate the age-dependent cardiomyopathy in ACE2 null mice," *Cardiovascular Research*, vol. 75, no. 1, pp. 29–39, 2007.

[33] M. A. Crackower, R. Sarao, G. Y. Oudit et al., "Angiotensin-converting enzyme 2 is an essential regulator of heart function," *Nature*, vol. 417, no. 6891, pp. 822–828, 2002.

[34] M. Yamazato, Y. Yamazato, C. Sun, C. Diez-Freire, and M. K. Raizada, "Overexpression of angiotensin-converting enzyme 2 in the rostral ventrolateral medulla causes long-term decrease in blood pressure in the spontaneously hypertensive rats," *Hypertension*, vol. 49, no. 4, pp. 926–931, 2007.

[35] Y. Feng, H. Xia, Y. Cai et al., "Brain-selective overexpression of human angiotensin-converting enzyme type 2 attenuates neurogenic hypertension," *Circulation Research*, vol. 106, no. 2, pp. 373–382, 2010.

[36] Y. Feng, X. Yue, H. Xia et al., "Angiotensin-converting enzyme 2 overexpression in the subfornical organ prevents the angiotensin II-mediated pressor and drinking responses and is associated with angiotensin II type 1 receptor downregulation," *Circulation Research*, vol. 102, no. 6, pp. 729–736, 2008.

[37] D. I. Diz, M. A. Garcia-Espinosa, S. Gegick et al., "Injections of angiotensin-converting enzyme 2 inhibitor MLN4760 into nucleus tractus solitarii reduce baroreceptor reflex sensitivity for heart rate control in rats," *Experimental Physiology*, vol. 93, no. 5, pp. 694–700, 2008.

[38] F. Pieruzzi, Z. A. Abassi, and H. R. Keiser, "Expression of renin-angiotensin system components in the heart, kidneys, and lungs of rats with experimental heart failure," *Circulation*, vol. 92, no. 10, pp. 3105–3112, 1995.

[39] M. Ruzicka, F. W. Keeley, and F. H. H. Leenen, "The renin-angiotensin system and volume overload-induced changes in cardiac collagen and elastin," *Circulation*, vol. 90, no. 4, pp. 1989–1996, 1994.

[40] A. T. Hirsch, C. E. Talsness, H. Schunkert, M. Paul, and V. J. Dzau, "Tissue-specific activation of cardiac angiotensin converting enzyme in experimental heart failure," *Circulation Research*, vol. 69, no. 2, pp. 475–482, 1991.

[41] S. Hokimoto, H. Yasue, K. Fujimoto et al., "Expression of angiotensin-converting enzyme in remaining viable myocytes of human ventricles after myocardial infarction," *Circulation*, vol. 94, no. 7, pp. 1513–1518, 1996.

[42] J. I. Sadoshima, Y. Xu, H. S. Slayter, and S. Izumo, "Autocrine release of angiotensin II mediates stretch-induced hypertrophy of cardiac myocytes in vitro," *Cell*, vol. 75, no. 5, pp. 977–984, 1993.

[43] M. Kakishita, K. Nakamura, M. Asanuma et al., "Direct evidence for increased hydroxyl radicals in angiotensin II-induced cardiac hypertrophy through angiotensin II type 1a receptor," *Journal of Cardiovascular Pharmacology*, vol. 42, supplement 1, pp. S67–S70, 2003.

[44] K. T. Weber, Y. Sun, and E. Guarda, "Structural remodeling in hypertensive heart disease and the role of hormones," *Hypertension*, vol. 23, no. 6, pp. 869–877, 1994.

[45] M. A. Pfeffer, E. Braunwald, L. A. Moye et al., "Effect of captopril on mortality and morbidity in patients with left ventricular dysfunction after myocardial infarction—results of the survival and ventricular enlargement trial," *The New England Journal of Medicine*, vol. 327, no. 10, pp. 669–677, 1992.

[46] P. J. Garabelli, J. G. Modrall, J. M. Penninger, C. M. Ferrario, and M. C. Chappell, "Distinct roles for angiotensin-converting enzyme 2 and carboxypeptidase A in the processing of angiotensins within the murine heart," *Experimental Physiology*, vol. 93, no. 5, pp. 613–621, 2008.

[47] J. A. Stewart Jr., E. Lazartigues, and P. A. Lucchesi, "The angiotensin converting enzyme 2/Ang-(1–7) axis in the heart: a role for mas communication?" *Circulation Research*, vol. 103, no. 11, pp. 1197–1199, 2008.

[48] Z. Kassiri, J. Zhong, D. Guo et al., "Loss of angiotensin-converting enzyme 2 accelerates maladaptive left ventricular remodeling in response to myocardial infarction," *Circulation*, vol. 2, no. 5, pp. 446–455, 2009.

[49] S. Bodiga, J. C. Zhong, W. Wang et al., "Enhanced susceptibility to biomechanical stress in ACE2 null mice is prevented by loss of the p47phox NADPH oxidase subunit," *Cardiovascular Research*, vol. 91, no. 1, pp. 151–161, 2011.

[50] K. Yamamoto, M. Ohishi, T. Katsuya et al., "Deletion of angiotensin-converting enzyme 2 accelerates pressure overload-induced cardiac dysfunction by increasing local angiotensin II," *Hypertension*, vol. 47, no. 4, pp. 718–726, 2006.

[51] K. Nakamura, N. Koibuchi, H. Nishimatsu et al., "Candesartan ameliorates cardiac dysfunction observed in angiotensin-converting enzyme 2-deficient mice," *Hypertension Research*, vol. 31, no. 10, pp. 1953–1961, 2008.

[52] T. N. Doan, N. Gletsu, J. Cole, and K. E. Bernstein, "Genetic manipulation of the renin-angiotensin system," *Current Opinion in Nephrology and Hypertension*, vol. 10, no. 4, pp. 483–491, 2001.

[53] N. Li, J. Zimpelmann, K. Cheng, J. A. Wilkins, and K. D. Burns, "The role of angiotensin converting enzyme 2 in the generation of angiotensin 1–7 by rat proximal tubules," *American Journal of Physiology*, vol. 288, no. 2, pp. F353–F362, 2005.

[54] G. Y. Oudit, A. M. Herzenberg, Z. Kassiri et al., "Loss of angiotensin-converting enzyme-2 leads to the late development of angiotensin II-dependent glomerulosclerosis," *American Journal of Pathology*, vol. 168, no. 6, pp. 1808–1820, 2006.

[55] J. Zhong, D. Guo, C. B. Chen et al., "Prevention of angiotensin II-mediated renal oxidative stress, inflammation, and fibrosis by angiotensin-converting enzyme 2," *Hypertension*, vol. 57, no. 2, pp. 314–322, 2010.

[56] M. J. Soler, J. Wysocki, M. Ye, J. Lloveras, Y. Kanwar, and D. Batlle, "ACE2 inhibition worsens glomerular injury in association with increased ACE expression in streptozotocin-induced diabetic mice," *Kidney International*, vol. 72, no. 5, pp. 614–623, 2007.

[57] D. W. Wong, G. Y. Oudit, H. Reich et al., "Loss of Angiotensin-converting enzyme-2 (Ace2) accelerates diabetic kidney injury," *American Journal of Pathology*, vol. 171, no. 2, pp. 438–451, 2007.

[58] A. T. Lely, I. Hamming, H. van Goor, and G. J. Navis, "Renal ACE2 expression in human kidney disease," *The Journal of Pathology*, vol. 204, no. 5, pp. 587–593, 2004.

[59] D. G. Kiely, R. I. Cargill, N. M. Wheeldon, W. J. Coutie, and B. J. Lipworth, "Haemodynamic and endocrine effects of type 1 angiotensin II receptor blockade in patients with hypoxaemic cor pulmonale," *Cardiovascular Research*, vol. 33, no. 1, pp. 201–208, 1997.

[60] Y. Imai, K. Kuba, S. Rao et al., "Angiotensin-converting enzyme 2 protects from severe acute lung failure," *Nature*, vol. 436, no. 7047, pp. 112–116, 2005.

[61] W. Li, M. J. Moore, N. Vasllieva et al., "Angiotensin-converting enzyme 2 is a functional receptor for the SARS coronavirus," *Nature*, vol. 426, no. 6965, pp. 450–454, 2003.

[62] K. Kuba, Y. Imai, S. Rao et al., "A crucial role of angiotensin converting enzyme 2 (ACE2) in SARS coronavirus-induced lung injury," *Nature Medicine*, vol. 11, no. 8, pp. 875–879, 2005.

[63] F. Lovren, Y. Pan, A. Quan et al., "Angiotensin converting enzyme-2 confers endothelial protection and attenuates atherosclerosis," *American Journal of Physiology*, vol. 295, no. 4, pp. H1377–H1384, 2008.

[64] B. Dong, C. Zhang, J. B. Feng et al., "Overexpression of ACE2 enhances plaque stability in a rabbit model of atherosclerosis," *Arteriosclerosis, Thrombosis, and Vascular Biology*, vol. 28, no. 7, pp. 1270–1276, 2008.

[65] C. Díez-Freire, J. Vázquez, M. F. Correa De Adjounian et al., "ACE2 gene transfer attenuates hypertension-linked pathophysiological changes in the SHR," *Physiological Genomics*, vol. 27, no. 1, pp. 12–19, 2006.

[66] B. Rentzsch, M. Todiras, R. Iliescu et al., "Transgenic angiotensin-converting enzyme 2 overexpression in vessels of SHRSP rats reduces blood pressure and improves endothelial function," *Hypertension*, vol. 52, no. 5, pp. 967–973, 2008.

[67] J. Wysocki, M. Ye, E. Rodriguez et al., "Targeting the degradation of angiotensin II with recombinant angiotensin-converting enzyme 2: prevention of angiotensin II-dependent hypertension," *Hypertension*, vol. 55, no. 1, pp. 90–98, 2010.

[68] G. Y. Oudit, G. C. Liu, J. Zhong et al., "Human recombinant ACE2 reduces the progression of diabetic nephropathy," *Diabetes*, vol. 59, no. 2, pp. 529–538, 2010.

[69] C. X. Liu, Q. Hu, Y. Wang et al., "Angiotensin-converting enzyme (ACE) 2 overexpression ameliorates glomerular injury in a rat model of diabetic nephropathy: a comparison with ACE inhibition," *Molecular Medicine*, vol. 17, no. 1-2, pp. 59–69, 2011.

[70] A. B. Goulter, M. Avella, K. M. Botham, and J. Elliott, "Chylomicron-remnant-like particles inhibit the basal nitric oxide pathway in porcine coronary artery and aortic endothelial cells," *Clinical Science*, vol. 105, no. 3, pp. 363–371, 2003.

[71] R. Studer, H. Reinecke, B. Muller, J. Holtz, H. Just, and H. Drexler, "Increased angiotensin-I converting enzyme gene expression in the failing human heart. Quantification by competitive RNA polymerase chain reaction," *The Journal of Clinical Investigation*, vol. 94, no. 1, pp. 301–310, 1994.

[72] L. S. Zisman, R. S. Keller, B. Weaver et al., "Increased angiotensin-(1–7)-forming activity in failing human heart ventricles: evidence for upregulation of the angiotensin-converting enzyme homologue ACE2," *Circulation*, vol. 108, no. 14, pp. 1707–1712, 2003.

[73] C. Communal, M. Singh, B. Menon, Z. Xie, W. S. Colucci, and K. Singh, "β1 integrins expression in adult rat ventricular myocytes and its role in the regulation of β-adrenergic receptor-stimulated apoptosis," *Journal of Cellular Biochemistry*, vol. 89, no. 2, pp. 381–388, 2003.

[74] M. Donoghue, H. Wakimoto, C. T. Maguire et al., "Heart block, ventricular tachycardia, and sudden death in ACE2 transgenic mice with downregulated connexins," *Journal of Molecular and Cellular Cardiology*, vol. 35, no. 9, pp. 1043–1053, 2003.

[75] R. Masson, S. A. Nicklin, M. A. Craig et al., "Onset of experimental severe cardiac fibrosis is mediated by overexpression of Angiotensin-converting enzyme 2," *Hypertension*, vol. 53, no. 4, pp. 694–700, 2009.

[76] S. Der Sarkissian, J. L. Grobe, L. Yuan et al., "Cardiac overexpression of angiotensin converting enzyme 2 protects the heart from ischemia-induced pathophysiology," *Hypertension*, vol. 51, no. 3, pp. 712–718, 2008.

[77] J. Zhong, R. Basu, D. Guo et al., "Angiotensin-converting enzyme 2 suppresses pathological hypertrophy, myocardial fibrosis, and cardiac dysfunction," *Circulation*, vol. 122, no. 7, pp. 717–728, 2010.

[78] N. Hayashi, K. Yamamoto, M. Ohishi et al., "The counter-regulating role of ACE2 and ACE2-mediated angiotensin 1–7 signaling against angiotensin II stimulation in vascular cells," *Hypertension Research*, vol. 33, no. 11, pp. 1182–1185, 2010.

[79] G. Raffai, M. J. Durand, and J. H. Lombard, "Acute and chronic angiotensin-(1–7) restores vasodilation and reduces oxidative stress in mesenteric arteries of salt-fed rats," *American Journal of Physiology*, vol. 301, no. 4, pp. H1341–H1352, 2011.

[80] J. C. Zhong, D. Y. Huang, Y. M. Yang et al., "Upregulation of angiotensin-converting enzyme 2 by all-trans retinoic acid in spontaneously hypertensive rats," *Hypertension*, vol. 44, no. 6, pp. 907–912, 2004.

[81] J. A. Hernández Prada, A. J. Ferreira, M. J. Katovich et al., "Structure-based identification of small-molecule angiotensin-converting enzyme 2 activators as novel antihypertensive agents," *Hypertension*, vol. 51, no. 5, pp. 1312–1317, 2008.

[82] A. J. Ferreira, V. Shenoy, Y. Qi et al., "Angiotensin-converting enzyme 2 activation protects against hypertension-induced cardiac fibrosis involving extracellular signal-regulated kinases," *Experimental Physiology*, vol. 96, no. 3, pp. 287–294, 2011.

[83] A. J. Ferreira, V. Shenoy, Y. Yamazato et al., "Evidence for angiotensin-converting enzyme 2 as a therapeutic target for the prevention of pulmonary hypertension," *American Journal of Respiratory and Critical Care Medicine*, vol. 179, no. 11, pp. 1048–1054, 2009.

[84] R. A. Fraga-Silva, B. S. Sorg, M. Wankhede et al., "ACE2 activation promotes antithrombotic activity," *Molecular Medicine*, vol. 16, no. 5-6, pp. 210–215, 2010.

8

Significant Increase in Salivary Substance P Level after a Single Oral Dose of Cevimeline in Humans

Yosuke Suzuki, Hiroki Itoh, Kohei Amada, Ryota Yamamura, Yuhki Sato, and Masaharu Takeyama

Department of Clinical Pharmacy, Oita University Hospital, Hasama-machi, Oita 879-5593, Japan

Correspondence should be addressed to Yosuke Suzuki; y-suzuki@oita-u.ac.jp

Academic Editor: Eva Ekblad

Cevimeline is a novel muscarinic acetylcholine receptor agonist currently being developed as a therapeutic agent for xerostomia. We examined the effects of cevimeline on salivary and plasma levels of substance-P- (SP-), calcitonin-gene-related-peptide- (CGRP-), and vasoactive-intestinal-polypeptide- (VIP-) like immunoreactive substances (ISs) in humans. An open-labeled crossover study was conducted on seven healthy volunteers. Saliva volume was measured, and saliva and venous blood samples were collected before and 30–240 min after a single oral dose of cevimeline or placebo. Salivary and plasma levels of SP-, CGRP-, and VIP-IS were measured using a highly sensitive enzyme immunoassay. A single oral dose of cevimeline resulted in significant increases in salivary but not plasma SP-IS level compared to placebo. Cevimeline administration did not alter the salivary or plasma levels of CGRP-IS or VIP-IS compared to placebo. Significant increases in salivary volume were observed after cevimeline administration compared to placebo. A significant correlation was observed between the total release of SP-IS and that of salivary volume. These findings suggest an association of SP with the enhancement of salivary secretion by cevimeline.

1. Introduction

The functions of the salivary glands are controlled by the autonomic nervous system and influenced by the sensory nervous system. When parasympathetic impulses dominate, salivary flow is greatly enhanced and the saliva has a low protein content. Studies of animal and human innervation have revealed that parasympathetic nerve fibers are present around acinar cells, ducts, and blood vessels in the major salivary glands [1]. A research has also shown that beside the classic transmitters noradrenaline and acetylcholine, neuropeptides such as substance P (SP), calcitonin gene-related peptide (CGRP), and vasoactive intestinal polypeptide (VIP) (Figure 1) are present in the nerve fibers of the autonomic nervous system as well as in the auriculotemporal nerve, facial nerve, and cervical dorsal root fibers [2]. These neuropeptides are known to cause salivation in rats [2–7]. In recent years, the mechanisms of actions of drugs that used to treat xerostomia have been elucidated pharmacologically from the viewpoint of salivary neuropeptide levels. Anethole trithione and pilocarpine have been shown to elevate SP and CGRP levels in human saliva [8–11].

Cevimeline hydrochloride hydrate (cevimeline) (Figure 2) is a novel muscarinic acetylcholine receptor agonist currently being developed as a therapeutic agent for Sjögren's syndrome. Sjögren's syndrome is a serious and chronic autoimmune disorder characterized by inflammation in the exocrine glands such as the salivary and lacrimal glands [12], leading to xerostomia (dry mouth) and xerophthalmia (dry eyes). Cevimeline acts as a stimulator of the M3 acetylcholine receptor expressed on salivary glands and has been shown to increase saliva secretion in patients with Sjögren's syndrome [13]. Although cevimeline is useful for the treatment of dry mouth, it only enhances saliva production in 60% of the patients [14], and the mechanism of the drug response is still unknown. It is possible that individual variability of neuropeptide nerve stimulation in response to cevimeline may be involved in the variable drug response to cevimeline. However, the effects of cevimeline in stimulating neuropeptide nerves have not been demonstrated.

SP   Arg-Pro-Lys-Pro-Gln-Gln-Phe-Phe-Gly-Leu-Met-NH$_2$

CGRP Ala-Cys-Asp-Thr-Ala-Thr-Cys-Val-Thr-His-Arg-Leu-Ala-Gly-
            Leu-Leu-Ser-Arg-Ser-Gly-Gly-Val-Val-Lys-Asn-Asn-Phe-Val-
            Pro-Thr-Asn-Val-Gly-Ser-Lys-Ala-Phe-NH$_2$

VIP   His-Ser-Asp-Ala-Val-Phe-Thr-Asp-Asn-Tyr-Thr-Arg-Leu-Arg-
            Lys-Gln-Met-Ala-Val-Lys-Lys-Tyr-Leu-Asn-Ser-Ile-Leu-Asn-NH$_2$

FIGURE 1: Structures of substance P (SP), calcitonin gene-related peptide (CGRP), and vasoactive intestinal polypeptide (VIP).

(and enantiomer)

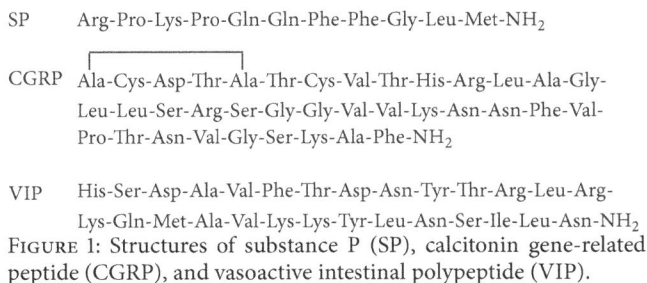

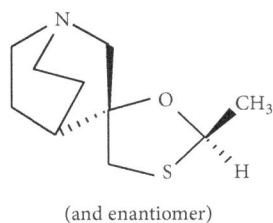

FIGURE 2: Structure of cevimeline.

The objective of the present study is to examine the effects of cevimeline on saliva and plasma levels of SP-, CGRP-, and VIP-like immunoreactive substances (ISs) in humans, as markers of nerve stimulation of these neuropeptides.

2. Materials and Methods

Cevimeline hydrochloride hydrate (Saligren Capsule 30 mg) was purchased from Nippon Kayaku Co. Ltd. (Tokyo, Japan). Lactose (Merck Hoei Co. Ltd., Osaka, Japan) was used as placebo. Synthetic human SP, CGRP and its fragment (8–37), and VIP were purchased from Peptide Institute, Inc. (Osaka, Japan). VIP fragment (11–28) was supplied by Professor Yajima (Kyoto University, Kyoto, Japan). Substance P antiserum (Y150) was purchased from Yanaihara Institute (Shizuoka, Japan), CGRP antiserum (14160) from Peptide Institute, Inc., and VIP antiserum (T-4116) from Peninsula Laboratories (California, USA). All other reagents were analytical reagent grade from commercial sources.

Seven healthy nonsmoking male volunteers aged 24–31 (median 27) years and weighing 56–70 (median 64) kg participated in this study. All subjects had no history of xerostomia, and their baseline fasting salivary and plasma levels of SP-, CGRP-, and VIP-IS were within the normal ranges for healthy subjects reported previously [8–11, 15, 16]. The study was approved by the Ethics Committee of Oita Medical University. Each subject gave informed consent after receiving explanation on the scientific purpose of the study. No subject received any medication during one month before the study. The subjects fasted for at least 2 hours before the study was commenced and during the experiments.

We performed an open-labeled, crossover study between May and October 2010. In each subject, cevimeline and placebo were studied in random order, in a crossover manner with an interval of one month between the two studies. On the day of study, all subjects finished lunch (standardized lunch of less than 800 kcal) before 12:00. Each study was conducted from 14:00 to 18:00 in a room with temperature controlled at 25°C, during which the subjects maintained a resting and relaxed state. A single dose of cevimeline 30 mg (cevimeline group) or placebo (placebo group) was administered orally with 100 mL water. At scheduled times after the test drug was administered, saliva production was measured, and saliva samples were collected for assaying salivary neuropeptide levels, while blood samples were collected for measuring plasma neuropeptide levels. The dose of cevimeline in this study was the normal daily dose used in clinical therapy. Saliva and venous blood samples were collected before and at 30, 60, 90, 120, 180, and 240 min after administration of cevimeline or placebo.

The volume of saliva produced in 5 min was measured by the Saxon test, an oral equivalent of the Schirmer test [17]. Two sterile absorbent cotton balls (no. 14, Kawamoto Houtai Zairyou, Osaka, Japan) and a polyethylene pouch were weighed. After swallowing to remove any existing oral fluid, saliva was collected by placing the two cotton balls onto the vestibule of the mouth for exactly 5 min. The subjects then expectorated the moist absorbent cotton balls into the polyethylene pouch. The weight of saliva was determined by subtracting the original weight of the pouch and cotton balls from the weight obtained after the cotton balls were placed in the mouth. The weight of the liquid was taken to be the salivary volume (mL) produced in 5 minutes.

Unstimulated whole saliva specimens were collected by the spitting method according to Navazesh and Christensen [18]. The subjects rinsed their mouth thoroughly with deionized water and rested for a few minutes before saliva collection began. After one minute practice collection, which was discarded, subsequently 3 mL of saliva was collected into a test tube containing 500 kallikrein inhibitor units/mL of aprotinin and 1.2 mg/mL of EDTA. Blood samples were collected into chilled tubes containing 500 kallikrein inhibitor units/mL of aprotinin and 1.2 mg/mL of EDTA.

The saliva samples were diluted 1:1 with 4% acetate buffer (pH 4.0), centrifuged at 3500 rpm for 5 min at 4°C, and then the supernatant was diluted 2:3 with 4% acetate buffer (pH 4.0) and loaded onto C18 reverse-phase cartridges (Sep-Pak C18; Millipore Corp., Milford, MA, USA). Blood samples were centrifuged, and the plasma samples were diluted 1:4 with 4% acetate buffer (pH 4.0) and loaded onto C18 reverse-phase cartridges. After washing with 4% acetate buffer, neuropeptides in the columns were eluted with 70% acetonitrile in 0.5% acetate buffer (pH 4.0). Eluates were concentrated by spin-vacuum evaporation, lyophilized, and stored at −40°C until use. The recovery of SP-, CGRP-, and VIP-IS in saliva and plasma was greater than 90% using this extraction procedure [19–21].

Neuropeptide levels in saliva and plasma were measured using highly sensitive enzyme immunoassays for SP [19], CGRP [20], and VIP [21] as described previously. The assays were performed by a delayed addition method. An immunoplate (Nunc-Immuno Module Maxisorp F8, InterMed, Denmark) coated with anti-rabbit IgG (55641, ICN Pharmaceuticals, Inc., Ohio, USA) was used to separate bound and free antigens. Human SP, CGRP fragment (8–37), or VIP

(a)

(b)

(c)

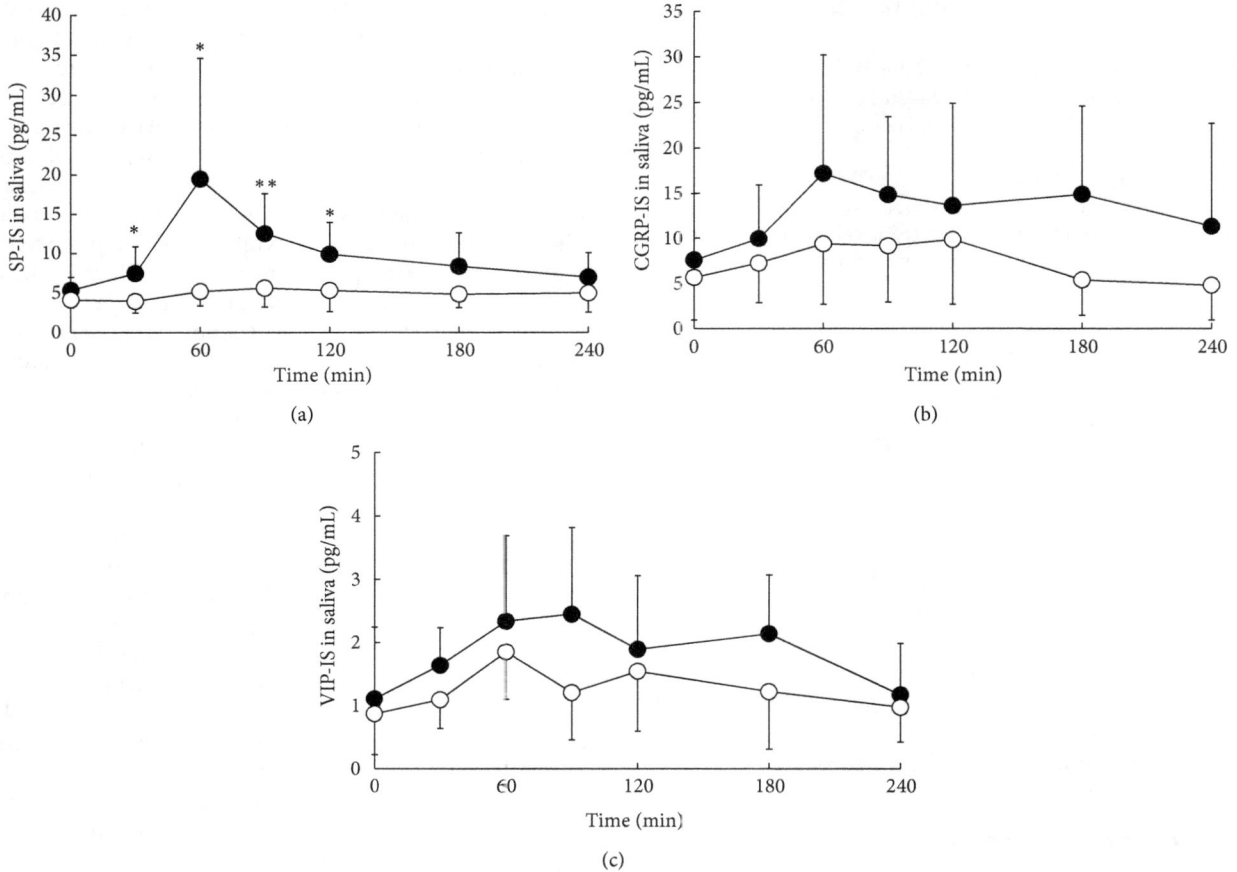

FIGURE 3: Effects of a single oral dose of cevimeline (•) or placebo (○) on salivary levels of substance P (SP) (a), calcitonin gene-related peptide (CGRP) (b), and vasoactive intestinal polypeptide (VIP) immunoreactive substance (c). Values are means ± SD, $n = 7$. $^{*}P < 0.05$, $^{**}P < 0.01$, versus placebo at the same time point.

TABLE 1: Total amounts of substance-P- (SP-), calcitonin-gene-related-peptide- (CGRP-), and vasoactive-intestinal-polypeptide- (VIP-) like immunoreactive substance in saliva released after administration of cevimeline or placebo to 7 healthy volunteers.

Drugs	AUC_{0-240} in saliva (pg min/mL)		
	SP	CGRP	VIP
Cevimeline	$2420.9 \pm 744.6^{**}$	3014.4 ± 2009.3	389.3 ± 120.4
Placebo	1185.8 ± 398.6	1644.1 ± 1094.7	288.3 ± 125.7

Data are expressed as mean ± SD, $^{**}P < 0.01$ versus placebo.

fragment (11–28) was conjugated with β-D-galactosidase by N-(ε-maleimido-caproyloxy)-succinimide according to the methods of Kitagawa et al. [22]. The enzyme immunoassays were specific and highly sensitive, with detection limits of 0.08, 0.40, and 1.00 fmol/well for SP-, CGRP-, and VIP-IS, respectively.

All values are expressed as means ± standard deviation (SD). Total release of each neuropeptide or saliva was calculated as the area under the level—or volume—time curve (AUC_{0-240}) using the trapezoidal method. Differences in neuropeptide-IS level, salivary volume, and their AUC_{0-240} between the cevimeline and placebo groups were analyzed by paired t-test or Mann-Whitney U test. The relationship between AUC_{0-240} of neuropeptide-IS level and AUC_{0-240} of salivary volume was analyzed by Pearson's product-moment correlation coefficient. A P value less than 0.05 was considered statistically significant. Statistical analyses were performed using the SPSS software package (version 17.0; SPSS Inc., IL, USA).

3. Results

The salivary SP-IS level-time profile and total release of SP-IS (AUC_{0-240}) after a single oral dose of cevimeline or placebo are shown in Figure 3(a) and Table 1. Oral administration of cevimeline resulted in significant increases in salivary SP-IS level at 30, 60, 90, and 120 min (7.5 ± 3.4, 19.1 ± 15.1, 12.5 ± 5.1, and 9.9 ± 4.1 pg/mL, resp.) compared with the corresponding levels after placebo administration (4.0 ± 1.5, 5.2 ± 1.8, 5.6 ± 2.4, and 5.3 ± 2.7 pg/mL). Furthermore, AUC_{0-240} was significantly higher after cevimeline administration (2420.9 ± 744.6 pg·min/mL) compared with placebo (1185.8±398.6 pg·min/mL). On the other hand, no significant changes in salivary CGRP- and VIP-IS levels and AUC_{0-240}

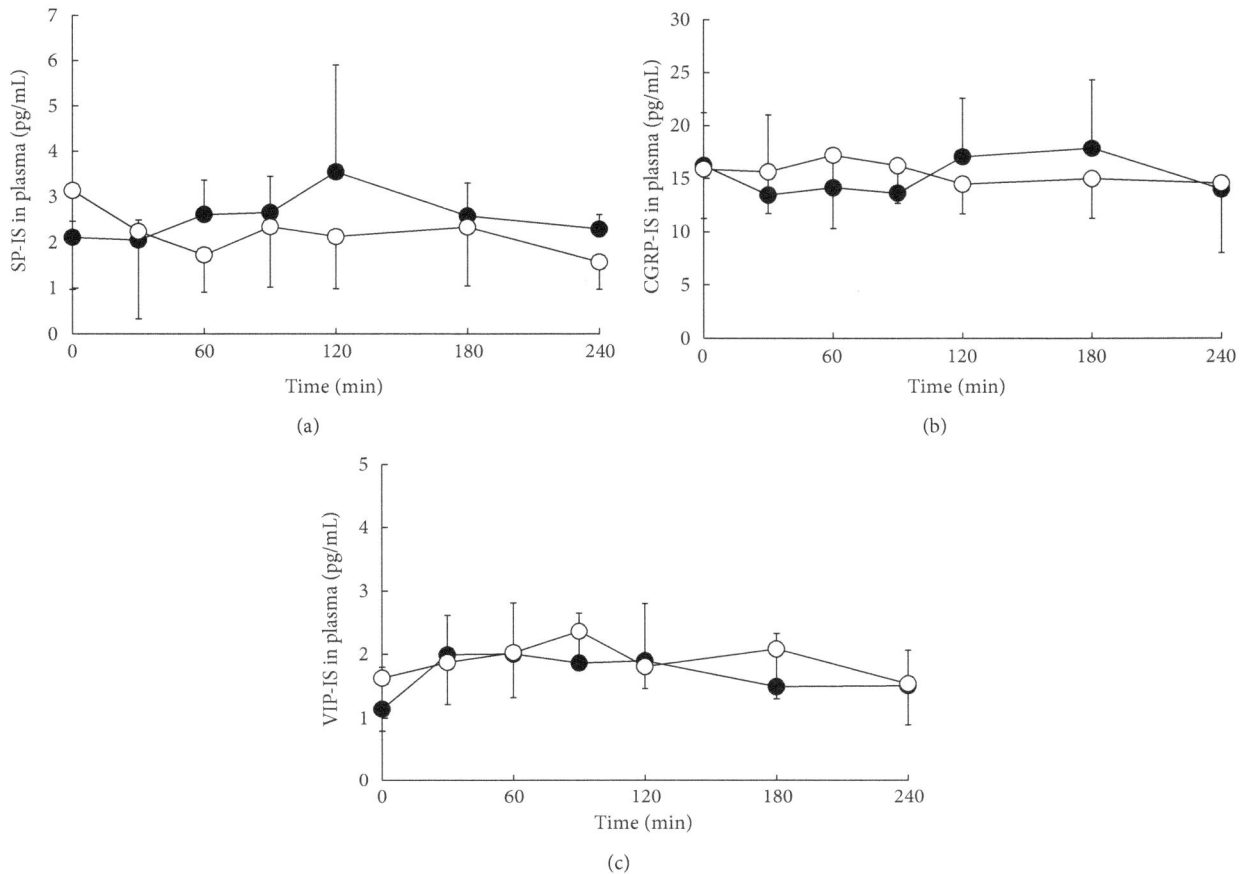

FIGURE 4: Effects of a single oral dose of cevimeline (•) or placebo (○) on plasma levels of substance P (SP) (a), calcitonin gene-related peptide (CGRP) (b), and vasoactive intestinal polypeptide (VIP) immunoreactive substance (c). Values are means ± SD, $n = 7$.

TABLE 2: Total amounts of substance-P- (SP-), calcitonin-gene-related-peptide- (CGRP-), and vasoactive-intestinal-polypeptide- (VIP-) like immunoreactive substance in plasma released after administration of cevimeline or placebo to 7 healthy volunteers.

Drugs	AUC$_{0-240}$ in plasma (pg min/mL)		
	SP	CGRP	VIP
Cevimeline	453.4 ± 343.6	2518.0 ± 1841.0	295.1 ± 210.4
Placebo	445.3 ± 316.2	2640.4 ± 1936.1	398.4 ± 189.3

Data are expressed as mean ± SD.

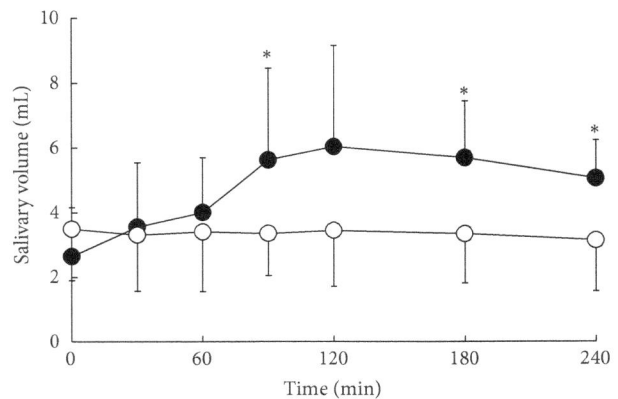

FIGURE 5: Effects of a single oral dose of cevimeline (•) or placebo (○) on salivary volume. Values are means ± SD, $n = 7$. *$P < 0.05$ versus placebo at the same time point.

were observed after the administration of cevimeline (Figures 3(b) and 3(c) and Table 1) compared with placebo.

The plasma SP-, CGRP-, and VIP-IS level-time profiles and total releases of SP-, CGRP-, and VIP-IS (AUC$_{0-240}$) after a single oral dose of cevimeline or placebo are shown in Figure 4 and Table 2. Cevimeline administration did not alter the plasma levels or AUC$_{0-240}$ of SP-, CGRP-, or VIP-IS compared with placebo.

The changes in salivary volume and total release of saliva (AUC$_{0-240}$) after cevimeline or placebo administration are shown in Figure 5 and Table 3. Cevimeline administration resulted in significant increases in salivary volume at 90, 180, and 240 min (5.6 ± 2.8, 5.7 ± 1.8, and 5.1 ± 1.2 mL, resp.)

compared with the corresponding levels after placebo administration (3.4 ± 1.3, 3.4 ± 1.5, and 3.2 ± 1.6 mL). The AUC$_{0-240}$ was also significantly higher after cevimeline administration (1200.8 ± 403.4 mL·min) compared with placebo (804.9 ± 369.8 mL·min).

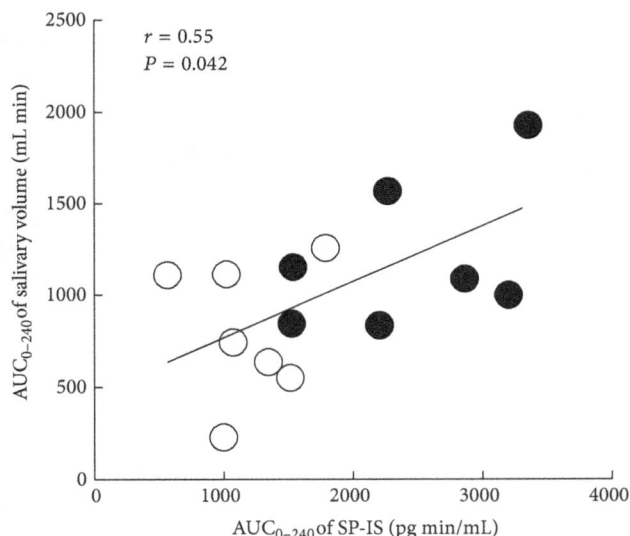

FIGURE 6: Relationship between the area under the time curve (AUC_{0-240}) of substance P (SP)-immunoreactive substance (IS) level and AUC_{0-240} of salivary volume after a single oral dose of cevimeline (•) or placebo (○).

TABLE 3: Total amount of saliva released after administration of cevimeline or placebo to 7 healthy volunteers.

Drugs	AUC_{0-240} of salivary volume (mL min)
Cevimeline	$1200.8 \pm 403.4^{**}$
Placebo	804.9 ± 369.8

Data are expressed as mean $\pm$ SD, $^{**}P < 0.01$ versus placebo.

The relationship between AUC_{0-240} of SP-IS level and salivary volume after administration of cevimeline or placebo is shown in Figure 6. A significant correlation was observed between AUC_{0-240} of SP-IS level and AUC_{0-240} salivary volume ($r = 0.55$, $P = 0.042$).

4. Discussion

In this study, we investigated the effects of cevimeline on saliva and plasma levels of SP-, CGRP-, and VIP-IS in healthy subjects. Past studies have established that salivary and plasma levels of SP-, CGRP-, and VIP-IS vary within 30 min after a meal and then maintain constant from 1 hour after a meal [8, 9, 19–21]. Furthermore, it is known that the absorption of cevimeline is little affected by a meal. These data support our study design, and the present study appropriately evaluates the effects of cevimeline on neuropeptide levels and saliva production without being affected by a meal.

Neuropeptides such as SP, CGRP, and VIP are important stimulators of salivation. SP is mainly localized in submandibular and parotid glands and increases blood flow via its vasodilatory effect in salivary glands, stimulates the production of saliva and amylase, and influences ionic flow in rats [23, 24]. Previous report indicates that CGRP also enhances the release of saliva and amylase in rats [3, 6],

and VIP induces alterations in salivary fluid and protein secretion [4, 25]. In the present study, a single oral dose of cevimeline resulted in significant increases in salivary SP-IS level at 30, 60, 90, and 120 min and in the AUC_{0-240} of SP-IS compared with placebo administration, whereas cevimeline did not alter the plasma levels or AUC_{0-240} of SP-IS. Anethole trithione and pilocarpine have also been reported to increase SP-IS in saliva but not in plasma [8–11]. These results indicate a close association of SP with the enhancement of salivary secretion by cevimeline, in the same manner as anethole trithione and pilocarpine. In addition, these findings suggest that cevimeline may mainly promote salivary secretion from submandibular and parotid glands by increasing SP. On the other hand, no significant changes in salivary and plasma levels and AUC_{0-240} of CGRP- and VIP-IS were observed after the administration of cevimeline. These findings suggest that pathways via CGRP and VIP nerves may not be involved in the stimulatory effect on salivation by cevimeline. On the other hand, anethole trithione and pilocarpine increase not only SP but also CGRP levels in human saliva [8–11]. Cevimeline acts as a selective stimulator of the M3 acetylcholine receptor expressed on salivary glands [13], and this selectivity may reflect the specificity of the cevimeline action on SP nerves in salivary glands.

Oral cevimeline administration resulted in significant increases in salivary volumes at 90, 180, and 240 min and in the AUC_{0-240} compared with placebo administration. Furthermore, a significant correlation was observed between AUC_{0-240} of SP-IS level and AUC_{0-240} of salivary volume, suggesting the possible involvement of SP in the cevimeline-enhanced saliva secretory activity. A lag time was observed between elevation of salivary SP level and increase in salivary volume, suggesting that SP secreted from the SP nerves stimulated by cevimeline may initially increase blood flow and cause vasodilatation in salivary glands, followed by a gradual increase in salivary production. However, some reports have suggested that human salivary glands are thought to lack an SP innervation of the acinar cells, and in vitro pieces of human submandibular glands do not respond with fluid secretion to the administration of SP, as judged by the release of potassium [26, 27]. Furthermore, other neuropeptides not tested in this study may also be involved in the mechanism of enhancement of salivary secretion by cevimeline. Therefore, this notion requires verification by further studies.

Cevimeline is known to enhance saliva production in only 60% of the treated patients [14], and the mechanism of drug response remains unknown. The present study shows a possibility that individual variability of SP nerve stimulation in response to cevimeline may account for the variable drug response to cevimeline, although it is uncertain whether this trend in healthy volunteers is also observed in patients. Therefore, further studies are required to investigate the effects of cevimeline in patients with conditions such as xerostomia.

5. Conclusions

This study demonstrated the effects of cevimeline on salivary and plasma levels of neuropeptides in humans. A single oral dose of cevimeline resulted in a significant increase in salivary but not plasma SP-IS level, and a significant correlation was observed between the total release of salivary SP-IS and of salivary volume. These findings suggest a close association of SP with the enhancement of salivary secretion by cevimeline. A large-scale controlled study evaluating multiple dosing conditions of cevimeline would help to better understand the effects of cevimeline.

Conflict of Interests

The authors declare that they have no conflict of interests to disclose.

References

[1] N. Emmelin, "Nerve interactions in salivary glands," *Journal of Dental Research*, vol. 66, no. 2, pp. 509–517, 1987.

[2] J. Ekström, "Autonomic control of salivary secretion," *Proceedings of the Finnish Dental Society*, vol. 85, no. 4-5, pp. 323–331, 1989.

[3] J. Ekström, "Neuropeptides and secretion," *Journal of Dental Research*, vol. 66, no. 2, pp. 524–530, 1987.

[4] J. Ekström, B. Månsson, and G. Tobin, "Vasoactive intestinal peptide evoked secretion of fluid and protein from rat salivary glands and the development of supersensitivity," *Acta Physiologica Scandinavica*, vol. 119, no. 2, pp. 169–175, 1983.

[5] J. Ekström, B. Månsson, and G. Tobin, "Non-adrenergic, non-cholinergic parasympathetic secretion in the rat submaxillary and sublingual glands," *Pharmacology & Toxicology*, vol. 60, no. 4, pp. 284–287, 1987.

[6] J. Ekström, R. Ekman, R. Håkanson, S. Sjogren, and F. Sundler, "Calcitonin gene-related peptide in rat salivary glands: neuronal localization, depletion upon nerve stimulation, and effects on salivation in relation to substance P," *Neuroscience*, vol. 26, no. 3, pp. 933–949, 1988.

[7] B. Lindh and T. Hökfelt, "Structural and functional aspects of acetylcholine peptide coexistence in the autonomic nervous system," *Progress in Brain Research*, vol. 84, pp. 175–191, 1990.

[8] M. Takeyama, T. Nagano, and K. Ikawa, "Anethole trithione raises levels of substance P in human saliva," *Pharmaceutical Sciences*, vol. 2, no. 12, pp. 581–584, 1996.

[9] T. Nagano, K. Ikawa, and M. Takeyama, "Anethole trithione raises levels of α-calcitonin gene-related peptide in saliva of healthy subjects," *Pharmacy and Pharmacology Communications*, vol. 4, no. 9, pp. 459–463, 1998.

[10] T. Nagano and M. Takeyama, "Enhancement of salivary secretion and neuropeptide (substance P, α-calcitonin gene-related peptide) levels in saliva by chronic anethole trithione treatment," *Journal of Pharmacy and Pharmacology*, vol. 53, no. 12, pp. 1697–1702, 2001.

[11] T. Nagano, H. Itoh, T. Hayashi, and M. Takeyama, "Effect of pilocarpine on levels of substance P and α-calcitonin gene-related peptide in human saliva," *Pharmacy and Pharmacology Communications*, vol. 5, no. 9, pp. 571–574, 1999.

[12] K. J. Bloch, W. W. Buchanan, M. J. Wohl, and J. J. Bunim, "Sjögren's syndrome. A clinical, pathological, and serological study of sixty-two cases. 1965," *Medicine*, vol. 71, no. 6, pp. 386–401, 1992.

[13] Y. Iwabuchi and T. Masuhara, "Sialogogic activities of SNI-2011 compared with those of pilocarpine and McN-A-343 in rat

salivary glands: identification of a potential therapeutic agent for treatment of Sjogren's syndrome," *General Pharmacology*, vol. 25, no. 1, pp. 123–129, 1994.

[14] R. S. Fife, W. F. Chase, R. K. Dore et al., "Cevimeline for the treatment of xerostomia in patients with Sjögren syndrome: a randomized trial," *Archives of Internal Medicine*, vol. 162, no. 11, pp. 1293–1300, 2002.

[15] I. Dawidson, M. Blom, T. Lundeberg, E. Theodorsson, and B. Angmar-Månsson, "Neuropeptides in the saliva of healthy subjects," *Life Sciences*, vol. 60, no. 4-5, pp. 269–278, 1996.

[16] T. Naito, H. Itoh, and M. Takeyama, "Effects of Hange-koboku-to (Banxia-houpo-tang) on neuropeptide levels in human plasma and saliva," *Biological and Pharmaceutical Bulletin*, vol. 26, no. 11, pp. 1609–1613, 2003.

[17] P. F. Kohler and M. E. Winter, "A quantitative test for xerostomia. The Saxon test, an oral equivalent of the Schirmer test," *Arthritis and Rheumatism*, vol. 28, no. 10, pp. 1128–1132, 1985.

[18] M. Navazesh and C. M. Christensen, "A comparison of whole mouth resting and stimulated salivary measurement procedures," *Journal of Dental Research*, vol. 61, no. 10, pp. 1158–1162, 1982.

[19] M. Takeyama, K. Mori, F. Takayama, K. Kondo, K. Kitagawa, and N. Fujii, "Enzyme immunoassay of a substance P-like immunoreactive substance in human plasma and saliva," *Chemical and Pharmaceutical Bulletin*, vol. 38, no. 12, pp. 3494–3496, 1990.

[20] T. Nagano, K. Ikawa, and M. Takeyama, "Enzyme immunoassay of calcitonin gene-related peptide in human plasma and saliva," *Japanese Journal of Hospital Pharmacy*, vol. 24, no. 1, pp. 363–369, 1998.

[21] M. Takeyama, K. Wakayama, F. Takayama, K. Kondo, N. Fujii, and H. Yajima, "Micro-enzyme immunoassay of vasoactive intestinal polypeptide (VIP)-like immunoreactive substance in bovine milk," *Chemical and Pharmaceutical Bulletin*, vol. 38, no. 4, pp. 960–962, 1990.

[22] T. Kitagawa, T. Shimozono, T. Aikawa, T. Yoshida, and H. Nishimura, "Preparation and characterization of hetero-bifunctional cross-linking reagents for protein modifications," *Chemical and Pharmaceutical Bulletin*, vol. 29, no. 2, pp. 1130–1135, 1981.

[23] I. L. Gibbins, "Target-related patterns of co-existence of neuropeptide Y, vasoactive intestinal peptide, enkephalin and substance P in cranial parasympathetic neurons innervating the facial skin and exocrine glands of guinea-pigs," *Neuroscience*, vol. 38, no. 2, pp. 541–560, 1990.

[24] D. L. Pikula, E. F. Harris, D. M. Desiderio, G. H. Fridland, and J. L. Lovelace, "Methionine enkephalin-like, substance P-like, and β-endorphin-like immunoreactivity in human parotid saliva," *Archives of Oral Biology*, vol. 37, no. 9, pp. 705–709, 1992.

[25] E. Bobyock and W. S. Chernick, "Vasoactive intestinal peptide interacts with alpha-adrenergic-, cholinergic-, and substance-P-mediated responses in rat parotid and submandibular glands," *Journal of Dental Research*, vol. 68, no. 11, pp. 1489–1494, 1989.

[26] C. Hauser-Kronberger, A. Saria, and G. W. Hacker, "Peptidergic innervation of the human parotid and submandibular salivary glands," *HNO*, vol. 40, no. 11, pp. 429–436, 1992.

[27] O. Larsson, M. Duner-Engstrom, and J. M. Lundberg, "Effects of VIP, PHM and substance P on blood vessels and secretory elements of the human submandibular gland," *Regulatory Peptides*, vol. 13, no. 3-4, pp. 319–326, 1986.

9

Leptin in Anorexia and Cachexia Syndrome

Diana R. Engineer[1, 2, 3] and Jose M. Garcia[1, 2, 4]

[1] Division of Diabetes, Endocrinology and Metabolism, Michael E DeBakey Veterans Affairs Medical Center, Houston, TX 77030, USA
[2] Baylor College of Medicine, 2002 Holcombe Boulevored, Building 109, Room 210, Houston, TX 77030, USA
[3] Division of Diabetes, Department of Medicine, Endocrinology and Metabolism, St Luke's Episcopal Hospital, Houston, TX 77030, USA
[4] Huffington Center of Aging, Baylor College of Medicine, Houston, TX 77030, USA

Correspondence should be addressed to Jose M. Garcia, jgarcia1@bcm.edu

Academic Editor: Lloyd D. Fricker

Leptin is a product of the obese (OB) gene secreted by adipocytes in proportion to fat mass. It decreases food intake and increases energy expenditure by affecting the balance between orexigenic and anorexigenic hypothalamic pathways. Low leptin levels are responsible for the compensatory increase in appetite and body weight and decreased energy expenditure (EE) following caloric deprivation. The anorexia-cachexia syndrome is a complication of many chronic conditions including cancer, chronic obstructive pulmonary disease, congestive heart failure, chronic kidney disease, and aging, where the decrease in body weight and food intake is not followed by a compensatory increase in appetite or decreased EE. Crosstalk between leptin and inflammatory signaling known to be activated in these conditions may be responsible for this paradox. This manuscript will review the evidence and potential mechanisms mediating changes in the leptin pathway in the setting of anorexia and cachexia associated with chronic diseases.

1. Introduction

Leptin was discovered in 1994 by Friedman and colleagues after cloning an obese (OB) gene responsible for obesity in *ob/ob* mice [1]. It is a 167 amino acid peptide produced by adipocytes and it is a member of the adipocytokine family. Leptin has been noted to play a major role in body mass regulation by acting in the central nervous system to both stimulate energy expenditure and decrease food intake [2–4]. Named after the Greek word *leptos*, meaning lean, leptin was the first adipocyte-secreted hormone discovered, proving the active role of adipocytes in metabolic signaling.

Leptin crosses the blood-brain barrier in a process that is highly regulated [5–8] and its receptors are found both centrally, in the hypothalamus, and peripherally, in pancreatic islets, liver, kidney, lung, skeletal muscle, and bone marrow [9]. Besides its key role on body weight regulation, leptin affects various metabolic pathways, including growth hormone (GH) signaling [10], insulin sensitivity, and lipogenesis [11]. While leptin levels are directly related to adiposity,

there are several other factors resulting in individual variability. Leptin secretion is regulated by insulin, glucocorticoids, and catecholamines [3, 12, 13]. Also, females have significantly higher levels of leptin than men, for any degree of fat mass [14]. Along with adiponectin, leptin assists in peripheral insulin sensitization independent of body weight [15–17]. In leptin-deficient (ob/ob) mice, leptin injections led to dose-dependent reductions in serum glucose levels compared to fed ob/ob controls, before any significant change in body weight occurred [1, 18, 19].

Inactivating mutations of leptin or leptin receptor gene result in the body's false perception of starvation and subsequent hyperphagia, decreased energy expenditure, and severe obesity [20, 21]. In the absence of these mutations and presence of diet-induced obesity, increased adipose tissue results in increased leptin levels. A 10% increase in body weight leads to a 300% increase in serum leptin concentrations [22]. These elevated leptin levels should lead to decreased food intake and increased energy expenditure via physical activity and thermogenesis. Although obese humans have high plasma leptin concentrations, these high levels do not result in the predicted decrease in appetite. This was initially thought to be due to leptin resistance. However, an increasing body of evidence now suggests central leptin insufficiency as a mechanism. The conventional idea of leptin resistance was thought to be due to inhibition of leptin signaling pathways in leptin-responsive neurons, defects downstream of leptin receptor, and blood-brain barrier (BBB) transport limitation for leptin [23–27]. Decreased leptin transfer via the BBB does not, however, compromise intracellular signaling in the hypothalamus. Furthermore, centrally administered leptin effectively decreases the rate of fat accumulation, hyperglycemia, insulin resistance, hyperinsulinemia, and progression to metabolic syndrome in obese rodents [28–30]. The validity of the concept of leptin resistance has, thus, been questioned.

Central leptin insufficiency due to dietary and lifestyle changes for extended periods of time has been shown to result in increased fat accrual, decreased energy expenditure, hyperinsulinemia, hyperglycemia, neuroendocrine disorders, osteoporosis, and impaired memory [6, 31, 32]. Leptin permeability across the BBB is modulated by various endogenous factors, including adiposity, daily mealtimes, intrinsic circadian rhythms governing ingestion behavior, and aging [5–8]. Transport of leptin to the brain is reduced by fasting and increased by pretreatment with glucose [33, 34]. Leptin binding proteins in the blood can affect leptin levels available for transport to the brain. For example, hepatic C-reactive protein (CRP), which is increased in obesity, binds leptin and limits leptin receptor binding and transport across the BBB [35, 36]. Central leptin gene therapy in obese mice resulted in multiple benefits, including normoinsulinemia, euglycemia, elimination of fatty liver, increased energy expenditure, and more than doubling of lifespan [32, 37, 38].

Cytokines, such as IL-6 and TNF-α, are increased in obesity and correlate with insulin resistance [39]. After three weeks of a very low-calorie diet, IL-6 levels decrease in adipose tissue as well as in serum. Furthermore, IL-6 knockout mice develop obesity at the young age of 6 months [40]. Unlike leptin, IL-6 has higher CSF concentrations than serum in some obese, but otherwise healthy, men [41]. The presence of increased cytokines, in addition to the hyperleptinemia and central leptin insufficiency, seen in obesity appears to play a key role in the pathophysiology of metabolic syndrome.

Taken together, the data suggest that the leptin system may be more efficient in signaling a decrease in fat mass and lack of nutrients (low leptin state) and triggering a compensatory increase in food intake and a decrease in energy expenditure than as a satiety signal when its serum levels are elevated. Moreover, recent evidence suggests that the neurobiology of leptin signaling in obesity appears to involve central leptin insufficiency, as opposed to the previously postulated notion of leptin resistance. Interestingly, central leptin administration and gene therapy has successfully improved energy homeostasis as well as prevented diet-induced obesity and metabolic syndrome in mice.

The arcuate nucleus (ARC), ventromedial (VMH), dorsomedial (DMH), and lateral (LH) hypothalamic nuclei are important regions regulating food intake and energy expenditure. Disrupting lesions in the ARC, VMH, and DMH of rats resulted in hyperphagia and obesity [42]. Moreover, lesions in the LH resulted in decreased food intake [43]. Binding of leptin to its hypothalamic receptors activates a signaling cascade in the ARC that results in inhibition of orexigenic pathways as indicated by decreased mRNA expression of neuropeptide Y (NPY) and agouti-related peptide (AgRP), and stimulation of anorexigenic pathways as suggested by increases in the mRNA levels of alpha-melanocyte-stimulating hormone (α-MSH) and cocaine and amphetamine-regulated transcript (CART) [44, 45]. Activation of POMC/CART-expressing neurons by leptin results in release of α-MSH, which subsequently binds to melanocotin receptors (MCRs) and leads to anorexia and increased energy expenditure. At the same time, leptin inhibits NPY/AgRP neurons, which stimulate orexigenic responses and directly inhibits POMC neuron expression as indicated by POMC mRNA expression [46, 47]. Interestingly, there is no feedback from POMC neurons to NPY/AgRP neurons, revealing that the default function of the circuit is to promote food intake [48, 49]. Loss of function mutations of the MC-4R, the most important melanocortin receptor (MCR), is the most common genetic etiology of obesity and accounts for 3–5% of severe human obesity [50, 51], highlighting the relevance of this pathway in humans. Leptin modifies postsynaptic action of orexigenic and anorectic signals via the JAK2 (Janus kinase 2)-STAT3 (signal transducer and activator of transcription 3) and PI3K-PDE3B (phosphatidylinositol-3 kinase-phosphodiesterase 3B-cAMP) pathways [52].

Pinto et al. hypothesized that leptin may cause rewiring of the ARC neural circuit when they found that the NPY/AgRP and POMC neurons in ob/ob and wild-type mice differed. Treatment with leptin normalized synaptic density within six hours, even before leptin levels affected food intake. These findings indicate that leptin may also function via neural plasticity in the hypothalamus [53]. Another anorectic hypothalamic pathway has been recently characterized involving the protein nesfatin-1 and it appears to be leptin-independent. Nesfatin-1 targets magnocellular and parvocellular oxytocin neurons as well as nesfatin-1 neurons to stimulate oxytocin release. Oxytocin then activates POMC neurons in the nucleus of the tractus solitaries (NTS) and induces melanocortin-dependent anorexia in leptin-resistant Zucker-fatty rats. Injecting nesfatin-1 was shown to activate the PVN and result in leptin-independent melanocortin-mediated anorexia [54].

In summary, the discovery of ob/ob mice and leptin has provided evidence that there is hormonal communication between adipose tissue and the hypothalamus, regulating food intake and energy metabolism. Leptin controls feeding via the ARC melanocortin system, by altering gene transcription and neural plasticity. The ARC then integrates all the information it receives and accordingly alters feeding and metabolism through hormonal and neural pathways. The hyperleptinemic state of obesity has been associated with leptin resistance or, more likely, central leptin deficiency.

Convincing evidence suggest that one of the main roles of leptin is to signal a state of nutrient deficiency and fat loss. Low leptin levels in this setting will trigger a centrally mediated, compensatory response leading to increased appetite and food intake, decreased energy expenditure, and, ultimately, weight regain. However, this mechanism does not appear to be preserved in most chronic diseases in spite of the weight loss seen. This manuscript will review the evidence and potential mechanisms mediating changes in this pathway in the setting of anorexia and cachexia associated with chronic diseases.

2. Anorexia-Cachexia Syndrome

Anorexia, defined as the loss of desire to eat, despite caloric deprivation, is frequently seen in patients with advanced chronic illness [55]. Cachexia, a term derived from Greek *kakos*, meaning bad, and *hexis*, meaning condition, describes a progressive loss of adipose tissue and lean body mass. Increased proteolysis, decreased protein synthesis, and accelerated lipolysis due to high energy demands result in a dramatic decline in lean body mass and fat mass and increase in mortality in this setting [56, 57]. Caloric restriction per se induces a less severe degree of weight loss and a different metabolic pattern characterized by decreased energy expenditure and preservation of lean mass at the expense of fat loss. This suggests that anorexia alone does not cause the extreme weight loss seen in cachexia. Moreover, nutritional support does not reverse cachexia [58]. Clinically, ACS presents with weight loss, decreased appetite, early satiety, muscle atrophy, and weakness. This process has been observed in various illnesses, including cancer, chronic heart disease, pulmonary disease, chronic kidney disease, and aging.

Anorexia-cachexia syndrome appears to be multifactorial, often associated with the underlying disease process, and related to both peripheral and central neurohormonal signals regulating both appetite and energy expenditure. Inflammatory cytokines, such as tumor necrosis factor- (TNF-) α, interleukin- (IL-) 1, IL-6, and interferon- (IFN-) γ have been postulated to play a key pathogenic role in the decreased food intake and increased energy expenditure seen in most chronic conditions associated with the anorexia and cachexia syndrome (ACS) [59]. Increased cytokines in the hypothalamus enhance serotoninergic tone through tryptophan, resulting in activation of POMC neurons and subsequent anorexia [60]. IL-1 inhibition in tumor-bearing animals has been shown to improve appetite and promote weight gain [61]. The somatomedin pathway, including GH and insulin-like growth factor-1 (IGF-1), stimulates skeletal muscle protein synthesis and is inhibited by inflammatory cytokines [62, 63]. In spite of the devastating effect that ACS has in patients, its pathophysiology is only partially understood and there are no approved treatments for this condition.

3. Crosstalk between Leptin Signaling and Inflammation

Leptin receptors belong to the class I cytokine receptor family and have similar structure to the signal-transducing subunits of the IL-6 receptors [64]. Leptin levels decrease with fasting and increase during the postprandial phase afterwards. These changes are directly correlated with changes in hypothalamic interleukin- (IL-) 1β mRNA levels, suggesting that leptin has

a proinflammatory role centrally [65]. This link between proinflammatory cytokines and leptin has been illustrated well in animal models. Fasted hamsters treated with the cytokines tumor necrosis factor- (TNF-) α and IL-1 showed increased levels of both circulating leptin and leptin mRNA in adipose tissue. These increases in leptin were associated with a decline in food intake [66]. This is also supported by experiments where peripheral leptin administration caused hypothalamic inflammation and central injection of IL-1 receptor (IL-1r) antagonist inhibited the suppression of food intake caused by central or peripheral injection of leptin [67]. Mice lacking the main IL-1 receptor responsible for IL-1 actions showed no reduction in food intake in response to leptin [68]. Increased inflammatory mediators have been shown to increase hypothalamic POMC mRNA expression [69]. Administering melanocortin receptor antagonist centrally results in blockade of inflammatory anorexia [70].

Leptin levels are significantly lower in patients with inflammatory states such as cancer [71] despite correction for body fat. These low levels of leptin, however, are not associated with greater appetite or lower energy expenditure, as might be expected. Disturbances in the feedback mechanism in the hypothalamus and/or release of pro-inflammatory cytokines, such as IL-1, IL-6, and TNF-α, are thought to be responsible for cachexia in this setting. These circulating cytokines result in insulin resistance, lipolysis, and loss of skeletal muscle mass [72]. IL-1 influences size, duration, and frequency of meals in rats via hypothalamic signaling [73]. Cytokines also suppress gastric production of the orexigenic peptide ghrelin that decreases production of inflammatory cytokines TNF-α, IL-6, and IL-1β.

Nuclear factor-κB (NF-κB), a transcription factor for inflammation-related proteins, is activated in the hypothalamus of animal models of infection-associated anorexia. These models are created by administration of bacterial and viral products such as lipopolysaccharide (LPS) and HIV-1 transactivator protein (Tat). In vitro, NF-κB activation stimulated POMC transcription, showing the connection of NF-κB in feeding regulation. Hypothalamic injection of LPS and Tat showed reductions in food intake and body weight, while inhibition of NF-κB and melanocortin cancelled these effects. Moreover, hypothalamic NF-κB is activated by leptin and is involved in leptin-stimulated POMC transcription, showing that it may serve as a downstream signaling pathway of leptin [74]. Paradoxically, inflammation is also thought to play a role in obesity [75]. Obesity in both human [76] and animal models [77] has been associated with increased inflammatory markers, including TNF-α and IL-6. In rats and mice with diet-induced obesity, inflammation of both peripheral tissues and the hypothalamus was noted [78–81]. Blocking hypothalamic inflammation signaling via pharmacological approach or gene therapy led to a reduction in food intake and lower body weight in these animals [79, 82].

In summary, the evidence suggests that the central effects of leptin in suppressing appetite and increasing energy expenditure via activation of POMC neurons is at least partially dependent upon inflammation. Moreover, inflammation may influence the same pathways affecting appetite and body weight independently of leptin.

4. Leptin in Cancer Cachexia

Cancer anorexia-cachexia syndrome (Cancer-ACS) is found in 80% of patients with advanced cancer. It has been shown to decrease performance status, quality of life, response to therapy, and survival [56, 57]. Cancer-ACS may account for up to 20% of cancer deaths [58]. Although the tumor itself is primarily responsible, treatments such as chemotherapy and radiation, and associated conditions such as depression, pain, gastrointestinal obstruction, and taste alterations can also contribute to weight loss [83]. Cancer-ACS appears to be multifactorial, involving tumor-host interactions that result in catabolism overwhelming anabolism.

Leptin is thought to play a major role in the pathophysiology of cancer-ACS. Animal studies have shown that circulating leptin levels are decreased in the setting of tumor-induced cachexia, as expected given the decrease in fat mass seen in this setting [84]. However, mRNA levels of NPY were decreased and for POMC were increased in the ARC, unlike what is seen in caloric restriction where low leptin levels cause activation of NPY and suppression of POMC pathways. Levels of phosphorylated signal transducer and activator of transcription-3 (P-Stat3), a central molecule activated via the leptin receptor signaling pathway, are upregulated in subsets of α-MSH and NPY positive neurons that are not responsive to leptin. This pathway appears to be induced by the cytokine macrophage inhibitory cytokine-1 (MIC-1) via activation of the transforming growth factor- (TGF-) β receptor II, suggesting a potential alternative pathway through which MIC-1 could regulate appetite independently of leptin. This is also supported by the fact that MIC-1 infusion can induce anorexia and weight loss in leptin-deficient mice [85].

Leptin levels are significantly decreased in cancer cachexia patients compared to both cancer noncachexia and healthy controls [86, 87]. Proinflammatory cytokines, such as TNF-α, IL-1, and IL-6, have been proposed to cause cachexia in spite of low circulating leptin due to increased expression of the hypothalamic leptin receptor [88]. This dysregulation of the normal feedback loop in cancer cachexia may explain why a decrease in leptin does not increase appetite or lower energy expenditure in patients with cancer cachexia. Interestingly, leptin was found to be directly associated with appetite and insulin resistance [87], suggesting that these patients are resistant to the orexigenic effects of hypoleptinemia. Leptin also has been postulated as an early marker of disease progression in advanced ovarian cancer [89]. Moreover, in these patients, there was a clear correlation between disease stage and performance status with markers of inflammation, such as IL-6 whereas low leptin levels were more closely associated with tumor stage and IL-6 levels than BMI.

Given the large role of proinflammatory cytokines in cancer cachexia, the use of anticytokine antibodies and cytokine receptor antagonists has been investigated as potential therapies. Unfortunately, despite promising experimental data, clinical trials have not been conclusive. In the Yoshida AH-130 model, anti-TNF therapy partially reversed metabolic abnormalities in cachexia [90]. Clinical trials, however, showed transient improvement at best. In a double-blinded, placebo-controlled, randomized study, pentoxyphylline, which inhibits TNFα transcription, failed to show any benefit on cancer cachexia [91]. In small, unrandomized clinical trials, thalidomide, a TNF-α inhibitor, has been shown to improve some cancer-ACS symptoms [92]. Anti-inflammatory cytokines, IL-12 and IL-15, have shown some improvement in cancer-ACS in tumor-bearing animals [93, 94]. Mantovani et al. performed a phase II clinical trial with cyclooxygenase-2 (COX-2) inhibitors in patients with cancer-ACS showing a significant increase in LBM, decrease in TNF-α, and improvement in overall performance status [95]. Whether these interventions that blocked inflammation had an effect on leptin or its pathway is not known given that leptin levels or changes in its downstream mediators were not reported in these studies.

In summary, cancer anorexia-cachexia syndrome is a major predictive factor of mortality. In both animal and human models, circulating leptin levels decrease in the setting of cancer-ACS. However, this decrease in leptin is not associated with a compensatory increase in appetite and food intake. Animal studies suggest that hypothalamic inflammation may account for the lack of response of leptin targets to the effects of hypoleptinemia. Although cytokines appear to play a major role in the development of cancer-ACS via central and peripheral effects, preliminary studies targeting this pathway have not shown convincing evidence of a beneficial effect.

5. Leptin in Chronic Heart Failure-Induced Cachexia

The incidence of chronic heart failure is steadily rising and carries a poor prognosis. Cardiac cachexia is defined as nonedematous weight loss of >6% of previous normal weight observed over a period of >6 months and is associated with poor prognosis [96]. CHF patients with cardiac cachexia have been noted to have a mortality of 50% at 18 months, versus 17% in noncachectic patients [97]. Various factors can contribute to the weight loss seen in CHF-induced cachexia, including malnutrition from medications, metabolic disturbances (i.e., hyponatremia, renal failure), and hepatic congestion; malabsorption from severe heart failure; or increased nutritional requirements. The basal metabolic rate in these patients is increased by 20% [98]. As in other chronic conditions, inflammation has been implicated as a key aspect of CHF-induced cachexia [99].

Some controversy exists regarding leptin levels in cachectic versus noncachectic CHF patients. While many studies show lower leptin levels in cachectic patients [10, 100, 101], as it would be expected with their decreased fat tissue, other studies report normal levels [102]. These differences may exist due to sex distribution and presence of cachexia in selected subjects. When fat tissue was normalized, leptin levels for both cachectic and noncachectic CHF patients were elevated in comparison to non-CHF controls [10]. Several groups have hypothesized that leptin has a cardioprotective role and that this increase in its levels in the setting of CHF may represent a compensatory response rather than simply a marker of fat atrophy [103–105]. This could also explain the fact that circulating leptin levels directly correlate with NYHA class and overall prognosis in this setting. It is also possible that this increase in leptin is at least partially responsible for the weight loss and anorexia in CHF patients since an increase in leptin would lead to activation of the melanocortin system that in turn would cause an increase in energy expenditure and a decrease in food intake [106]. This hypothesis remains to be tested. It is also postulated that hyperleptinemia in these patients may be a result of insulin resistance [107, 108]. Regardless of the reason, this hyperleptinemia in both cachectic and noncachectic CHF patients suggests that leptin-mediated decrease in appetite and food intake is not particularly important in the development of CHF-induced cachexia.

Leptin may also contribute to CHF-cachexia and obesity-related cardiomyopathy by various cardiovascular mechanisms, including increasing sympathetic activity and producing vasodilation by an endothelium-dependent mechanism peripherally. It also promotes inflammation, calcification, proliferation, and thrombosis in the vasculature [109]. Animal models, however, do not show any increase in blood pressure, despite this increase in sympathetic activity [110, 111]. It is hypothesized that this may be due to leptin's vasodilatory effects via unclear mechanisms [109, 112].

The presence of proinflammatory cytokines in this setting suggests that inflammation plays an important role in the pathogenesis of CHF. Increasing levels of TNF-α, IL-1, and IL-6, lead to activation of the renin-angiotensin-aldosterone-system, improving renal and organ perfusion early on. However, TNF-α also induces apoptosis and activates protein breakdown in various tissues, including striate muscle. It also contributes to endothelial dysfunction and subsequent decreased blood flow to skeletal muscle, which results in decreased exercise endurance and nutrient supply [113]. Plasma TNF-α receptor levels in CHF patients have been associated with poor- long and short-term prognosis [114, 115]. Importantly, TNF-α increases the expression of leptin [66, 116]. No current therapy is approved to target cardiac cachexia. Standard treatment of CHF including ACE-inhibitors and beta-blockers has been shown to increase weight in CHF patients [96]. However, it appears that increases in leptin in noncachectic patients with CHF are primarily driven by body weight increases [117].

Although CHF is associated with elevated leptin levels, these are closely related to the amount of fat tissue; hence, levels are lower in cachectic individuals compared to noncachectic, CHF controls. This elevation in leptin may be due in part to insulin resistance but a direct cardioprotective effect of leptin also has been proposed. Taken together, the data suggests that leptin's role in this setting is not entirely related to body weight regulation and that the decreased appetite and increased energy expenditure seen in CHF-induced cachexia are more likely due to other factors. There is a scarcity of therapeutic data on patients with CHF-cachexia, so little is known about leptin responses to treatment at this time.

6. Leptin in Pulmonary Cachexia

Chronic obstructive pulmonary disease (COPD), including chronic bronchitis, emphysema, asthmatic bronchitis, and bronchiolitis obliterans, is a leading cause of morbidity and mortality worldwide. Cachexia has been reported in 20–40% of COPD patients and is associated with negative prognosis [118]. Although the increased work of breathing may be partly responsible for increased energy expenditure, this alone does not explain the reported weight loss [119, 120]. It is hypothesized that many different pathways are involved in the pathophysiology of COPD cachexia, including anorexia, nutritional deficiency, hypoxia, increased metabolic rate, inactivity, sympathetic upregulation, inflammation, and anabolic hormone deficiency.

Circulating levels of TNF-α and TNF-α production by peripheral monocytes are increased in patients with pulmonary cachexia [121, 122]. In both animal and human models, endotoxin or cytokine (TNF-α or IL-1) administration produces a dose-dependence increase in serum leptin levels [66, 123, 124]. However, COPD patients were reported to have lower leptin levels compared to healthy controls and these levels correlated well with BMI and percentage body fat [125]. Moreover, there is a loss of circadian variation in leptin levels that may represent alterations in the autonomic nervous system tone. Conversely, serum TNF-α levels were significantly higher in COPD patients compared to healthy controls and did not correlate with leptin levels [126], suggesting that in pulmonary cachexia, leptin levels are physiologically regulated and are independent of inflammatory markers, such as TNF-α.

In addition to the effects of chronic inflammation, it appears that hypoxia plays a major role in COPD-cachexia. Leptin-deficient animals show CO_2 retention and respiratory depression; leptin administration to these animals increases minute ventilation and improves lung mechanics suggesting that an increase in leptin levels in patients with lung disease may represent a compensatory response to hypoxia [127–130]. Consistent with this hypothesis, elevated leptin has been described in hypoxic patients compared to BMI-matched controls [131]. It has been shown that expression of the human leptin gene is induced by hypoxia through the hypoxia-inducible factor-1 (HIF-1) pathway. The introduction of noninvasive positive airway pressure ventilation (NIPPV) in patients with severe COPD serves to both decrease energy expenditure and hypoxia. Moreover, body weight significantly improved (increased 12%) in patients with severe COPD and cachexia placed on NIPPV for 1 year [132]. This elevation in leptin was reversed by improving hypoxia, although this was associated with a decrease in fat accumulation in some studies that may have accounted at least partially for the changes in leptin.

In summary, pulmonary cachexia likely involves multiple pathways, including hypoxia and inflammation. Circulating leptin levels are decreased in patients with pulmonary cachexia suggesting that the physiologic regulation of leptin is maintained despite weight loss. Hypoxia also appears to play a role in leptin expression via the HIF-1 pathway that is reversible by correction of hypoxia. Furthermore, correction of hypoxia is associated with weight gain in spite of a decrease in leptin levels. These findings are consistent with the hypothesis that leptin plays a role in regulating the respiratory drive besides its usual role as a metabolic sensor.

7. Leptin in CKD Cachexia

Chronic kidney disease (CKD) is a common illness associated with a state of chronic inflammation and, oftentimes, cachexia. CKD-associated cachexia is linked to higher morbidity and mortality [133]. Uremic anorexia appears to be multifactorial. High plasma levels of insulin, leptin, and uremic toxins induce MC4-R stimulation to increase energy expenditure and decrease food intake [134]. Leptin levels are significantly elevated in CKD and ESRD patients and are associated with markers of poor nutritional status, such as low serum albumin and hypercatabolism as well as decline in renal function [135]. Serum leptin concentrations have been shown to inversely associate with survival in some studies [136]. In others, leptin was shown to correlate with fat mass rather than independently affecting food intake or mortality [137]. The hyperleptinemia seen in dialysis patients may be due to poor renal clearance, overproduction, or both. It has been postulated that uremic patients may have an acquired leptin receptor disorder resulting in central insensitivity or resistance, similar to obese individuals. Leptin reduces hypothalamic NPY levels and increases sympathetic activity with hyperinsulinemia, resulting in appetite suppression [138]. Supporting this hypothesis is the observation of increased sympathetic activity, via elevated dopamine, norepinephrine, and serotonin levels, found in uremic patients [139]. Elevated serum acute phase reactants, including C-reactive protein (CRP) and several cytokines, most prominently IL-6 and TNF-α, are found in CKD patients and may be associated with reduced appetite in dialysis patients [140]. Increased inflammation in renal failure is multifactorial, and possible factors include decreased renal clearance and increased production of proinflammatory cytokines [141, 142]. The mediators of inflammation act on the central nervous system to alter both appetite and metabolic rate [143, 144].

The administration of AgRP to mice with CKD resulted in increased food intake, normalization of basal metabolic rate, and increases in total body weight and lean body mass, independent of caloric or protein intake. Also, studies in db/db mice show that the lack of leptin receptor is protective against CKD-induced cachexia. Furthermore, manipulation of leptin's downstream mediators in the hypothalamus, MC4 by either gene deletion or by using antagonists of its receptor, confirms the relevance of this pathway in mediating anorexia and weigh loss in the setting of CKD [145].

A cross-sectional study of 217 hemodialysis patients followed for 31 months showed that those in the lowest tertile of ghrelin levels were the oldest and had the highest BMI, and highest CRP and leptin levels. These patients all had increased mortality risk, despite adjustment for age, gender, and dialysis history. Moreover, those in this group with protein-energy wasting had the highest all-cause and cardiovascular mortality risk (hazards ratios 3.34 and 3.54, resp.) [151]. In the setting of CKD, there is the opportunity to manipulate leptin levels not only by administering recombinant leptin but also by removing leptin from circulation using super-flux polysulfone dialyzers. van Tellingen et al. tried such approach and although leptin levels were significantly reduced, no other parameters such as appetite or body composition were examined [152, 153]. Therefore, the effectiveness of this intervention remains unknown.

Taken together, the evidence shows that CKD and ESRD-induced cachexia are associated with poor prognosis. Elevated levels of inflammatory mediators and leptin are likely results of decreased renal clearance and disease-related inflammation. Activation of the melanocortin system by leptin is key in the pathophysiology of CKD cachexia. Further studies to explore the efficacy of therapeutic options, including polysulfone dialyzers to lower leptin levels, are needed to determine the role of leptin in this setting.

8. Leptin in Aging

Weight loss in the geriatric population is a strong predictor of morbidity and mortality [154]. Normal aging involves a decline in appetite, decrease in lean body mass, increase in fat mass, and decrease in energy expenditure [155]. Aging has significant effects on energy homeostasis and dysregulation of adipokines, including leptin. These effects appear to be mediated at least in part by a decrease in the tone of the orexigenic AgRP/NPY pathway, an increase of the anorexigenic CART/POMC pathway, and failure of these pathways in responding to caloric restriction.

Elevated circulating leptin levels with decreased hypothalamic leptin responsiveness have been found in both animal and human models of aging [146]. Using a rodent model of aging, Wolden-Hanson showed that caloric restriction in aged Brown Norway rats failed to induce a compensatory increase in appetite after refeeding, unlike what is seen in young animals [147]. Leptin levels are increased in aged animals paralleling changes in fat mass but fail to decrease in response to fasting, suggesting that hyperleptinemia may contribute to this energy balance dysregulation and play a causative role in the poor tolerance of aged individuals to catabolic conditions. Also, leptin resistance has been proposed as one of the alterations seen in the elderly [156]. Hence, hyperleptinemia may be a compensatory mechanism to overcome the impaired leptin action in the brain. Uptake of leptin in the hypothalamus is significantly lower in old animals [157]. In aged rats, leptin administration does not suppress appetite, hypothalamic NPY expression, circulating leptin levels, or ob mRNA levels in white adipose tissue to the same extent as in young animals [148]. Although expression of the leptin receptor was not investigated in this report, others have shown a decrease in expression of the long form of this receptor during aging [157]. Downstream of leptin, abnormalities in other hypothalamic neuropeptides have been reported as well. Transcript mRNA expression of the orexigenic peptides NPY and AgRP decreases and fails to increase in response to caloric restriction [149, 158]; while CART mRNA expression increases with aging. Although response to exogenous AgRP appears to be maintained, response to NPY administration was significantly blunted in aged animals [159].

Previous studies have shown protective effects of fat mass on morbidity and mortality in the geriatric population [160]. Lipoatrophy and lipodystrophy in aging have been associated with dysregulation of adipokines and, subsequently, metabolic derangements such as insulin resistance,

TABLE 1: Summary of markers of appetite regulation in various cachectic states.

Condition	Appetite	Body Weight	Circulating Leptin Levels	POMC/α-MSH hypothalamic levels	NPY/AgRP hypothalamic levels	Circulating inflammatory markers	Hypothalamic inflammatory markers	References
Cancer cachexia	↓*#	↓*#	↓*#	↑*	↓*	↑*#	↑*	[60, 84–87]
CHF-induced cachexia	↓*#	↓*#	↑*#	unknown	unknown	↑*#	unknown	[10, 99–102, 115]
Pulmonary cachexia	↓*#	↓*#	↓*#	unknown	unknown	↑*#	unknown	[121, 122, 124–126]
CKD cachexia	↓*#	↓*#	↑*#	unknown	unknown	↑*#	unknown	[135, 137, 139, 140, 142, 145]
Aging cachexia	↓*#	↓*#	↓*#	↑*	↓*	↑*#	unknown	[146–150]

#—supported by human model data; *—supported by animal model data.

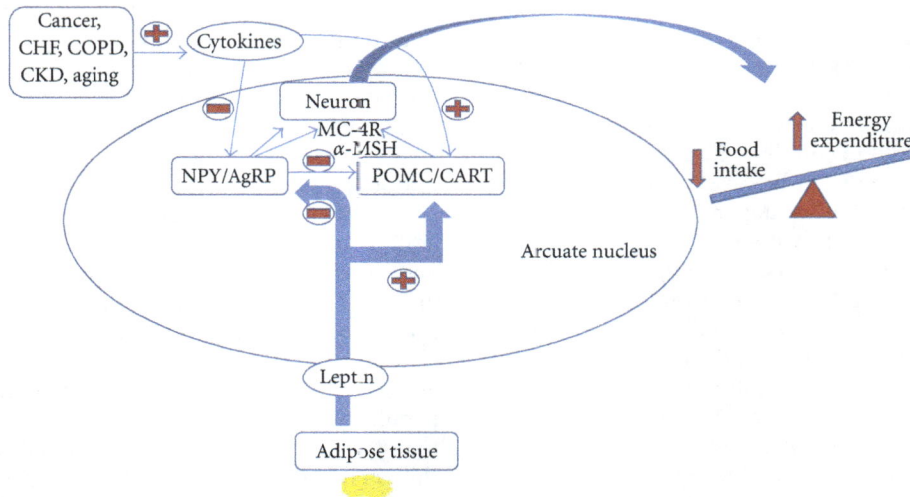

FIGURE 1: *Summary of the effects of peripheral hormones on hypothalamic regulation of food intake and energy expenditure.* NPY = neuropeptide Y; AgRP = Agouti-related peptide; POMC = pro-opiomelacortin; CART = cocaine-amphetamine-related peptide; α-MSH = alpha-melanocyte-stimulating hormone; MC-4R = type-4 melanocortin receptor.

dyslipidemia, metabolic syndrome, hypertension, and hyperglycemia. Centenarians, models of health and longevity, had been reported to exhibit preserved insulin sensitivity and intact adipokine profiles. Studying this population revealed that poor prognosis was associated with dysregulated adipokines, including leptin levels inappropriately low for fat mass [161]. Leptin concentrations in 19 elderly patients with protein-energy malnutrition were significantly lower than their age-matched controls. However, others have reported an increase in leptin levels in aged individuals even after adjusting for fat mass [162]. It has also been shown that the there are increased levels of IL-6 and CRP in aging [150]. Studies in obese patients have suggested an association between hypothalamic inflammation and decreased leptin action via persistence activation of the melanocortin system; no studies, however, have been performed in the elderly [69, 163]. Functional status, anthropometry, and serum markers of nutrition and inflammation, including leptin and CRP, in

seventy elderly patients versus controls revealed that those with the lowest functional status and highest frailty indices displayed features of cachexia. Moreover, they had low leptin levels, appropriate for their low body fat, as well as high CRP and IL-6 levels [164]. This suggests that the mechanism for cachexia in the elderly may involve disrupted hypothalamic feedback of leptin from the effects of proinflammatory cytokines like other chronic inflammatory states.

In summary, weight loss in the geriatric population is associated with higher mortality. In normal aging, fat mass is increased and hyperleptinemia arises. Despite these high levels of circulating leptin, however, there appears to be decreased leptin action and subsequently no decrease in appetite, similar to obese individuals. Moreover, a failed hypothalamic response to caloric restriction appears to be responsible for the poor tolerance of the elderly to catabolic stress. Elderly patients with cachexia tend to have elevated inflammatory markers and low leptin levels, both correlating with worsened prognosis.

9. Conclusion

Leptin, a product of the obese gene secreted by adipose tissue, acts centrally to suppress appetite and increase thermogenesis by activating the POMC neurons in the arcuate nucleus and triggering the release of α-MSH from POMC axon terminals and subsequently activating MC4-R (Figure 1). Moreover, the NPY and AgRP-producing neurons of the arcuate nucleus, which are suppressed by leptin, can antagonize these anorexigenic melanocortin cells. Consequently, low levels of leptin cause an increase in appetite and reduce energy expenditure.

Cachexia is a unique process characterized by depletion of adipose tissue and lean body mass found in various chronic diseases often accompanied by anorexia. Anorexia-cachexia can be seen in cancer, CHF, COPD, CKD, and aging (Table 1). All of these conditions are associated with elevated inflammatory markers such as TNF-α, IL-6, IL-2, and IL-1β. These inflammatory markers may regulate hypothalamic feedback mechanisms and are thought to contribute to the development of cachexia. Leptin receptors belong to the class I cytokine family and there is crosstalk between leptin signaling and inflammation. This crosstalk could explain why, despite low levels of leptin in chronic inflammatory processes such as cancer, COPD, and aging, patients do not have the expected increased appetite or lower energy expenditure.

Cancer, COPD, and aging-associated cachexia are all associated with low leptin levels, in spite of low appetite and elevated energy expenditure suggesting a state of resistance to the effects of hypoleptinemia. On the contrary, elevated levels have been noted in CKD- and CHF-induced cachexia. In CHF-induced cachexia, the reason for this elevation is unclear but there is no association with weight or fat mass change or with appetite suggesting that leptin may have a different function in this setting and that it likely does not play a major role in the ensuing cachexia. In CKD and ESRD, circulating levels of leptin and inflammatory agents are likely elevated due to poor renal clearance but there is no association with the degree of weight loss or anorexia.

Given the key role that inflammation appears to play in the pathogenesis of leptin-mediated cachexia, therapeutic intervention with anti-inflammatory drugs may prove to be beneficial in restoring sensitivity to the effect of hypoleptinemia in ACS. COX-2 inhibitors have shown some promise in patients with cancer cachexia. In the setting of uremic cachexia, polysulfone dialysers decrease leptin levels but more studies are needed to evaluate the effect of this intervention on appetite and weight parameters. As we gain more insight into the pathophysiology of cachexia, the therapeutic possibilities increase. Further investigations into anti-inflammatory drugs, appetite stimulants, and immunomodulators in these various conditions are warranted.

Acknowledgment

J. Garcia received research support from a M. MERIT grant from the Department of Veterans Affairs (I01-BX000507).

References

[1] J. L. Halaas, K. S. Gajiwala, M. Maffei et al., "Weight-reducing effects of the plasma protein encoded by the obese gene," *Science*, vol. 269, no. 5223, pp. 543–546, 1995.

[2] J. M. Friedman, "A tale of two hormones," *Nature Medicine*, vol. 16, no. 10, pp. 1100–1106, 2010.

[3] J. M. Friedman, "Modern science versus the stigma of obesity," *Nature Medicine*, vol. 10, no. 6, pp. 563–569, 2004.

[4] J. S. Flier and E. Maratos-Flier, "Lasker lauds leptin," *Cell*, vol. 143, no. 1, pp. 9–12, 2010.

[5] W. A. Banks, "Enhanced leptin transport across the blood-brain barrier by α1-adrenergic agents," *Brain Research*, vol. 899, no. 1-2, pp. 209–217, 2001.

[6] W. A. Banks, "Is obesity a disease of the blood-brain barrier? Physiological, pathological, and evolutionary considerations," *Current Pharmaceutical Design*, vol. 9, no. 10, pp. 801–809, 2003.

[7] W. A. Banks and C. L. Farrell, "Impaired transport of leptin across the blood-brain barrier in obesity is acquired and reversible," *American Journal of Physiology*, vol. 285, no. 1, pp. E10–E15, 2003.

[8] A. J. Kastin and W. Pan, "Dynamic regulation of leptin entry into brain by the blood-brain barrier," *Regulatory Peptides*, vol. 92, no. 1–3, pp. 37–43, 2000.

[9] S. Margetic, C. Gazzola, G. G. Pegg, and R. A. Hill, "Leptin: a review of its peripheral actions and interactions," *International Journal of Obesity*, vol. 26, no. 11, pp. 1407–1433, 2002.

[10] W. Doehner, C. D. Pflaum, M. Rauchhaus et al., "Leptin, insulin sensitivity and growth hormone binding protein in chronic heart failure with and without cardiac cachexia," *European Journal of Endocrinology*, vol. 145, no. 6, pp. 727–735, 2001.

[11] W. Doehner and S. D. Anker, "Cardiac cachexia in early literature: a review of research prior to Medline," *International Journal of Cardiology*, vol. 85, no. 1, pp. 7–14, 2002.

[12] W. M. Mueller, F. M. Gregoire, K. L. Stanhope et al., "Evidence that glucose metabolism regulates leptin secretion from cultured rat adipocytes," *Endocrinology*, vol. 139, no. 2, pp. 551–558, 1998.

[13] P. Leroy, S. Dessolin, P. Villageois et al., "Expression of ob gene in adipose cells: regulation by insulin," *Journal of Biological Chemistry*, vol. 271, no. 5, pp. 2365–2368, 1996.

[14] C. S. Mantzoros and S. J. Moschos, "Leptin: in search of role(s) in human physiology and pathophysiology," *Clinical Endocrinology*, vol. 49, no. 5, pp. 551–567, 1998.

[15] A. Khan, S. Narangoda, B. Ahren, C. Holm, F. Sundler, and S. Efendic, "Long-term leptin treatment of ob/ob mice improves glucose-induced insulin secretion," *International Journal of Obesity*, vol. 25, no. 6, pp. 816–821, 2001.

[16] C. Koch, R. A. Augustine, J. Steger et al., "Leptin rapidly improves glucose homeostasis in obese mice by increasing hypothalamic insulin sensitivity," *Journal of Neuroscience*, vol. 30, no. 48, pp. 16180–16187, 2010.

[17] J. W. Lee and D. R. Romsos, "Leptin administration normalizes insulin secretion from islets of Lep^{ob}/Lep^{ob} mice by food intake-dependent and -independent mechanisms," *Experimental Biology and Medicine*, vol. 228, no. 2, pp. 183–187, 2003.

[18] M. W. Schwartz, D. G. Baskin, T. R. Bukowski et al., "Specificity of leptin action on elevated blood glucose levels and hypothalamic neuropeptide Y gene expression in ob/ob mice," *Diabetes*, vol. 45, no. 4, pp. 531–535, 1996.

[19] M. A. Pelleymounter, M. J. Cullen, M. B. Baker et al., "Effects of the obese gene product on body weight regulation in ob/ob mice," *Science*, vol. 269, no. 5223, pp. 540–543, 1995.

[20] C. T. Montague, I. S. Farooqi, J. P. Whitehead et al., "Congenital leptin deficiency is associated with severe early-onset obesity in humans," *Nature*, vol. 387, no. 6636, pp. 903–908, 1997.

[21] K. Clément, C. Vaisse, N. Lahlou et al., "A mutation in the human leptin receptor gene causes obesity and pituitary dysfunction," *Nature*, vol. 392, no. 6674, pp. 398–401, 1998.

[22] R. V. Considine, M. K. Sinha, M. L. Heiman et al., "Serum immunoreactive-leptin concentrations in normal-weight and obese humans," *New England Journal of Medicine*, vol. 334, no. 5, pp. 292–295, 1996.

[23] B. Burguera, M. E. Couce, G. L. Curran et al., "Obesity is associated with a decreased leptin transport across the blood-brain barrier in rats," *Diabetes*, vol. 49, no. 7, pp. 1219–1223, 2000.

[24] J. F. Caro, J. W. Kolaczynski, M. R. Nyce et al., "Decreased cerebrospinal-fluid/serum leptin ratio in obesity: a possible mechanism for leptin resistance," *The Lancet*, vol 348, no. 9021, pp. 159–161, 1996.

[25] S. P. Kalra, "Circumventing leptin resistance for weight control," *Proceedings of the National Academy of Sciences of the United States of America*, vol. 98, no. 8, pp. 4279–4281, 2001.

[26] S. P. Kalra, M. G. Dube, S. Pu, B. Xu, T. L. Horvath, and P. S. Kalra, "Interacting appetite-regulating pathways in the hypothalamic regulation of body weight," *Endocrine Reviews*, vol. 20, no. 1, pp. 68–100, 1999.

[27] M. W. Schwartz, E. Peskind, M. Raskind, E. J. Boyko, and D. Porte Jr., "Cerebrospinal fluid leptin levels: relationship to plasma levels and to adiposity in humans," *Nature Medicine*, vol. 2, no. 5, pp. 589–593, 1996.

[28] L. A. Campfield, F. J. Smith, Y. Guisez, R. Devos, and P. Burn, "Recombinant mouse OB protein: evidence for a peripheral signal linking adiposity and central neural networks," *Science*, vol. 269, no. 5223, pp. 546–549, 1995.

[29] M. G. Dube, E. Beretta, H. Dhillon, N. Ueno, P. S. Kalra, and S. P. Kalra, "Central leptin gene therapy blocks high-fat diet-induced weight gain, hyperleptinemia, and hyperinsulinemia: increase in serum ghrelin levels," *Diabetes*, vol. 51, no. 6, pp. 1729–1736, 2002.

[30] J. L. Halaas, C. Boozer, J. Blair-West, N. Fidahusein, D. A. Denton, and J. M. Friedman, "Physiological response to long-term peripheral and central leptin infusion in lean and obese mice," *Proceedings of the National Academy of Sciences of the United States of America*, vol. 94, no. 16, pp. 8878–8883, 1997.

[31] P. A. Baldock, A. Sainsbury, S. Allison et al., "Hypothalamic control of bone formation: distinct actions of leptin and Y2 receptor pathways," *Journal of Bone and Mineral Research*, vol. 20, no. 10, pp. 1851–1857, 2005.

[32] H. Dhillon, S. P. Kalra, V. Prima et al., "Central leptin gene therapy suppresses body weight gain, adiposity and serum insulin without affecting food consumption in normal rats: a long-term study," *Regulatory Peptides*, vol. 99, no. 2-3, pp. 69–77, 2001.

[33] A. J. Kastin and V. Akerstrom, "Fasting, but not adrenalectomy, reduces transport of leptin into the brain," *Peptides*, vol. 21, no. 5, pp. 679–682, 2000.

[34] A. J. Kastin and V. Akerstrom, "Glucose and insulin increase the transport of leptin through the blood-brain barrier in normal mice but not in streptozotocin-diabetic mice," *Neuroendocrinology*, vol. 73, no. 4, pp. 237–242, 2001.

[35] K. Chen, F. Li, J. Li et al., "Induction of leptin resistance through direct interaction of C-reactive protein with leptin," *Nature Medicine*, vol. 12, no. 4, pp. 425–432, 2006.

[36] H. Florez, S. Castillo-Florez, A. Mendez et al., "C-reactive protein is elevated in obese patients with the metabolic syndrome," *Diabetes Research and Clinical Practice*, vol. 71, no. 1, pp. 92–100, 2006.

[37] E. Beretta, M. G. Dube, P. S. Kalra, and S. P. Kalra, "Long-term suppression of weight gain, adiposity, and serum insulin by central leptin gene therapy in prepubertal rats: effects on serum ghrelin and appetite-regulating genes," *Pediatric Research*, vol. 52, no. 2, pp. 189–198, 2002.

[38] S. Boghossian, N. Ueno, M. G. Dube, P. Kalra, and S. Kalra, "Leptin gene transfer in the hypothalamus enhances longevity in adult monogenic mutant mice in the absence of circulating leptin," *Neurobiology of Aging*, vol. 28, no. 10, pp. 1594–1604, 2007.

[39] J. P. Bastard, C. Jardel, E. Bruckert et al., "Elevated levels of interleukin 6 are reduced in serum and subcutaneous adipose tissue of obese women after weight loss," *Journal of Clinical Endocrinology and Metabolism*, vol. 85, no. 9, pp. 3338–3342, 2000.

[40] I. Wernstedt, B. Olsson, M. Jernås et al., "Increased levels of acylation-stimulating protein in interleukin-6- deficient (IL-6$^{-/-}$) mice," *Endocrinology*, vol. 147, no. 6, pp. 2690–2695, 2006.

[41] K. Stenlöf, I. Wernstedt, T. Fjällman, V. Wallenius, K. Wallenius, and J. O. Jansson, "Interleukin-6 levels in the central nervous system are negatively correlated with fat mass in overweight/obese subjects," *Journal of Clinical Endocrinology and Metabolism*, vol. 88, no. 9, pp. 4379–4383, 2003.

[42] G. A. Bray and D. A. York, "Hypothalamic and genetic obesity in experimental animals: an autonomic and endocrine hypothesis," *Physiological Reviews*, vol. 59, no. 3, pp. 719–809, 1979.

[43] B. K. Anand and J. R. Brobeck, "Localization of a "feeding center" in the hypothalamus of the rat," *Proceedings of the Society for Experimental Biology and Medicine*, vol. 77, no. 2, pp. 323–324, 1951.

[44] M. A. Cowley, J. L. Smart, M. Rubinstein et al., "Leptin activates anorexigenic POMC neurons through a neural network in the arcuate nucleus," *Nature*, vol. 411, no. 6836, pp. 480–484, 2001.

[45] N. Ibrahim, M. A. Bosch, J. L. Smart et al., "Hypothalamic proopiomelanocortin neurons are glucose responsive and express KATP channels," *Endocrinology*, vol. 144, no. 4, pp. 1331–1340, 2003.

[46] T. L. Horvath, F. Naftolin, S. P. Kalra, and C. Leranth, "Neuropeptide-Y innervation of β-endorphin-containing cells in the rat mediobasal hypothalamus: a light and electron microscopic double immunostaining analysis," *Endocrinology*, vol. 131, no. 5, pp. 2461–2467, 1992.

[47] T. L. Horvath, L. M. Garcia-Segura, and F. Naftolin, "Control of gonadotropin feedback: the possible role of estrogen-induced hypothalamic synaptic plasticity," *Gynecological Endocrinology*, vol. 11, no. 2, pp. 139–143, 1997.

[48] T. L. Horvath, Z. B. Andrews, and S. Diano, "Fuel utilization by hypothalamic neurons: roles for ROS," *Trends in Endocrinology and Metabolism*, vol. 20, no. 2, pp. 78–87, 2009.

[49] S. Diano, "New aspects of melanocortin signaling: a role for PRCP in α-MSH degradation," *Frontiers in Neuroendocrinology*, vol. 32, no. 1, pp. 70–83, 2011.

[50] A. Hinney, A. Schmidt, K. Nottebom et al., "Several mutations in the melanocortin-4 receptor gene including a nonsense and a frameshift mutation associated with dominantly

inherited obesity in humans," *Journal of Clinical Endocrinology and Metabolism*, vol. 84, no. 4, pp. 1483–1486, 1999.

[51] C. Vaisse, K. Clement, E. Durand, S. Hercberg, B. Guy-Grand, and P. Froguel, "Melanocortin-4 receptor mutations are a frequent and heterogeneous cause of morbid obesity," *Journal of Clinical Investigation*, vol. 106, no. 2, pp. 253–262, 2000.

[52] A. Sahu, "Minireview: a hypothalamic role in energy balance with special emphasis on leptin," *Endocrinology*, vol. 145, no. 6, pp. 2613–2620, 2004.

[53] S. Pinto, A. G. Roseberry, H. Liu et al., "Rapid rewiring of arcuate nucleus feeding circuits by leptin," *Science*, vol. 304, no. 5667, pp. 110–115, 2004.

[54] Y. Maejima, U. Sedbazar, S. Suyama et al., "Nesfatin-1-regulated oxytocinergic signaling in the paraventricular nucleus causes anorexia through a leptin-independent melanocortin pathway," *Cell Metabolism*, vol. 10, no. 5, pp. 355–365, 2009.

[55] D. Walsh, S. Donnelly, and L. Rybicki, "The symptoms of advanced cancer: relationship to age, gender, and performance status in 1,000 patients," *Supportive Care in Cancer*, vol. 8, no. 3, pp. 175–179, 2000.

[56] W. D. Dewys, C. Begg, P. T. Lavin et al., "Prognostic effect of weight loss prior to chemotherapy in cancer patients," *American Journal of Medicine*, vol. 69, no. 4, pp. 491–497, 1980.

[57] P. O'Gorman, D. C. McMillan, and C. S. McArdle, "Impact of weight loss, appetite, and the inflammatory response on quality of life in gastrointestinal cancer patients," *Nutrition and Cancer*, vol. 32, no. 2, pp. 76–80, 1998.

[58] M. J. Tisdale, "Cachexia in cancer patients," *Nature Reviews Cancer*, vol. 2, no. 11, pp. 862–871, 2002.

[59] C. R. Plata-Salamán, "Cytokines and feeding," *International Journal of Obesity*, vol. 25, supplement 5, pp. S48–S52, 2001.

[60] A. Laviano, M. M. Meguid, and F. Rossi-Fanelli, "Cancer anorexia: clinical implications, pathogenesis, and therapeutic strategies," *The Lancet Oncology*, vol. 4, no. 11, pp. 686–694, 2003.

[61] M. E. Martignoni, P. Kunze, and H. Friess, "Cancer cachexia," *Molecular Cancer*, vol. 2, article 36, 2003.

[62] S. R. Broussard, R. H. Mccusker, J. E. Novakofski et al., "Cytokine-hormone interactions: tumor necrosis factor α impairs biologic activity and downstream activation signals of the insulin-like growth factor I receptor in myoblasts," *Endocrinology*, vol. 144, no. 7, pp. 2988–2996, 2003.

[63] R. A. Frost and C. H. Lang, "Alteration of somatotropic function by proinflammatory cytokines," *Journal of Animal Science*, vol. 82, pp. E100–E109, 2004.

[64] G. H. Lee, R. Proenca, J. M. Montez et al., "Abnormal splicing of the leptin receptor in diabetic mice," *Nature*, vol. 379, no. 6566, pp. 632–635, 1996.

[65] M. W. Schwartz and G. J. Morton, "Keeping hunger at bay," *Nature*, vol. 418, no. 6898, pp. 595–597, 2002.

[66] C. Grunfeld, C. Zhao, J. Fuller et al., "Endotoxin and cytokines induce expression of leptin, the ob gene product, in hamsters," *Journal of Clinical Investigation*, vol. 97, no. 9, pp. 2152–2157, 1996.

[67] B. E. Wisse, K. Ogimoto, G. J. Morton et al., "Physiological regulation of hypothalamic IL-1β gene expression by leptin and glucocorticoids: implications for energy homeostasis," *American Journal of Physiology*, vol. 287, no. 6, pp. E1107–E1113, 2004.

[68] G. N. Luheshi, J. D. Gardner, D. A. Rushforth, A. S. Loudon, and N. J. Rothwell, "Leptin actions on food intake and body temperature are mediated by IL-1," *Proceedings of the National Academy of Sciences of the United States of America*, vol. 96, no. 12, pp. 7047–7052, 1999.

[69] V. Sergeyev, C. Broberger, and T. Hökfelt, "Effect of LPS administration on the expression of POMC, NPY, galanin, CART and MCH mRNAs in the rat hypothalamus," *Molecular Brain Research*, vol. 90, no. 2, pp. 93–100, 2001.

[70] D. L. Marks, N. Ling, and R. D. Cone, "Role of the central melanocortin system in cachexia," *Cancer Research*, vol. 61, no. 4, pp. 1432–1438, 2001.

[71] G. Mantovani, A. Macciò, L. Mura et al., "Serum levels of leptin and proinflammatory cytokines in patients with advanced-stage cancer at different sites," *Journal of Molecular Medicine*, vol. 78, no. 10, pp. 554–561, 2000.

[72] H. Baumann and J. Gauldie, "The acute phase response," *Immunology Today*, vol. 15, no. 2, pp. 74–80, 1994.

[73] A. Laviano, M. M. Meguid, Z. J. Yang, J. R. Gleason, C. Cangiano, and F. R. Fanelli, "Cracking the riddle of cancer anorexia," *Nutrition*, vol. 12, no. 10, pp. 706–710, 1996.

[74] P. G. Jang, C. Namkoong, G. M. Kang et al., "NF-κB activation in hypothalamic pro-opiomelanocortin neurons is essential in illness- and leptin-induced anorexia," *Journal of Biological Chemistry*, vol. 285, no. 13, pp. 9706–9715, 2010.

[75] J. P. Thaler, S. J. Choi, M. W. Schwartz, and B. E. Wisse, "Hypothalamic inflammation and energy homeostasis: resolving the paradox," *Frontiers in Neuroendocrinology*, vol. 31, no. 1, pp. 79–84, 2010.

[76] V. Mohamed-Ali, S. Goodrick, A. Rawesh et al., "Subcutaneous adipose tissue releases interleukin-6, but not tumor necrosis factor-α, in vivo," *Journal of Clinical Endocrinology and Metabolism*, vol. 82, no. 12, pp. 4196–4200, 1997.

[77] F. Kim, M. Pham, E. Maloney et al., "Vascular inflammation, insulin resistance, and reduced nitric oxide production precede the onset of peripheral insulin resistance," *Arteriosclerosis, Thrombosis, and Vascular Biology*, vol. 28, no. 11, pp. 1982–1988, 2008.

[78] C. T. De Souza, E. P. Araujo, S. Bordin et al., "Consumption of a fat-rich diet activates a proinflammatory response and induces insulin resistance in the hypothalamus," *Endocrinology*, vol. 146, no. 10, pp. 4192–4199, 2005.

[79] K. A. Posey, D. J. Clegg, R. L. Printz et al., "Hypothalamic proinflammatory lipid accumulation, inflammation, and insulin resistance in rats fed a high-fat diet," *American Journal of Physiology*, vol. 296, no. 5, pp. E1003–E1012, 2009.

[80] X. Zhang, G. Zhang, H. Zhang, M. Karin, H. Bai, and D. Cai, "Hypothalamic IKKβ/NF-κB and ER stress link overnutrition to energy imbalance and obesity," *Cell*, vol. 135, no. 1, pp. 61–73, 2008.

[81] L. Ozcan, A. S. Ergin, A. Lu et al., "Endoplasmic reticulum stress plays a central role in development of leptin resistance," *Cell Metabolism*, vol. 9, no. 1, pp. 35–51, 2009.

[82] M. Milanski, G. Degasperi, A. Coope et al., "Saturated fatty acids produce an inflammatory response predominantly through the activation of TLR4 signaling in hypothalamus: implications for the pathogenesis of obesity," *Journal of Neuroscience*, vol. 29, no. 2, pp. 359–370, 2009.

[83] M. Burstow, T. Kelly, S. Panchani et al., "Outcome of palliative esophageal stenting for malignant dysphagia: a retrospective analysis," *Diseases of the Esophagus*, vol. 22, no. 6, pp. 519–525, 2009.

[84] H. Suzuki, H. Hashimoto, M. Kawasaki et al., "Similar changes of hypothalamic feeding-regulating peptides mRNAs and plasma leptin levels in PTHrP-, LIF-secreting tumors-induced cachectic rats and adjuvant arthritic rats," *International Journal of Cancer*, vol. 128, no. 9, pp. 2215–2223, 2011.

[85] Q. Ding, T. Mracek, P. Gonzalez-Muniesa et al., "Identification of macrophage inhibitory cytokine-1 in adipose tissue and its secretion as an adipokine by human adipocytes," *Endocrinology*, vol. 150, no. 4, pp. 1688–1696, 2009.

[86] B. Weryńska, M. Kosacka, M. Gołecki, and R. Jankowska, "Leptin serum levels in cachectic and non-cachectic lung cancer patients," *Pneumonologia i Alergologia Polska*, vol. 77, no. 6, pp. 500–506, 2009.

[87] J. Smiechowska, A. Utech, G. Taffet, T. Hayes, M. Marcelli, and J. M. Garcia, "Adipokines in patients with cancer anorexia and cachexia," *Journal of Investigative Medicine*, vol. 58, no. 3, pp. 554–559, 2010.

[88] C. Bing, S. Taylor, M. J. Tisdale, and G. Williams, "Cachexia in MAC16 adenocarcinoma: suppression of hunger despite normal regulation of leptin, insulin and hypothalamic neuropeptide Y," *Journal of Neurochemistry*, vol. 79, no. 5, pp. 1004–1012, 2001.

[89] A. Macciò, C. Madeddu, D. Massa et al., "Interleukin-6 and leptin as markers of energy metabolic changes in advanced ovarian cancer patients," *Journal of Cellular and Molecular Medicine*, vol. 13, no. 9, pp. 3951–3959, 2009.

[90] P. Costelli, N. Carbo, L. Tessitore et al., "Tumor necrosis factor-α mediates changes in tissue protein turnover in a rat cancer cachexia model," *Journal of Clinical Investigation*, vol. 92, no. 6, pp. 2783–2789, 1993.

[91] R. M. Goldberg, C. L. Loprinzi, J. A. Mailliard et al., "Pentoxifylline for treatment of cancer anorexia and cachexia? A randomized, double-blind, placebo-controlled trial," *Journal of Clinical Oncology*, vol. 13, no. 11, pp. 2856–2859, 1995.

[92] E. Bruera, C. M. Neumann, E. Pituskin, K. Calder, G. Ball, and J. Hanson, "Thalidomide in patients with cachexia due to terminal cancer: preliminary report," *Annals of Oncology*, vol. 10, no. 7, pp. 857–859, 1999.

[93] K. Mori, K. Fujimoto-Ouchi, T. Ishikawa, F. Sekiguchi, H. Ishitsuka, and Y. Tanaka, "Murine interleukin-12 prevents the development of cancer cachexia in a murine model," *International Journal of Cancer*, vol. 67, no. 6, pp. 849–855, 1996.

[94] N. Carbó, J. López-Soriano, P. Costelli et al., "Interleukin-15 antagonizes muscle protein waste in tumour-bearing rats," *British Journal of Cancer*, vol. 83, no. 4, pp. 526–531, 2000.

[95] G. Mantovani, A. Macciò, C. Madeddu et al., "Phase II nonrandomized study of the efficacy and safety of COX-2 inhibitor celecoxib on patients with cancer cachexia," *Journal of Molecular Medicine*, vol. 88, no. 1, pp. 85–92, 2010.

[96] S. D. Anker, A. Negassa, A. J. S. Coats et al., "Prognostic importance of weight loss in chronic heart failure and the effect of treatment with angiotensin-converting-enzyme inhibitors: an observational study," *The Lancet*, vol. 361, no. 9363, pp. 1077–1083, 2003.

[97] S. D. Anker, P. Ponikowski, S. Varney et al., "Wasting as independent risk factor for mortality in chronic heart failure," *The Lancet*, vol. 349, no. 9058, pp. 1050–1053, 1997.

[98] C. R. Gibbs, G. Jackson, and G. Y. H. Lip, "ABC of heart failure: non-drug management," *British Medical Journal*, vol. 320, no. 7231, pp. 366–369, 2000.

[99] B. Levine, J. Kalman, L. Mayer, H. M. Fillit, and M. Packer, "Elevated circulating levels of tumor necrosis factor in severe chronic heart failure," *New England Journal of Medicine*, vol. 323, no. 4, pp. 236–241, 1990.

[100] G. S. Filippatos, K. Tsilias, K. Venetsanou et al., "Leptin serum levels in cachectic heart failure patient. Relationship with tumor necrosis factor-α system," *International Journal of Cardiology*, vol. 76, no. 2-3, pp. 117–122, 2000.

[101] D. R. Murdoch, E. Rooney, H. J. Dargie, D. Shapiro, J. J. Morton, and J. J. V. McMurray, "Inappropriately low plasma leptin concentration in the cachexia associated with chronic heart failure," *Heart*, vol. 82, no. 3, pp. 352–356, 1999.

[102] M. J. Toth, S. S. Gottlieb, M. L. Fisher, A. S. Ryan, B. J. Nicklas, and E. T. Poehlman, "Plasma leptin concentrations and energy expenditure in heart failure patients," *Metabolism*, vol. 46, no. 4, pp. 450–453, 1997.

[103] G. Paolisso, M. R. Rizzo, G. Mazziotti et al., "Lack of association between changes in plasma leptin concentration and in food intake during the menstrual cycle," *European Journal of Clinical Investigation*, vol. 29, no. 6, pp. 490–495, 1999.

[104] M. W. Nickola, L. E. Wold, P. B. Colligan, G. J. Wang, W. K. Samson, and J. Ren, "Leptin attenuates cardiac contraction in rat ventricular myocytes role of NO," *Hypertension*, vol. 36, no. 4, pp. 501–505, 2000.

[105] K. R. McGaffin, C. S. Moravec, and C. F. McTiernan, "Leptin signaling in the failing and mechanically unloaded human heart," *Circulation*, vol. 2, no. 6, pp. 676–683, 2009.

[106] P. C. Schulze, J. Kratzsch, A. Linke et al., "Elevated serum levels of leptin and soluble leptin receptor in patients with advanced chronic heart failure," *European Journal of Heart Failure*, vol. 5, no. 1, pp. 33–40, 2003.

[107] F. Leyva, S. D. Anker, K. Egerer, J. C. Stevenson, W. J. Kox, and A. J. S. Coats, "Hyperleptinaemia in chronic heart failure. Relationships with insulin," *European Heart Journal*, vol. 19, no. 10, pp. 1547–1551, 1998.

[108] T. Tsutamoto, T. Hisanaga, A. Wada et al., "Interleukin-6 spillover in the peripheral circulation increases with the severity of heart failure, and the high plasma level of interleukin-6 is an important prognostic predictor in patients with congestive heart failure," *Journal of the American College of Cardiology*, vol. 31, no. 2, pp. 391–398, 1998.

[109] V. Sharma and J. H. McNeill, "The emerging roles of leptin and ghrelin in cardiovascular physiology and pathophysiology," *Current Vascular Pharmacology*, vol. 3, no. 2, pp. 169–180, 2005.

[110] W. G. Haynes, W. I. Sivitz, D. A. Morgan, S. A. Walsh, and A. L. Mark, "Sympathetic and cardiorenal actions of leptin," *Hypertension*, vol. 30, no. 3, pp. 619–623, 1997.

[111] W. G. Haynes, D. A. Morgan, S. A. Walsh, W. I. Sivitz, and A. L. Mark, "Cardiovascular consequences of obesity: role of leptin," *Clinical and Experimental Pharmacology and Physiology*, vol. 25, no. 1, pp. 65–69, 1998.

[112] G. Lembo, C. Vecchione, L. Fratta et al., "Leptin induces direct vasodilation through distinct endothelial mechanisms," *Diabetes*, vol. 49, no. 2, pp. 293–297, 2000.

[113] S. D. Anker and S. Von Haehling, "Inflammatory mediators in chronic heart failure: an overview," *Heart*, vol. 90, no. 4, pp. 464–470, 2004.

[114] R. Ferrari, T. Bachetti, R. Confortini et al., "Tumor necrosis factor soluble receptors in patients with various degrees of congestive heart failure," *Circulation*, vol. 92, no. 6, pp. 1479–1486, 1995.

[115] M. Rauchhaus, W. Doehner, D. P. Francis et al., "Plasma cytokine parameters and mortality in patients with chronic heart failure," *Circulation*, vol. 102, no. 25, pp. 3060–3067, 2000.

[116] T. G. Kirchgessner, K. T. Uysal, S. M. Wiesbrock, M. W. Marino, and G. S. Hotamisligil, "Tumor necrosis factor-α contributes to obesity-related hyperleptinemia by regulating leptin release from adipocytes," *Journal of Clinical Investigation*, vol. 100, no. 11, pp. 2777–2782, 1997.

[117] D. Kovačić, M. Marinšek, L. Gobec, M. Lainščak, and M. Podbregar, "Effect of selective and non-selective β-blockers on body weight, insulin resistance and leptin concentration in chronic heart failure," *Clinical Research in Cardiology*, vol. 97, no. 1, pp. 24–31, 2008.

[118] Y. Schutz and V. Woringer, "Obesity in switzerland: a critical assessment of prevalence in children and adults," *International Journal of Obesity*, vol. 26, supplement 2, pp. S3–S11, 2002.

[119] M. K. Sridhar, R. Carter, M. E. J. Lean, and S. W. Banham, "Resting energy expenditure and nutritional state of patients with increased oxygen cost of breathing due to emphysema, scoliosis and thoracoplasty," *Thorax*, vol. 49, no. 8, pp. 781–785, 1994.

[120] O. Hugli, Y. Schutz, and J. W. Fitting, "The cost of breathing in stable chronic obstructive pulmonary disease," *Clinical Science*, vol. 89, no. 6, pp. 625–632, 1995.

[121] M. Di Francia, D. Barbier, J. L. Mege, and J. Orehek, "Tumor necrosis factor-alpha levels and weight loss in chronic obstructive pulmonary disease," *American Journal of Respiratory and Critical Care Medicine*, vol. 150, no. 5, pp. 1453–1455, 1994.

[122] I. De Godoy, M. Donahoe, W. J. Calhoun, J. Mancino, and R. M. Rogers, "Elevated TNF-α production by peripheral blood monocytes of weight-losing COPD patients," *American Journal of Respiratory and Critical Care Medicine*, vol. 153, no. 2, pp. 633–637, 1996.

[123] P. Sarraf, R. C. Frederich, E. M. Turner et al., "Multiple cytokines and acute inflammation raise mouse leptin levels: potential role in inflammatory anorexia," *Journal of Experimental Medicine*, vol. 185, no. 1, pp. 171–175, 1997.

[124] M. S. Zumbach, M. W. J. Boehme, P. Wahl, W. Stremmel, R. Ziegler, and P. P. Nawroth, "Tumor necrosis factor increases serum leptin levels in humans," *Journal of Clinical Endocrinology and Metabolism*, vol. 82, no. 12, pp. 4080–4082, 1997.

[125] N. Takabatake, H. Nakamura, S. Abe et al., "Circulating leptin in patients with chronic obstructive pulmonary disease," *American Journal of Respiratory and Critical Care Medicine*, vol. 159, no. 4, pp. 1215–1219, 1999.

[126] N. Takabatake, H. Nakamura, O. Minamihaba et al., "A novel pathophysiologic phenomenon in cachexic patients with chronic obstructive pulmonary disease: the relationship between the circadian rhythm of circulating leptin and the very low-frequency component of heart rate variability," *American Journal of Respiratory and Critical Care Medicine*, vol. 163, no. 6, pp. 1314–1319, 2001.

[127] C. P. O'Donnell, C. D. Schaub, A. S. Haines et al., "Leptin prevents respiratory depression in obesity," *American Journal of Respiratory and Critical Care Medicine*, vol. 159, no. 5, pp. 1477–1484, 1999.

[128] C. G. Tankersley, C. O'Donnell, M. J. Daood et al., "Leptin attenuates respiratory complications associated with the obese phenotype," *Journal of Applied Physiology*, vol. 85, no. 6, pp. 2261–2269, 1998.

[129] H. Groeben, S. Meier, R. H. Brown, C. P. O'Donnell, W. Mitzner, and C. G. Tankersley, "The effect of leptin on the ventilatory response to hyperoxia," *Experimental Lung Research*, vol. 30, no. 7, pp. 559–570, 2004.

[130] C. Tankersley, S. Kleeberger, B. Russ, A. Schwartz, and P. Smith, "Modified control of breathing in genetically obese (ob/ob) mice," *Journal of Applied Physiology*, vol. 81, no. 2, pp. 716–723, 1996.

[131] A. Grosfeld, J. André, S. H.-D. Mouzon, E. Berra, J. Pouysségur, and M. Guerre-Millo, "Hypoxia-inducible factor 1 transactivates the human leptin gene promoter," *Journal of Biological Chemistry*, vol. 277, no. 45, pp. 42953–42957, 2002.

[132] S. Budweiser, F. Heinemann, K. Meyer, P. J. Wild, and M. Pfeifer, "Weight gain in cachectic COPD patients receiving noninvasive positive-pressure ventilation," *Respiratory Care*, vol. 51, no. 2, pp. 126–132, 2006.

[133] K. J. Tracey, "The inflammatory reflex," *Nature*, vol. 420, no. 6917, pp. 853–859, 2002.

[134] R. H. Mak, W. Cheung, R. D. Cone, and D. L. Marks, "Leptin and inflammation-associated cachexia in chronic kidney disease," *Kidney International*, vol. 69, no. 5, pp. 794–797, 2006.

[135] W. W. Cheung, K. H. Paik, and R. H. Mak, "Inflammation and cachexia in chronic kidney disease," *Pediatric Nephrology*, vol. 25, no. 4, pp. 711–724, 2010.

[136] A. Scholze, D. Rattensperger, W. Zidek, and M. Tepel, "Low serum leptin predicts mortality in patients with chronic kidney disease stage 5," *Obesity*, vol. 15, no. 6, pp. 1617–1622, 2007.

[137] I. Beberashvili, I. Sinuani, A. Azar et al., "Longitudinal study of leptin levels in chronic hemodialysis patients," *Nutrition Journal*, vol. 10, article 68, no. 1, 2011.

[138] P. Stenvinkel, "Leptin—a new hormone of definite interest for the nephrologist," *Nephrology Dialysis Transplantation*, vol. 13, no. 5, pp. 1099–1101, 1998.

[139] A. Aguilera, J. A. Sánchez-Tomero, and R. Selgas, "Brain activation in uremic anorexia," *Journal of Renal Nutrition*, vol. 17, no. 1, pp. 57–61, 2007.

[140] A. F. Suffredini, G. Fantuzzi, R. Badolato, J. J. Oppenheim, and N. P. O'Grady, "New insights into the biology of the acute phase response," *Journal of Clinical Immunology*, vol. 19, no. 4, pp. 203–214, 1999.

[141] R. Pecoits-Filho, L. C. Sylvestre, and P. Stenvinkel, "Chronic kidney disease and inflammation in pediatric patients: from bench to playground," *Pediatric Nephrology*, vol. 20, no. 6, pp. 714–720, 2005.

[142] K. Kalantar-Zadeh, P. Stenvinkel, L. Pillon, and J. D. Kopple, "Inflammation and nutrition in renal insufficiency," *Advances in Renal Replacement Therapy*, vol. 10, no. 3, pp. 155–169, 2003.

[143] R. H. Mak and W. Cheung, "Energy homeostasis and cachexia in chronic kidney disease," *Pediatric Nephrology*, vol. 21, no. 12, pp. 1807–1814, 2006.

[144] R. H. Mak and W. Cheung, "Adipokines and gut hormones in end-stage renal disease," *Peritoneal Dialysis International*, vol. 27, supplement 2, pp. S298–S302, 2007.

[145] W. Cheung, P. X. Yu, B. M. Little, R. D. Cone, D. L. Marks, and R. H. Mak, "Role of leptin and melanocortin signaling in uremia-associated cachexia," *Journal of Clinical Investigation*, vol. 115, no. 6, pp. 1659–1665, 2005.

[146] P. J. Scarpace, M. Matheny, R. L. Moore, and N. Tümer, "Impaired leptin responsiveness in aged rats," *Diabetes*, vol. 49, no. 3, pp. 431–435, 2000.

[147] T. Wolden-Hanson, "Mechanisms of the anorexia of aging in the Brown Norway rat," *Physiology and Behavior*, vol. 88, no. 3, pp. 267–276, 2006.

[148] E. W. Shek and P. J. Scarpace, "Resistance to the anorexic and thermogenic effects of centrally administered leptin in obese aged rats," *Regulatory Peptides*, vol. 92, no. 1–3, pp. 65–71, 2000.

[149] D. A. Gruenewald, M. A. Naai, B. T. Marck, and A. M. Matsumoto, "Age-related decrease in neuropeptide-Y gene

expression in the arcuate nucleus of the male rat brain is independent of testicular feedback," *Endocrinology*, vol. 134, no. 6, pp. 2383–2389, 1994.

[150] D. Horrillo, J. Sierra, C. Arribas et al., "Age-associated development of inflammation in Wistar rats: effects of caloric restriction," *Archives of Physiology and Biochemistry*, vol. 117, no. 3, pp. 140–150, 2011.

[151] J. J. Carrero, A. Nakashima, A. R. Qureshi et al., "Protein-energy wasting modifies the association of ghrelin with inflammation, leptin, and mortality in hemodialysis patients," *Kidney International*, vol. 79, no. 7, pp. 749–756, 2011.

[152] A. van Tellingen, M. P. C. Grooteman, M. Schoorl et al., "Enhanced long-term reduction of plasma leptin concentrations by super-flux polysulfone dialysers," *Nephrology Dialysis Transplantation*, vol. 19, no. 5, pp. 1198–1203, 2004.

[153] E. D. Javor, E. K. Cochran, C. Musso, J. R. Young, A. M. DePaoli, and P. Gorden, "Long-term efficacy of leptin replacement in patients with generalized lipodystrophy," *Diabetes*, vol. 54, no. 7, pp. 1994–2002, 2005.

[154] A. B. Newman, D. Yanez, T. Harris, A. Duxbury, P. L. Enright, and L. P. Fried, "Weight change in old age and its association with mortality," *Journal of the American Geriatrics Society*, vol. 49, no. 10, pp. 1309–1318, 2001.

[155] G. Hauser and M. Neumann, "Aging with quality of life—a challenge for society," *Journal of Physiology and Pharmacology*, vol. 56, supplement 2, pp. 35–48, 2005.

[156] Z. Kmiec, "Central regulation of food intake in ageing," *Journal of Physiology and Pharmacology*, vol. 57, supplement 6, pp. 7–16, 2006.

[157] C. Fernández-Galaz, T. Fernández-Agulló, F. Campoy et al., "Decreased leptin uptake in hypothalamic nuclei with ageing in Wistar rats," *Journal of Endocrinology*, vol. 171, no. 1, pp. 23–32, 2001.

[158] D. A. Gruenewald, B. T. Marck, and A. M. Matsumoto, "Fasting-induced increases in food intake and neuropeptide Y gene expression are attenuated in aging male brown Norway rats," *Endocrinology*, vol. 137, no. 10, pp. 4460–4467, 1996.

[159] T. Wolden-Hanson, B. T. Marck, and A. M. Matsumoto, "Blunted hypothalamic neuropeptide gene expression in response to fasting, but preservation of feeding responses to AgRP in aging male Brown Norway rats," *American Journal of Physiology*, vol. 287, no. 1, pp. R138–R146, 2004.

[160] O. Bouillanne, C. Dupont-Belmont, P. Hay, B. Hamon-Vilcot, L. Cynober, and C. Aussel, "Fat mass protects hospitalized elderly persons against morbidity and mortality," *American Journal of Clinical Nutrition*, vol. 90, no. 3, pp. 505–510, 2009.

[161] Y. Arai, M. Takayama, Y. Abe, and N. Hirose, "Adipokines and aging," *Journal of Atherosclerosis and Thrombosis*, vol. 18, no. 7, pp. 545–550, 2011.

[162] M. Zamboni, E. Zoico, F. Fantin et al., "Relation between leptin and the metabolic syndrome in elderly women," *Journals of Gerontology*, vol. 59, no. 4, pp. 396–400, 2004.

[163] P. J. Enriori, A. E. Evans, P. Sinnayah et al., "Diet-induced obesity causes severe but reversible leptin resistance in arcuate melanocortin neurons," *Cell Metabolism*, vol. 5, no. 3, pp. 181–194, 2007.

[164] R. E. Hubbard, M. S. O, B. L. Calver, and K. W. Woodhouse, "Nutrition, inflammation, and leptin levels in aging and frailty," *Journal of the American Geriatrics Society*, vol. 56, no. 2, pp. 279–284, 2008.

10

Peptide-Modulated Activity Enhancement of Acidic Protease Cathepsin E at Neutral pH

Masayuki Komatsu,[1,2] **Madhu Biyani,**[1] **Sunita Ghimire Gautam,**[1,2] **and Koichi Nishigaki**[1,2]

[1] *Department of Functional Materials Science, Graduate School of Science and Engineering, Saitama University, 255 Shimo-okubo, Sakura-ku, Saitama-shi, Saitama 338-8570, Japan*
[2] *Rational Evolutionary Design of Advanced Biomolecules, Saitama (REDS), Saitama Small Enterprise Promotion Corporation, No. 552, Saitama Industrial Technology Center, 3-12-18 Kami-Aoki, Kawaguchi, Saitama 333-0844, Japan*

Correspondence should be addressed to Koichi Nishigaki, koichi@fms.saitama-u.ac.jp

Academic Editor: Weihong Pan

Enzymes are regulated by their activation and inhibition. Enzyme activators can often be effective tools for scientific and medical purposes, although they are more difficult to obtain than inhibitors. Here, using the paired peptide method, we report on protease-cathepsin-E-activating peptides that are obtained at neutral pH. These selected peptides also underwent molecular evolution, after which their cathepsin E activation capability improved. Thus, the activators we obtained could enhance cathepsin-E-induced cancer cell apoptosis, which indicated their potential as cancer drug precursors.

1. Introduction

Although a number of enzyme inhibitors, such as kinase inhibitors and protease inhibitors, have been successfully used in cancer therapy, very few enzyme activators have been successfully applied [1–3]. This discrepancy is partly because enzyme activators are difficult to identify as there are currently no rational design principles or effective screening methods that can be used [4]. Because various diseases are caused by reducing the activities of endogenous enzymes [5, 6], a general method for identifying enzyme activators is highly desirable. In particular, a method for obtaining enzyme-activating peptides is attractive because of the potential of activators to reveal phenomena that cannot be elucidated by inhibitors [7].

A preliminary approach used an *in vitro* evolution method called the evolutionary rapid panning system (eRA-PANSY) [8], and peptides that moderately enhanced cathepsin E activity were successfully identified after secondary library selection [9]. Cathepsin E, which usually operates at acidic pH [10], has been shown to induce cancer cell apoptosis [11, 12] and inhibit tumor angiogenesis [13] at neutral pH, which promotes the finding of its activators for

cancer therapy. Therefore, in this study, we sought to obtain peptides with sufficiently high biological activity that would be suitable for medical purposes.

To achieve this, we adopted the systemic *in vitro* evolution method (eRAPANSY) along with the paired peptide method (PPM), in which selected peptides were arbitrarily combined by linking through a definite length of a spacer sequence. This resulted in a paired peptide library containing a set of peptides consisting of two peptide moieties, each of which was per se functional. Thus, linking of these peptide aptamers via a spacer is highly promising for obtaining far more advanced peptides. This strategy has already been successfully applied to obtain cathepsin E-inhibitory peptides. Thus, this study confirms the effectiveness of systematic *in vitro* evolution combined with the progressive library method.

The paired peptide library was subjected to a conventional cDNA display [14] and function-based selection could identify cathepsin E-activating peptides with greater activating capability. We also examined the biological activities of these peptides by examining the induction of cancer cell apoptosis. This showed that these peptides were biologically active.

2. Materials and Methods

2.1. Preparation of Cathepsin E and Its Substrate. Cathepsin E was isolated from rat spleens as previously described [15]. The fluorogenic substrate for cathepsin E at neutral pH (pH 7.4) was the same as previously reported [14]. The fluorogenic substrate (10 mM) was prepared by dissolving in 100% DMSO and diluting in reaction buffer immediately before use.

2.2. Construction of the Paired Peptide Library. The paired peptide library was generated by the Y-Ligation-based block shuffling (YLBS) method [16]. In brief, 10 species of DNA blocks were synthesized corresponding to the peptides that were based on information for cathepsin E-activating peptide aptamers obtained from the secondary library selection [9]. This resulted in construction of a library with about 400 different variants (the actual diversity was assumed to be much higher due to substitutions and indel mutations during DNA construct generation [17]). This DNA library was integrated into the cDNA display construct after the cDNA display procedure (Figure 1(b)).

2.3. Selection of Cathepsin-E-Activating Peptides. To select cathepsin-E-activating peptides, the selection-by-function method was used [14] with minor modifications. After completing the final cDNA display construct (Figure 1(b)) by transcription, puromycin-linker ligation, translation, and reverse transcription, the selection-by-function method was employed. First, paired peptide-coding DNA was incubated in Selection buffer (50 mM Tris-HCl, 100 mM NaCl, and 5 mM MgCl$_2$, pH 7.4) containing 5 pmol of cathepsin E-immobilizing sepharose beads (GE Healthcare, USA) at 25°C for 10 min. Unbound DNA molecules were rapidly removed by washing with Selection buffer. DNA molecules that were bound to cathepsin E immobilizing beads were incubated at the temperature (37°C) optimum for cathepsin E proteolytic reaction. DNA molecules released from the beads were collected for the next selection step, as most could be assumed to be generated by the enhanced proteolytic reaction of cathepsin E. This procedure was repeated three times with increasing selection stringency (i.e., shorter reaction periods). The final product was subjected to cloning and sequencing to identify cathepsin E-activating peptides.

2.4. Cathepsin E Protease Activity Assay. Activation of cathepsin E by the selected peptides was determined as previously described [9]. The selected peptides were prepared by *in vitro* translation or chemical synthesis. A solution containing 20 nM cathepsin E was pre incubated with 20 nM of selected peptides in Selection buffer (50 mM Tris-HCl, 100 mM NaCl, pH 7.4) at 25°C for 10 min. Next, the fluorogenic substrate was added (5 μM) and the mixture was incubated at 37°C for 1 h for the enzyme reaction. The fluorescent reaction product was monitored at 440 nm

(excitation at 340 nm) using Infinite 200 (TECAN, Japan). The percent of cathepsin E activation (*A*) was calculated by:

$$A = 100 \times \frac{\left(S_f - B_f\right)}{\left(C_f - B_f\right)} [\% \text{ activation}], \qquad (1)$$

where S_f was the fluorescence intensity of the cathepsin E reaction in the presence of a selected peptide, C_f was the fluorescence intensity of the control in the absence of a selected peptide, and B_f was background fluorescence of the solution containing the fluorogenic substrate only.

2.5. In Vitro Translation or Chemical Synthesis of Selected Peptides. The selected peptides were prepared by *in vitro* translation or chemical synthesis, depending on their intended use. To rapidly estimate cathepsin E-activating capability, peptides were prepared by *in vitro* translation using the DNA coding sequence for the selected peptide. The coding sequence was integrated into the DNA construct for *in vitro* translation [8]. The *in vitro* translated products consisted of a streptavidin-binding peptide for molecular fishing, a protease Factor Xa recognition site for removing the unnecessary peptide portion, and a functional peptide coding region. Finally, peptides were synthesized using a translation kit (PURESYSTEM, Wako, Japan) following the manufacturer's protocol. Peptides T4 (IEGRGCPCIDFMVEVQVEVAEALLTALSLSPGS) and T11 (IEGRLLSGGAGACSVRTVDDSFDCG) were chemically synthesized by commercial vendors (SCRUM Corporation, Ltd., Japan and Operon Biotechnologies, Inc., Japan) with certification of more than 95% purity for the precise assays of cathepsin E activity/affinity/*in vitro* or *in vivo*.

2.6. Affinity Determinations. The dissociation constant (K_d) of each selected peptide with cathepsin E was determined by the SPR method using a Biacore2000 (GE Healthcare, UK). Cathepsin E was immobilized on a CM5 Biacore sensor chip (GE Healthcare, UK) by the general amine coupling method. A small quantity of acetate buffer (50 mM sodium acetate, 100 mM NaCl, pH 4.5) containing cathepsin E (150 μg/mL) was injected into the flow cell for the sample lane. The reference lane was prepared similarly but without cathepsin E. Different concentrations (10, 20, 30, and 40 nM) of the selected peptide candidates, T4 and T11 were injected into both lanes at a flow rate of 20 ul/min to measure the interaction between cathepsin E and each peptide candidate. For all experiments, a neutral pH buffer (50 mM Tris-HCl, 100 mM NaCl, pH 7.4) was used as the running buffer and a solution of 50 mM NaOH was used to remove the peptides. The resulting sensorgram curves were fit to a 1:1 Langmuir binding model, and the K_d values were calculated using BIA evaluation software (GE Healthcare, UK).

2.7. Cell-Based Assay. HeLa cells were used for an *in vitro* cell-based bioassay. Approximately 10,000 HeLa cells were seeded into the wells of a 96-well culture plate (Corning, USA) containing 100 uL of DMEM medium supplemented with 10% FBS and 2% penicillin-streptomycin and incubated

(a)

(b)

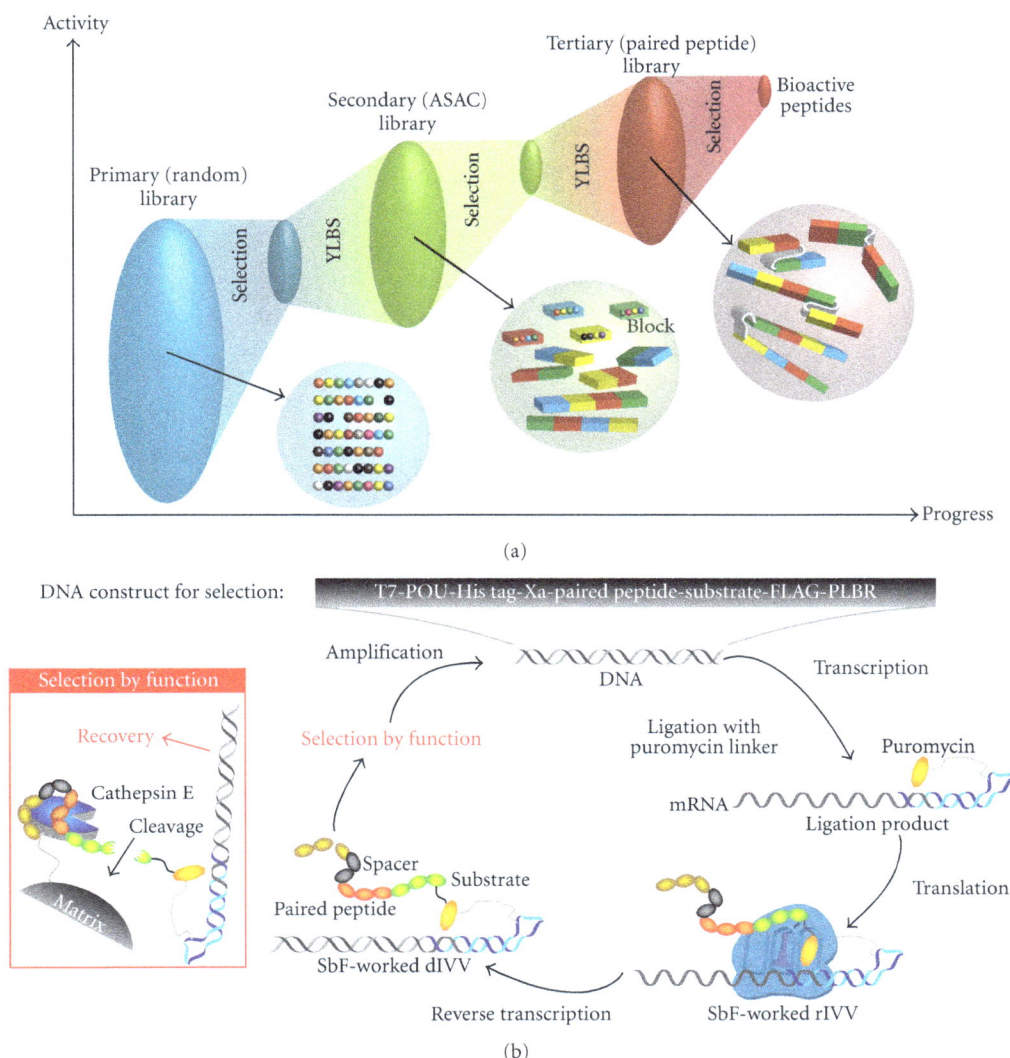

FIGURE 1: Systematic of the *in vitro* evolution strategy for obtaining enzyme activators. (a) Progressive library. Functional peptides were identified from the first random peptide library. The second library was generated by combining peptide blocks selected from the primary library and subjecting them to the next round of selection. The third library was constructed by pairing two peptides selected from the second library. (b) Schematic representation of cDNA display-based selection-by-function. SbF-worked r/d-IVV was an RNA/DNA-type *in vitro* virus construct augmented with the selection-by-function construct. SbF-worked dIVV could be cleaved by cathepsin E if its binding activated cathepsin E, as shown in the box.

overnight at 37°C under 5% CO_2. The cells were treated with cathepsin E (77 nM) alone or together with two different concentrations (770 nM, 7.7 μM) of the selected peptide T11 in 100 μL of serum-free Opti-MEM at 37°C for 20 h. After incubation, viable cells were determined with a Cell Counting Kit-F (Dojindo Molecular Technologies, Japan) according to the manufacturer's protocol. Viable cells were identified by their fluorescent emissions intensity using an Infinite 200 (TECAN, Japan) with excitation at 490 nm and emission at 515 nm. To detect apoptotic cells, cells were treated with Annexin V-Cy3 (BioVision, USA) following the manufacturer's protocol. Apoptotic cells were detected by their fluorescent emissions with excitation at 543 nm and emission at 570 nm. To detect caspase-3/7 activity induced

by cathepsin E, an Apo-ONE Homogeneous Caspase-3/7 Assay kit (Promega, USA) was used according to the manufacturer's protocol. Activity was detected by fluorescent emissions with excitation at 485 nm and emission at 530 nm. The cathepsin E and selected peptide concentrations used for apoptosis and caspase assays were the same as those used for cell viability assays.

3. Results

We sought to develop a method to obtain cathepsin E-activating peptides and identify a sufficiently strong enzyme activator with verifiable biological activity, such as inducing cancer cell apoptosis.

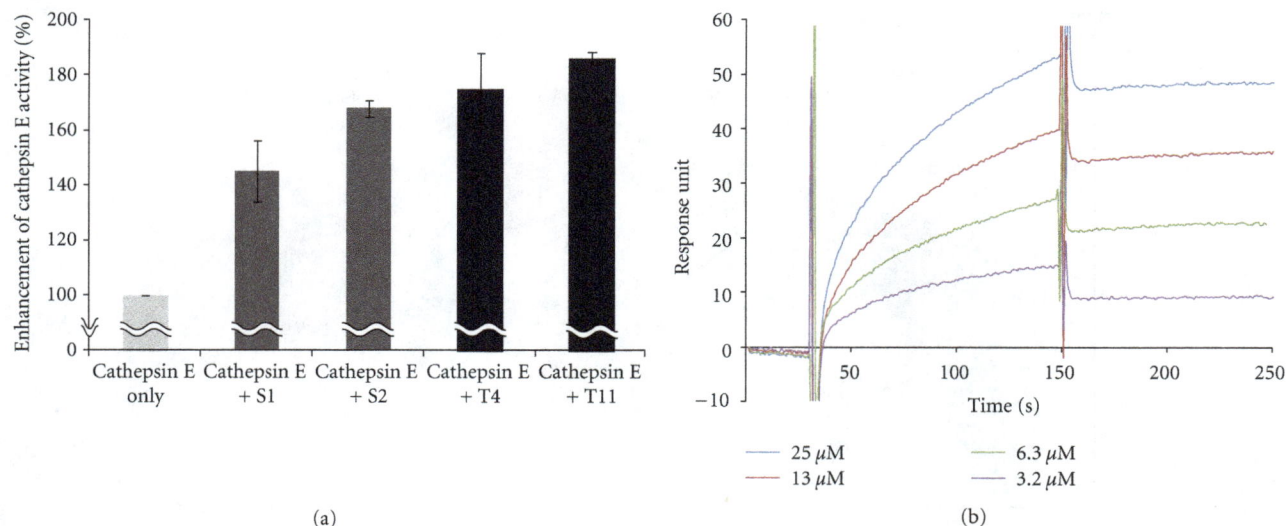

(a)

(b)

FIGURE 2: Activities and affinities of selected peptides for cathepsin E activation. (a) Cathepsin E activity enhancement by chemically synthesized peptides selected from the secondary ASAC library (S1 and S2) and the third paired peptide library (T4 and T11). Cathepsin E and peptide concentrations were 20 nM. Error bars indicate the standard deviations of triplicate experiments. (b) Typical SPR sensorgram obtained from the interaction between paired peptide T4 and cathepsin E. To determine the dissociation constant (K_d), four different peptide concentrations were injected. The range from 40 s to 150 s corresponded to association, while that from 150 s to 250 s corresponded to dissociation.

3.1. Overall Scheme to Obtain Activating Peptides.

Cathepsin-E-activating peptides that were previously developed (Table 1) were used as starting materials for this study. These peptides moderately enhanced cathepsin-E activity (~60% using our assay system) and had a maximum binding affinity of 400 nM. These activity-enhancing peptides were selected using the selection-by-function method [17] and a block shuffling method: ASAC (all-steps all-combinations; Figure 1). Among various selection methods, the selection-by-function method was found to select activating peptides and the ASAC method provided a proven library, from which we could identify peptides with improved activities [8, 9, 18]. These methods enabled the selection of the desired peptides that otherwise would have been difficult, as there is no general approach to effectively identify protease-activating reagents (including peptides).

Thus, we introduced a novel method that exploited previously acquired molecular information, that is, pairing of selected peptides that have affinity for a target protein, specifically for different epitopes, to increase activity in a cooperative manner. The library that was generated by arbitrarily pairing the second library selection products (Table 1) was the third library (i.e., paired peptide library). This library was screened by the selection-by-function method (Table 2). For the technical reasons (see [8, 18] for details), the initial paired peptide library contained substantial amounts of unintended molecules in addition to the whole set of intended molecules.

3.2. Activity Enhancing Peptides Acquired by the Paired Peptide Method.

A few selected clones were assayed for

TABLE 1: Cathepsin-E-activating peptides and spacer peptides used for paired peptide library construction.

Name[1]	Size (a.a.)	Amino acid sequence (N → C)	Activity (%)[2]
S1	13	IEGRVGCDFMYVG	**130**
S2	8	GSPCIGII	*133*
S3	8	IVIHQQLL	—
S4	8	PGIKIIIG	*151*
S5	9	IGPQFGMCG	—
S6	10	PGFEERSSEG	—
S7	16	SPIISHIVGCDPPSCG	*160*
S8	16	IGCEERSFPNIIIIG	**168**
S9	13	SGIKVGCDPPSCG	*140*
S10	13	PGIKPPPCIIIG	**145**
s1	10	GGGSGGGSGG	—
s2	10	GGGPGGGPGG	—
s3	15	GGGSGGGSGGGSGGG	—
s4	15	GGGPGGGPGGGPGGG	—

[1] Cathepsin E-activating peptides (S1–S10) obtained from primary library selection [9] and spacer peptides (s1–s4) are shown.
[2] Activities were determined using peptides produced by the *in vitro* translation system (in italics) or chemical synthesis (in bold). Note that the former is less reliable than the latter and sometimes exhibits a higher activity than the latter and sometimes adversely systematically [19]. Cathepsin E activity alone was considered as to be 100%.

cathepsin E-activating capability using *in vitro* translation products (Supplementary Figure 1 available online at doi:10.1155/2012/316432). Two peptides selected by this method were analyzed further using chemically synthesized peptides and were found to be clearly superior with regard to

TABLE 2: Amino acid sequences of selected peptides.

Name	Paired blocks[1]	Amino acid sequence (N → C)[2]	Size (a.a.)	Activity (%)[3]	Frequency[4]
T1	S2-(s1)-α	GCPCIDFMVEVQVEVAEALLTALSLSPGL GMTATKGEFQHTGGRY	45	—	6
T2	S2-(s1)-α	SCPCIDFMVEVQVEVAEALLTALSLSPGL GMTATKGEFQHTGGRY	45	—	1
T3	S2-(s1)-α	SCPCIDFMVEVQVEVAEALLTALSLSPGL GMTAT	34	—	6
T4	S2-(s1)-α	IEGRGCPCIDFMVEVQVEVAEALLTALSL SPGS	33	175 (122)	1
T5	S2-(s1)-α	SCPCIDFMVEVQVEVAEALLTALSLSPGL GM	32	—	1
T6	S10-(s1)-α	SDDKSTTLVEVQVEVAEELWRHYHYLLHG	29	118	2
T7	β-(s2)-γ	SYKDSCIGGRGSGGGPGGIPGRIGYIG	25	111	1
T8	β-(s2)-γ	NYKDSCIGGRGSGGGPGGIPGRIGYIG	25	—	1
T9	δ	VFVVGRSCLRLARGRVHFVSG	21	—	1
T10	ε-(s1)-S2	AVDAVLGGDPNLGGHSIGSCG	21	—	1
T11	ζ-(s1)-S1	IEGRLLSGGAGACSVRTVDDSFDCG	26	186 (138)	1

[1] See Table 1 for block names detail. Novel blocks were temporarily labeled as α, β, γ, δ, ε, and ζ.
[2] Bold regions were contained in the original blocks.
[3] Peptides for activity assay were obtained by *in vitro* translation (in italic) or chemical synthesis (in bold). In this study, the discrepancy is unexpectedly large and adverse to our previous experiences due to unknown reason [19].
[4] Copy numbers found in sequenced clones.

their activity or binding affinity than the previously selected products (Figure 2). Peptide T11 had 1.3 times greater activity than the most improved activity peptide (S2) in the second library of selected products. Another peptide, T4, had a very high affinity for cathepsin E (K_d = 2 nM).

It is worth noting that, although the affinity of peptide T11 to cathepsin E was too weak to be determined by the conventional SPR method (two failed trials), peptide T11 did have a high activating capability in a preliminary test. This situation sometimes occurs, as reported for PDK1 activators [20]. Considering that the members of the third library, including those that were unintentionally generated, were a relatively enriched population and they comprised moieties that had already been selected for their cathepsin E activation, the remarkable improvement shown here was expected.

3.3. Biological Activity of a Cathepsin E Activator: Apoptosis Induction. The peptide T11 exhibited the highest cathepsin-E-activating ability; thus, we used it in an apoptosis induction assay. As shown in Figures 3(a) and 3(b), the percentage of dying cells in the presence of T11 was significantly higher than without it. This apoptosis phenomenon was confirmed by detecting of increased levels of apoptosis-associated protein caspase-3/7 (Figure 3(c)). TRAIL-induced-apoptosis, which is assumed to be the pathway responsible for cathepsin-E-mediated cell death [11], results in caspase-mediated apoptosis, including the activity of caspase-3/7. Caspase-3/7 activity was found to be activated by the addition of peptide T11 to a much higher level of apoptosis (188%) than with cathepsin E alone (155%). This phenomenon was not observed when T11 was added without cathepsin E.

Although peptide T11 could be assumed to induce cancer cell apoptosis through its activation of cathepsin E, the actual mechanism may not be so simple. It was previously proposed that cathepsin E cleaved off the soluble TRAIL ligand from the TRAIL precursor protein and that this ligand switched on the apoptosis pathway of cancer cells. There may also be another pathway for cathepsin E-induced apoptosis of cancer cells (i.e., different target for cathepsin E than the TRAIL precursor protein). Our preliminary data may support this proposition, as we attempted to find the cleaved TRAIL molecule when HeLa cells were treated with cathepsin E and the cathepsin E-activating peptide (T11). With these conditions, apoptosis could be induced, although the cathepsin E concentration was significantly lower (77 nM) than that previously reported (1 μM) [11].

4. Discussion

In this study, we attempted to establish and confirm an approach for identifying protease-activating peptides. In general, it is more difficult to identify protease activity-enhancing molecules than activity-inhibiting molecules because no general principles have been established, whereas activity-inhibiting molecules are rather easily obtained by fabricating substrate-analog molecules. Yet, increasing the activity of a particular protease often contributes not only to elucidating molecular systems within cells, such as caspase signal transduction and metabolic pathway regulation, but can also enhance activities, such as antimicrobial or anti-cancer activities [21].

It is likely that diminished protease activity, which leads to the accumulation of unprocessed proteins or a shortage of necessary processed proteins, plays a causative role in various

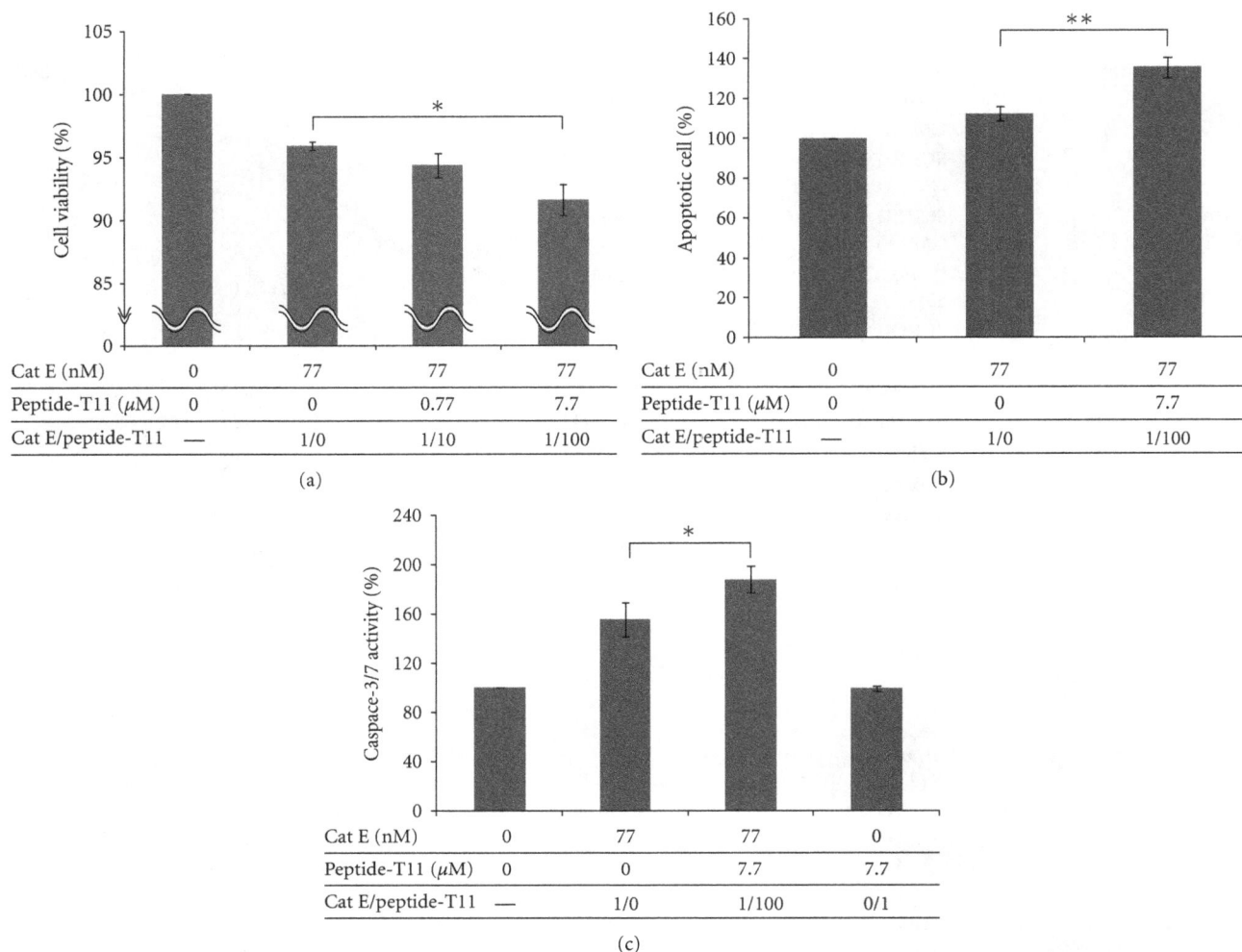

FIGURE 3: Biological effects of a paired peptide on HeLa cells. (a) Induction of cancer cell death by cathepsin E and its enhancement by a peptide. Cell viability was determined using a cell counting kit after treating HeLa cells for 20 h with cathepsin E and peptide T11 at different molar ratios (cathepsin E: peptide = 1 : 0, 1 : 10, 1 : 100). (b) Effect of a cathepsin E-activating peptide on cancer cell apoptosis. Apoptotic cells were stained with Annexin V-Cy3 for 24 h after incubating HeLa cells in the presence of 77 nM cathepsin E and 7.7 μM peptide T11. (c) Assessment of caspase activity induced by cathepsin E-activating peptide. Caspase-3 and/or -7 was measured at 24 h after incubating HeLa cells in the presence of 77 nM cathepsin E and 7.7 μM peptide T11. Error bars indicate the standard deviations of three independent experiments. Statistical significance is denoted by the symbols (* < 0.05, ** < 0.01) and was based on comparison by Student's t-test.

diseases. To treat these diseases, either the amount of the protease needs be increased or the protease activity needs be enhanced. The latter is much easier to accomplish because of the small size and high stability of proteases. Therefore, the aim of this study, to identify cathepsin E-activating peptides was reasonable.

The peptide that we identified here, T11, was associated with inducting cancer cell apoptosis, as shown in Figures 3(a) and 3(b). Cancer cells apoptosis was further confirmed by increased levels of apoptosis-related factors (caspase-3 and/or -7) that were observed in the presence of cathepsin E and its activator T11 (Figure 3(c)). As expected, the peptide alone had no effect on cells. Because tumor growth was suppressed by cathepsin E treatment in a mouse xenograft model [11], T11 administration might be expected to enhance therapeutic efficacy.

Another important finding of this study was the consistent, high performance of *in vitro* evolution reported in a previous study [9]. This was confirmed here by the observed steady improvement in activities of the selected products as the stage of the library progressed from first to second, and from second to third (Table 2 and [9]). It is worth noting that some of the mutation-derived peptides had a higher capability for cathepsin E activation than the non-mutated peptides and ultimately survived the selection-by-function processes. Because this approach could actually find more functional peptides than what was expected, this library that contained unintentionally mutated molecules was effective.

Although this demonstration was rather phenomenological and will require many confirmatory examples before it is sufficiently reliable, the present data together with previously reported data (7 trials; 4 trials regarding cathepsin E inhibition and activation at acidic pH ([8] and Kitamura et al.), [21]), 2 trials regarding cathepsin E activation at neutral pH ([9] and this study) and one about Aβ42-binding peptides [18]) strongly support the idea that the progressive library method is an effective means for identifying activity-enhancing peptides.

5. Conclusions

By adopting the systemic *in vitro* evolution method (eRAPANSY) augmented by function-based screening (selection-by-function), we could identify peptides that had high activating capability for the protease cathepsin E. We also demonstrated the availability of these activators for the induction of cancer cell apoptosis. The same strategy is most likely to be applicable to develop the clinically available peptide activators which are targeted to other proteases.

Acknowledgments

This study was performed as part of the Saitama-Bio Project 3 (REDS3), Central Saitama Area in the Program for Fostering Regional Innovation (City Area Type), supported by MEXT.

References

[1] S. K. Grant, "Therapeutic protein kinase inhibitors," *Cellular and Molecular Life Sciences*, vol. 66, no. 7, pp. 1163–1177, 2009.

[2] B. Turk, "Targeting proteases: successes, failures and future prospects," *Nature Reviews Drug Discovery*, vol. 5, no. 9, pp. 785–799, 2006.

[3] C. Ottmann, P. Hauske, and M. Kaiser, "Activation instead of inhibition: targeting proenzymes for small-molecule intervention," *ChemBioChem*, vol. 11, no. 5, pp. 637–639, 2010.

[4] J. A. Zorn and J. A. Wells, "Turning enzymes on with small molecules," *Nature Chemical Biology*, vol. 6, no. 3, pp. 179–188, 2010.

[5] C. S. Thakur, B. K. Jha, B. Dong et al., "Small-molecule activators of RNase L with broad-spectrum antiviral activity," *Proceedings of the National Academy of Sciences of the United States of America*, vol. 104, no. 23, pp. 9585–9590, 2007.

[6] B. B. Zhang, G. Zhou, and C. Li, "AMPK: an emerging drug target for diabetes and the metabolic syndrome," *Cell Metabolism*, vol. 9, no. 5, pp. 407–416, 2009.

[7] T. Ishii, K. Fukano, K. Shimada et al., "Proinsulin C-peptide activates α-enolase: implications for C-peptide cell membrane interaction," *Journal of Biochemistry*, vol. 152, no. 1, pp. 53–62, 2012.

[8] K. Kitamura, C. Yoshida, Y. Kinoshita et al., "Development of systemic in vitro evolution and its application to generation of peptide-aptamer-based inhibitors of cathepsin E," *Journal of Molecular Biology*, vol. 387, no. 5, pp. 1186–1198, 2009.

[9] M. Biyani, M. Futakami, K. Kitamura et al., "In vitro selection of cathepsin E-activity-enhancing peptide aptamers at neutral pH," *International Journal of Peptides*, vol. 2012, Article ID 834525, 10 pages, 2012.

[10] M. Chlabicz, M. Gacko, A. Worowska, and R. Lapinski, "Cathepsin E (EC 3. 4. 23. 34)-a review," *Folia Histochem. Cytobiol*, vol. 49, pp. 547–557, 2011.

[11] T. Kawakubo, K. Okamoto, J. I. Iwata et al., "Cathepsin E prevents tumor growth and metastasis by catalyzing the proteolytic release of soluble TRAIL from tumor cell surface," *Cancer Research*, vol. 67, no. 22, pp. 10869–10878, 2007.

[12] A. Yasukochi, T. Kawakubo, S. Nakamura, and K. Yamamoto, "Cathepsin E enhances anticancer activity of doxorubicin on human prostate cancer cells showing resistance to TRAIL-mediated apoptosis," *Biological Chemistry*, vol. 391, no. 8, pp. 947–958, 2010.

[13] M. Shin, T. Kadowaki, J. I. Iwata et al., "Association of cathepsin E with tumor growth arrest through angiogenesis inhibition and enhanced immune responses," *Biological Chemistry*, vol. 388, no. 11, pp. 1173–1181, 2007.

[14] I. Tabuchi, S. Soramoto, N. Nemoto, and Y. Husimi, "An in vitro DNA virus for in vitro protein evolution," *FEBS Letters*, vol. 508, no. 3, pp. 309–312, 2001.

[15] D. Bromme and K. Okamoto, "Human cathepsin O2, a novel cysteine protease highly expressed in osteoclastomas and ovary molecular cloning, sequencing and tissue distribution," *Biological Chemistry Hoppe-Seyler*, vol. 376, no. 6, pp. 379–384, 1995.

[16] K. Kitamura, Y. Kinoshita, S. Narasaki, N. Nemoto, Y. Husimi, and K. Nishigaki, "Construction of block-shuffled libraries of DNA for evolutionary protein engineering: Y-ligation-based block shuffling," *Protein Engineering*, vol. 15, no. 10, pp. 843–853, 2003.

[17] M. Naimuddin, K. Kitamura, Y. Kinoshita et al., "Selection-by-function: efficient enrichment of cathepsin E inhibitors from a DNA library," *Journal of Molecular Recognition*, vol. 20, no. 1, pp. 58–68, 2007.

[18] S. Tsuji-Ueno, M. Komatsu, K. Iguchi et al., "Novel High-affinity Aβ-binding peptides identified by an advanced in vitro evolution, progressive library method," *Protein and Peptide Letters*, vol. 18, no. 6, pp. 642–650, 2011.

[19] K. Kitamura, C. Yoshida, M. Salimullah et al., "Rapid in vitro synthesis of pico-mole quantities of peptides," *Chemistry Letters*, vol. 37, no. 12, pp. 1250–1251, 2008.

[20] A. Stroba, F. Schaeffer, V. Hindie et al., "3,5-Diphenylpent-2-enoic acids as allosteric activators of the protein kinase PDK1: structure-activity relationships and thermodynamic characterization of binding as paradigms for PIF-binding pocket-targeting compounds," *Journal of Medicinal Chemistry*, vol. 52, no. 15, pp. 4683–4693, 2009.

[21] E. Leung, A. Datti, M. Cossette et al., "Activators of cylindrical proteases as antimicrobials: identification and development of small molecule activators of ClpP protease," *Cell*, vol. 18, no. 9, pp. 1167–1178, 2011.

11

Large Scale Solid Phase Synthesis of Peptide Drugs: Use of Commercial Anion Exchange Resin as Quenching Agent for Removal of Iodine during Disulphide Bond Formation

K. M. Bhaskara Reddy,[1] **Y. Bharathi Kumari,**[2] **Dokka Mallikharjunasarma,**[1] **Kamana Bulliraju,**[1] **Vanjivaka Sreelatha,**[1] **and Kuppanna Ananda**[1]

[1] Chemical Research Division, Mylan Laboratories Ltd., Anrich Industrial Estate, Bollaram, Hyderabad 502325, India
[2] Department of Chemistry, College of Engeenering, Jawaharlal Nehru Technological University Hyderabad, Kukatpally, Hyderabad 500085, India

Correspondence should be addressed to Kuppanna Ananda, ananda.kuppanna@mylan.in

Academic Editor: John D. Wade

The S-acetamidomethyl (Acm) or trityl (Trt) protecting groups are widely used in the chemical synthesis of peptides that contain one or more disulfide bonds. Treatment of peptides containing S-Acm protecting group with iodine results in simultaneous removal of the sulfhydryl protecting group and disulfide formation. However, the excess iodine needs to be quenched or adsorbed as quickly as possible after completion of the disulfide bond formation in order to minimize side reactions that are often associated with the iodination step. We report here a simple method for simultaneous quenching and removal of iodine and isolation of disulphide bridge peptides. The use of excess inexpensive anion exchange resin to the oxidized peptide from the aqueous acetic acid/methanol solution affords quantitative removal of iodine and other color impurities. This improves the resin life time of expensive chromatography media that is used in preparative HPLC column during the purification of peptide using preparative HPLC. Further, it is very useful for the conversion of TFA salt to acetate in situ. It was successfully applied commercially, to the large scale synthesis of various peptides including Desmopressin, Oxytocin, and Octreotide. This new approach offers significant advantages such as more simple utility, minimal side reactions, large scale synthesis of peptide drugs, and greater cost effectiveness.

1. Introduction

Many naturally occurring peptides contain intradisulphide bridges, which play an important role in biological activities. Disulfide bond-containing peptides have long presented a particular challenge for their chemical synthesis [1]. There are many ways to form a disulfide bridge in solution, and solid phase synthesis is well known and widely used in peptide community. Solution phase cyclization is most commonly carried out using air oxidation and or mild basic conditions [2]. The cyclization of free thiols by air oxygen usually leads to low yields of target peptides (10–15%) [3]. The application of potassium ferricyanide [4] or dimethyl sulphoxide [5] usually results in homogeneous reaction mixtures, and yields of cyclic peptides are considerably higher.

However, a multistage purification of peptide is necessary for the removal of excess of these oxidative reagents [6]. Beginning with the initial discovery by Kamber et al. [7] that in peptides, where thiols are protected by Acm groups, iodine offered the potential to carry out removal and simultaneous oxidative disulphide bond formation in one-single step, several disulphide-bonded peptides have been synthesized using this strategy [8]. Unfortunately, side reactions are often associated with this reaction including, for example, iodination of some sensitive residues such as Tyr, Met, and Trp [9].

To help limit these, excess iodine should be quenched or adsorbed as quickly as possible after completion of the disulphide bond formation by addition of sodium bisulfite [9], sodium thiosulfate [10], ascorbic acid [11], powdered zinc dust, activated charcoal [12], or by dilution with

Large Scale Solid Phase Synthesis of Peptide Drugs: Use of Commercial Anion Exchange Resin as Quenching Agent for Removal of Iodine during Disulphide Bond Formation

75

water followed by extraction with carbon tetrachloride [13]. However, the quenching reagents themselves can sometimes cause additional side reactions including formation of thiosulfate adduct of the peptide [11]. The large volumes of highly diluted aqueous peptide solutions, post-Acm removal, can also encumber the subsequent RP-HPLC purification. Further, lyophilization of the peptide solution can result in concentration of unremoved iodine leading to additional side reactions. Thus, improved method for quenching of the iodine and subsequent isolation of disulphide peptides are desirable.

Ion exchange resins are insoluble polymers that contain acidic or basic functional groups and have the ability to exchange counter ions within aqueous solutions surrounding them. Anion exchange resins have positively charged functional groups, and there exchanges negatively charged ions. While the best-known usage of ion exchange resins is in water treatment, pharmaceutical applications were recognized in the early 1950's when amberlite IRC-50 was used in successful purification of streptomycin. Over the years, ion exchange resins have found use as pharmacologically active ingredients in drug formulations. However, the ion exchange resins have also been used in pharmaceutical manufacturing for the isolation and purification of drugs and catalysis of reactions [14, 15].

Anionexchange resins are widely used to remove acidic molecules from solution. They are used for scavenging excess reagents, and byproducts have been exploited in the solution phase synthesis of chemical libraries [16]. Ion exchange resins are synthetic polymeric materials that contain basic or acidic groups that are able to interact with ionizable molecules to create insoluble complexes. A typical strongly basic anion exchange resin will have quaternary ammonium groups attached to a styrene and divinylbenzene copolymer [17]. It has been reported by many workers that polyiodide ions, such as triiodide and pentaiodide, exhibit extremely high affinity for quaternary ammonium strong-base anionexchange resins in aqueous solutions [18]. Here we report, the utility of the indions 830-S [19] commercially available, strongly basic quaternary ammonium anion exchange resin [20] as quenching agent to remove excess of iodine and iodide during disulphide bridge formation. It is further used for the salt exchange (TFA salt to Acetate) simultaneously in a single step.

2. Materials and Methods

All amino acid derivatives and Rink amide resin were purchased from GL Biochem china.Scavengers, coupling agents and cleavage reagents [N, N-diisopropylcarbodiimide (DIC), N-diisopropyl-ethylamine (DIEA), 1-Hydroxybenzotriazole (HOBt), triisopropylsilane (TIS), piperidine, trifluoroacetic acid (TFA)] were Fluka products. Oxyma and COMU were procured from Sigma-Aldrich. Indion 830-S resin was purchased from Ion Exchange India Pvt Limited. The solvents used were all commercial grade.

Peptide Synthesis. Peptides were synthesized manually by solid phase method by stepwise coupling of Fmoc-amino

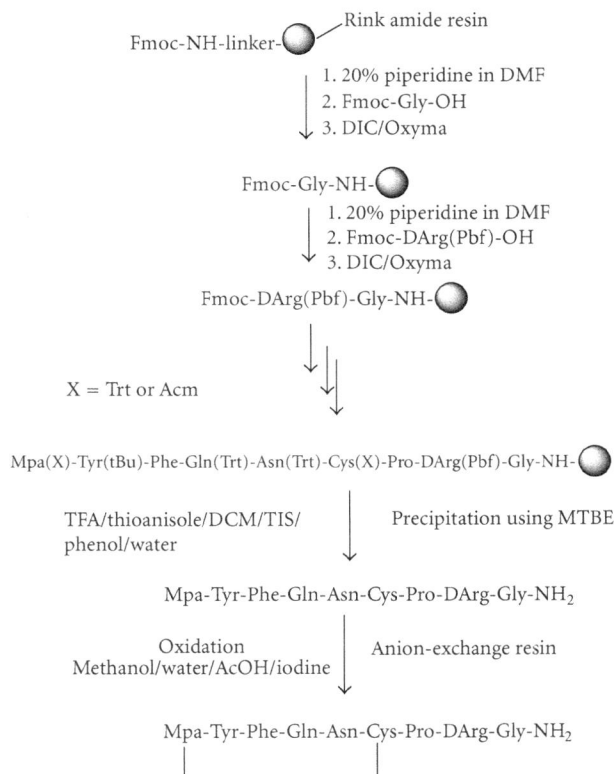

FIGURE 1: Scheme for solid phase synthesis of Desmopressin.

acids to the growing chain on a Rink amide resin (0.9–1.1 mmol/g, Figures 1 and 3 for Desmopressin and Oxytocin) and 2-chlorotrityl chloride resin (Figure 2, for Octreotide). The Fmoc-amino acids were used with the Trt and the Pbf as side chain protecting groups for Asn, Gln, Cys, and Arg, respectively. The couplings were done using DIC/HOBt, in a mixture of DMF/NMP (1 : 1), in most of the cases, the amino acids were coupled two fold excess. The completeness of each coupling reaction was monitored by Kaiser or Chloranil test. The individual coupling steps, if showing low coupling efficiency must be repeated prior to proceeding for deprotection and coupling with next amino acid of the sequence.

Octreotide was synthesized manually on 2-chlorotrityl chloride resin (substitution 0.90 mmol/g) by standard Fmoc solid phase synthesis strategy (Figure 2). Fmoc-Thr(tBu)-OL/Fmoc-Thr-OL was treated with the swelled 2-CTC resin in DCM in the presence of DIEA, and substitution level was determined by weight gain measurements and also by UV Method. It was found that Fmoc-Thr-OL gives better loading than Fmoc-Thr(tBu)-OL. This may be due to steric hindrance of t-butyl group. After the coupling of the first amino acid onto the resin, the unreacted linkers on the resin (polymer) are protected, to avoid the undesired peptide chain formation, with a solution of 5% DIEA and 10% methanol in DCM. This process of capping was performed after anchoring the first protected amino acid to the resin.

The complete synthesis was achieved by stepwise coupling of Fmoc-Amino acids to the growing peptide chain on

2-Chlorotrityl chloride resin

Fmoc-Thr(tBu)-OL
DIEA/DCM, 2 hrs

Fmoc-Thr(tBu)-O-resin

1. MDC/MeOH/DIEA(17: 2: 1)
2. Piperidine/DMF
3. Amino acid, DIC/Oxyma
4. Repeat steps 2 and 3

Boc-D-Phe-Cys(Acm)-Phe-D-Trp(Boc)-Thr(tBu)-Cys(Acm)-Thr(tBu)-O-resin

90% TFA, 5% water, 5% TIS

D-Phe-Cys(Acm)-Phe-D-Trp-Lys-Thr-Cys(Acm)-Thr-OL

Oxidation   Precipitation using MTBE
MeOH/water/AcOH/   anion exchange resin
iodine in MeOH

D-Phe-Cys-Phe-D-Trp-Lys-Thr-Cys-Thr-OL
         | |
         S——————————————S
                    Octreotide

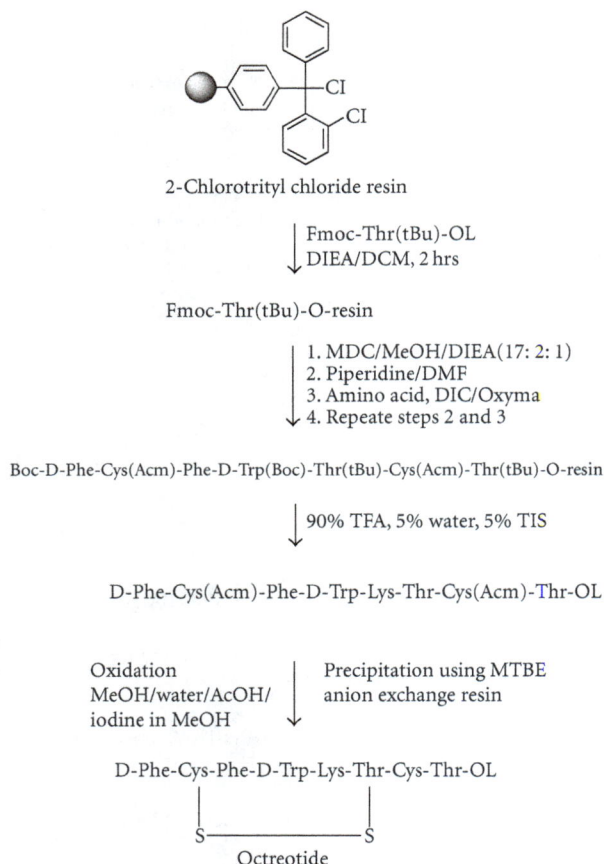

FIGURE 2: Solid phase synthesis of Octreotide.

Fmoc-NH-linker-○ ⟵ Rink amide resin

1. 20% piperidine inDMF
2. Fmoc-Gly
3. DIC/HOBt

Fmoc-Gly-NH-○

1. 20% piperidine in DMF
2. Fmoc-Leu
3. DIC/Oxyma

Fmoc-Leu-Gly-NH-○

Cys(Acm)-Tyr(tBu)-Ile-Gln(Trt)-Asn(Trt)-Cys(Acm)-Pro-Leu-Gly-NH-○
TFA/EDT/thioanisole/DCM/TIS | Precipitation using MTBE
2 hrs RT

Cys(Acm)-Tyr-Ile-Gln-Asn-Cys(Acm)-Pro-Leu-Gly-NH₂

Oxidation
MeOH/AcOH/iodine | Precipitation using MTBE

Cys-Tyr-Ile-Gln-Asn-Cys-Pro-Leu-Gly-NH₂
  | |
  S————————————————S

FIGURE 3: Solid phase synthesis of Oxytocin.

the resin. All the couplings were carried out in DMF. The couplings were performed by dissolving the Fmoc-Amino acid (2 eq.) and HOBt (2 eq.) in DMF, and then DIC (2 eq.) was added. The reaction mixture was added to the resin and allowed to stir for 2 hrs. The efficiency of the coupling was monitored using the Kaiser Ninhydrin test. The coupling step was repeated if Kaiser test was found positive.

2.1. Cleavage of Peptide from Resin Along with Global Deprotection. The Octreotide resin (20 g) was swelled in DCM (50 mL) for 15 to 20 minutes under nitrogen at 25–30°C. The cocktail mixture (TFA (180 mL), water (10 mL), and TIS (10 mL)) was added to the resin and stirred for 2.5 hours at room temperature under nitrogen atmosphere. The reaction mixture was filtered and washed the resin with TFA (25 mL). The obtained filtrate was added into cold MTBE (500 ml, precooled to a temperature of 0–5°C) under stirring and allowed the temperature to rise more than 5°C. The reaction mixture was stirred for 45 minutes at 0–5°C. The obtained suspension was filtered, washed the solid with MTBE (500 mL), and dried the solid under nitrogen.

2.2. Purification of Mpa-Tyr-Phe-Gln-Asn-Cys-Pro-DArg-Gly-NH₂. Peptide thiol (5 g) was slurried in a mixture of ethyl acetate as follows: ethanol (95 : 5) at 0°C for 1 hour. The

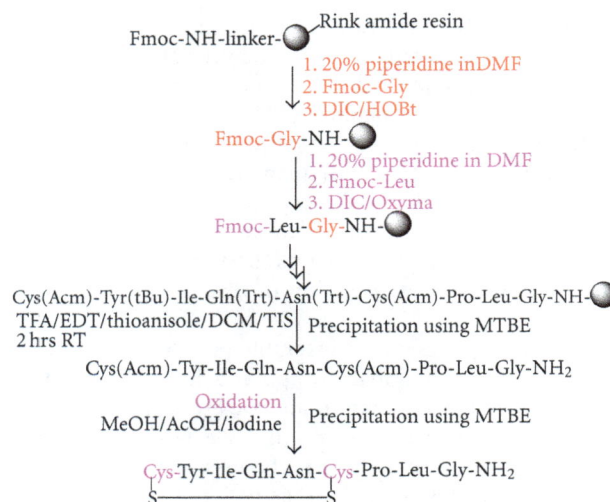

reaction mixture was filtered and washed with ethyl acetate to afford 4.5 g of pure peptide thiol (~95%).

2.3. Conversion of Thiol to Disulphide. A 2% solution of iodine in methanol (12 mL) was added portion wise to a solution of the peptide free thiol (1 g) in 10% aq AcOH (10 mL), and the mixture was stirred at room temperature for 60 min. Reaction completion was confirmed by RP-HPLC. Anion exchange resin (10 g, Cl⁻ form) was added, and stirring was continued for 30 min. The suspension was then filtered to remove the resin, and the resin was washed further with additional small amounts of water. The combined filtrates were evaporated; the residue was suspended in water and lyophilized. The obtained products were analyzed by RP-HPLC and ESI-MS.

2.4. Disulphide Bridge Formation with Acm Protecting Groups. To a solution of semiprotected peptide in 80% aq methanol, was added slowly 10% solution of iodine in methanol till yellow color persists. The reaction is stirred at room temperature for 60 min. Reaction completion was confirmed by RP-HPLC. The excess of iodine was quenched with anion exchange resin, and stirring was continued for 30 min. The suspension was then filtered to remove the resin, and the resin was washed further with additional small amounts of water. The combined filtrates were evaporated; the residue was suspended in water and lyophilized. The obtained products were analyzed by RP-HPLC and ESI-MS.

2.5. Procedure for Conversion of Ion Exchange Resin to Acetate Form. Indion 830-S anion exchange resin (Chloride form, Cl⁻) was converted to acetate form by washing with IM aq NaOH, water, acetic acid, water, and methanol. The resin was dried and stored.

2.6. Procedure for Salt Exchange. The peptide in its trifluoroacetate form was dissolved in methanol water and added the anion exchange resin (acetate form) and stirred for

Large Scale Solid Phase Synthesis of Peptide Drugs: Use of Commercial Anion Exchange Resin as Quenching Agent for
Removal of Iodine during Disulphide Bond Formation

77

15 min. The resin is filtered and washed with methanol. The methanol is evaporated, and the product is lyophilized and monitored the acetate content by HPLC (Assay). The acetate content usually in the range of 4–6%.

2.7. Preparative HPLC Purification of Disulphide Bridge Peptides. The crude peptides were loaded on to preparative C18 column (50 × 250 mm, 100 A°). The peptide was purified on a preparative HPLC using gradient method by elution with a gradient comprising of buffer A: Water/Acetic acid (0.05%) and buffer B: Methanol/Acetic acid (0.05%). The peptides were eluted at around 30% of methanol, during the elution the fractions are collected at regular intervals. The collected fractions are assayed by HPLC to determine the purities, and fractions with desired purities are pooled together. The methanol was evaporated and the aqueous layer was lyophilized to obtain the peptides as white solid. The purified peptides were analyzed by RP-HPLC and mass determined by mass spectrometer. The purification achieved by this method utilizes the desired salt in a single purification step avoids the additional desalting step. Analytical HPLC was carried out on a 150 × 4.6 mm, 5 micron Purospher RP-18 column, flow rate 1 mL/min using a gradient conducted over 50 min. of 0.1% aqueous TFA and 0.1% TFA-CH_3CN from 10–90%.

3. Results and Discussion

Disulphide bridges play a crucial role in the folding and structural stabilization of many peptide and protein molecules, including many enzymes and hormones. The simplest approach to prepare disulphide-containing peptides involves complete deprotection and then careful oxidation to the properly folded product (Figure 1). The crude cleaved peptide generally needs to be treated with suitable reducing agents to ensure disulphide bridge is formed.

The resulting dithiols after cleavage of the peptide from resin was cyclized with a 0.1M I_2 in methanol using a standard procedure. The reaction was monitored by RP-HPLC, and after completion of disulphide bond formation as evidenced by the disappearance of the starting peptide, the reaction is quenched with an anion exchange resin and stirred for 30 min's. The reaction turns colorless; the liberated iodide ions are adsorbed or exchanged to quaternary ammonium exchange resin [R-N^+(CH_3)$_3I^-$]. The resin is filtered and washed with methanol/water. The solvent is evaporated under reduced pressure, and the resulting residue is directly taken for preparative HPLC purification or dissolved in water and lyophilized.

The peptides (crude) were isolated (Table 2) in ~80% yield by iodine oxidation followed by quenching with an anion exchange resin. The yield of peptides after preparative HPLC purification was 50–60% recovery of the peptides. There is no yield loss as confirmed by washing the used anion exchange resin with a mild base.

The ion exchange resin was washed thoroughly with water and methanol before use. The resin prewash step is necessary to ensure that no extractable material from resin contaminated the product when the later was mixed with the resin. Indion 830-S was stirred with water and methanol for 30 min's and filtered through a Buchner funnel. It was further washed with water and methanol and dried. The free thiol is added portion wise to a cooled mixture composed of (methanol : acetic acid : water) and stirred for 15 min at RT. Add iodine solution in methanol to the reaction slowly up to yellow color persists. The reaction is stirred for another 1 hr at room temperature and monitored by HPLC for disulphide bridge formation. No dimmer was detected as confirmed by HPLC analysis. After completion, the washed anion exchange resin was added into the iodine solution of the disulphide forming reaction mixture and stirred for 15–30 min's. The reaction mixture shows colorless suggesting quantitative removal of iodine and iodide ions [21]. It was filtered through Buchner funnel, and the solvent was evaporated (Residue). The residue was directly injected into preparative HPLC or by dissolving the residue with water and isolated the peptide by lyophilization.

Iodine has proven its utility in forming disulphide bonds in Acm or trityl protected peptides. However, as soon as the disulphide bonds have been made, a safe practice is to quench the reactive iodine so as not to allow any side reaction on the peptide. One such common side reaction is the iodination of the phenolic side chain of tyrosine. Iodine can be quenched with thiosulphate or ascorbic acid, the quenched reaction mixture was lyophilized, and salts removed by gel filtration or RP-HPLC. However, caution must be exercised since unexpected side reactions have been observed following the use of quenching agents.

Our initial attempts to use ascorbic acid as quenching agent lead to less pure peptides compared to the present method. The yields are low (Table 2), and the main issue is that during preparative HPLC purification, due to ascorbic acid, the separation was not good for higher loading of the peptide to preparative column.

In the present study, it has been established that an ion exchange resin was used to quench the excess of iodine used for disulphide bridge formation, there is no need to remove the salts by gel filtration, the reaction is neat and clean circumvents, the formation of impurities. Since an intramolecular disulphide must be made in conditions of high dilution, the postreaction handling of iodine and the cyclized peptides needs special consideration. A conventional gel filtration experiment is impractical. The resin quenching method is especially useful for the scale up of peptide drugs on a large scale. The method is utilized in the large scale manufacture of peptide drugs like desmopressin, Octreotide, and Oxytocin.

Cyclization was performed by additions of portions of the crude peptide to aqueous acetic acid methanol and stirred for 10 min's. 0.1M I_2 in methanol was added slowly till a light yellow solution was obtained, and the peptide concentration was adjusted to 350 mL/g of the peptide with methanol. The cyclizations were tried with different concentrations and found that 350 mL/g was ideal as there was no dimmer/isomer formation. After the disulphide bridge was formed as monitored by HPLC. Anion exchange resin (chloride form) was added and stirring was continued for 30 min to remove iodide and excess of iodine. It was filtered

TABLE 1: Resins of different loading capacity tried during Desmopressin scale up.

Resin	Particle size (μm)	Matrix	Loading (mmol/g)
Tentagel SRAM	90	Poly(oxyethylene)-RAM Polymer bound	0.24
Rink amide resin	100–200	Amino methyl polystyrene crosslinked with 1% DVB	1.1
Rink amide resin	100–200	Amino methyl polystyrene crosslinked with 1% DVB	0.43
2-chlorotrityl chloride resin	100–200	Polystyrene crosslinked with 1% DVB	0.7–0.9

TABLE 2: Recovery yield of disulphide peptide quenched with anion exchange resin.

Peptide	Peptide conc (mL/g)	MeOH (%)	Crude purity (%)	Yield (%) crude	Yield (%) purified	Quenching agent
Desmopressin (Thiol)	350	85	90	80	50	Anion exchange resin
Desmopressin (Acm)	300	80	85	78	48	Anion exchange resin
Octreotide	200	95	88	85	58	Anion exchange resin
Oxytocin	450	80	82	82	47	Anion exchange resin
Desmopressin (Acm)	300	80	69	70 (including ascorbic acid)	30	Ascorbic acid
Octreotide	200	95	70	72 (including ascorbic acid)	32	Ascorbic acid

through Buckner funnel, and the filtrate is evaporated. The residue is dissolved in water and lyophilized or directly taken for preparative HPLC purification. The purity of the peptide was assayed by HPLC and characterized by LCMS.

The absence of iodine peak as monitored by HPLC (Figure 5) was found to be less than 0.5%. In an experiment that was carried out without the complete removal of iodine in the crude peptide, we found that even after purification using preparative HPLC the iodine peak was present (Figure 4).

Further, exchange resin is stirred with 0.1M iodine in methanol for 10 min's. The TLC analysis showed there are not even traces of iodine present; the resin beads were brown in color and completely adsorbed by the resin.

The utility of other ion exchange resins, like Indion 860-S and Deuolite A-68, were tried with limited successes. It is found that a strongly, basic anion exchange resin is required for complete removal of iodine. Weak basic anion exchange resins were found to be less effective for the removal of iodine.

The peptides synthesized on solid phase were cleaved using TFA leading to the formation of trifluoroacetate salt. Most of the peptide drugs in the market are having acetate salt. The salt exchange of trifluoroacetate to acetate was done using the anion exchange resin. The anion exchange resin after converting to its acetate form was added to peptide trifluoroactate salt dissolved in methanol/water and stirred for 30 mins and filtered, and the filtrate was lyophilized. The acetate was confirmed by HPLC (assay) and the IR for the absence of C-F stretching mode (1110–1200 cm^{-1}). It is further confirmed by F NMR on the TFA peptide salt and compared the signal with an equivalent amount of TFA. The acetate peptide salt was at same concentration and compared the fluorine strength (Figure 6). Thus, the salt exchange is

100% as there is no signal corresponding to fluorine in acetate peptide.

The resin used in the process of the preparation of desmopressin acts as support material and is selected from Tentagel SRAM, 2-Chlorotrityl chloride resin (2-CTC), Rink amide resin (0.43 mmol/g), and Rink amide resin (1.1 mmol/g). The selection of polymeric support and attached linker is very critical for overall outcome of the solid phase peptide synthesis. The Rink amide resin (1.1 mmol/g) with amino methyl polystyrene type linker used in present application provides additional advantages over the other resins. Tentagel resin is also found to be very effective for the preparation of desmopressin and is comprising of grafted copolymers consisting of a low cross-linked polystyrene.

The advantage of using high load resin is that significantly more peptide for unit measure of beads could be produced with high load resins. This is a consequence of the fact that higher concentrations of reagents and reactants can be achieved with high load resins. Smaller vessel sizes could be employed to generate a given amount of peptide and at least 50% less wash solvents needed while using high loaded resins. For the scale up of the solid phase it is important to reduce the large amounts of reagents typically employed in solid phase peptide synthesis.

The different resins (Table 1) were tried in desmopressin scale up optimization on large scale synthesis and found that Rink amide resin with 0.9–1.1 mmol/g was more useful for scale up in terms of yield and cost.

An ion exchange resin is a cost effective and used extensively for purifying water and food processing. Therefore, the use of ion exchange resin in the large scale peptide drug process is an advantageous. In comparison with the reported methods for the large scale synthesis of desmopressin, our method found to be advantageous in terms of scale up as

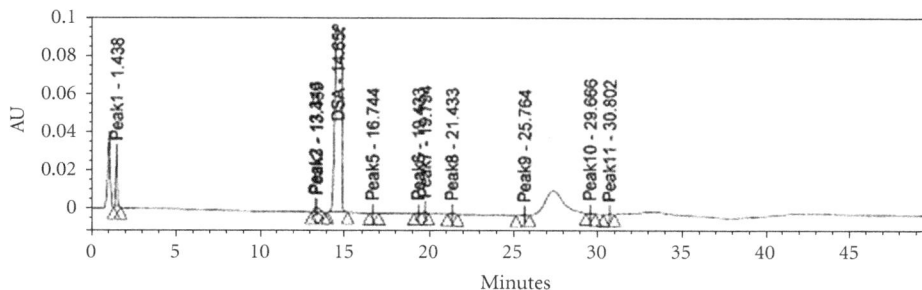

Peak results

	Name	RT	Area	Area (%)	RT ratio
1	Peak1	1.438	181013	1.77	0.1
2	Peak2	13.348	34169	0.33	0.91
3	Peak3	13.439	27111	0.27	0.92
4	Peak4	14.656	9907360	97.09	
5	Peak5	16.744	2369	0.02	1.14
6	Peak6	19.433	3664	0.04	1.33
7	Peak7	19.794	17915	0.18	1.35
8	Peak8	21.433	10123	0.1	1.46
9	Peak9	25.764	8748	0.09	1.76
10	Peak10	29.666	6926	0.07	2.02
11	Peak11	30.802	5073	0.05	2.1
Sum				100	

FIGURE 4: HPLC Chromatograph of desmopressin with incomplete removal of iodine In such cases, the peptide dissolved in water added the resin and stirred for 5 min's filtered and Lyophilized. The HPLC analysis (Figure 5) showed not even traces of iodine present.

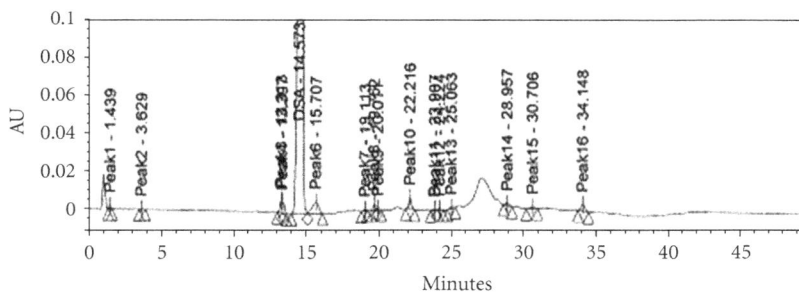

Peak results

	Name	RT	Area	Area (%)	RT ratio
1	Peak1	1.439	11308	0.05	0.1
2	Peak2	3.629	2367	0.01	0.25
3	Peak3	13.313	71514	0.3	0.921
4	Peak4	13.397	57166	0.24	0.92
5	Peak5	14.573	23378791	97.45	
6	Peak6	15.707	189734	0.79	1.08
7	Peak7	19.113	7438	0.03	1.31
8	Peak8	19.762	34954	0.15	1.36
9	Peak9	20.011	10188	0.04	1.37
10	Peak10	22.216	79715	0.33	1.52
11	Peak11	23.907	7154	0.03	1.64
12	Peak12	24.224	11554	0.05	1.68
13	Peak13	25.063	10247	0.04	1.72
14	Peak14	28.957	23374	0.1	1.99
15	Peak15	30.706	23344	0.1	2.11
16	Peak16	34.148	71760	0.3	2.34
Sum				100	

FIGURE 5: HPLC Chromatograph of desmopressin with complete removal of iodine.

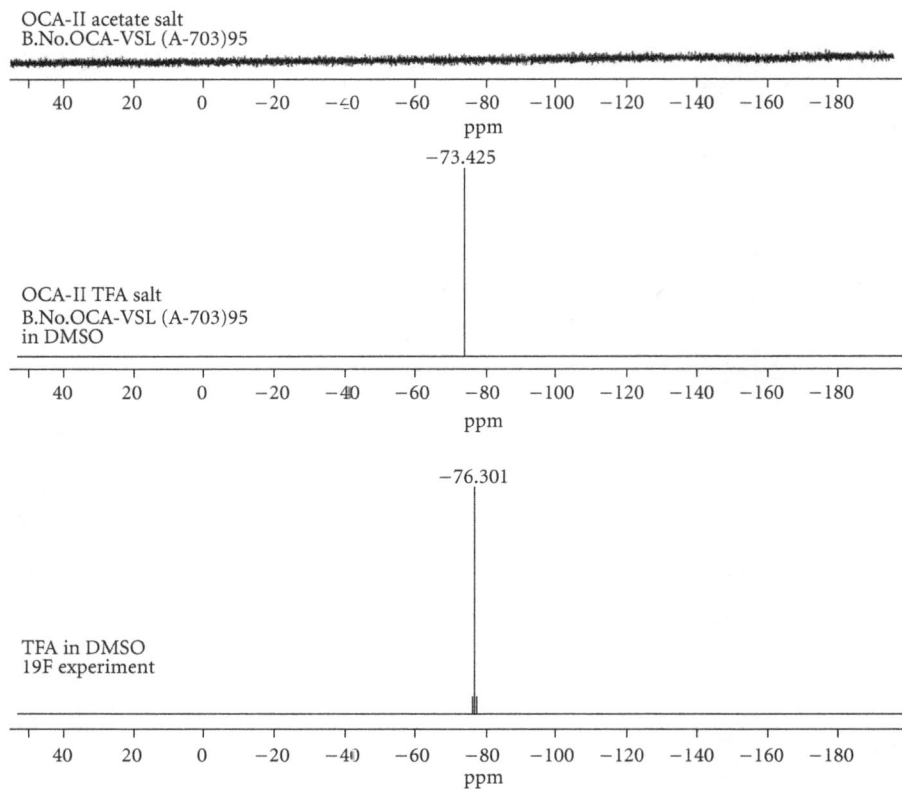

FIGURE 6: F NMR comparision of Octreotide acetate and Octreotide Trifluoro acetate.

well as isolation of peptide. The reported method utilizes two step purification process for crude peptide and a separate step for salt exchange. The present method describes the use of high loading resin for the synthesis, and anion exchange resin for iodine quenching makes the method more robust for the synthesis of peptide drugs. The yields as well as purities of the peptides were good when compared with other quenching agents. It offers a number of attractive advantages such as ease of application, minimal side reactions, and cost effectiveness.

4. Conclusion

A new and simple method for simultaneous quenching and removal of excess of iodine and iodide ions from aqueous solution containing iodine and iodide ions and peptide isolation has been developed. The strong base anion exchange resin is more effective than the weak anion exchange resins. This simple method of quenching of iodine has been successfully applied to the synthesis of several peptides including Desmopressin, Oxytocin, and Octreotide. This method is very useful in large scale synthesis especially for disulphide bridge peptide API's to produce on a commercial scale.

Abbreviations

Acm: Acetimidomethyl
Trt: Trityl
TFA: Trifluoroacetic acid

Pbf: 2, 2, 4, 6, 7-pentamethyl-dihydrobenzofuran-5-sulfonyl
MTBE: Methyltert-butyl ether
CTC: Chlorotrityl chloride resin
Tyr: Tyrosine
Met: Methionine
Trp: Tryptophan.

Acknowledgments

The authors express their sincere thanks to Mylan laboratories Limited, Hyderabad, India, for supporting this research work. Authors, also thankful to authorities of Jawaharlal Nehru Technological University, Hyderabad.

References

[1] I. Annis, B. Hargittai, and G. Barany, "Disulfide bond formation in peptides," *Methods in Enzymology*, vol. 289, pp. 198–221, 1997.

[2] C. Kellenberger, H. Hietter, and B. Luu, "Regioselective formation of the three disulfide bonds of a 35-residue insect peptide," *Peptide Research*, vol. 8, no. 6, pp. 321–327, 1995.

[3] E. V. Kudryavtseva, M. V. Sidorova, and R. P. Evstigneeva, "Some peculiarities of synthesis of cysteine-containing peptides," *Russian Chemical Reviews*, vol. 67, no. 7, pp. 545–562, 1998.

[4] R. Eritja, J. P. Ziehler-Martin, P. A. Walker et al., "On the use of s-t-butylsulphenyl group for protection of cysteine in

solid-phase peptide synthesis using fmoc-amino acids," *Tetrahedron*, vol. 43, no. 12, pp. 2675–2680, 1987.

[5] J. P. Tam, C. R. Wu, W. Liu, and J. W. Zhang, "Disulfide bond formation in peptides by dimethyl sulfoxide. Scope and applications," *Journal of the American Chemical Society*, vol. 113, no. 17, pp. 6657–6662, 1991.

[6] D. Andreu, F. Albericio, N. A. Sole, M. C. Munson, M. Ferrer, and G. Barany, "In methods in molecular biology, formation of disulphide bonds in synthetic peptides and proteins," in *Peptide Synthesis Protocols*, M. W. Penington and B. M. Dunn, Eds., vol. 35, pp. 91–169, Humana Press, Totowa, NJ, USA, 1994.

[7] B. Kamber, A. Hartmann, K. Eisler et al., "Synthesis of cystiene peptides by iodine oxidation of S-trityl-cystiene," *Helvetica Chimica Acta*, vol. 63, no. 4, pp. 899–915, 1980.

[8] S. Zhang, F. Lin, M. A. Hossain, F. Shabanpoor, G. W. Tregear, and J. D. Wade, "Simultaneous post-cysteine(S-Acm) group removal quenching of iodine and isolation of peptide by one step ether precipitation," *International Journal of Peptide Research and Therapeutics*, vol. 14, no. 4, pp. 301–305, 2008.

[9] M. Engebretsen, E. Agner, J. Sandosham, and P. M. Fischer, "Unexpected lability of cysteine acetamidomethyl thiol protecting group: tyrosine ring alkylation and disulfide bond formation upon acidolysis," *Journal of Peptide Research*, vol. 49, no. 4, pp. 341–346, 1997.

[10] W. L. Mendelson, A. M. Tickner, M. M. Holmes, and I. Lantos, "Efficient solution phase synthesis of {1-(β-mercapto-β,β-cyclopentamethylenepropionic acid}-2-(O-ethyl-D-tyrosine)-4-valine-9-desglycine]arginine vasopressin," *International Journal of Peptide and Protein Research*, vol. 35, no. 3, pp. 249–257, 1990.

[11] M. Pohl, D. Ambrosius, J. Grotzinger et al., "Cyclic disulfide analogues of the complement component C3a. Synthesis and conformational investigations," *International Journal of Peptide and Protein Research*, vol. 41, no. 4, pp. 362–375, 1993.

[12] E. V. Kudryavtseva, M. V. Sidorova, M. V. Ovchinnikov, Z. D. Bespalova, and V. N. Bushuev, "Comparative evaluation of different methods for disulfide bond formation in synthesis of the HIV-2 antigenic determinant," *Journal of Peptide Research*, vol. 49, no. 1, pp. 52–58, 1997.

[13] D. Sahal, "Removal of iodine by solid phase adsorption to charcoal following iodine oxidation of acetamidomethyl-protected peptide precursors to their disulfide bonded products: oxytocin and a Pre-S1 peptide of hepatitis B virus illustrate the method," *Journal of Peptide Research*, vol. 53, no. 1, pp. 91–97, 1999.

[14] L. Chen, H. Bauerová, J. Slaninová, and G. Barany, "Syntheses and biological activities of parallel and antiparallel homo and hetero bis-cystine dimers of oxytocin and deamino-oxytocin," *Peptide Research*, vol. 9, no. 3, pp. 114–121, 1996.

[15] K. Kunin, *In Amber-Hi-Lites. Fifty Years of Ion-Exchange Technology*, Rohm and Hass, Philadephia, Pa, USA, 1996.

[16] D. P. Elder, "Pharmaceutical applications of ion-exchange resins," *Journal of Chemical Education*, vol. 82, no. 4, pp. 575–587, 2005.

[17] L. M. Gayo and M. J. Suto, "Ion-exchange resins for solution phase parallel synthesis of chemical libraries," *Tetrahedron Letters*, vol. 38, no. 4, pp. 513–516, 1997.

[18] G. L. Hatch, J. L. Lambert, and L. R. Fina, "Some properties of the quaternary ammonium anion-exchange resin-triiodide disinfectant for water," *Industrial & Engineering Chemistry Product Research and Development*, vol. 19, no. 2, pp. 259–264, 1980.

[19] H. A. Ezzeldin, A. Apblett, and G. L. Foutch, "Synthesis and properties of anion exchangers derived from chloromethyl styrene codivinylbenzene and their use in water treatment," *International Journal of Polymer Science*, vol. 2010, Article ID 684051, 9 pages, 2010.

[20] Ion exchange India pvt limited.

[21] V. C. Gerald, "Process for removing iodine/iodide from aqueous solution," US Patent 5624567, 1997.

12

Amyloid Beta Peptides Differentially Affect Hippocampal Theta Rhythms *In Vitro*

Armando I. Gutiérrez-Lerma,[1,2] **Benito Ordaz,**[1] **and Fernando Peña-Ortega**[1]

[1] *Departamento de Neurobiología del Desarrollo y Neurofisiología, Instituto de Neurobiología, UNAM, Boulevard Juriquilla 3001, 16230 Querétaro, Mexico*

[2] *Departamento de Farmacobiología, Cinvestav-IPN, Calzada de los Tenorios 235, Col. Granjas Coapa, 14330 México, DF, Mexico*

Correspondence should be addressed to Fernando Peña-Ortega; jfpena@unam.mx

Academic Editor: John D. Wade

Soluble amyloid beta peptide ($A\beta$) is responsible for the early cognitive dysfunction observed in Alzheimer's disease. Both cholinergically and glutamatergically induced hippocampal theta rhythms are related to learning and memory, spatial navigation, and spatial memory. However, these two types of theta rhythms are not identical; they are associated with different behaviors and can be differentially modulated by diverse experimental conditions. Therefore, in this study, we aimed to investigate whether or not application of soluble $A\beta$ alters the two types of theta frequency oscillatory network activity generated in rat hippocampal slices by application of the cholinergic and glutamatergic agonists carbachol or DHPG, respectively. Due to previous evidence that oscillatory activity can be differentially affected by different $A\beta$ peptides, we also compared $A\beta_{25-35}$ and $A\beta_{1-42}$ for their effects on theta rhythms *in vitro* at similar concentrations (0.5 to 1.0 μM). We found that $A\beta_{25-35}$ reduces, with less potency than $A\beta_{1-42}$, carbachol-induced population theta oscillatory activity. In contrast, DHPG-induced oscillatory activity was not affected by a high concentration of $A\beta_{25-35}$ but was reduced by $A\beta_{1-42}$. Our results support the idea that different amyloid peptides might alter specific cellular mechanisms related to the generation of specific neuronal network activities, instead of exerting a generalized inhibitory effect on neuronal network function.

1. Introduction

Alzheimer's disease (AD) is a dementia of increasing prevalence [1], which is produced, at least in its early stages, by the extracellular accumulation of amyloid beta protein ($A\beta$) [2–4]. Early deterioration of hippocampal function, likely induced by soluble $A\beta$, contributes to the initial memory deficits observed in AD patients [4–8]. Interestingly, $A\beta$ encompasses several peptide species which differ in their length, solubility, biological activity, toxicity, and aggregation propensity [3, 4, 9]. $A\beta_{1-40}$ and $A\beta_{1-42}$ are the most abundant $A\beta$ peptides found in senile plaques and vascular deposits of AD patients [10, 11]; however, these deposits also contain $A\beta$ peptides with shorter sequences such as $A\beta_{25-35}$ [12–14]. $A\beta_{25-35}$ can be produced in AD patients by enzymatic cleavage of $A\beta_{1-40}$ at its hydrophobic C-terminus [14, 15],

and it has been proposed that $A\beta_{25-35}$ constitutes one of the biologically active fragments of $A\beta$ [12, 16, 17]. Despite the extensive literature showing that the effects produced by $A\beta_{25-35}$ are mostly reproduced by the full-length sequence [3, 12, 16, 18–28], other reports indicate that this is not always the case. For instance, it has been shown that the reduction in long term potentiation (LTP) produced by $A\beta_{1-40}$ is not reproduced by $A\beta_{25-35}$ [29]. In contrast, whereas $A\beta_{25-35}$ induces intracellular actin aggregation and alters axonal transport, $A\beta_{1-42}$ does not [30]. The same pattern occurs with the increase in intracellular calcium observed after application of $A\beta_{25-35}$, which is not reproduced by $A\beta_{1-42}$ [31]. It has already been proposed that different forms of soluble $A\beta$ alter cognitively related, synchronized electrical oscillatory activity in neural circuits [9, 18, 19, 32–34]. Normal hippocampal function is strongly dependent on an oscillatory activity called theta rhythm [35–38] which, in lower mammals, includes oscillations ranging from 3 to 12 Hz [35–38]. Hippocampal theta rhythm participates in memory consolidation during REM sleep, in synaptic plasticity, in neural coding of place, and in spatial memory [8, 37, 39–43]. Interestingly, AD patients show alterations in theta rhythm activity [5, 44, 45]. Some reports indicate that cognitive dysfunction correlates with an increase of resting theta rhythm [5, 8, 44, 45], but other studies show a reduction of cognitive-induced theta rhythm [45]. Similar findings have been observed in transgenic mice that overproduce $A\beta$ and exhibit AD-like symptoms [8, 46–50]. The complex relationship between AD pathology and theta rhythms can be explained by the theta rhythm heterogeneity that exists both in humans and in mice [51, 52]. Theta rhythms require the activation of either cholinergic or glutamatergic pathways, or both [53, 54], which are profoundly damaged in AD patients [55, 56] and severely compromised in transgenic AD mouse models [57–59]. Moreover, it is known that $A\beta$ can affect both cholinergic [60, 61] and glutamatergic transmission [19, 62]. Thus, the application of $A\beta$ in either the medial septum or in the hippocampus reduces theta rhythm both *in vivo* and *in vitro* [19, 33, 34, 63–65] and simultaneously induces cognitive deficits [33, 65]. Moreover, carbachol-induced theta rhythm generation is impaired in slices obtained from triple-transgenic AD mice [48]. Despite all this evidence, it is still unknown if $A\beta$ can alter cholinergically and/or glutamatergically induced theta rhythms to the same extent *in vitro*. We have addressed this question and also evaluated the potential differences in biological activity between $A\beta_{25-35}$ and $A\beta_{1-42}$ on both cholinergically and glutamatergically induced theta rhythms, as was done earlier for beta rhythms [9]. We tested these peptides at concentrations of $1\,\mu M$ or less because most studies report $A\beta$ concentrations in the low nM range both in AD patients [66–70] and in AD transgenic mice [71–74]. However, some reports show that $A\beta$ in AD patients can reach high nM [75–77] or even μM concentrations [66, 78, 79] (for a review see [80]). Our results show that $A\beta_{25-35}$ reduces, with less potency than $A\beta_{1-42}$, carbachol-induced field theta oscillatory activity. In contrast, DHPG-induced oscillatory activity was not affected by $A\beta_{25-35}$ but was reduced by $A\beta_{1-42}$.

2. Material and Methods

2.1. Animals.
All experiments were performed using 7- to 10-week-old male Wistar rats. Animals used in this study were housed at $22°C$ and maintained on a 12-h:12-h light/dark cycle with free access to food and water. All the experimental protocols were approved by the Local Committees of Ethics on Animal Experimentation (CICUAL-Cinvestav and INB-UNAM). Experiments were performed according to the Mexican Official Norm for the Use and Care of Laboratory Animals (NOM-062-ZOO-1999).

2.2. Hippocampal Slice Preparation.
Hippocampal slices were obtained as follows: the animals were anesthetized intraperitoneally with sodium pentobarbital (63 mg/Kg) and perfused transcardially with cold modified artificial cerebrospinal fluid (maCSF) of the following composition (in mM): 238 sucrose, 3 KCl, 2.5 $MgCl_2$, 25 $NaHCO_3$, and 30 D-glucose, pH 7.4, bubbled with carbogen (95% O_2 and 5% CO_2). After a maximum of 1 min of transcardial perfusion, the animals were decapitated, and the brains were removed and dissected in ice-cold artificial cerebrospinal fluid (aCSF) containing (in mM) 119 NaCl, 3 KCl, 1.5 $CaCl_2$, 1 $MgCl_2$, 25 $NaHCO_3$, and 30 D-glucose, pH 7.4, continuously bubbled with carbogen. One cerebral hemisphere was glued to an agar block with a $30°$ inclination and mounted on a vibratome (The Vibratome Company, St. Louis, MO, USA). Horizontal slices (400 μm thick) containing the hippocampal formation were cut and left to recover at room temperature for at least 90 min in aCSF continuously bubbled with carbogen.

2.3. Electrophysiological Recordings.
The slices were transferred to a submerged recording chamber continuously perfused at 20 mL/min with oxygenated aCSF. The temperature was kept constant at $29 \pm 2°C$. Extracellular field recordings were obtained with suction electrodes filled with aCSF and positioned over hippocampal area CA1, *stratum pyramidale* [62, 81]. The signal was amplified and filtered (highpass, 0.5 Hz; lowpass, 1.5 KHz) with a wide-band AC amplifier (Grass Instruments, Quincy, MA, USA). The experimental protocol consisted of recording the spontaneous basal activity for 10 min before adding either the general cholinergic agonist carbachol (Cch) [20 μM] or the metabotropic glutamate group I agonist (S)-3,5-dihydroxyphenylglycine (DHPG) [10 μM] to the perfusion system in order to generate stable and persistent oscillatory activity [53]. Then, we added freshly dissolved $A\beta_{25-35}$ to the perfusion system at two different concentrations: 0.5 μM and 1 μM [19]. In a different set of experiments, the freshly dissolved full peptide $A\beta_{1-42}$ [0.5 μM] was also tested [62]. We used the freshly dissolved inverse sequences $A\beta_{35-25}$ [1 μM] and $A\beta_{42-1}$ [0.5 μM] as negative controls. In some experiments, atropine [1 μM], a broad-spectrum muscarinic acetylcholine antagonist, was added to the perfusion system to block carbachol-induced oscillations. The metabotropic glutamate receptor group 5 antagonist 2-methyl-6-(phenylethynyl) pyridine (MPEP) [25 μM] was added to block the DHPG-induced oscillations. All drugs were dissolved in distilled water and were obtained from Sigma (Sigma-RBI, St. Louis, MO). The stock solutions were prepared at concentrations of at least 1000X. In most cases 1 μL of distilled water, containing the drug, was applied to 1 mL of aCSF which, in our hands, does not affect its osmolarity.

2.4. Data Analysis.
The recordings were stored on a personal computer with an acquisition system from National Instruments (Austin, TX, USA) by using custom-made software designed in the LabView environment. The recordings obtained were analyzed off-line using the program Clampfit (Molecular Devices). The full experiments were analyzed, and for quantification purposes, segments of 100 sec were analyzed every 10 min, using a Fast Fourier Transform Algorithm with a Hamming window. The power spectra

of all segments were normalized to the basal spontaneous activity of each individual experiment or normalized to the control cholinergic or glutamatergic oscillatory activity. We also performed an Autocorrelation Function Estimate to test for the self cross-correlation (rhythmicity) of the signal on short (1 sec window) traces. All data are expressed as mean ± standard error of the mean (SEM). In most cases the data distribution was markedly skewed, and hence we used a Mann-Whitney Rank Sum Test with statistical significance denoted by $P < 0.05$, or a Kruskal-Wallis One-Way Analysis of Variance on Ranks followed by either Dunn's Method for Multiple Comparisons versus Control Group or for All Pairwise Multiple Comparison. Differences with statistical significance are denoted by $P < 0.001$.

3. Results

3.1. $A\beta$ Peptides Inhibit Carbachol-Induced Hippocampal Theta Oscillatory Activity. In order to study the effect of $A\beta$ on cholinergically induced theta oscillatory activity, we first generated such activity by applying carbachol [20 μM] to the perfusion system [82]. In these conditions, most of the slices (10 out of 13) generated field oscillatory activity with frequency components that fall into the theta range (Figure 1(a), middle trace) and have a peak frequency of 9.8 ± 0.39 Hz (Figure 1(a), middle power spectrum). The insets in the power spectra show the corresponding autocorrelation estimates of the hippocampal population activity, indicating that in the presence of carbachol [20 μM] rhythmic population activity emerges. The rhythmicity exhibits a high degree of self-correlation (Figure 1(a)). Quantification of power in the classical theta range (4 to 12 Hz) [48, 53] shows that application of carbachol [20 μM] increases the power of hippocampal population activity to 204.06 ± 13.79% of control basal activity (Figure 1(a), right graph). The activity described maintains stability for at least 100 min ($n = 10$). As expected, addition of atropine [1 μM] completely abolished the oscillatory activity generated by carbachol and even reduced the power of the hippocampal activity below control levels, to 58.18 ± 3.64% of control basal activity (Figure 1(a)).

To test whether or not the amyloid peptides were capable of disrupting population theta oscillatory activity, we applied them to the bath after the application of carbachol. Addition of $A\beta_{25-35}$ [0.5 μM] to the perfusion system did not significantly change carbachol-induced theta oscillations (Figure 1(b)). However, it is important to notice that $A\beta_{25-35}$ [0.5 μM] shows a slight tendency to increase such activity (to 140.01 ± 19.86% of the cholinergic control (Figure 1(b); $n = 8$)). In the presence of $A\beta_{25-35}$ [0.5 μM] the peak oscillatory frequency remained unaffected (10.17 ± 0.78 Hz). Increasing $A\beta_{25-35}$ concentration, in an independent group of slices, to 1 μM reduced cholinergically induced theta oscillations to 47.93 ± 8.77% of the cholinergic control (Figure 1(b); $n = 9$). Likewise, the peak oscillatory frequency diminished significantly to 5.57 ± 1.26 Hz. Similarly, application of the full peptide $A\beta_{1-42}$ [0.5 μM] reduced the cholinergically induced theta oscillations to 42.60 ± 13.28% of the cholinergic control (Figure 1(c); $n = 4$); this inhibition in power

was accompanied by a significant reduction in the peak oscillatory frequency to 6.45 ± 1.43 Hz.

As an internal control, we washed out the $A\beta_{25-35}$ [1 μM] with aCSF containing carbachol [20 μM] and found that the population theta oscillatory activity returned to nearly the same power level as before application of $A\beta_{25-35}$ [1 μM] (89.40 ± 5.95% of the cholinergic control; Figure 2(a), graph and lower trace; $n = 4$). The peak oscillatory frequency was also restored upon washout (8.56 ± 1.06 Hz; Figure 2(a), lower power spectrum; $n = 4$), and the rhythmicity of the theta oscillatory activity was identical to that observed before the application of $A\beta_{25-35}$ [1 μM] (Figure 2(a), upper inset). To test for the specificity of the effect produced by $A\beta_{25-35}$, we used its inverse sequence $A\beta_{35-25}$ [1 μM] and found no effect on the carbachol-induced theta oscillations (Figure 2(b)). The power of the activity remained unaltered in the presence of $A\beta_{35-25}$ (97.00 ± 9.16% of the cholinergic control, $n = 10$; Figure 2(b), graph and middle trace), and the peak theta frequency remained unaltered as well (8.49 ± 0.49 Hz, $n = 10$, Figure 2(b), middle power spectrum). Of course, subsequent application of $A\beta_{25-35}$ [1 μM] in the continued presence of $A\beta_{35-25}$ (1 μM) reduced the carbachol-induced theta oscillations to 33.71 ± 10.19% of the cholinergic control (Figure 2(b), lower trace and power spectrum). We also tested the effect of the inverse sequence $A\beta_{42-1}$ [0.5 μM] and found that this peptide sequence does not significantly alter carbachol-induced activity (81.57 ± 26.67% of the cholinergic control, $n = 5$).

3.2. $A\beta$ Peptides Differentially Affect DHPG-Induced Delta/Theta Oscillatory Activity. As mentioned earlier, theta rhythm can also be induced by glutamatergic activation, mainly through metabotropic group I receptors [53, 83]. In order to study the effect of $A\beta$ on glutamatergically induced theta oscillatory activity, we first generated such activity by applying DHPG [10 μM] to the perfusion system. In these conditions, most of the slices (8 out of 10) generated oscillatory activity with frequency components slower than those present in cholinergically induced oscillatory activity, ranging from 2 to 10 Hz in the mixed delta/theta rhythm range (Figure 3(a); $n = 8$) with a mean peak frequency of 5.71 ± 0.92 Hz (Figure 3(a), middle power spectrum). The insets in the power spectra show the corresponding autocorrelations of the hippocampal population activity indicating that, in the presence of DHPG [10 μM], rhythmic population activity emerges and is characterized by a high degree of self-correlation (Figure 3(a), inset). Quantification of power in the delta/theta range (2 to 10 Hz) shows that application of DHPG [10 μM] increases the power of hippocampal population activity to 207.47 ± 13.66% of the control basal activity (Figure 3(a), right graph). This activity was stable for at least 100 min ($n = 8$). As expected, addition of MPEP [25 μM] completely abolished the oscillatory activity generated by DHPG and reduced the power of the hippocampal activity to 110 ± 8.24% of control basal activity (Figure 3(a), lower trace and power spectrum). To test whether or not the amyloid peptides were capable of disrupting the glutamatergic population theta oscillatory activity, we added them after the application of DHPG.

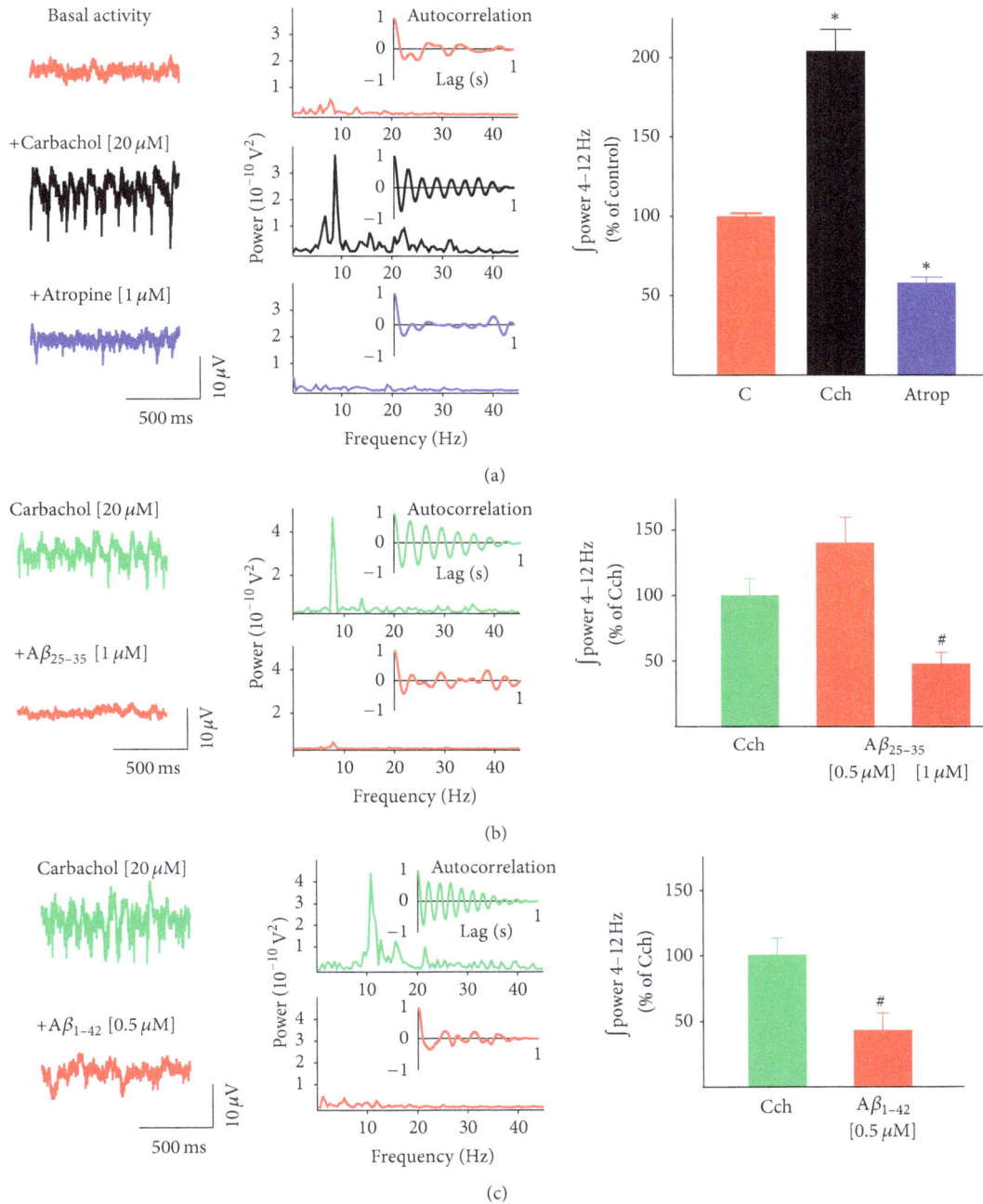

FIGURE 1: Amyloid beta peptides inhibit carbachol-induced hippocampal theta oscillatory activity. (a) Representative recordings (left) and the corresponding power spectra (right) of hippocampal population activity recorded in control conditions (upper trace and power spectrum) after bath application of carbachol (Cch 20 μM; middle trace and power spectrum) and after the application of atropine 1 μM (lower trace and power spectrum). The graph on the right shows the quantification of the integrated spectral power from 4 to 12 Hz. Note that the addition of Cch induces atropine-sensitive rhythmic oscillations that increase the power of the hippocampal population activity. (b) Representative recordings (left) and the corresponding power spectra (right) of hippocampal population activity recorded in the presence of carbachol 20 μM (upper trace and power spectrum) and after bath application of $A\beta_{25-35}$ 1 μM (lower trace and power spectrum). The graph on the right shows the quantification of the integrated spectral power from 4 to 12 Hz before and after bath application of $A\beta_{25-35}$ [0.5 μM] and [1 μM]. Note that a high concentration of $A\beta_{25-35}$ inhibits Cch-induced rhythmic theta oscillations and that a lower concentration does not affect this activity. (c) Representative recordings (left) and their corresponding power spectra (right) of hippocampal population activity recorded in the presence of carbachol 20 μM (upper trace and power spectrum) and after bath application of $A\beta_{1-42}$ 0.5 μM (lower trace and power spectrum). The graph on the right shows the quantification of the integrated spectral power from 4 to 12 Hz before and after bath application of $A\beta_{1-42}$. Note that $A\beta_{1-42}$ inhibits Cch-induced rhythmic theta oscillations. The inset shown on each power spectrum is an autocorrelogram obtained from the corresponding trace. *indicates a significant difference with respect to control ($P < 0.001$), and #indicates a significant difference with respect to the carbachol-induced oscillatory activity.

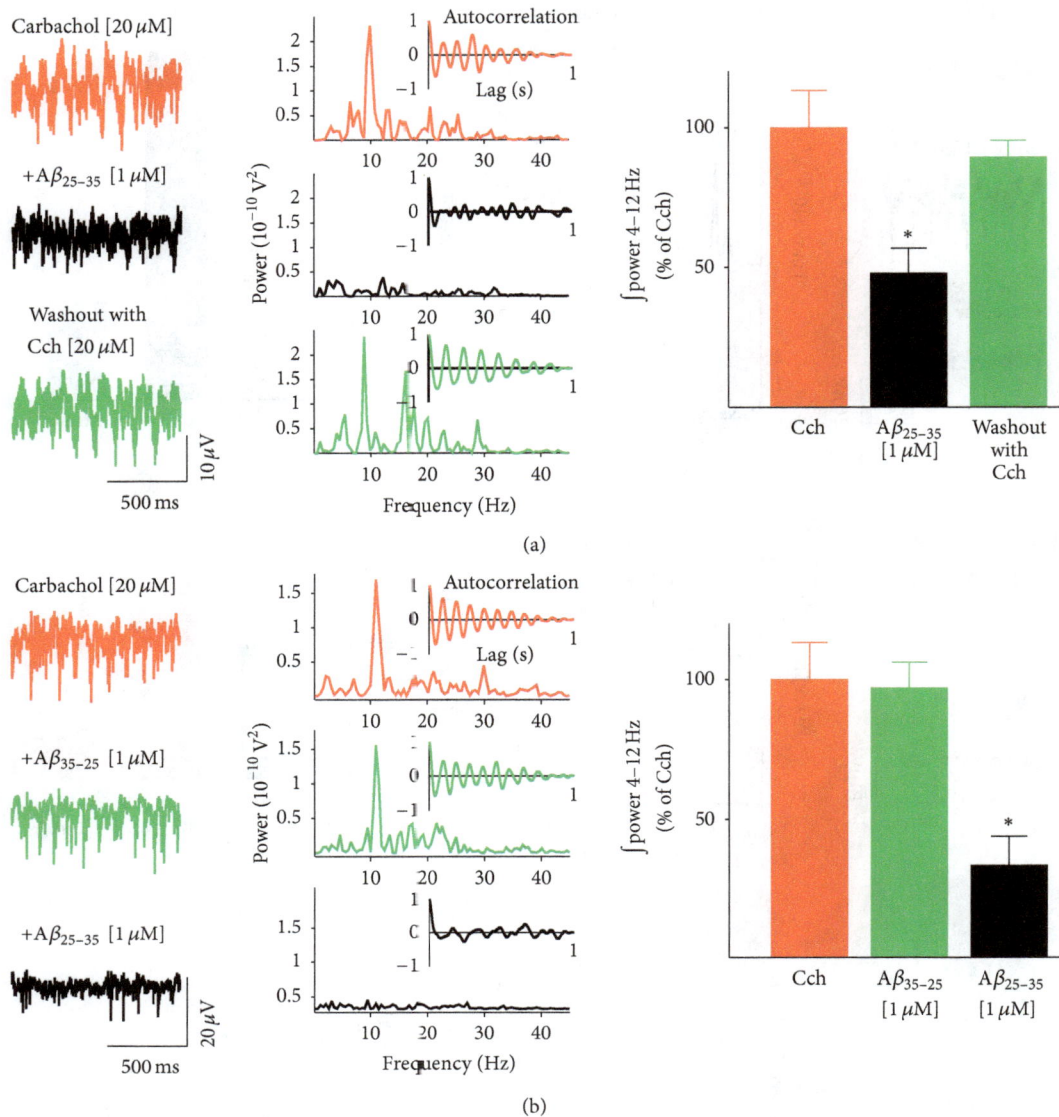

FIGURE 2: Amyloid beta inhibition of carbachol-induced hippocampal theta oscillatory activity is reversible and specific. (a) Representative recordings (left) and the corresponding power spectra (right) of hippocampal population activity recorded in the presence of carbachol 20 μM (upper trace and power spectrum) after bath application of $A\beta_{25-35}$ 1 μM (middle trace and power spectrum) and after washout of $A\beta$ (in the presence of carbachol 20 μM; lower trace and power spectrum). The graph on the right shows the quantification of the integrated spectral power from 4 to 12 Hz. Note that the $A\beta_{25-35}$-induced inhibition of theta activity is reversible upon washout. (b) Representative recordings (left) and the corresponding power spectra (right) of hippocampal population activity recorded in the presence of carbachol 20 μM (upper trace and power spectrum) after bath application of the inverse sequence of $A\beta_{25-35}$, $A\beta_{35-25}$ [1 μM] (middle trace and power spectrum) and after bath application of $A\beta_{25-35}$ 1 μM (in the presence of carbachol; lower trace and power spectrum). The inset shown on each power spectrum is an autocorrelogram obtained from the corresponding trace. The graph on the right shows the quantification of the integrated spectral power from 4 to 12 Hz. Note that the inverse sequence $A\beta_{35-25}$ has no effect on the theta rhythm. *indicates a significant difference with respect to the carbachol-induced oscillatory activity ($P < 0.001$).

Addition of $A\beta_{25-35}$ [1 μM] to the perfusion system did not significantly change DHPG-induced delta/theta oscillations (120.52 ± 12.92% of the glutamatergic control; Figure 3(b), graph, lower trace and power spectrum). In the presence of $A\beta_{25-35}$ [1 μM], the peak oscillatory frequency remained unaffected (7.57 ± 0.64 Hz; $n = 10$). In contrast, application

of full length $A\beta_{1-42}$ [0.5 μM] reduced DHPG-induced theta oscillations to 33.27 ± 8.92% of the glutamatergic control (Figure 3(c), graph, lower trace and power spectrum; $n = 4$). However, such inhibition in power was not accompanied by a significant reduction of the peak oscillatory frequency, which was 5.77 ± 1.12 Hz.

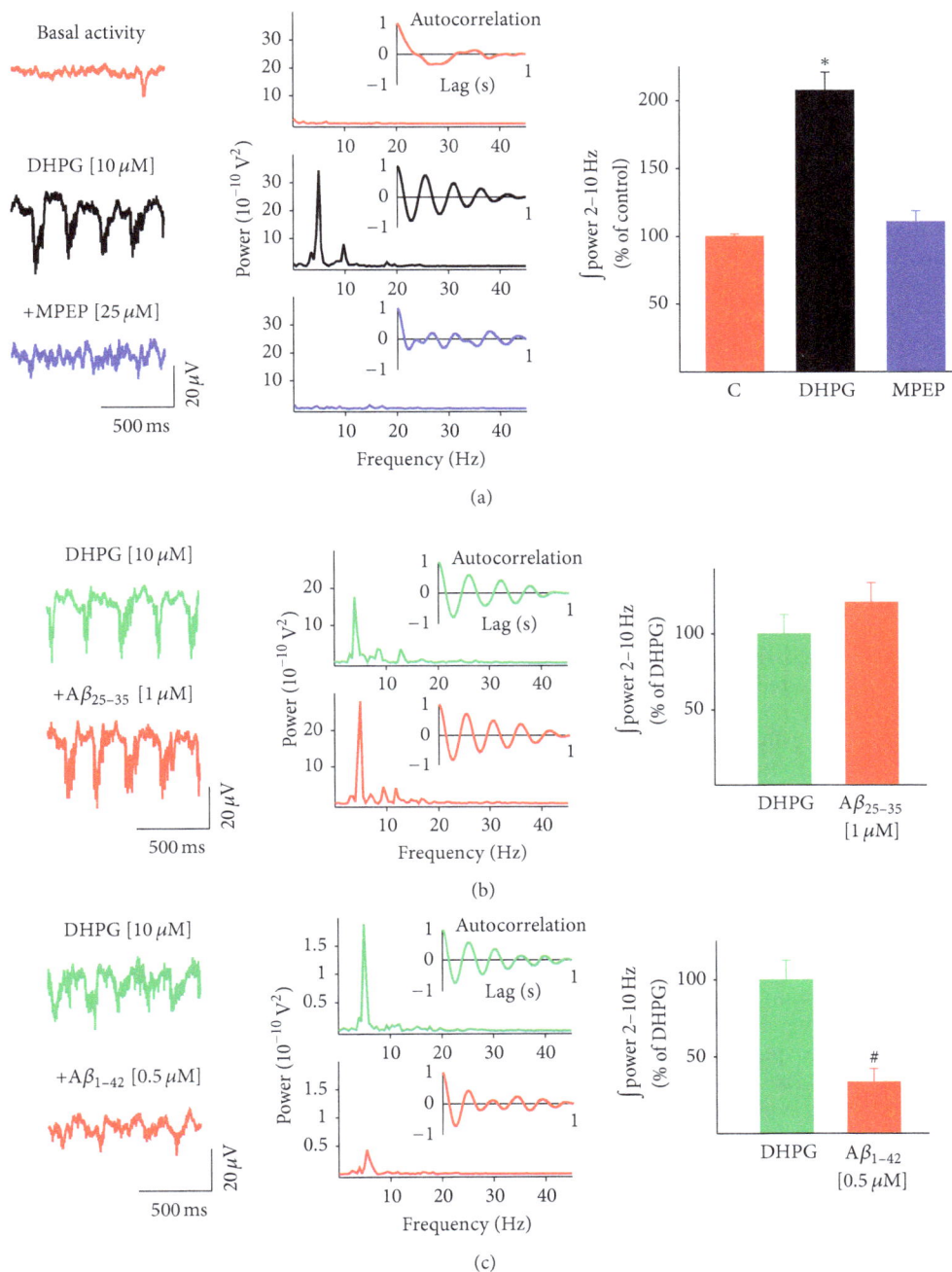

FIGURE 3: Amyloid beta peptides differentially inhibit DHPG-induced hippocampal theta oscillatory activity. (a) Representative recordings (left) and their corresponding power spectra (right) of hippocampal population activity recorded in control conditions (upper trace and power spectrum), after bath application of DHPG 10 μM (middle trace and power spectrum) and after the application of MPEP 25 μM (lower trace and power spectrum). The graph on the right shows the quantification of the integrated spectral power from 2 to 10 Hz. Note that DHPG induces MPEP-sensitive rhythmic oscillations that increase the power of the hippocampal population activity. (b) Representative recordings (left) and the corresponding power spectra (right) of hippocampal population activity recorded in the presence of DHPG 10 μM (upper trace and power spectrum) and after bath application of $A\beta_{25-35}$ 1 μM (lower trace and power spectrum). The graph on the right shows the quantification of the integrated spectral power from 2 to 10 Hz. Note that $A\beta_{25-35}$ does not affect DHPG-induced oscillatory activity. (c) Representative recordings (left) and the corresponding power spectra (right) of hippocampal population activity recorded in the presence of DHPG 10 μM (upper trace and power spectrum) and after bath application of $A\beta_{1-42}$ 0.5 μM (lower trace and power spectrum). The inset shown on each power spectrum is an autocorrelogram obtained from the corresponding trace. The graph on the right shows the quantification of the integrated spectral power from 2 to 10 Hz. Note that $A\beta_{1-42}$ inhibits DHPG-induced rhythmic theta oscillations. * indicates a significant difference with respect to control ($P < 0.001$), and # indicates a significant difference with respect to the DHPG-induced oscillatory activity.

4. Discussion

The study of theta rhythm generation as well as its alterations in several neurological pathologies has received a great deal of attention [8, 84]. Such research has revealed that instead of a single theta rhythm, there actually exist a variety of rhythms that are involved in different behaviors and rely on different cellular mechanisms [52, 85]. In general, it has been proposed that theta rhythms are evoked through either cholinergic or glutamatergic means [8, 53, 84]. Our finding that $A\beta$ affects both types of theta rhythms is of particular importance, given that both cholinergic and glutamatergic neurotransmission are necessary to generate cognitive-related theta network oscillatory activity (as reviewed by [84]), and therefore any pathological disruption in the normal function of either neurotransmitter system will generate cognitive dysfunction and, eventually, dementia. Previous research has shown that soluble $A\beta$ affects the power, frequency, and structure of synchronized oscillatory activity, mainly theta and gamma rhythms, which are strongly related to cognition [9, 18, 19, 32–34, 64, 65] and that such activity is disrupted in AD patients [5, 44, 86, 87] and in AD transgenic mice [46–50]. As we have previously shown that theta generation *in vivo* is affected by both $A\beta_{25-35}$ [19] and $A\beta_{1-42}$ [34], we took this knowledge a step further and tested whether or not freshly dissolved $A\beta$ affected the power and frequency of either cholinergically or glutamatergically induced hippocampal population theta oscillatory activity in slices *in vitro*. We found that different sequences of $A\beta$ induce, with some differences in potency, a statistically significant reduction in power of cholinergic population theta oscillatory activity. In contrast, glutamatergically induced oscillatory activity was not sensitive to the application of $A\beta_{25-35}$, whereas it was strongly inhibited by $A\beta_{1-42}$. The mechanisms involved in the $A\beta$-induced reduction of both cholinergically and glutamatergically generated population theta oscillatory activity are not clear, and there are several cellular and molecular targets through which $A\beta$ can pathologically interact to produce synaptic and cellular dysfunction (as reviewed in [3, 88]). However, very recent evidence indicates that $A\beta$ affects the oscillatory activity of the hippocampus mainly in two ways: the first one is a generalized reduction of both glutamatergic and GABAergic transmission of presynaptic origin [19, 62, 89, 90], and the second involves alterations in specific subpopulations of GABAergic interneurons [33, 89–91]. Additionally, carbachol induces rhythmic oscillations of membrane potential in hippocampal neurons that allow them to engage in the theta oscillatory rhythm [92]. It has been shown that such subthreshold oscillations are abolished by $A\beta$ when they are induced by carbachol induction [18] or through application of depolarizing DC current [19]. Regarding the biochemical mechanisms involved in this effect, both muscarinic acetylcholine M1/M3 receptors and group I metabotropic glutamate receptors modulate several cellular mechanisms by activating phospholipase C and protein kinase C (PKC) [83, 93]. It has been shown that the activation of PKC mediated by carbachol is disrupted by $A\beta$ [94], whereas the activation of PKC induced by ACPD, a general metabotropic glutamate agonist, is completely unaffected by

$A\beta$ [94]. Another important factor to take into account is that, as it has already been suggested, carbachol and DHPG most probably generate oscillatory activity through the activation of different subsets of hippocampal interneurons [53], and, therefore, $A\beta$ may pathologically alter the proper function of some, but not of other kinds of interneurons, which may also explain why DHPG-induced oscillatory activity is resistant to $A\beta_{25-35}$. Another possible explanation for the higher sensitivity of carbachol-induced theta rhythm to bath application of $A\beta$ compared to DHPG-induced theta rhythms has emerged from the recent finding that $A\beta$ directly disrupts the M1 muscarinic receptor/G-protein interaction [95]. In contrast, to our knowledge, such direct disruption has not been demonstrated for metabotropic glutamatergic receptors. To finish with this point, it is important to mention that, just like the theta oscillatory activity generated with DHPG, potassium-induced hippocampal beta rhythm was reduced by $A\beta_{1-42}$ but was resistant to $A\beta_{25-35}$ [9]. In our previous studies, we had tested the changes in hippocampal population activity induced by $A\beta_{25-35}$ at concentrations ranging from $300\,nM$ to $3\,\mu M$ [9, 19], whereas $A\beta_{1-42}$ has been tested in the range from $10\,nM$ to $0.5\,\mu M$ [19, 62]. Based on our previous observations, and those from others, we decided to use a common concentration of both peptides of $0.5\,\mu M$, which had already been shown to produce similar effects on hippocampal population activity [19].

In another series of experiments, we washed out the peptide $A\beta_{25-35}$ [$1\,\mu M$] from the system with aCSF containing carbachol, and both peak frequency and power of the theta field oscillatory activity were completely restored; therefore, the reduction of field oscillatory activity induced by $A\beta$ is completely reversible. We have shown previously that inhibition of hippocampal population activity induced by $A\beta_{25-35}$ and $A\beta_{1-42}$ is equally reversible [19, 62]. The reversibility of the effect of $A\beta_{25-35}$ on theta rhythm observed here corroborates these previous findings and gives us confidence that the effects induced by $A\beta_{1-42}$ are reversible as well [62]. The reversibility of the effects induced by $A\beta$ suggests that soluble $A\beta$ does not permanently damage the neural circuitry needed to generate theta oscillatory activity *in vitro*, and secondly, it strongly supports the idea that soluble $A\beta$ produces synaptic impairment rather than irreversible synaptic loss *in vitro*. Finally, we also tested the $A\beta$ inverse sequences as a control for the specificity of the biological activity of the $A\beta$ peptides used; as expected, the inverse sequences had no effect at all on either the peak frequency or the power of the cholinergically induced population theta oscillatory activity. As additional proof that $A\beta$ has biological effects that are sequence and conformation specific, in the presence of the inverse sequence $A\beta_{35-25}$ we added into the perfusion system the regular sequence $A\beta_{25-35}$ [$1\,\mu M$] which, unsurprisingly, significantly reduced the power of the cholinergic theta oscillatory activity to approximately one-third of control cholinergic values. Finally, in relation to the differences in biological activity between $A\beta_{25-35}$ and $A\beta_{1-42}$, the latter always produced a reduction of theta oscillatory activity, and, furthermore, it was more effective at a lower concentration than $A\beta_{25-35}$, which is consistent with the idea that the most toxic $A\beta$ species is soluble $A\beta_{1-42}$ [96, 97], especially in its oligomerized form [98, 99]. In this study, both peptides were used in their soluble forms, which have been shown to be mainly composed of monomers and a few oligomers [100], although the precise concentrations of each aggregation species are hard to determine. Considering that the inhibition of beta rhythm hippocampal population oscillations produced by $A\beta_{25-35}$ did not exhibit a clear concentration-dependent relationship in the low μM range [9], we decided not to study higher concentrations. Given that $A\beta_{1-42}$ was the most effective peptide used in terms of disrupting theta rhythm power, it would be of interest to test the effect of chemically oligomerized $A\beta_{1-42}$ on population theta oscillatory activity induced both cholinergically and glutamatergically, since the oligomers should be even more potent at reducing the power of the theta rhythm at low concentrations.

5. Conclusions

In conclusion, $A\beta$ peptides, but more potently $A\beta_{1-42}$, can differentially reduce population theta oscillatory activity induced through either cholinergic or glutamatergic means, which suggests that instead of having a widespread effect, $A\beta$ disrupts some particular mechanism of generation and maintenance of this activity. Importantly, this effect is sequence specific and completely reversible, suggesting that $A\beta$ produces synaptic and cognitive impairments that could be potentially delayed or reversed.

Conflict of Interests

The authors declare that they have no conflict of interests.

Authors' Contribution

F. P. Ortega designed the research. A. I. Gutiérrez-Lerma and B. Ordaz carried out the recordings and analyzed the data. A. I. Gutiérrez-Lerma and F. P. Ortega wrote the paper. All the authors read and approved the final paper.

Acknowledgments

This work was funded by Consejo Nacional de Ciencia y Tecnología (CONACyT) Grant no. 151261 and 181323 (to F. Peña-Ortega) and a scholarship 195367 (to A. I. Gutiérrez-Lerma), by Dirección General de Asuntos del Personal Académico-UNAM Grant no. IACODI1201511 and IB200212-RR280212 (to F. Peña-Ortega), and by Alzheimer's Association Grant NIRG-11-205443. This study was performed in partial fulfillment of the requirements for the Ph.D. degree in Biomedical Sciences of the Programa de Doctorado en Ciencias Biomédicas (INB, Universidad Nacional Autónoma de México) for A. I. Gutiérrez-Lerma. The authors would like to thank Dr. Dorothy Pless for editorial comments.

References

[1] L. E. Hebert, P. A. Scherr, J. L. Bienias, D. A. Bennett, and D. A. Evans, "Alzheimer disease in the US population: prevalence estimates using the 2000 census," *Archives of Neurology*, vol. 60, no. 8, pp. 1119–1122, 2003.

[2] H. Braak and E. Braak, "Frequency of stages of Alzheimer-related lesions in different age categories," *Neurobiology of Aging*, vol. 18, no. 4, pp. 351–357, 1997.

[3] F. Peña, A. I. Gutiérrez-Lerma, R. Quiroz-Baez, and C. Arias, "The role of β-amyloid protein in synaptic function: implications for Alzheimer's disease therapy," *Current Neuropharmacology*, vol. 4, no. 2, pp. 149–163, 2006.

[4] D. J. Selkoe, "Alzheimer's disease is a synaptic failure," *Science*, vol. 298, no. 5594, pp. 789–791, 2002.

[5] C. Babiloni, M. Pievani, F. Vecchio et al., "White-matter lesions along the cholinergic tracts are related to cortical sources of eeg rhythms in amnesic mild cognitive impairment," *Human Brain Mapping*, vol. 30, no. 5, pp. 1431–1443, 2009.

[6] W. L. Klein, G. A. Krafft, and C. E. Finch, "Targeting small A β oligomers: the solution to an Alzheimer's disease conundrum?" *Trends in Neurosciences*, vol. 24, no. 4, pp. 219–224, 2001.

[7] T. Ondrejcak, I. Klyubin, N.-W. Hu, A. E. Barry, W. K. Cullen, and M. J. Rowan, "Alzheimer's disease amyloid β-protein and synaptic function," *NeuroMolecular Medicine*, vol. 12, no. 1, pp. 13–26, 2010.

[8] F. Peña-Ortega, "Amyloid β-protein and neural network dysfunction," *Journal of Neurodegenerative Diseases*, vol. 2013, Article ID 657470, 8 pages, 2013.

[9] A. Adaya-Villanueva, B. Ordaz, H. Balleza-Tapia, A. Márquez-Ramos, and F. Peña-Ortega, "β-like hippocampal network activity is differentially affected by amyloid β peptides," *Peptides*, vol. 31, no. 9, pp. 1761–1766, 2010.

[10] T. Iwatsubo, A. Odaka, N. Suzuki, H. Mizusawa, N. Nukina, and Y. Ihara, "Visualization of Aβ42(43) and Aβ40 in senile plaques with end-specific Aβ monoclonals: evidence that an initially deposited species is Aβ42(43)," *Neuron*, vol. 13, no. 1, pp. 45–53, 1994.

[11] A. Güntert, H. Döbeli, and B. Bohrmann, "High sensitivity analysis of amyloid-β peptide composition in amyloid deposits from human and PS2APP mouse brain," *Neuroscience*, vol. 143, no. 2, pp. 461–475, 2006.

[12] C. J. Pike, A. J. Walencewicz-Wasserman, J. Kosmoski, D. H. Cribbs, C. G. Glabe, and C. W. Cotman, "Structure-activity analyses of β-amyloid peptides: contributions of the β25–35 region to aggregation and neurotoxicity," *Journal of Neurochemistry*, vol. 64, no. 1, pp. 253–265, 1995.

[13] T. Kubo, S. Nishimura, Y. Kumagae, and I. Kaneko, "In vivo conversion of racemized βamyloid ([D-Ser^{26}]Aβ_{1-40}) to truncated and toxic fragments ([D-Ser^{26}]Aβ25–35/40) and fragment presence in the brains of Alzheimer's patients," *Journal of Neuroscience Research*, vol. 70, no. 3, pp. 474–483, 2002.

[14] M. A. Gruden, T. B. Davudova, M. Mališauskas et al., "Autoimmune responses to amyloid structures of Aβ_{25-35} peptide and human lysozyme in the serum of patients with progressive Alzheimer's disease," *Dementia and Geriatric Cognitive Disorders*, vol. 18, no. 2, pp. 165–171, 2004.

[15] I. Kaneko, K. Morimoto, and T. Kubo, "Drastic neuronal loss in vivo by β-amyloid racemized at Ser26 residue: conversion of non-toxic [D-Ser26]β-amyloid 1–40 to toxic and proteinase-resistant fragments," *Neuroscience*, vol. 104, no. 4, pp. 1003–1011, 2001.

[16] B. A. Yankner, L. K. Duffy, and D. A. Kirschner, "Neurotrophic and neurotoxic effects of amyloid β protein: reversal by tachykinin neuropeptides," *Science*, vol. 250, no. 4978, pp. 279–282, 1990.

[17] M. P. Mattson, B. Cheng, D. Davis, K. Bryant, I. Lieburg, and R. E. Rydel, "β-Amyloid peptides destabilize calcium homeostasis and render human cortical neurons vulnerable to excitotoxicity," *Journal of Neuroscience*, vol. 12, no. 2, pp. 376–389, 1992.

[18] M.-K. Sun and D. L. Alkon, "Impairment of hippocampal CA1 heterosynaptic transformation and spatial memory by β-amyloid25–35," *Journal of Neurophysiology*, vol. 87, no. 5, pp. 2441–2449, 2002.

[19] F. Peña, B. Ordaz, H. Balleza-Tapia et al., "β-amyloid protein (25–35) disrupts hippocampal network activity: role of Fyn-kinase," *Hippocampus*, vol. 20, no. 1, pp. 78–96, 2010.

[20] S. Delobette, A. Privat, and T. Maurice, "In vitro aggregation facilitates β-amyloid peptide-(25–35)-induced amnesia in the rat," *European Journal of Pharmacology*, vol. 319, no. 1, pp. 1–4, 1997.

[21] D. B. Freir and C. E. Herron, "Nicotine enhances the depressive actions of Aβ1–40 on long-term potentiation in the rat hippocampal CA1 region in vivo," *Journal of Neurophysiology*, vol. 89, no. 6, pp. 2917–2922, 2003.

[22] D. B. Freir, D. A. Costello, and C. E. Herron, "Aβ_{25-35}-induced depression of long-term potentiation in area CA1 in vivo and in vitro is attenuated by verapamil," *Journal of Neurophysiology*, vol. 89, no. 6, pp. 3061–3069, 2003.

[23] E. A. Grace, C. A. Rabiner, and J. Busciglio, "Characterization of neuronal dystrophy induced by fibrillar amyloid β: implications for Alzheimer's disease," *Neuroscience*, vol. 114, no. 1, pp. 265–273, 2002.

[24] C. Holscher, S. Gengler, V. A. Gault, P. Harriott, and H. A. Mallot, "Soluble β-amyloid[25–35] reversibly impairs hippocampal synaptic plasticity and spatial learning," *European Journal of Pharmacology*, vol. 561, no. 1–3, pp. 85–90, 2007.

[25] T. Maurice, B. P. Lockhart, and A. Privat, "Amnesia induced in mice by centrally administered β-amyloid peptides involves cholinergic dysfunction," *Brain Research*, vol. 706, no. 2, pp. 181–193, 1996.

[26] M. Y. Stepanichev, I. M. Zdobnova, I. I. Zarubenko et al., "Aβ_{25-35}-induced memory impairments correlate with cell loss in rat hippocampus," *Physiology and Behavior*, vol. 80, no. 5, pp. 647–655, 2004.

[27] C. Tohda, N. Matsumoto, K. Zou, M. R. Meselhy, and K. Komatsu, "Aβ_{25-35}-induced memory impairment, axonal atrophy, and synaptic loss are ameliorated by MI, A metabolite of protopanaxadiol-type saponins," *Neuropsychopharmacology*, vol. 29, no. 5, pp. 860–868, 2004.

[28] Y. Yamaguchi and S. Kawashima, "Effects of amyloid-β-(25–35) on passive avoidance, radial-arm maze learning and choline acetyltransferase activity in the rat," *European Journal of Pharmacology*, vol. 412, no. 3, pp. 265–272, 2001.

[29] R. Rönicke, A. Klemm, J. Meinhardt, U. H. Schröder, M. Fändrich, and K. G. Reymann, "AB mediated diminution of MTT reduction—an artefact of single cell culture?" *PLoS ONE*, vol. 3, no. 9, Article ID e3236, 2008.

[30] H. Hiruma, T. Katakura, S. Takahashi, T. Ichikawa, and T. Kawakami, "Glutamate and amyloid β-protein rapidly inhibit fast axonal transport in cultured rat hippocampal neurons by different mechanisms," *Journal of Neuroscience*, vol. 23, no. 26, pp. 8967–8977, 2003.

[31] A. R. Korotzer, E. R. Whittenmore, and C. W. Cotman, "Differential regulation by β-amyloid peptides of intracellular free Ca^{2+} concentration in cultured rat microglia," *European Journal of Pharmacology*, vol. 288, no. 2, pp. 125–130, 1995.

[32] J. E. Driver, C. Racca, M. O. Cunningham et al., "Impairment of hippocampal gamma (γ)-frequency oscillations in vitro in mice overexpressing human amyloid precursor protein (APP)," *European Journal of Neuroscience*, vol. 26, no. 5, pp. 1280–1288, 2007.

[33] V. Villette, F. Poindessous-Jazat, A. Simon et al., "Decreased rhythmic GABAergic septal activity and memory-associated θ oscillations after hippocampal amyloid-β pathology in the rat," *Journal of Neuroscience*, vol. 30, no. 33, pp. 10991–11003, 2010.

[34] F. Peña-Ortega and R. Bernal-Pedraza, "Amyloid β peptide slows down sensory-induced hippocampal oscillations," *International Journal of Peptides*, vol. 2012, Article ID 236289, 8 pages, 2012.

[35] B. H. Bland and L. V. Colom, "Extrinsic and intrinsic properties underlying oscillation and synchrony in limbic cortex," *Progress in Neurobiology*, vol. 41, no. 2, pp. 157–208, 1993.

[36] M. J. Kahana, D. Seelig, and J. R. Madsen, "Theta returns," *Current Opinion in Neurobiology*, vol. 11, no. 6, pp. 739–744, 2001.

[37] M. J. Kahana, "The cognitive correlates of human brain oscillations," *Journal of Neuroscience*, vol. 26, no. 6, pp. 1669–1672, 2006.

[38] W. Klimesch, "EEG alpha and theta oscillations reflect cognitive and memory performance: a review and analysis," *Brain Research Reviews*, vol. 29, no. 2-3, pp. 169–195, 1999.

[39] A. L. Griffin, Y. Asaka, R. D. Darling, and S. D. Berry, "Theta-contingent trial presentation accelerates learning rate and enhances hippocampal plasticity during trace eyeblink conditioning," *Behavioral Neuroscience*, vol. 118, no. 2, pp. 403–411, 2004.

[40] N. McNaughton, M. Ruan, and M.-A. Woodnorth, "Restoring theta-like rythmicity in rats restores initial learning in the Morris water maze," *Hippocampus*, vol. 16, no. 12, pp. 1102–1110, 2006.

[41] T. Nakashiba, D. L. Buhl, T. J. McHugh, and S. Tonegawa, "Hippocampal CA3 output is crucial for ripple-associated reactivation and consolidation of memory," *Neuron*, vol. 62, no. 6, pp. 781–787, 2009.

[42] M. J. Kahana, R. Sekuler, J. B. Caplan, M. Kirschen, and J. R. Madsen, "Human theta oscillations exhibit task dependence during virtual maze navigation," *Nature*, vol. 399, no. 6738, pp. 781–784, 1999.

[43] B. R. Cornwell, L. L. Johnson, T. Holroyd, F. W. Carver, and C. Grillon, "Human hippocampal and parahippocampal theta during goal-directed spatial navigation predicts performance on a virtual Morris water maze," *Journal of Neuroscience*, vol. 28, no. 23, pp. 5983–5990, 2008.

[44] C. Babiloni, E. Cassetta, G. Binetti et al., "Resting EEG sources correlate with attentional span in mild cognitive impairment and Alzheimer's disease," *European Journal of Neuroscience*, vol. 25, no. 12, pp. 3742–3757, 2007.

[45] T. D. R. Cummins, M. Broughton, and S. Finnigan, "Theta oscillations are affected by amnestic mild cognitive impairment and cognitive load," *International Journal of Psychophysiology*, vol. 70, no. 1, pp. 75–81, 2008.

[46] J. Wang, S. Ikonen, K. Gurevicius, T. Van Groen, and H. Tanila, "Alteration of cortical EEG in mice carrying mutated human

APP transgene," *Brain Research*, vol. 943, no. 2, pp. 181–190, 2002.

[47] J. P. Wisor, D. M. Edgar, J. Yesavage et al., "Sleep and circadian abnormalities in a transgenic mouse model of Alzheimer's disease: a role for cholinergic transmission," *Neuroscience*, vol. 131, no. 2, pp. 375–385, 2005.

[48] M. Akay, K. Wang, Y. M. Akay, A. Dragomir, and J. Wu, "Nonlinear dynamical analysis of carbachol induced hippocampal oscillations in mice," *Acta Pharmacologica Sinica*, vol. 30, no. 6, pp. 859–867, 2009.

[49] B. Platt, B. Drever, D. Koss et al., "Abnormal cognition, sleep, eeg and brain metabolism in a novel knock-in alzheimer mouse, plb1," *PLoS ONE*, vol. 6, no. 11, Article ID e27068, 2011.

[50] L. Scott, J. Feng, T. Kiss, E. Needle, K. Atchison, and T. T. Kawabe, "Age-dependent disruption in hippocampal theta oscillation in amyloid-β overproducing transgenic mice," *Neurobiol Aging*, vol. 33, pp. e13–e23, 2012.

[51] L. V. Colom, "Septal networks: relevance to theta rhythm, epilepsy and Alzheimer's disease," *Journal of Neurochemistry*, vol. 96, no. 3, pp. 609–623, 2006.

[52] J. Shin, "Theta rhythm heterogeneity in humans," *Clinical Neurophysiology*, vol. 121, no. 3, pp. 456–457, 2010.

[53] J. Pálhalmi, O. Paulsen, T. F. Freund, and N. Hájos, "Distinct properties of carbachol- and DHPG-induced network oscillations in hippocampal slices," *Neuropharmacology*, vol. 47, no. 3, pp. 381–389, 2004.

[54] C. G. Reich, M. A. Karson, S. V. Karnup, L. M. Jones, and B. E. Alger, "Regulation of IPSP theta rhythm by muscarinic receptors and endocannabinoids in hippocampus," *Journal of Neurophysiology*, vol. 94, no. 6, pp. 4290–4299, 2005.

[55] K. M. Cullen, G. M. Halliday, K. L. Double, W. S. Brooks, H. Creasey, and G. A. Broe, "Cell loss in the nucleus basalis is related to regional cortical atrophy in Alzheimer's disease," *Neuroscience*, vol. 78, no. 3, pp. 641–652, 1997.

[56] D. S. Auld, T. J. Kornecook, S. Bastianetto, and R. Quirion, "Alzheimer's disease and the basal forebrain cholinergic system: relations to β-amyloid peptides, cognition, and treatment strategies," *Progress in Neurobiology*, vol. 68, no. 3, pp. 209–245, 2002.

[57] M. Klingner, J. Apelt, A. Kumar et al., "Alterations in cholinergic and non-cholinergic neurotransmitter receptor densities in transgenic Tg2576 mouse brain with β-amyloid plaque pathology," *International Journal of Developmental Neuroscience*, vol. 21, no. 7, pp. 357–369, 2003.

[58] H.-J. Lüth, J. Apelt, A. O. Ihunwo, T. Arendt, and R. Schliebs, "Degeneration of β-amyloid-associated cholinergic structures in transgenic APPSW mice," *Brain Research*, vol. 977, no. 1, pp. 16–22, 2003.

[59] Y. Ikarashi, Y. Harigaya, Y. Tomidokoro et al., "Decreased level of brain acetylcholine and memory disturbance in APPsw mice," *Neurobiology of Aging*, vol. 25, no. 4, pp. 483–490, 2004.

[60] X. Ma, W. Ye, and Z. Mei, "Change of cholinergic transmission and memory deficiency induced by injection of β-amyloid protein into NBM of rats," *Science in China Series C*, vol. 44, no. 4, pp. 435–442, 2001.

[61] L. Fang, J. Duan, D. Ran, Z. Fan, Y. Yan, and N. Huang, "Amyloid-β depresses excitatory cholinergic synaptic transmission in *Drosophila*," *Neuroscience Bulletin*, vol. 28, pp. 585–594, 2012.

[62] H. Balleza-Tapia, A. Huanosta-Gutiérrez, A. Márquez-Ramos, N. Arias, and F. Peña, "Amyloid β oligomers decrease hippocampal spontaneous network activity in an age-dependent manner," *Current Alzheimer Research*, vol. 7, no. 5, pp. 453–462, 2010.

[63] R. N. Leão, L. V. Colom, L. Borgius, O. Kiehn, and A. Fisahn, "Medial septal dysfunction by Aβ-induced KCNQ channel-block in glutamatergic neurons," *Neurobiology of Aging*, vol. 33, pp. 2046–2061, 2012.

[64] L. V. Colom, M. T. Castañeda, C. Bañuelos et al., "Medial septal β-amyloid 1–40 injections alter septo-hippocampal anatomy and function," *Neurobiology of Aging*, vol. 31, no. 1, pp. 46–57, 2010.

[65] E. A. Mugantseva and I. Y. Podolski, "Animal model of Alzheimer's disease: characteristics of EEG and memory," *Central European Journal of Biology*, vol. 4, no. 4, pp. 507–514, 2009.

[66] Y.-M. Kuo, M. R. Emmerling, C. Vigo-Pelfrey et al., "Water-soluble A$^\beta$ (N-40, N-42) oligomers in normal and Alzheimer disease brains," *Journal of Biological Chemistry*, vol. 271, no. 8, pp. 4077–4081, 1996.

[67] W. E. Klunk, B. J. Lopresti, M. D. Ikonomovic et al., "Binding of the positron emission tomography tracer Pittsburgh Compound-B reflects the amount of amyloid-β in Alzheimer's Disease brain but not in transgenic mouse brain," *Journal of Neuroscience*, vol. 25, no. 46, pp. 10598–10606, 2005.

[68] T. Matsui, M. Ingelsson, H. Fukumoto et al., "Expression of APP pathway mRNAs and proteins in Alzheimer's disease," *Brain Research*, vol. 1161, no. 1, pp. 116–123, 2007.

[69] M. D. Ikonomovic, W. E. Klunk, E. E. Abrahamson et al., "Post-mortem correlates of in vivo PiB-PET amyloid imaging in a typical case of Alzheimer's disease," *Brain*, vol. 131, no. 6, pp. 1630–1645, 2008.

[70] J. R. Steinerman, M. Irizarry, N. Scarmeas et al., "Distinct pools of β-amyloid in Alzheimer disease-affected brain: a clinicopathologic study," *Archives of Neurology*, vol. 65, no. 7, pp. 906–912, 2008.

[71] I. Dewachter, J. Van Dorpe, L. Smeijers et al., "Aging increased amyloid peptide and caused amyloid plaques in brain of old APP/V717I transgenic mice by a different mechanism than mutant presenilin1," *Journal of Neuroscience*, vol. 20, no. 17, pp. 6452–6458, 2000.

[72] D. Praticó, K. Uryu, S. Sung, S. Tang, J. Q. Trojanowski, and V. M.-Y. Lee, "Aluminum modulates brain amyloidosis through oxidative stress in APP transgenic mice," *The FASEB Journal*, vol. 16, no. 9, pp. 1138–1140, 2002.

[73] J.-Y. Lee, J. E. Friedman, I. Angel, A. Kozak, and J.-Y. Koh, "The lipophilic metal chelator DP-109 reduces amyloid pathology in brains of human β-amyloid precursor protein transgenic mice," *Neurobiology of Aging*, vol. 25, no. 10, pp. 1315–1321, 2004.

[74] J. L. Jankowsky, L. H. Younkin, V. Gonzales et al., "Rodent Aβ modulates the solubility and distribution of amyloid deposits in transgenic mice," *Journal of Biological Chemistry*, vol. 282, no. 31, pp. 22707–22720, 2007.

[75] Y. Shinkai, M. Yoshimura, M. Morishima-Kawashima et al., "Amyloid β-protein deposition in the leptomeninges and cerebral cortex," *Annals of Neurology*, vol. 42, no. 6, pp. 899–908, 1997.

[76] J. Wang, D. W. Dickson, J. Q. Trojanowski, and V. M.-Y. Lee, "The levels of soluble versus insoluble brain aβ distinguish Alzheimer's disease from normal and pathologic aging," *Experimental Neurology*, vol. 158, no. 2, pp. 328–337, 1999.

[77] J. Fonte, J. Miklossy, C. Atwood, and R. Martins, "The severity of cortical Alzheimer's type changes is positively correlated with increased amyloid-β levels: resolubilization of amyloid-β with

transition metal ion chelators," *Journal of Alzheimer's Disease*, vol. 3, no. 2, pp. 209–219, 2001.

[78] R. L. Patton, W. M. Kalback, C. L. Esh et al., "Amyloid-β peptide remnants in AN-1792-immunized Alzheimer's disease patients: a biochemical analysis," *American Journal of Pathology*, vol. 169, no. 3, pp. 1048–1063, 2006.

[79] K. A. Bates, G. Verdile, Q.-X. Li et al., "Clearance mechanisms of Alzheimer's amyloid-B peptide: implications for therapeutic design and diagnostic tests," *Molecular Psychiatry*, vol. 14, no. 5, pp. 469–486, 2009.

[80] G. C. Gregory and G. M. Halliday, "What is the dominant aβ species in human brain tissue? A review," *Neurotoxicity Research*, vol. 7, no. 1-2, pp. 29–41, 2005.

[81] F. Peña and N. Alavez-Pérez, "Epileptiform activity induced by pharmacologic reduction of M-current in the developing hippocampus in vitro," *Epilepsia*, vol. 47, no. 1, pp. 47–54, 2006.

[82] J. M. Fellous and T. J. Sejnowski, "Cholinergic induction of oscillations in the hippocampal slice in the slow (0.5–2 Hz), theta (5–12 Hz), and gamma (35–70 Hz) bands," *Hippocampus*, vol. 10, pp. 187–197, 2000.

[83] K. Winiewski and H. Car, "(S)-3,5-DHPG: a review," *CNS Drug Reviews*, vol. 8, no. 1, pp. 101–116, 2002.

[84] G. Buzsáki, "Theta oscillations in the hippocampus," *Neuron*, vol. 33, no. 3, pp. 325–340, 2002.

[85] J. Shin, D. Kim, R. Bianchi, R. K. S. Wong, and H.-S. Shin, "Genetic dissection of theta rhythm heterogeneity in mice," *Proceedings of the National Academy of Sciences of the United States of America*, vol. 102, no. 50, pp. 18165–18170, 2005.

[86] F. Nobili, F. Copello, P. Vitali et al., "Timing of disease progression by quantitative EEG in Alzheimer's patients," *Journal of Clinical Neurophysiology*, vol. 16, no. 6, pp. 566–573, 1999.

[87] A. Jyoti, A. Plano, G. Riedel, and B. Platt, "EEG, activity, and sleep architecture in a transgenic AβPPswe/PSEN1A246E Alzheimer's disease mouse," *Journal of Alzheimer's Disease*, vol. 22, pp. 873–887, 2010.

[88] F. M. LaFerla, K. N. Green, and S. Oddo, "Intracellular amyloid-β in Alzheimer's disease," *Nature Reviews Neuroscience*, vol. 8, no. 7, pp. 499–509, 2007.

[89] L. Verret, E. O. Mann, G. B. Hang et al., "Inhibitory interneuron deficit links altered network activity and cognitive dysfunction in alzheimer model," *Cell*, vol. 149, no. 3, pp. 708–721, 2012.

[90] G. J. Chen, Z. Xiong, and Z. Yan, "Aβ impairs nicotinic regulation of inhibitory synaptic transmission and interneuron excitability in prefrontal cortex," *Molecular Neurodegeneration*, vol. 8, article 3, 2013.

[91] S. E. Rubio, G. Vega-Flores, A. Martinez, C. Bosch, A. Perez-Mediavilla, and J. del Rio, "Accelerated aging of the GABAergic septohippocampal pathway and decreased hippocampal rhythms in a mouse model of Alzheimer's disease," *The FASEB Journal*, vol. 26, pp. 4458–4467, 2012.

[92] C. A. Chapman and J.-C. Lacaille, "Cholinergic induction of theta-frequency oscillations in hippocampal inhibitory interneurons and pacing of pyramidal cell firing," *Journal of Neuroscience*, vol. 19, no. 19, pp. 8637–8645, 1999.

[93] C. C. Felder, "Muscarinic acetylcholine receptors: signal transduction through multiple effectors," *The FASEB Journal*, vol. 9, no. 8, pp. 619–625, 1995.

[94] H.-M. Huang, H.-C. Ou, and S.-J. Hsieh, "Amyloid β peptide impaired carbachol but not glutamate-mediated phosphoinositide pathways in cultured rat cortical neurons," *Neurochemical Research*, vol. 25, no. 2, pp. 303–312, 2000.

[95] H. Janickova, V. Rudajev, P. Zimcik, J. Jakubik, H. Tanila, and E. E. El-Fakahany, "Uncoupling of M1 muscarinic receptor/G-protein interaction by amyloid β_{1-42}," *Neuropharmacology*, vol. 67, pp. 272–283, 2013.

[96] K. N. Dahlgren, A. M. Manelli, W. Blaine Stine Jr., L. K. Baker, G. A. Krafft, and M. J. Ladu, "Oligomeric and fibrillar species of amyloid-β peptides differentially affect neuronal viability," *Journal of Biological Chemistry*, vol. 277, no. 35, pp. 32046–32053, 2002.

[97] E. McGowan, F. Pickford, J. Kim et al., "Aβ42 is essential for parenchymal and vascular amyloid deposition in mice," *Neuron*, vol. 47, no. 2, pp. 191–199, 2005.

[98] G. Bitan, M. D. Kirkitadze, A. Lomakin, S. S. Vollers, G. B. Benedek, and D. B. Teplow, "Amyloid β-protein (Aβ) assembly: Aβ40 and Aβ42 oligomerize through distinct pathways," *Proceedings of the National Academy of Sciences of the United States of America*, vol. 100, no. 1, pp. 330–335, 2003.

[99] A. Sandberg, L. M. Luheshi, S. Söllvander et al., "Stabilization of neurotoxic Alzheimer amyloid-β oligomers by protein engineering," *Proceedings of the National Academy of Sciences of the United States of America*, vol. 107, no. 35, pp. 15595–15600, 2010.

[100] H. Lin, R. Bhatia, and R. Lal, "Amyloid β protein forms ion channels: implications for Alzheimer's disease pathophysiology," *The FASEB Journal*, vol. 15, no. 13, pp. 2433–2444, 2001.

13

Phage Display Screening for Tumor Necrosis Factor-α-Binding Peptides: Detection of Inflammation in a Mouse Model of Hepatitis

Coralie Sclavons,[1] Carmen Burtea,[1] Sébastien Boutry,[2] Sophie Laurent,[1] Luce Vander Elst,[1] and Robert N. Muller[1,2]

[1] Department of General, Organic and Biomedical Chemistry, NMR and Molecular Imaging Laboratory, University of Mons, Mendeleïev Building, 19 Avenue Maistriau, 7000 Mons, Belgium
[2] Center for Microscopy and Molecular Imaging (CMMI), 8 Rue Adrienne Bolland, 6041 Gosselies, Belgium

Correspondence should be addressed to Robert N. Muller; robert.muller@umons.ac.be

Academic Editor: Tzi Bun Ng

TNF-α is one of the most abundant cytokines produced in many inflammatory and autoimmune conditions such as multiple sclerosis, chronic hepatitis C, or neurodegenerative diseases. These pathologies remain difficult to diagnose and consequently difficult to treat. The aim of this work is to offer a new diagnostic tool by seeking new molecular probes for medical imaging. The target-specific part of the probe consists here of heptameric peptides selected by the phage display technology for their affinity for TNF-α. Several affinity tests allowed isolating 2 peptides that showed the best binding capacity to TNF-α. Finally, the best peptide was synthesized in both linear and cyclic forms and tested on the histological sections of concanavalin-A-(ConA-)treated mice liver. In this well-known hepatitis mouse model, the best results were obtained with the cyclic form of peptide 2, which allowed for the staining of inflamed areas in the liver. The cyclic form of peptide 2 (2C) was, thus, covalently linked to iron oxide nanoparticles (magnetic resonance imaging (MRI) contrast agent) and tested in the ConA-induced hepatitis mouse model. The vectorized nanoparticles allowed for the detection of inflammation as well as of the free peptide. These *ex vivo* results suggest that phage display-selected peptides can direct imaging contrast agents to inflammatory areas.

1. Introduction

Tumor necrosis factor alpha (TNF-α) is a proinflammatory cytokine produced in many inflammatory and autoimmune diseases such as rheumatoid arthritis, Crohn's disease, multiple sclerosis, or chronic hepatitis C [1, 2]. TNF-α is produced by different cell types including macrophages, monocytes, T-cells, smooth muscle cells, adipocytes, and fibroblasts. This cytokine is also implicated in the diseases of the central nervous system like Alzheimer's and Parkinson's diseases [3], where it can be produced by several cell populations, including microglia, astrocytes, endothelial cells, Th1 lymphocytes and neurons.

Mature TNF-α is secreted as a 157-amino acid form [4] with a molecular weight of 17 kDa [5]. Before being released from cells, TNF-α is anchored in the plasma membrane as a 26 kDa precursor containing both hydrophobic and hydrophilic regions [6]. The 17 kDa form of TNF-α is excised from the integral transmembrane precursor by proteolytic cleavage mediated by the tumor necrosis factor alpha converting enzyme (TACE) [7]. Soluble and transmembrane TNF-α are produced by cells as homotrimers that bind to two kinds of receptors, TNF-RI and TNF-RII (tumor necrosis factor receptor type I, p55; type II, p75, resp.), which are present in the membrane of all cell types except erythrocytes.

TNF-α, which is one of the most abundant cytokines involved in apoptotic and inflammatory pathways, has both desirable and undesirable effects. Tumor necrosis induction and mediation of the immune response to bacterial, viral, and parasitic invasions are beneficial functions of TNF-α

[6]. It is also an acute phase protein that initiates a cascade of cytokines and increases vascular permeability, thereby recruiting macrophages and neutrophils to a site of infection. However, TNF-α can also have pathological consequences such as promoting the growth of some tumor cell types. It also plays an important role in the chronic inflammation that occurs in various pathologies and has been identified as the major mediator in various autoimmune diseases [8, 9]. TNF-α thus represents a good marker of inflammatory events.

Phage display is a high-throughput screening (HTS) method. It is an effective way of selecting target-specific proteins and peptides that can be synthesized and linked to an imaging reporter for diagnostic use. This technique can be used to identify peptides or antibodies capable of interacting with inflammatory mediators [10, 11].

In the present work, a heptapeptide phage display library was screened against TNF-α, as a first step in the development of new molecular imaging tools for detecting inflammation in chronic inflammatory pathologies and diseases of uncertain diagnosis. Peptide specificity for the cytokine was tested on histological liver sections of mice treated with concanavalin A (ConA), which is a mitogenic lectin known to stimulate lymphocytes to produce lymphokines [12] mediating chronic inflammation processes [13] and inducing severe injury to hepatocytes [14]. ConA induces an autoimmune hepatitis due to massive CD4+ lymphocyte activation and infiltration into the liver parenchyma leading to the secretion of the proinflammatory cytokines TNF-α, interferon-γ (IFNγ), interleukin (IL)-2, IL-6, granulocyte macrophage-colony stimulating factor, and IL-1 [15]. The phage display-selected heptapeptide was then grafted to iron oxide nanoparticles (known for their use as MRI contrast agents). This new peptide was thus set up as an iron oxide-based probe and used to detect inflammatory process on histological liver sections prior to further *in vivo* MRI tests.

2. Methods

2.1. Phage Display

2.1.1. The Biopanning of PhD-C7C Phage Display Library against TNF-α. The well of an ELISA microtiter plate was coated with 100 μL of target solution (0.1 mg/mL of human TNF-α (GenScript Corporation, Piscataway, USA) in 0.1 M NaHCO$_3$ buffer, pH 8.6) by overnight incubation at 4°C in a humid chamber. The next day, the target solution was removed and replaced by the blocking buffer (Bovine Serum Albumin, 5 mg/mL; 0.1 M NaHCO$_3$, pH 8.6, NaN$_3$ 0.02%) for 2 hours and finally washed with Tris-buffered saline (TBS) supplemented with 0.1% Tween-20 (TBS-T, 50 mM Tris-HCl, 150 mM NaCl, pH 7.4). After negative selection on a BSA-coated well, the phage library (2 × 10^{11} phages in 100 μL of TBS-T 0.1%; PhD-C7C, New England Biolabs Inc., Westburg B. X., Leusden, The Netherlands) was incubated during 1 hour at 37°C with the target. After rinsing, TNF-α-bound phages were eluted with 0.2 M glycine-HCl buffer (pH 2.2) complemented with 0.1% BSA and then neutralized with 1 M Tris-HCl buffer (pH 9.1).

Eluted phages were amplified during 4 h30 via *Escherichia coli* (ER2738 host strain, New England Biolabs Inc.) infection. Amplified phages were collected by two precipitations at 4°C in PEG-NaCl solution (20% polyethylene glycol-8000, 2.5 M NaCl). The phage pellet was finally solubilized in a TBS buffer solution (50 mM Tris-HCl, 150 mM NaCl, pH 7.5). This succession of steps was repeated 4 times and constitutes a biopanning round. The selective pressure was increased during the third and the fourth rounds of biopanning by increasing the Tween-20 concentration in the incubation and rinsing buffers to 0.3% and 0.5%, respectively, and by reducing the incubation time to 45 min and 30 min, respectively.

E. coli was grown on a selective medium containing isopropyl-beta-D-thiogalactoside (IPTG) (ICN Biomedical Inc., Brussels, Belgium) and 5-bromo-4-chloro-3-indolyl-beta-D-galactopyranoside (Xgal) (Sigma-Aldrich, Bornem, Belgium). The phage genome contains a part of the LacZ gene that confers to bacteria the ability to produce β-galactosidase, which reacts with the X-gal substrate and results in a blue plaque staining. Phage titer is, thus, achieved by counting phage-infected *E. coli* (blue-colored) colonies after each biopanning round.

2.1.2. Sequencing of Selected Phage Clones. The genome sequencing of selected phage clones was based on the Sanger method which uses dideoxynucleotides triphosphate as DNA chain terminators. Briefly, DNA is extracted by the phenol/chloroform extraction procedure [16] and denatured by several heating cycles. Virus genome is sequenced by using a Start Mix solution (Beckman Coulter, Analis, Namur, Belgium) and a 20-base primer (5′-CCCTCATAGTTAGCGTAACG-3′, New England Biolabs Inc.) located 96 nucleotides upstream to the inserted peptide-encoding sequence. The Start Mix solution is the sequencing reaction buffer containing 4 ddNTPs, 4 dNTPs, and the DNA polymerase enzyme.

The DNA sequence was analyzed on a CEQ 2000 XL DNA Analysis System (Beckman Coulter, Analis). The sequence reading was performed automatically using the JaMBW 1.1 software (http://bioinformatics.org/JaMBW/).

2.1.3. Evaluation of the Affinity of Selected Clones for the Target. (a) *Estimation of the Apparent Dissociation Constant* (K_d^*) *between the Phage Clones and TNF-α.* The K_d^* of phage clones against their target was assessed by the ELISA (enzyme-linked immunosorbent assay) method, using a range of phage dilutions incubated with the specific target.

ELISA plates were treated overnight with the target solution (human TNF-α, 0.01 mg/mL, at 4°C). The next day, the target solution was replaced by the blocking solution as described above. Plates were then incubated for 2 h at 37°C with serial phage dilutions (6.5 × 10^{-11} – 3.3 × 10^{-8} M) prepared in PBS containing 0.3% Tween 20. After rinsing the plate 6 times with the same buffer, phages were detected with a peroxidase-conjugated anti-M13 monoclonal antibody (Amersham Pharmacia Biotech Benelux, Roosendaal, The Netherlands) diluted 1 : 5000 in the blocking buffer. The

Phage Display Screening for Tumor Necrosis Factor-α-Binding Peptides: Detection of Inflammation in
a Mouse Model of Hepatitis

95

peroxidase detection was performed using 2,2′-azino-bis[3-ethylbenztiazoline-6-sulfonic acid, diammonium salt] (ABTS, Sigma-Aldrich; 22 mg in 100 mL of 50 mM sodium citrate, pH 4.0) supplemented with 0.05% H_2O_2. After 30–60 min of incubation at room temperature, the OD_{405} values were measured on a microplate reader (Stat-Fax-2100, Awareness Technology Inc., Fisher Bioblock Scientific, Tournai, Belgium) for K_d^* evaluation.

(b) Competition Experiments with an Anti-TNF-α Antibody. Competition experiments were performed in the same manner as the K_d^* experiments except that clones were set in competition with an anti-TNF-α antibody (R&D Systems Inc., Abington, UK) in dilutions ranging from 1.3×10^{-9} to 6.66×10^{-7} M in a PBS solution. Plates were first incubated for 30 min with 50 μL/well of antibody dilutions before a one-hour incubation with 50 μL of phage solutions at the K_d^* concentration. The IC_{50} of the anti-TNF-α antibody is the antibody concentration required to inhibit by 50% the interaction between phages and TNF-α.

2.1.4. Synthesis of Biotinylated Peptides. Peptides displayed on the surface of selected phage clones were synthesized in a biotinylated form by NeoMPS (Polypeptide group, Strasbourg, France). Biotinylated peptides were synthesized in two forms: a linear form (L) and a cyclic (C) form.

2.1.5. The Evaluation of the Affinity of Selected Peptides for the Target. The affinity of the selected peptides was evaluated by ELISA as for the phage clones (see Section 2.1.3). The target was immobilized overnight on a 96-well plate at 4°C before a 2-hour incubation with a protein-free blocking buffer solution (PFBB, Pierce, Aalst, Belgium) and then with a range of peptide dilutions: 10^{-3} M to 10^{-6} M. The target-bound biotinylated peptides were detected using ABC (Avidin Biotin Complex) and ABTS solutions as described above. The dissociation constant was estimated from the inflection point of the curve.

2.2. The Physicochemical Characterization of the 2C Peptide-Grafted Nanoparticles. The phage display-selected 2C peptide was grafted to ultrasmall nanoparticles of iron oxide (USPIO). The particles were pegylated with the amino-PEG 750 (Fluka, Bornem, Belgium) to avoid the rapid uptake of USPIO by macrophages and to prolong the blood circulation time.

Peptide-USPIO-PEG was prepared in our laboratory from previously described nanoparticles bearing carboxylated groups on their surface [17–19]. USPIOs were functionalized in two successive steps: first, they were functionalized with the peptide, and, second, they were coupled to an amino-PEG 750. Briefly, 19 mg of the peptide and 6 mg of EDCI (1-ethyl-3-(3-dimethylaminopropyl) carbodiimide) were added to 15 mL of nanoparticles ([Fe] = 0.175 Mol. Peptide/Fe molar percent is 0.02%), and the mixture was stirred overnight. The nanoparticle suspension was ultrafiltrated on a 30 kD membrane. Amino-PEG 750 (0.503 g) and EDCI (0.327 g) were added to the nanoparticles (PEG/Fe molar percent is

2%) and the reaction was stirred during 17 h. The mixture was ultrafiltrated on a 30 kD membrane to remove low-molecular material.

NMRD (Nuclear Magnetic Resonance Dispersion) profiles were recorded at 37°C on a Fast Field Cycling Relaxometer (Stelar, Mede, Italy) over a magnetic field range from 0.24 mT to 0.24 T. Additional longitudinal (R_1) and transverse (R_2) relaxation rate measurements at 0.47 T and 1.41 T were obtained on Minispec MQ 20 and MQ 60 spin analyzers (Bruker, Karlsruhe, Germany), respectively. The fitting of the NMRD profiles by a theoretical relaxation model [20] allowed the determination of the crystal radius (r) and the specific magnetization (M_S).

Hydrodynamic size measurement was carried out by photon correlation spectroscopy (PCS) on a Zêtasizer Nanoseries ZEN 3600 (Malvern, UK).

Total iron concentration was determined by proton (^{1}H) relaxometry at 20 MHz and 310 K after microwave digestion (MLS-1200 MEGA, MILESTONE, Analis, Namur, Belgium).

2.3. Animals and Treatments. Adult male mice C57BL/6JOlaHsd (6 months old, Harlan Laboratories B.V., The Netherlands) were used.

All animals were housed in plastic cages under a 12 h light/12 h dark cycle and had free access to food and water. Ambient temperature was maintained at 25 ± 2°C. Animals were maintained and treated in compliance with the guidelines specified by the Belgian Ministry of Trade and Agriculture.

Adult mice ($n = 6$) received a single i.p. injection of 20 mg/Kg of ConA [21], a lymphocyte T activator producing inflammation in mice liver.

2.4. Immunohistochemistry

2.4.1. Tissue Preparation. Mice were sacrificed 2 hours after intravenous injection of ConA, and their livers were fixed for 24 h in 4% paraformaldehyde. After fixation, the liver was dehydrated by several alcohol and butanol soakings and the organ was subsequently embedded in paraffin according to standard procedures. Sections of 4-micrometer thickness were cut for histological tests.

2.4.2. TNF-α Detection by a Polyclonal Antibody in Liver Histological Sections. Liver sections were dewaxed and rehydrated by several washes in toluene and alcohol. They were subsequently treated for 5 minutes with 2% H_2O_2 in order to block endogenous peroxidases. Endogenous biotins and nonspecific epitopes were blocked successively with an avidin/biotin and casein solution before proceeding to incubate the liver sections overnight with a rabbit anti-mouse TNF-α polyclonal antibody (1 : 100, Abcam, Paris, France). The next day, liver sections were incubated for 1 hour with a biotinylated goat anti-rabbit antibody (Vector Labs, Brussels, Belgium). Reactive sites were detected with the avidin-biotin peroxidase complex (ABC, Vector Labs) and revealed with DAB (3,3′-diaminobenzidine, Sigma-Aldrich, Bornem, Belgium) and 0.02% of H_2O_2 prepared in PBS.

Sections were washed in distilled water to stop the reaction. Counterstaining of liver tissue was performed with Luxol fast blue before mounting in a permanent medium.

2.4.3. TNF-α Detection by 2C and 2L Peptides in Liver Histological Sections.
TNF-α detection by the phage display-selected 2C and 2L peptides was carried out with the same protocol as for anti-TNF-α polyclonal antibody (see Section 2.4.2). Briefly, liver sections were incubated overnight with the biotinylated peptides (20 μM). Reactive sites were detected with the ABC complex and DAB staining.

2.4.4. TNF-α Detection by 2C Peptide-USPIO-PEG in Liver Histological Sections.
The selected peptide (2C) was grafted to USPIO-PEG in order to produce a vectorized MRI contrast agent for the specific targeting of injured areas. The specificity of the 2C peptide-USPIO-PEG was tested on liver sections. Slices were deparaffinated in toluene and rehydrated in alcohol. Sections were then incubated overnight with 30 mM of 2C peptide-USPIO-PEG at 4°C before being immersed for 30 minutes in Prussian blue reagent (5% potassium ferrocyanide in PBS-HCl 2% solution) for the iron staining. Slides were counterstained with eosin during 8 minutes before mounting in a permanent medium.

3. Results

3.1. Characterization of the Selected Phage Clones

3.1.1. TNF-α Affinity.
The screening of a cyclic heptamer peptide library displayed by the phage coat allowed for the selection of 48 phage clones from round 3 and 50 phage clones from round 4 of the selection-amplification procedure. Therefore, 98 phages presenting an affinity for the target were collected.

Nineteen clones from round 3 and 23 clones from round 4 were chosen for their higher affinity for TNF-α and their lower affinity for BSA coated plates (TNF-α/BSA = 2.9 ± 0.7 [mean value ± SEM] for round 3 clones and TNF-α/BSA = 3.1 ± 0.5 [mean value ± SEM] for round 4 clones).

Based on these results, 28 clones with higher affinity were, thus, selected for further characterization (Figure 1).

3.1.2. Peptide Sequence and Biochemical Parameters of the Selected TNF-α Binding Clones.
The DNA of the 28 phage clones selected for their affinity to the TNF-α cytokine was sequenced. Fifteen different peptide inserts were identified. Peptide sequences mainly contained hydrophobic amino acids (61%), that is, glycine, leucine, proline, and tryptophan, occupying the first 5 positions in the peptide (Figure 2). Hydrophilic amino acids with a hydroxyl function (i.e., serine and threonine) are less well represented (39%) than hydrophobic amino acids and are most predominant in the last two positions.

Among the fifteen different peptides that were identified from the 28 determined peptide sequences, the cyclic peptides C-GLPWLST-C and C-GPAVYMK-C were found 13 and 2 times, respectively (Table 1).

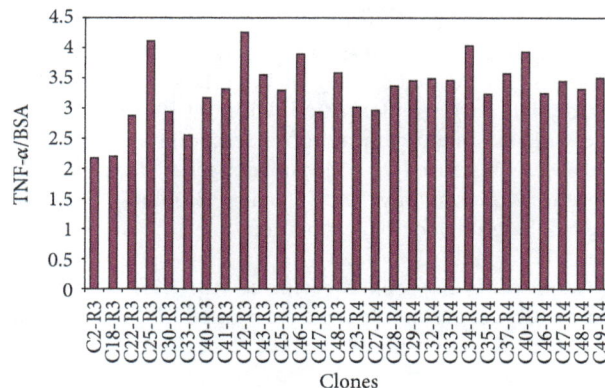

FIGURE 1: Affinity of the 28 selected clones for the TNF-α cytokine. Results are presented as a ratio between the clones' affinity for TNF-α and that for BSA. Clones show 4-times higher affinity for TNF-α than for BSA. The phage clones were selected after the biopanning rounds 3 and 4.

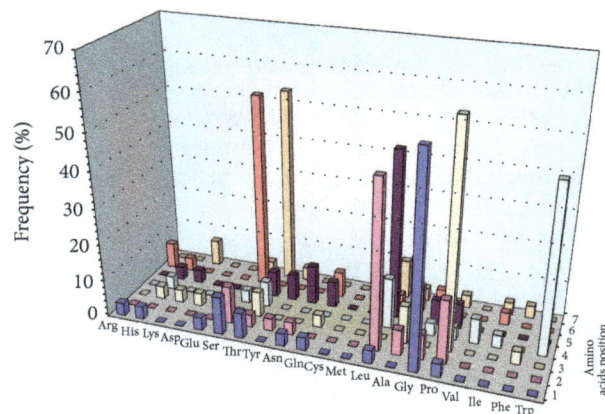

FIGURE 2: The analysis of the amino acid frequency in each position of the heptapeptide sequences. The 5 first positions are mainly occupied by hydrophobic amino acids while the last two positions are mainly occupied by hydrophilic amino acids.

Two biochemical parameters of these peptides were theoretically estimated by using the EXPASY ProtParam tools (http://www.expasy.ch/tools/protparam.html).

The predicted half-life is the time required for half of the intravenously injected peptide dose to be degraded. The peptides' half-life must be long enough to allow specific interactions between peptides and their targets to take place (e.g., during MRI scans with the peptides linked to contrast agents). The half-life of peptides ranges between 0.8 and 30 hours. The longest predicted half-lives (20–30 hours) were found for the following peptides: C-PATLTSL-C, C-GPAVYMK-C, C-GLPWLST-C, and C-GSKTQAP-C (Table 1).

The isoelectric point (pI) provides information that is important for further applications in physiological conditions. Indeed, an electrical charge borne by the peptide at physiological pH can change its biodistribution in vivo. A change of the peptide charge can also affect the affinity for

Phage Display Screening for Tumor Necrosis Factor-α-Binding Peptides: Detection of Inflammation in
a Mouse Model of Hepatitis

97

TABLE 1: Biochemical properties of the selected peptide clones (from http://www.expasy.ch/tools/protparam.html).

Clone	Sequence	$T_{1/2}$ (hours)	pI
C2-R3	**C-GSKTQAP-C**	**30**	**8.06**
C18-R3	C-GLPWLST-C	30	5.51
C22-R3	**C-STPHNLG-C**	**1.9**	**6.72**
C25-R3	C-LATGNQI-C	5.5	5.51
C30-R3	C-PATLTSL-C	20	5.51
C33-R3	**C-HGAPNRL-C**	**3.5**	**8.08**
C41-R3	**C-RPPIGAF-C**	**1**	**8.07**
C42-R3	**C-TSQSQHM-C**	**7.2**	**6.72**
C45-R3	**C-GPAVYMK-C**	**30**	**8.05**
C46-R3	C-QGDLPGY-C	0.8	3.80
C32-R4	**C-SPHTTIA-C**	**1.9**	**6.72**
C33-R4	C-EPFAGRS-C	1	5.99
C35-R4	C-TSPLPGT-C	7.2	5.51
C46-R4	C-SYEAHQT-C	1.9	5.24
C47-R4	**C-NNPLKSL-C**	**1.4**	**8.06**

$T_{1/2}$: half-life theoretically estimated in mammalian reticulocytes *in vitro* and based on the N-end rule [20].

pI: isoelectric point which is the pH at which a particular molecule or surface carries no net electrical charge.

the target and consequently that of the conjugated imaging agent. The pI values of the 28 selected peptides were found to vary from 3.8 to 8.08. Eight of them would be in an ionized state at a pH near the physiological pH of 7.4 (Table 1, bold text).

3.1.3. Estimation of the K_d^* between the Phage Clones and TNF-α.
The K_d^* value was estimated for 15 phage clones. One clone was chosen to represent the sequences that were found several times. The clone that showed the best K_d^* value is C30-R3 (clone 30 from round 3, C-PATLTSL-C, $K_d^* = 2.05 \times 10^{-11}$ M). The lowest affinity was found for clone C47-R4 (clone 47 from round 4, C-NNPLKSL-C, $K_d^* = 9.89 \times 10^{-6}$ M). All the other clones were found to have a K_d^* value lower than 10^{-8} M.

The best clones were chosen on the basis of the ratio K_d^* BSA/K_d^* TNF-α to be tested in competition with an anti-TNF-α antibody (Figure 3). Seven phage clones were thus selected for these competition tests: C2-R3 (C-GSKTQAP-C), C22-R3 (C-STPHNLG-C), C30-R3 (C-PATLTSL-C), C41-R3 (C-RPPIGAF-C), C42-R3 (C-TSQSQHM-C), C45-R3 (C-GPAVYMK-C), and C46-R3 (C-QGDLPGY-C). The highest affinity clones were C2-R3 and C30-R3, with a K_d BSA/K_d TNF-α ratio of 6×10^6 and 4.2×10^6, respectively; the clone showing the lowest affinity was again C47-R4, with a K_d BSA/K_d TNF-α ratio of 0.02.

3.1.4. Competition Experiments with an Anti-TNF-α Antibody.
The objective of *this* test is to set up a competition between every phage clone and another ligand of the target to calculate an inhibition constant value (IC$_{50}$). A high IC$_{50}$ value indicates that a significant concentration of antibody is

FIGURE 3: Affinity of the 15 selected phage clones. Clones having the best affinity are the C2-R3 and C30-R3 clones with a K_d BSA/K_d TNF-α of 6×10^6 and 4.2×10^6, respectively.

necessary to displace phages from their binding sites on the target.

The best phage clone-target interaction was found for C2-R3 and C30-R3 (IC$_{50}$ values of 1.12×10^{-7} M and 1.27×10^{-7} M, resp.).

Subsequently, the best phage clones were selected on the basis of their IC$_{50}^{\text{Ab anti-TNF-α}}$/$K_d^{\text{clone}}$ ratio, which reflects the highest target affinity of a phage clone (Figure 4). The clones C2-R3 and C30-R3 showed the highest ratio (1087 and 6195, resp.).

3.2. The Characterization of the Phage Display-Selected Peptides

3.2.1. The Estimation of K_d^* of the Selected Peptides for TNF-α.
The K_d^* values of cyclic and linear biotinylated peptides (C-)GSKTQAP(-C) (peptide 1) and (C-)PATLTSL(-C) (peptide 2) were evaluated. (C-)PATLTSL(-C) showed a better affinity for the target than (C-)GSKTQAP(-C). Linear and cyclic forms of GSKTQAP weakly interact with TNF-α and the curve does not reach saturation. In comparison with the GSKTQAP peptide, the linear and cyclic forms of the PATLTSL peptide (resp., 2L and 2C peptides) have a good ability to interact with the target (K_d^* values of 1.79×10^{-4} and 1.12×10^{-4}, resp.). Both linear and cyclic forms show only a weak interaction with the PFBB coated plate; the K_d^* values could not be estimated.

3.3. The Physicochemical Characterization of the 2C Peptide Grafted to the USPIO-PEG Nanoparticles.
The 2C peptide-USPIO-PEG was prepared from iron oxide nanoparticles that were functionalized first with the peptide and then with an amino-PEG 750. These grafting steps do not change significantly the relaxometric properties of the nanoparticle. NMRD profiles showing the changes in relaxivity (1/T1) versus the applied magnetic field strength are given in Figure 5.

The physicochemical properties of the nanoparticles included a hydrodynamic size of 33 nm (determined by PCS),

$$IC_{50}^{AC\ anti\text{-}TNF\alpha}/K_d^{clone}$$

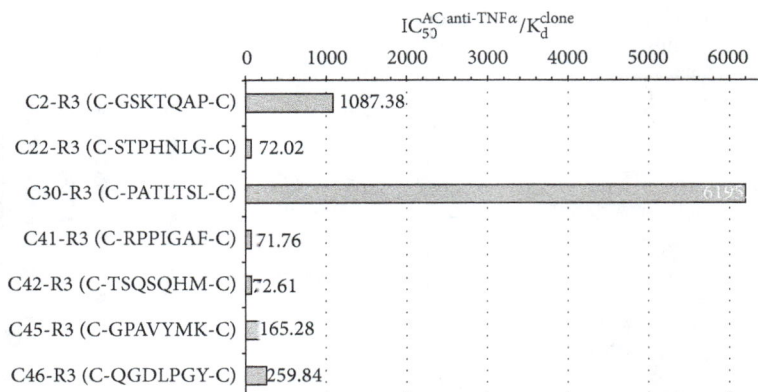

FIGURE 4: $IC_{50}^{AC\ anti\text{-}TNF\text{-}\alpha}/K_d^{clone}$ ratio of the 7 phage clones selected for their high K_d BSA/K_d TNF-α ratio. Clones C2-R3 and C30-R3 have the best affinity for the target ($IC_{50}^{AC\ anti\text{-}TNF\text{-}\alpha}/K_d^{clone}$ ratio values of 1087 and 6195, resp.). These results are in agreement with the previous ones.

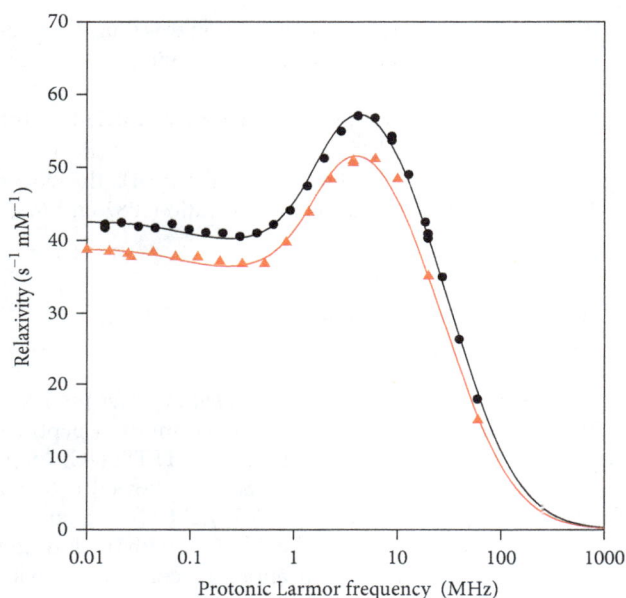

FIGURE 5: NMRD profiles of 2C peptide-USPIO-PEG (black circles) and USPIO-PEG (red triangles).

an r_1 of 40.2 s^{-1} mM^{-1} and r_2 of 81.6 s^{-1} mM^{-1} at 20 MHz, 37°C; an r_1 of 17.9 s^{-1} mM^{-1} and r_2 of 82.9 s^{-1} mM^{-1} at 60 MHz, 37°C.

The theoretical fitting of the NMRD profile according to the superparamagnetic relaxation model [20] gave a magnetization Msat of 55.82 Am^2/kg and a radius of 5.76 nm. The vectorization with peptides and PEG does not change significantly the magnetic properties of the nanoparticles.

3.4. Immunohistochemical Assay

3.4.1. TNF-α Immunodetection with a Polyclonal Antibody in a Mouse Model of Hepatitis. To confirm the specific affinity of the phage display-selected peptide PATLTSL, called peptide 2, to TNF-α, the proinflammatory cytokine was targeted in a well-known mouse model of hepatitis.

TNF-α was detected with a polyclonal anti-TNF-α antibody on liver tissue sections (Figures 6(a) and 6(b)) and was mainly detected around hepatic sinusoids 2 hours after i.v. ConA injection.

3.4.2. TNF-α Detection by the Phage Display-Selected 2C and 2L Peptides. The affinity of the phage display-selected 2C and 2L peptides for TNF-α was evaluated by immunohistochemistry on liver sections obtained from mice treated with ConA. The cyclic form of peptide 2 produced a similar staining to that obtained with the polyclonal anti-TNF-α antibody (Figures 6(c) and 6(d)). On the contrary, the linear form did not yield such a result.

3.4.3. TNF-α Detection with the Phage Display-Selected 2C Peptide Grafted to USPIO-PEG. After proving the affinity of the biotinylated 2C peptide on liver sections, this phage display-selected peptide was grafted to USPIO-PEG to study the affinity of the vectorized probe for the target. USPIO was detected with the Prussian blue iron staining method. A blue coloration indicated the presence of iron around sinusoids and blood vessels on ConA-treated liver sections (Figure 6(e)), similar to the results obtained with the biotinylated form of the 2C peptide. Healthy liver did not show any blue coloration (Figure 6(f)), suggesting that the vectorized nanoparticles were able to bind their target in the liver sections of ConA-treated mice. The selected peptide did not lose its affinity for TNF-α after grafting to USPIO.

4. Discussion

Inflammation is the first response of the immune system to infection, injury, or irritation. During the acute phase of the inflammatory process, some molecular and cellular

Phage Display Screening for Tumor Necrosis Factor-α-Binding Peptides: Detection of Inflammation in
a Mouse Model of Hepatitis

99

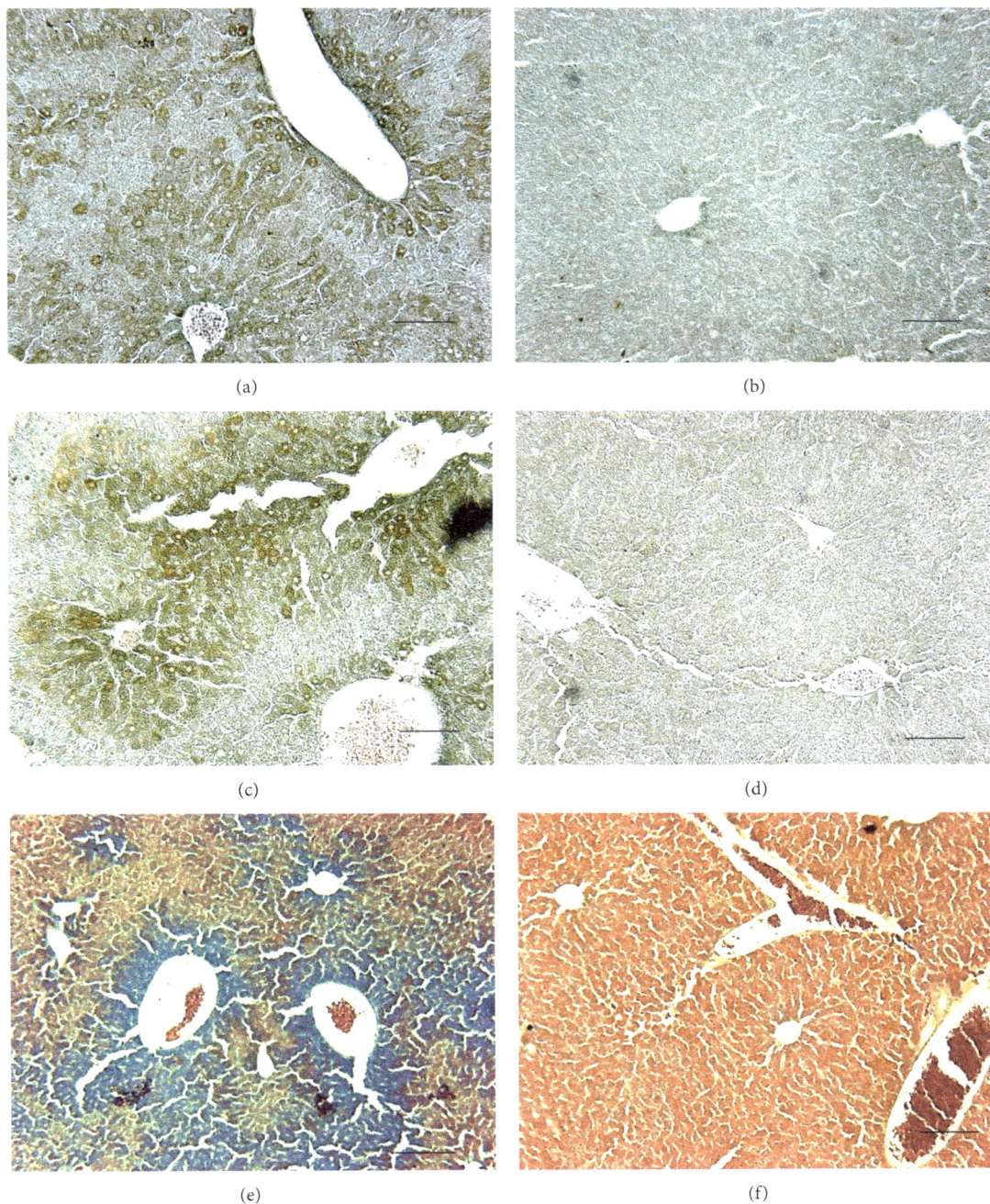

FIGURE 6: TNF-α detection in liver sections of ConA-treated mice. TNF-α was first detected with a polyclonal anti-TNF-α antibody in liver sections of ConA-treated mice by a brown coloration mainly around sinusoids and blood vessels (a) as compared to untreated-liver sections (b). TNF-α was then detected with the biotinylated 2C peptide (c) by a similar brown coloration located around sinusoids and blood vessels on liver sections of treated mice. The sections of healthy liver did not show such a staining (d). The 2C peptide-USPIO-PEG also allowed for the detection of the TNF-α cytokine on liver sections of treated mice (e), while it was not detectable on healthy liver sections (f). Scale bar: 100 μm.

changes occur like the accumulation of fluid, inflammatory cells, and soluble mediators. Many of these soluble mediators regulate the activation of resident cells such as monocytes, lymphocytes, or neutrophils, which results in the systemic response to the inflammatory process. These mediators include cytokines, inflammatory lipid metabolites, arachidonic acid derivatives (prostaglandin, leukotrienes), vasoactive amines like histamine or serotonin, and cascades of soluble proteases/substrates (clotting, complement, and kinin pathways), which generate numerous proinflammatory peptides. Among these factors, cytokines play key roles in mediating inflammatory reactions, especially IL and TNF, which are potent and primary inflammatory molecules mediating acute inflammation.

TNF-α is the most common proinflammatory cytokine secreted during the inflammatory process. It is involved in both acute and chronic inflammation and has a key position in the pathogenesis of various infectious and inflammatory diseases. It is abundantly present in inflammatory areas in two forms, a soluble and a transmembrane precursor forms. These characteristics make TNF-α a good candidate for inflammation targeting.

In the present study, we used the phage display technology to identify small peptide vectors for MRI detection of inflammation.

Results showed two phage displayed peptides having a better affinity for the target (TNF-α) among a selection of 28 phage clones. To optimize peptide target specificity, phage displayed peptides were incubated with TNF-α with an increasing concentration of destabilizing detergent during each round of selection. A competition test with an anti-TNF-α antibody was also performed to validate the specific interaction between peptides and their target.

Finally, the two best peptides were synthesized in a linear and a cyclic form, and K_d^* values were evaluated. The lowest K_d^* values of free peptides compared to phage displayed peptides might be explained by a multivalency effect allowed by the five peptide copies displayed by phages [22, 23]. However, it has been reported that a conformation constraint induced by disulfide cyclization of peptides could attenuate this affinity-decreasing effect [24]. The biotinylated linear and cyclic forms of peptide 2 were then chosen for histological studies on a well-known mouse model of hepatitis TNF-α was mainly detected around hepatic sinusoids 2 hours after i.v. ConA injection, in accordance with previously reported data [12, 21].

The best result was obtained with the cyclic form of peptide 2 (called 2C peptide) which allowed detecting TNF-α as efficiently as the anti-TNF-α antibody on liver histological sections. The cyclization gives the peptide a spatial conformation that may facilitate its interaction with the target, thus allowing better TNF-α staining results than those obtained with the linear form of the same peptide.

Finally, the 2C peptide was grafted to iron oxide nanoparticles that were evaluated on liver sections. Results obtained with the new specific iron oxide nanoparticles were compared to anti-TNF-α antibody and to the synthesized biotinylated 2C peptide experiments performed on the ConA-treated mice liver sections. The specific 2C peptide-USPIO-PEG showed the same labeling as the antibody and the biotinylated 2C peptide on histological liver sections, showing that the USPIO-PEG, thanks to the TNF-α specific 2C peptide linked to their surface, allowed for the staining of this proinflammatory cytokine in a ConA-induced mouse model of hepatitis.

At present, some inflammatory diseases like arthritis, asthma, atherosclerosis, Crohn's disease, or neurodegenerative diseases are not well understood and consequently not well diagnosed. Molecular imaging, which uses small probes (e.g., MRI contrast agents) vectorized to an injured area, offers new tools for diagnosing idiopathic diseases [25–27]. Increasingly, studies are focusing on new targeted approaches to the detection and treatment of inflammation [28–30].

The aim of this work was thus to develop specific iron oxide probes by vectorization with phage display-selected heptapeptides to offer a new tool for the MRI diagnosis of pathologies with an abundant inflammatory process. In this molecular approach of MRI, magnetic nanoparticles are optimized to reach a specific target, known to be a characteristic and abundant molecular feature of a certain disease. In this work, TNF-α was chosen as a target to detect inflammation. The phage display-selected 2C peptide and the specific 2C peptide-USPIO-PEG seemed to be able to detect inflammation by interacting with TNF-α in ConA-treated mice liver sections.

According to these results, the 2C peptide is appealing for molecular imaging applications. It can, for instance, be used as a vector for magnetic probes like iron oxide nanoparticles or gadolinium complexes, providing a new targeted agent for the detection of inflammation in MRI.

Acknowledgments

The authors thank Mrs. Patricia de Francisco for her help in preparing the paper and the following Contract/Grant sponsors: the European Cooperation in the Field of Scientific and Technical Research (COST Action D38); the European Network for Cell Imaging and Tracking Expertise (ENCITE) (Seventh Framework Program), FP7-HEALTH-2007-A; Action de Recherches Concertées of the French Community of Belgium, Convention AUWB-2010–10/15-UMONS-5; Fonds de la Recherche Scientifique (FNRS); the University of Mons; the Center for Microscopy and Molecular Imaging (CMMI) which is supported by the European Regional Development Fund and the Walloon Region.

References

[1] H. Knobler and A. Schattner, "TNF-α, chronic hepatitis C and diabetes: a novel triad," *QJMed*, vol. 98, no. 1, pp. 1–6, 2005.

[2] E. Larrea, N. Garcia, C. Qian, M. P. Civeira, and J. Prieto, "Tumor necrosis factor α gene expression and the response to interferon in chronic hepatitis C," *Hepatology*, vol. 23, no. 2, pp. 210–217, 1996.

[3] P. L. McGeer and E. G. McGeer, "Inflammation and the degenerative diseases of aging," *Annals of the New York Academy of Sciences*, vol. 1035, pp. 104–116, 2004.

[4] W. Fiers, "Precursor structures and structure-function analysis of TNF and lymphotoxin," *Immunology Series*, vol. 56, pp. 79–92, 1992.

[5] B. B. Aggarwal, B. Moffat, and R. N. Harkins, "Human lymphotoxin. Production by a lymphoblastoid cell line, purification, and initial characterization," *Journal of Biological Chemistry*, vol. 259, no. 1, pp. 686–691, 1984.

[6] J. Vilcek and T. H. Lee, "Tumor necrosis factor: new insights into the molecular mechanisms of its multiple actions," *Journal of Biological Chemistry*, vol. 266, no. 12, pp. 7313–7316, 1991.

[7] R. A. Black, C. T. Rauch, C. J. Kozlosky et al., "A metalloproteinase disintegrin that releases tumour-necrosis factor-Ø from cells," *Nature*, vol. 385, no. 6618, pp. 729–733, 1997.

[8] B. Beutler, I. W. Milsark, and A. C. Cerami, "Passive immunization against cachectin/tumor necrosis factor protects mice from lethal effect of endotoxin," *Science*, vol. 229, no. 4716, pp. 869–871, 1985.

[9] P. Vassalli, "The pathophysiology of tumor necrosis factors," *Annual Review of Cell and Developmental Biology*, vol. 10, pp. 411–452, 1992.

[10] T. J. M. Molenaar, C. C. M. Appeldoorn, S. A. M. De Haas et al., "Specific inhibition of P-selectin-mediated cell adhesion by phage display-derived peptide antagonists," *Blood*, vol. 100, no. 10, pp. 3570–3577, 2002.

[11] C. L. Chirinos-Rojas, M. W. Steward, and C. D. Partidos, "A peptidomimetic antagonist of TNF-α-mediated cytotoxicity identified from a phage-displayed random peptide library," *Journal of Immunology*, vol. 161, no. 10, pp. 5621–5626, 1998.

[12] E. Pick, J. Brostoff, J. Krejci, and J. L. Turk, "Interaction between "sensitized lymphocytes" and antigen in vitro. II. Mitogen-induced release of skin reactive and macrophage migration inhibitory factors," *Cellular Immunology*, vol. 1, no. 1, pp. 92–109, 1970.

[13] E. Pick and J. L. Turk, "The biological activities of soluble lymphocyte products," *Clinical and Experimental Immunology*, vol. 10, no. 1, pp. 1–23, 1972.

[14] G. Tiefs, J. Hentschel, and A. Wendel, "A T cell-dependent experimental liver injury in mice inducible by concanavalin A," *Journal of Clinical Investigation*, vol. 90, no. 1, pp. 196–203, 1992.

[15] A. Nonomura, M. Tanino, H. Kurumaya, G. Ohta, Y. Kato, and K. Kobayashi, "Studies of immune functions of patients with chronic hepatitis," *The Tohoku Journal of Experimental Medicine*, vol. 137, no. 2, pp. 163–177.

[16] J. Sambrook, E. F. Fritsch, and T. Maniatis, *Molecular Cloning: A Laboratory Manual*, Cold Spring Harbor Laboratory Press, New York, NY, USA, 1989.

[17] M. Port, C. Corot, I. Raynal, and O. Rousseaux, "Novel compositions magnetic particles covered with gem-bisphosphonate derivatives," http://www.patentlens.net/patentlens/patents.html?patnums=US_2004/0253181_A1&returnTo=quick.html.

[18] V. Rerat, S. Laurent, C. Burtéa et al., "Ultrasmall particle of iron oxide-RGD peptidomimetic conjugate: synthesis and characterisation," *Bioorganic and Medicinal Chemistry Letters*, vol. 20, no. 6, pp. 1861–1865, 2010.

[19] K. A. Radermacher, N. Beghein, S. Boutry et al., "In vivo detection of inflammation using pegylated iron oxide particles targeted at e-selectin a multimodal approach using mr imaging and epr spectroscopy," *Investigative Radiology*, vol. 44, no. 7, pp. 398–404, 2009.

[20] A. Roch, R. N. Muller, and P. Gillis, "Theory of proton relaxation induced by superparamagnetic particles," *Journal of Chemical Physics*, vol. 110, p. 5403, 1999.

[21] C. Trautwein, T. Rakemann, N. P. Malek, J. Plümpe, G. Tiegs, and M. P. Manns, "Concanavalin A-induced liver injury triggers hepatocyte proliferation," *Journal of Clinical Investigation*, vol. 101, no. 9, pp. 1960–1969, 1998.

[22] M. Mammen, S. K. Choi, and G. M. Whitesides, "Polyvalent interactions in biological systems: implications for design and use of multivalent ligands and inhibitors," *Angewandte Chemie International Edition*, vol. 37, no. 20, pp. 2754–2794, 1998.

[23] J. E. Gestwicki, C. W. Cairo, L. E. Strong, K. A. Oetjen, and L. L. Kiessling, "Influencing receptor-ligand binding mechanisms with multivalent ligand architecture," *Journal of the American Chemical Society*, vol. 124, no. 50, pp. 14922–14933, 2002.

[24] L. B. Giebel, R. T. Cass, D. L. Milligan, D. C. Young, R. Arze, and C. R. Johnson, "Screening of cyclic peptide phage libraries identifies ligands that bind streptavidin with high affinities," *Biochemistry*, vol. 34, no. 47, pp. 15430–15435, 1995.

[25] A. Chopra, "99 mTc-labeled murine IgM monoclonal antibody, fanolesomab, that targets the CD15 glycoprotein antigen," in *Molecular Imaging and Contrast Agent Database (MICAD)*, National Library of Medicine (US) NCBI, Bethesda, Md, USA, 2004–2011.

[26] C. Burtea, S. Laurent, E. Lancelot et al., "Peptidic targeting of phosphatidylserine for the MRI detection of apoptosis in atherosclerotic plaques," *Molecular Pharmaceutics*, vol. 6, no. 6, pp. 1903–1919, 2009.

[27] L. Larbanoix, C. Burtea, E. Ansciaux et al., "Design and evaluation of a 6-mer amyloid-beta protein derived phage display library for molecular targeting of amyloid plaques in Alzheimer's disease: comparison with two cyclic heptapeptides derived from a randomized phage display library," *Peptides*, vol. 32, no. 6, pp. 1232–1243, 2011.

[28] M. Kopf, M. F. Bachmann, and B. J. Marsland, "Averting inflammation by targeting the cytokine environment," *Nature Reviews Drug Discovery*, vol. 9, pp. 703–718, 2010.

[29] E. Bell, "Inflammation: targeting TNF," *Nature Reviews Immunology*, vol. 9, pp. 390–391, 2009.

[30] P. Laverman, C. P. Bleeker-Rovers, F. H. M. Corstens, O. C. Boerman, and W. J. G. Oyen, "Development of infection and inflammation targeting compounds," *Current Radiopharmaceuticals*, vol. 1, no. 1, pp. 42–48, 2008.

14

Infection by CXCR4-Tropic Human Immunodeficiency Virus Type 1 Is Inhibited by the Cationic Cell-Penetrating Peptide Derived from HIV-1 Tat

Shawn Keogan, Shendra Passic, and Fred C. Krebs

Department of Microbiology and Immunology, Center for Molecular Virology and Translational Neuroscience, and Center for Sexually Transmitted Disease, Institute for Molecular Medicine and Infectious Disease, Drexel University College of Medicine, Philadelphia, PA 19102, USA

Correspondence should be addressed to Fred C. Krebs, fkrebs@drexelmed.edu

Academic Editor: Lloyd D. Fricker

Cell-penetrating peptides (CPP), which are short peptides that are capable of crossing the plasma membrane of a living cell, are under development as delivery vehicles for therapeutic agents that cannot themselves enter the cell. One well-studied CPP is the 10-amino acid peptide derived from the human immunodeficiency virus type 1 (HIV-1) Tat protein. In experiments to test the hypothesis that multiple cationic amino acids within Tat peptide confer antiviral activity against HIV-1, introduction of Tat peptide resulted in concentration-dependent inhibition of HIV-1 IIIB infection. Using Tat peptide variants containing arginine substitutions for two nonionic residues and two lysine residues, HIV-1 inhibition experiments demonstrated a direct relationship between cationic charge and antiviral potency. These studies of Tat peptide as an antiviral agent raise new questions about the role of Tat in HIV-1 replication and provide a starting point for the development of CPPs as novel HIV-1 inhibitors.

1. Introduction

Cell penetrating peptides (CPP) are short peptides that can efficiently cross the plasma membrane, which is otherwise a formidable barrier to many extracellular molecules [1–3]. CPPs are capable of not only traversing the cell membrane, but also serving as a vehicle for transporting a variety of cargos, including nucleic acids, polymers, nanoparticles, and drugs that cannot otherwise gain entry to the cell [3]. Although the functions of various CPPs have been repeatedly verified in a variety of cells and conditions, the mechanism of CPP uptake is not yet fully understood and may involve energy-dependent and -independent mechanisms [4].

Of the numerous peptides shown to have cell penetrating properties, a 10-amino acid (aa) peptide derived from the human immunodeficiency virus type 1 (HIV-1) Tat protein has been well studied as an effective CPP and an attractive drug delivery agent [5]. The Tat peptide has received particular emphasis as a CPP due to its simplicity and capacity for modification to suit the delivery context or cargo [5, 6]. The core peptide is a 10-aa sequence comprised of six arginine and two lysine residues, as well as two non-ionic amino acids (Table 1). However, numerous Tat peptides of varied lengths and terminal sequences have been investigated with the goals of modifying activity or attaching different cargo [6]. A multitude of studies have determined that the activity of the Tat peptide as a CPP involves interactions with the cellular membrane and cytoskeleton [7], and is influenced by numerous variables related to the peptide, the cargo, and extracellular conditions [4].

CPPs such as the Tat peptide, the 16-aa penetratin peptide derived from the Drosophila melanogaster Antennapedia homeodomain protein, and nona-arginine contain numerous cationic arginine (R) and lysine (K) residues [2]. Interestingly, cationic charge is a feature also shared by molecules identified as inhibitors of HIV-1 infection. Multiple cationic charges are prominent features of molecules shown to have activity against HIV-1, including ALX40-4C [8], NeoR6 (an aminoglycoside-arginine conjugate) [9],

Infection by CXCR4-Tropic Human Immunodeficiency Virus Type 1 Is Inhibited by the Cationic Cell-Penetrating
Peptide Derived from HIV-1 Tat

103

TABLE 1: Sequences of peptides examined. Peptide sequences are shown relative to the primary amino acid sequence of the Tat peptide. Position numbers are derived from the full-length Tat protein amino acid sequence (HIV-1 strain SF2) [21].

Peptide	Sequence										Charge
Tat peptide	G	R	K	K	R	R	Q	R	R	R	+8
TPvar1	R	—	—	—	—	—	—	—	—	—	+9
TPvar2	—	—	—	—	—	—	R	—	—	—	+9
TPvar3	R	—	—	—	—	—	R	—	—	—	+10
R-10	R	—	R	R	—	—	R	—	—	—	+10
aa position	48	49	50	51	52	53	54	55	56	57	

the lysozyme-derived HL9 peptide [10], the cathelicidin LL-37 [11], the biguanide-based molecule NB325 [12–16], and compounds that incorporate multiple guanide groups [17]. Cationic peptides found in both semen and cervicovaginal fluids were shown to effectively inhibit HIV-1 infection [18, 19]. Indeed, full-length HIV-1 Tat protein, from which the cationic Tat peptide was derived, was shown to inhibit HIV-1 infection as a CXCR4 antagonist [20].

The present studies were conducted to test the hypothesis that Tat peptide, because of the numerous cationic amino acids contained within its primary sequence, can effectively inhibit HIV-1 infection. In vitro experiments involving Tat peptide and an HIV-1-susceptible indicator cell line demonstrated concentration-dependent inhibition of the X4 HIV-1 strain IIIB, which uses CXCR4 as a coreceptor. Additional experiments involving variants of Tat peptide with increased cationic charge suggested a direct relationship between charge magnitude and antiviral potency. These results provide further insights into a potential role for Tat as an HIV-1 inhibitor and suggest a novel anti-HIV-1 activity attributed to the family of CPPs.

2. Materials and Methods

2.1. Synthesis of Tat Peptide and Variants. Tat peptide (Table 1) was derived from residues 48–57 (numbering from HIV-1 strain SF2) of the full-length Tat protein [5, 6, 21]. Three arginine-enriched Tat peptide variants (Table 1) were designed by substituting arginine for G48, Q54, or both amino acids. A decaarginine peptide was also included in these studies as a Tat peptide variant with all four nonarginine residues converted to arginines. All peptides were synthesized commercially by liquid phase peptide synthesis (GenScript, Pascataway, NJ) and provided at >95% purity as determined by mass spectrometry and high-performance liquid chromatography analysis performed by the manufacturer (GenScript). Lyophilized peptides were suspended in 1 mL of sterile deionized water upon receipt and stored at $-20°C$ prior to use.

2.2. Cell Line Maintenance. P4-R5 MAGI indicator cells (NIH AIDS Research and Reference Reagent Program number 3580) were maintained in Dulbecco's modified eagle's media (DMEM) supplemented with 10% fetal bovine serum (FBS), 0.05% sodium bicarbonate, antibiotics (penicillin, streptomycin, and kanamycin at $40 \mu g/mL$), and $1 \mu g/mL$ puromycin (Cellgro, Manassas, VA).

2.3. Assessing Inhibition of HIV-1 Infection by Tat Peptide and Its Variants. Peptide effectiveness was determined in an HIV-1 infection inhibition assay using P4-R5 MAGI indicator cells. P4-R5 MAGI cells were plated at a concentration of 1.5×10^4 cells/well in a flat-bottom 96-well plate (BD Biosciences, Bedford, MA). The cells were then infected with HIV-1 strain IIIB (Advanced Biotechnologies, Inc., Columbia, MD; $10^{7.8}$ TCID$_{50}$/mL) at multiplicities of infection (MOI) of 0.6, 0.05, or 0.03 in the presence or absence of peptide or dextran sulfate (DS) (Sigma, St. Louis, MO). Following a 2 h incubation at $37°C$, the cells were washed with PBS, provided with $200 \mu L$ of new media, and incubated for an additional 46 h. Levels of infection were measured using the Galacto-Star-β-Galactosidase Reporter Gene Assay System for Mammalian Cells (Applied Biosystems, Carlsbad, CA) as described by the manufacturer. Chemiluminescence was measured using a Glomax Luminometer plate reader (Promega, Madison, WI).

2.4. Assessing the Effect of Tat Peptide on In Vitro Cell Viability. P4-R5 MAGI cells were plated at a concentration of 1.5×10^4 cells/well in a flat-bottom 96-well plate. The cells were then exposed to the indicated half log concentrations of peptide and incubated at $37°C$ for 2 h. Following exposure, the cells were washed with PBS and then assayed for cell viability immediately or at 24 h or 48 h after exposure. Viability was measured using a 3-(4,5-dimethylthiazol-2-yl)-2,5-diphenyltetrazolium bromide (MTT) assay as previously described [13].

2.5. Data Analyses. Mean values and standard deviations were calculated from two independent assays in which each concentration was examined in quadruplicate. Calculations of EC$_{50}$ (concentrations that resulted in 50% reductions in infection relative to mock-treated, HIV-1-infected cells) were calculated using the Forecast function of Microsoft Excel.

3. Results

3.1. Tat Peptide Inhibits HIV-1 Infection. Initial experiments were performed to test the hypothesis that Tat peptide, by virtue of its cationic charge, was capable of inhibiting HIV-1 infection. HIV-1-susceptible P4-R5 MAGI indicator cells were exposed to HIV-1 strain IIIB (0.6 MOI) for 2 h while in the presence of half log concentrations of Tat peptide up to 1 mg/mL. The presence of Tat peptide inhibited HIV-1 infection in a concentration-dependent manner (Figure 1),

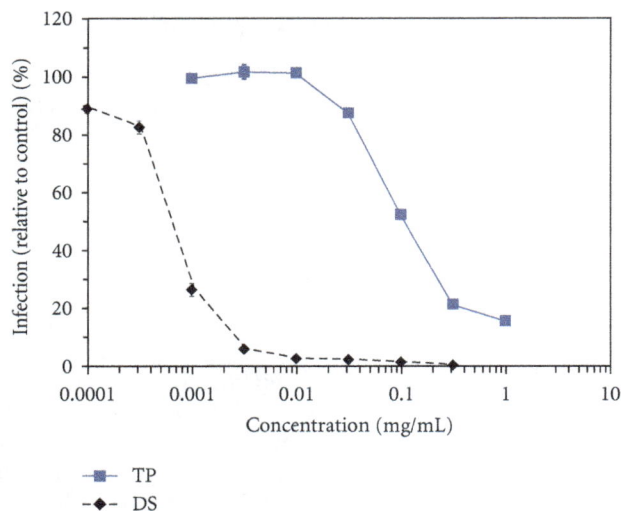

FIGURE 1: Tat peptide inhibits infection by HIV-1 strain IIIB. P4-R5 MAGI cells were exposed to half log concentrations of Tat peptide (TP) or dextran sulfate (DS) in the presence of HIV-1 strain IIIB for 2 h. Reductions in HIV-1 infection (%) were calculated relative to mock-exposed HIV-1 infected cells. The graph represents data from two independent assays in which infections at each concentration were repeated in quadruplicate. Error bars represent standard deviations.

TABLE 2: Viral titer does not affect the antiviral activity of Tat peptide. EC_{50} values were calculated from the results of antiviral assays (as described in Section 2) involving infection by HIV-1 IIIB at three different multiplicities of infection (MOI).

Virus concentration during infection (10^3 infectious virions/mL)	MOI	EC_{50}
88	0.6	0.094 mg/mL
8.8	0.05	0.14 mg/mL
4.4	0.03	0.10 mg/mL

with an EC_{50} of 0.094 mg/mL (50 μM). No inhibition was apparent at or below 0.0316 mg/mL. At the highest concentration tested (1 mg/mL), the presence of Tat peptide was insufficient to completely inhibit HIV-1 infection (15.6% infection relative to mock-exposed, HIV-1-infected cells). In comparison, the anionic compound dextran sulfate, which was included as a known inhibitor, blocked HIV-1 infection with an EC_{50} of 0.0007 mg/mL. To determine the potential effect of virus titer on Tat peptide antiviral activity, similar experiments were performed using reduced concentrations of input virus (0.05 and 0.03 MOI). In these experiments (Table 2), reductions in input virus had no effect on Tat peptide antiviral activity (EC_{50} values of 0.14 mg/mL and 0.10 mg/mL, resp.).

To confirm that any adverse effects of Tat peptide on reporter cell viability had not compromised the antiviral assays, MTT cytotoxicity assays were performed using conditions identical to those used in the antiviral assays. In these assays, 2 h exposures to Tat peptide at concentrations below 1 mg/mL had no effect on P4-R5 MAGI cell viability,

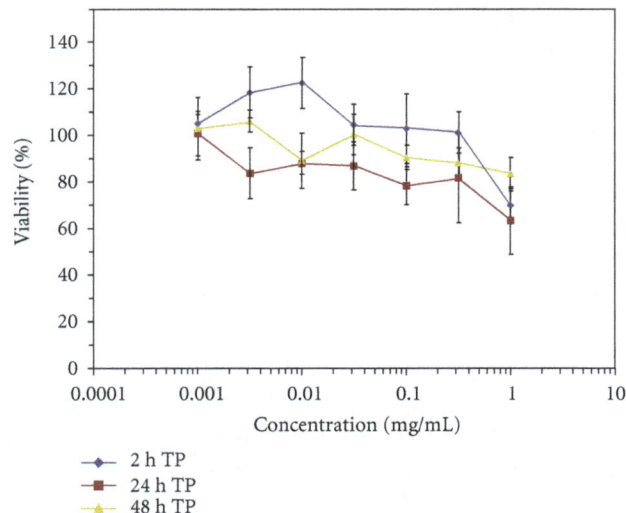

FIGURE 2: Tat peptide has no effect on reporter cell viability. P4-R5 MAGI cells were exposed to half log concentrations of Tat peptide for 2 h, washed, and assessed immediately for changes in cell viability or after extended maintenance (24 h or 48 h after exposure) in the absence of Tat peptide. Percent changes in cell viability were calculated relative to mock-exposed cells. The graph represents data from two independent assays in which exposure to each concentration of peptide was repeated in quadruplicate. Error bars represent standard deviations.

as measured immediately after exposure or after extended postexposure maintenance (24 h or 48 h) in the absence of Tat peptide (Figure 2). These results indicated that measurements of antiviral activity were not biased by reductions in P4-R5 MAGI cell viability. These results are also consistent with previous studies [22], in which Tat peptide alone (but not peptide conjugated to payload) had no effect on cell viability at concentrations up to 100 μM and exposure durations as long as 48 h.

3.2. Additional Cationic Charges Increase the Antiviral Potency of Tat Peptide. Having demonstrated the anti-HIV-1 activity of the Tat peptide, additional experiments were performed to investigate the role of charge in determining antiviral efficacy. Of the 10 aa residues in Tat peptide, eight are cationic (six arginine and two lysine residues) and the remaining two are uncharged (G48, nonpolar and aliphatic; Q54, polar). To increase the net peptide charge, arginine residues were substituted for one or both of the noncationic residues in the native Tat peptide sequence (Table 1). These substitutions increased the net positive side chain charge of the Tat peptide from +8 to +9 (TPvar1 and TPvar2) or +10 (TPvar3). An additional peptide, decaarginine (R-10), was also included in these studies. R-10 also had a net side chain charge of +10, but differed from TPvar3 in that all ten positive charges were contributed by the arginine guanidinium groups. R-10 was, in effect, a Tat peptide variant with arginine residues substituted into all nonarginine positions. Like the Tat peptide, none of the variants had any effect on P4-R5 MAGI cell viability after a 2 h exposure (data not shown).

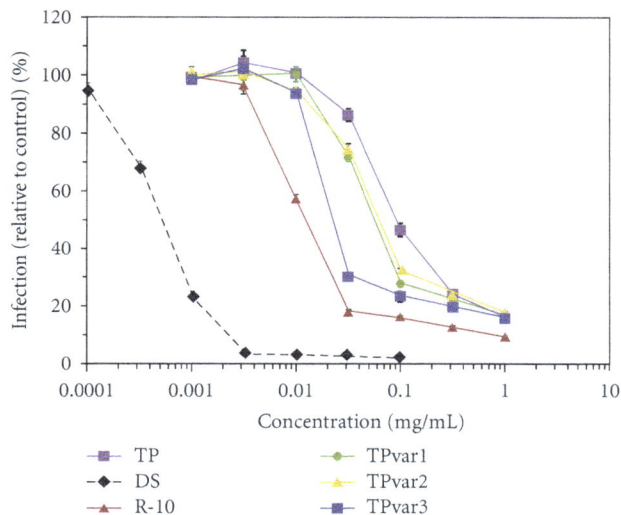

FIGURE 3: Increased peptide antiviral potency is associated with increased peptide cationic charge. P4-R5 MAGI cells were exposed to half log concentrations of Tat peptide (TP), three Tat peptide variants (TPvar1-3), decaarginine (R-10), or dextran sulfate (DS) in the presence of HIV-1 strain IIIB for 2 h. Peptide sequences are depicted in Table 1. Reductions in HIV-1 infection (%) were calculated relative to mock-exposed HIV-1 infected cells. The graph represents data from two independent assays in which infections at each concentration were repeated in quadruplicate. Error bars represent standard deviations.

Concurrent incubation of HIV-1 IIIB and each peptide with P4-R5 MAGI cells again resulted in concentration-dependent inhibition of HIV-1 infection (Figure 3). However, the Tat peptide, the Tat peptide variants, and R-10 differed in antiviral potency, with EC_{50} values ranging from 0.094 mg/mL (Tat peptide) to 0.014 mg/mL (R-10). Single substitutions of arginine into the Tat peptide sequence (G48R or Q54R) resulted in small increases in antiviral activity relative to Tat peptide (TPvar1 $EC_{50} = 0.065$ mg/mL; TPvar2 $EC_{50} = 0.071$ mg/mL). Arginine substitutions at both positions (TPvar3) further increased peptide antiviral activity ($EC_{50} = 0.025$ mg/mL). However, despite having the same number of positive charges, TPvar3 was less active than R-10. No peptide was active below 0.00316 mg/mL or 100% inhibitory at the highest concentration examined (1 mg/mL).

4. Discussion

Results presented in this paper demonstrate that Tat peptide, a CPP that is capable of delivering molecules intracellularly [1, 5] across living membranes, also contains intrinsic antiviral activity against HIV-1 infection. Nontoxic concentrations of Tat peptide inhibited CXCR4-mediated infection by the X4 virus HIV-1 IIIB in a concentration-dependent manner. In experiments designed to explore the contribution of cationic charge to antiviral activity, Tat peptide variants with arginine substitutions for non-ionic and lysine residues were also assessed for antiviral activity. Increases in antiviral potency with increased net peptide positive charge confirmed

our original hypothesis and suggest a role for peptide charge in the mechanism of action against HIV-1 infection.

In a broader virologic context, observations reported herein provide new information about Tat protein and its contributions to HIV-1-associated pathogenesis. In previous studies, soluble full-length Tat protein, which is secreted from infected cells [23], specifically inhibited an HIV-1 strain that used CXCR4 as a co-receptor (designated X4) but not a CCR5-tropic (R5) strain [20]. Our preliminary results using the Tat peptide are consistent with this observation (data not shown). The authors speculated that, during the course of HIV-1 infection, this mechanism attributed to extracellular Tat could favor the replication and spread of R5 virus by inhibiting X4 virus infection. They further suggested that Tat protein may interact with acidic regions of CXCR4 through the high concentration of basic residues scattered throughout the Tat protein primary sequence [20]. However, this study did not identify the specific sequences within full-length Tat that were the source of the antiviral activity. This antiviral activity was also complicated by apparent cytotoxicity associated with exposure to extracellular full-length Tat protein [24–27]. In contrast, the present studies demonstrated that the Tat peptide had no adverse effect on cell viability.

These studies also provide two starting points for the development of novel inhibitors of HIV-1. First, Tat peptide can serve as a prototype for the development of novel agents effective against HIV-1. Such agents may take the form of cationic peptides or small molecule inhibitors that mimic a peptide structure. Second, Tat peptide provides the basis for multifunctional therapeutic agents that combine the intrinsic and specific anti-HIV-1 activity of Tat peptide with its ability to deliver therapeutic agents that by themselves do not readily penetrate cells and tissues [3, 6]. For example, Tat peptide could be linked to an HIV-1 protease inhibitor to form a dual-activity antiretroviral agent that combines entry inhibition, increased drug penetration, and a second, distinct mechanism of antiretroviral activity.

Our experiments also indicate that this intrinsic antiviral activity is not limited to the Tat-derived CPP alone. In the present studies, the R-10 peptide was also an effective HIV-1 inhibitor, despite changes in four out of ten amino acids with respect to the Tat peptide. Preliminary studies have also demonstrated anti-HIV-1 activity associated with the well-studied CPP nona-arginine (R-9) and a 20-aa peptide consisting solely of alternating arginine and glycine residues (data not shown). The finding that antiviral activity is not limited to Tat peptide suggests that a key characteristic common to these molecules (i.e., multiple cationic charges) confers activity against HIV-1.

The involvement of cationic charge in CPP antiviral activity is also supported by the results of experiments involving the Tat peptide variants. Those results indicated a direct relationship between charge and antiviral activity. Tat peptide (+8 charge) was the least active while R-10 (+10 charge) was the most active, and variants with intermediate levels of cationic charge had intermediate levels of antiviral activity. Despite the fact that R-10 and TPvar3 had the same charge, these two peptides differed in their effects on HIV-1 infection, likely due to the replacement of two lysine residues with two arginine residues. Lysine has a single positive charge associated with a terminal amino group while arginine has a single positive charge associated with a terminal guanidinium group. The charge in arginine is delocalized across the guanidinium group, supporting the formation of multiple hydrogen bonds [2, 28]. Polar and charged interactions supported by the arginine end group may favor mechanisms that are responsible for antiviral activity and, perhaps, cell penetrating activity [5, 28]. Interestingly, peptides with multiple guanidinium groups, such as R-10 and nona-arginine, are in the same family with other demonstrated HIV-1 inhibitors ALX40-4C, NB325, and LL-37, which are also cationic molecules with multiple guanidinium groups.

Related studies have also indicated the importance of charge in cationic HIV-1 inhibitors and provided further evidence for a mechanism of CPP antiviral activity. We previously demonstrated that charge distribution plays a key role in the antiviral activity of biguanide-based molecules [13]. These molecules are oligomeric, cationic compounds characterized by the presence of alternating biguanide groups and hydrocarbon linkers [15]. Using rational compound design and structure-function screening of biguanide-containing synthetic molecules, studies identified polyethylene hexamethylene biguanide (PEHMB; also known as NB325) as a molecule with minimal cytotoxicity and considerable activity against HIV-1 [12, 15]. More recent work identified NB325 as an HIV-1 entry inhibitor [14, 15] that antagonizes CXCR4 through epitope-specific interactions with extracellular loop 2 (ECL2). Further experiments demonstrated relationships between charge density, cytotoxicity, and antiviral activity [13]. These findings add to a collective understanding of cationic HIV-1 inhibitors and can be used to guide further investigations focused on defining mechanisms of action and optimizing the antiviral potency of cationic inhibitors such as the CPPs.

These results provide the basis for further basic science and translational studies. Expanded studies will be necessary to investigate the antiviral effect of Tat and Tat peptide on HIV-1 replication in natively HIV-1-susceptible immune cell populations and to better understand the contribution of the potential bias toward R5 virus replication to viral pathogenesis and disease progression. Related efforts will be directed toward the development of novel CPP-based antiviral agents that can serve as multifunctional HIV-1 inhibitors. These efforts will address CPP potency, stability, mechanism of action, and combined activity as these agents are advanced into preclinical investigations and clinical trials.

Acknowledgments

These studies were supported by faculty development funds provided by the Department of Microbiology and Immunology, Drexel University College of Medicine and the Institute for Molecular Medicine and Infectious Disease. The authors would like to thank Dr. Brian Wigdahl for lively discussions and insightful contributions relevant to this work, and the critical review of this paper prior to its submission.

References

[1] C. Foerg, K. M. Weller, H. Rechsteiner et al., "Metabolic cleavage and translocation efficiency of selected cell penetrating peptides: a comparative study with epithelial cell cultures," *American Association of Pharmaceutical Scientists Journal*, vol. 10, no. 2, pp. 349–359, 2008.

[2] I. Nakase, T. Takeuchi, G. Tanaka, and S. Futaki, "Methodological and cellular aspects that govern the internalization mechanisms of arginine-rich cell-penetrating peptides," *Advanced Drug Delivery Reviews*, vol. 60, no. 4-5, pp. 598–607, 2008.

[3] V. P. Torchilin, "Tat peptide-mediated intracellular delivery of pharmaceutical nanocarriers," *Advanced Drug Delivery Reviews*, vol. 60, no. 4-5, pp. 548–558, 2008.

[4] A. T. Jones, "Gateways and tools for drug delivery: endocytic pathways and the cellular dynamics of cell penetrating peptides," *International Journal of Pharmaceutics*, vol. 354, no. 1-2, pp. 34–38, 2008.

[5] J. M. Gump and S. F. Dowdy, "TAT transduction: the molecular mechanism and therapeutic prospects," *Trends in Molecular Medicine*, vol. 13, no. 10, pp. 443–448, 2007.

[6] H. Brooks, B. Lebleu, and E. Vivès, "Tat peptide-mediated cellular delivery: back to basics," *Advanced Drug Delivery Reviews*, vol. 57, no. 4, pp. 559–577, 2005.

[7] A. Mishra, G. H. Lai, N. W. Schmidt et al., "Translocation of HIV TAT peptide and analogues induced by multiplexed membrane and cytoskeletal interactions," *Proceedings of the National Academy of Sciences of the United States of America*, vol. 108, no. 41, pp. 16883–16888, 2011.

[8] B. J. Doranz, K. Grovit-Ferbas, M. P. Sharron et al., "A small-molecule inhibitor directed against the chemokine receptor CXCR4 prevents its use as an HIV-1 coreceptor," *Journal of Experimental Medicine*, vol. 186, no. 8, pp. 1395–1400, 1997.

[9] A. Lapidot, A. Peled, A. Berchanski et al., "NeoR6 inhibits HIV-1-CXCR4 interaction without affecting CXCL12 chemotaxis activity," *Biochimica et Biophysica Acta*, vol. 1780, no. 6, pp. 914–920, 2008.

[10] S. Lee-Huang, V. Maiorov, P. L. Huang et al., "Structural and functional modeling of human lysozyme reveals a unique nonapeptide, HL9, with anti-HIV activity," *Biochemistry*, vol. 44, no. 12, pp. 4648–4655, 2005.

[11] P. Bergman, L. Walter-Jallow, K. Broliden, B. Agerberth, and J. Söderlund, "The antimicrobial peptide LL-37 inhibits HIV-1 replication," *Current HIV Research*, vol. 5, no. 4, pp. 410–415, 2007.

[12] F. C. Krebs, S. R. Miller, M. L. Ferguson, M. Labib, R. F. Rando, and B. Wigdahl, "Polybiguanides, particularly polyethylene hexamethylene biguanide, have activity against human immunodeficiency virus type 1," *Biomedicine and Pharmacotherapy*, vol. 59, no. 8, pp. 438–445, 2005.

[13] S. R. Passic, M. L. Ferguson, B. J. Catalone et al., "Structure-activity relationships of polybiguanides with activity against human immunodeficiency virus type 1," *Biomedicine and Pharmacotherapy*, vol. 64, no. 10, pp. 723–732, 2010.

[14] N. Thakkar, V. Pirrone, S. Passic et al., "Persistent interactions between biguanide-based compound NB325 and CXCR4 result in prolonged inhibition of human immunodeficiency virus type 1 infection," *Antimicrobial Agents and Chemotherapy*, vol. 54, no. 5, pp. 1965–1972, 2010.

[15] N. Thakkar, V. Pirrone, S. Passic et al., "Specific interactions between the viral coreceptor CXCR4 and the biguanide-based compound NB325 mediate inhibition of human immunodeficiency virus type 1 infection," *Antimicrobial Agents and Chemotherapy*, vol. 53, no. 2, pp. 631–638, 2009.

[16] K. Lozenski, T. Kish-Catalone, V. Pirrone et al., "Cervi-covaginal safety of the formulated biguanide-based human immunodeficiency virus type 1 (HIV-1) inhibitor NB325 in a murine model of microbicide application," *Journal of Biomedicine and Biotechnology*, vol. 2011, Article ID 941061, 10 pages, 2011.

[17] R. A. Wilkinson, S. H. Pincus, J. B. Shepard et al., "Novel com-pounds containing multiple guanide groups that bind the HIV coreceptor CXCR4," *Antimicrobial Agents and Chemotherapy*, vol. 55, no. 1, pp. 255–263, 2011.

[18] J. A. Martellini, A. L. Cole, N. Venkataraman et al., "Cationic polypeptides contribute to the anti-HIV-1 activity of human seminal plasma," *FASEB Journal*, vol. 23, no. 10, pp. 3609–3618, 2009.

[19] N. Venkataraman, A. L. Cole, P. Svoboda, J. Pohl, and A. M. Cole, "Cationic polypeptides are required for anti-HIV-1 activity of human vaginal fluid," *Journal of Immunology*, vol. 175, no. 11, pp. 7560–7567, 2005.

[20] H. Xiao, C. Neuveut, H. L. Tiffany et al., "Selective CXCR4 antagonism by Tat: implications for in vivo expansion of coreceptor use by HIV-1," *Proceedings of the National Academy of Sciences of the United States of America*, vol. 97, no. 21, pp. 11466–11471, 2000.

[21] M. Kuppuswamy, T. Subramanian, A. Srinivasan, and G. Chinnadurai, "Multiple functional domains of Tat, the trans-activator of HIV-1, defined by mutational analysis," *Nucleic Acids Research*, vol. 17, no. 9, pp. 3551–3561, 1989.

[22] A. K. Cardozo, V. Buchillier, M. Mathieu et al., "Cell-permeable peptides induce dose- and length-dependent cyto-toxic effects," *Biochimica et Biophysica Acta*, vol. 1768, no. 9, pp. 2222–2234, 2007.

[23] B. Ensoli, G. Barillari, S. Z. Salahuddin, R. C. Gallo, and F. Wong-Staal, "Tat protein of HIV-1 stimulates growth of cells derived from Kaposi's sarcoma lesions of AIDS patients," *Nature*, vol. 345, no. 6270, pp. 84–86, 1990.

[24] A. Nath, K. Psooy, C. Martin et al., "Identification of a human immunodeficiency virus type 1 Tat epitope that is neuroexcitatory and neurotoxic," *Journal of Virology*, vol. 70, no. 3, pp. 1475–1480, 1996.

[25] G. R. Campbell, E. Pasquier, J. Watkins et al., "The glutamine-rich region of the HIV-1 Tat protein is involved in T-cell apoptosis," *Journal of Biological Chemistry*, vol. 279, no. 46, pp. 48197–48204, 2004.

[26] A. Chauhan, A. Tikoo, A. K. Kapur, and M. Singh, "The taming of the cell penetrating domain of the HIV Tat: myths and realities," *Journal of Controlled Release*, vol. 117, no. 2, pp. 148–162, 2007.

[27] B. Romani, S. Engelbrecht, and R. H. Glashoff, "Functions of Tat: the versatile protein of human immunodeficiency virus type 1," *Journal of General Virology*, vol. 91, part 1, pp. 1–12, 2010.

[28] J. B. Rothbard, T. C. Jessop, R. S. Lewis, B. A. Murray, and P. A. Wender, "Role of membrane potential and hydrogen bonding in the mechanism of translocation of guanidinium-rich peptides into cells," *Journal of the American Chemical Society*, vol. 126, no. 31, pp. 9506–9507, 2004.

15

Antimicrobial Lactoferrin Peptides: The Hidden Players in the Protective Function of a Multifunctional Protein

Mau Sinha, Sanket Kaushik, Punit Kaur, Sujata Sharma, and Tej P. Singh

Department of Biophysics, All India Institute of Medical Sciences, Ansari Nagar, New Delhi 110029, India

Correspondence should be addressed to Sujata Sharma; afrank2@gmail.com and Tej P. Singh; tpsingh.aiims@gmail.com

Academic Editor: Severo Salvadori

Lactoferrin is a multifunctional, iron-binding glycoprotein which displays a wide array of modes of action to execute its primary antimicrobial function. It contains various antimicrobial peptides which are released upon its hydrolysis by proteases. These peptides display a similarity with the antimicrobial cationic peptides found in nature. In the current scenario of increasing resistance to antibiotics, there is a need for the discovery of novel antimicrobial drugs. In this context, the structural and functional perspectives on some of the antimicrobial peptides found in N-lobe of lactoferrin have been reviewed. This paper provides the comparison of lactoferrin peptides with other antimicrobial peptides found in nature as well as interspecies comparison of the structural properties of these peptides within the native lactoferrin.

1. Introduction

The innate immune system or the nonspecific immune system is the first and the oldest line of defense in organisms [1, 2]. It was the most dominant form of immunity before the evolution of the more sophisticated adaptive immunity. It is comprised of various mechanisms which are responsible for rapid defense of the host organism against invasion by other factors in a nonspecific manner. The innate immune system differs from the adaptive immune system in a way that while it is able to defend the body against pathogens, it is not able to impart long-lasting immunity to the host, unlike the latter [3]. Despite the evolution of the more complex and specific adaptive immunity, innate immunity still continues to function as the primary line of defense for most organisms [4]. The antimicrobial action in the innate immunity is mediated by various antimicrobial proteins and peptides, which have been evolutionary conserved. Antimicrobial peptides are small peptides which demonstrate broad-spectrum antibiotic activity against various gram-positive and gram-negative bacteria, fungi, protozoa, and viruses [5–8]. While the most common mechanism of action deployed by these peptides is perturbation of microbial cell membrane [9–11], there are other mechanisms which are also prevalent [12–16]. Due to increasing resistance to antibiotics, there is an urgent requirement of novel antimicrobial drugs [17–19]. Use of antimicrobial peptides is one of the promising approaches which may lead to potential antimicrobial drugs [16, 20–24]. It has been observed that peptides which are predominantly cationic and hydrophobic in nature show potent antimicrobial activity [5, 25–29]. Many of these peptides including indolicidin from bovine neutrophiles [30], tripticin from porcine neutrophil granules [31], puroindoline from wheat seeds [32], combi-1, a synthesized antimicrobial peptide [33], and Lys H and Lys C from lysozyme [34] have been extensively studied. Although these peptides adopt various conformations the alpha-helical conformation with polar and nonpolar groups on opposite sides of the helix tends to be the most abundant [35–37]. The antimicrobial property of these amphipathic alpha-helical peptides increases sequentially with increase in net charge [38].

Lactoferrin is an iron-binding glycoprotein which is found in most of the exocrine secretions such as milk, tears, nasal secretions, saliva, urine, uterine secretions, and amniotic fluids [39–41] as well as in secondary granules of neutrophils [42]. It exerts a wide antimicrobial activity against a number of bacterial, viral, and fungal pathogens in vitro [43–48]. Lactoferrin exerts its antimicrobial action not just in the form of the intact molecule but the monoferric lobes and active peptides of lactoferrin also have a role in the

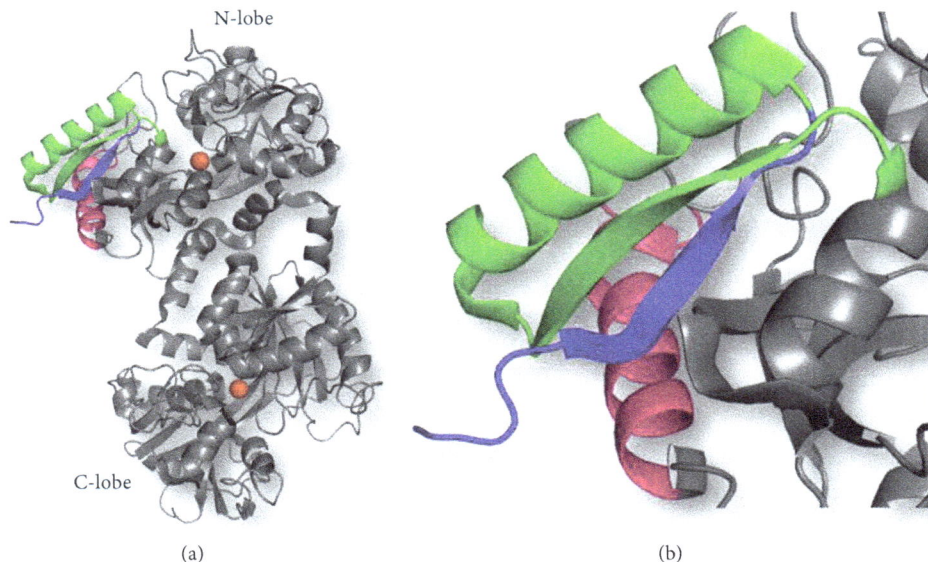

FIGURE 1: (a) Overall structure of lactoferrin showing positions of LF1-11 (blue), lactoferrampin (pink), and lactoferricin (green) peptides in the N-terminal lobe. (b) The zoomed structure showing the position of peptides in detail.

host defense against microbial disease [49–52]. Lactoferrin is a rich source of cationic and hydrophobic antimicrobial peptides, which may be used against microbes [53, 54]. These antibacterial peptides which are a part of the polypeptide chain of lactoferrin and are released upon the proteolysis of this molecule by various proteolytic enzymes can be developed into clinically useful lead molecules for antimicrobial therapeutics [55, 56].

Although it has been shown that native lactoferrin exerts its antimicrobial action through sequestration of iron [57–59], it is still unclear how the antimicrobial lactoferrin peptides act against the microbes. Though a number of studies have implicated these peptides in the binding to the outer membrane proteins of various bacteria or binding to microbial proteases [60–62], the structure-function interrelationships of these peptides have not yet been established.

A number of functional peptides are produced from lactoferrin by the action of proteolytic enzymes. It is expected that these enzymes are present in the gastrointestinal tract as well as the site of microbial infection, and hence, they may contribute in the natural function of lactoferrin in the human body. This paper reports the comparison of these peptides with other antimicrobial peptides found in nature as well as cross-species comparison of the sequences of these antimicrobial peptides from the native sequences of lactoferrin with the intent to draw evolutionary inferences of their function. Although many antimicrobial peptides from lactoferrin have been isolated and characterized, only three of them have been studied in detail. These are LF1-11, lactoferrampin, and lactoferricin. The sequences of these peptides indicate that these peptides belong to the N-terminal half of lactoferrin (Figure 1). Hydrophobicity, cationicity, and helical conformation of these antimicrobial peptides are the important characteristics that determine their antimicrobial

potency [9, 63, 64]. All these peptides have high pI values (>9) and is expected to interact with negatively charged elements. All three have different sequences, structural elements, and modes of action (Table 1). An attempt to analyze and decipher the structural and functional characteristics of three peptides is made in this paper. Their overall structural comparison as observed in intact lactoferrin is depicted in Figure 2.

2. LF1-11

LF1-11, as its name suggests, is the N-terminal peptide of lactoferrin, comprised of the first eleven residues of the molecule. This peptide has been shown to be highly effective against five multidrug-resistant *Acinetobacter baumannii* strains [65] and methicillin-resistant *Staphylococcus aureus* [66] and various *Candida* species [67, 68]. The potent antimicrobial effect of LF1-11 was attributed to the first two arginines at the N-terminus of human lactoferrin [69]. This conclusion was based on the fact that when the second or third arginines were replaced by alanine, the candidacidal activity of the LF1-11 was observed. Additionally, while LF1-11, LF2-11, and LF3-11 showed comparable candidacidal activities, the same was found compromised in the case of LF4-11.

The importance of the three arginines (R2–R4) for the potent antimicrobial activity of this peptide was established when synthetic peptides lacking the first three N-terminal residues were found to be less effective [70] in the killing of bacteria. Also, mutant lactoferrin lacking the first five N-terminal residues displayed decreased binding to bacterial lipopolysaccharide [71].

In yet another study, Stallmann et al. studied the efficacy of local prophylactic treatment with human LF1-11 in a rabbit model of femur infection and observed that hLF1-11 effectively reduced the development of osteomyelitis in a rabbit model [72].

TABLE 1: Amino acid sequences of LF1-11 from lactoferrin from six species.

	LF1-11	Lactoferrampin	Lactoferricin
Sequence	GRRRSVQWCAV	WNLLRQAQEKFGKDKSP	KCFQWQRNMRKVRGPPVSCIKRDS
pI	11.70	9.70	10.95
Secondary structure in the intact lactoferrin [X-ray crystallography]	Loop followed by β-strand	Amphipathic helix with a C-terminal tail	N-terminal amphipathic helix connected to a β-strand with a loop. The structural assembly is held together with a disulphide bond.
Secondary structure when isolated [NMR]	Not known	N-terminal amphipathic alpha-helical conformation across the first 11 residues and random C-terminus	N-terminal amphipathic helix and a random coil

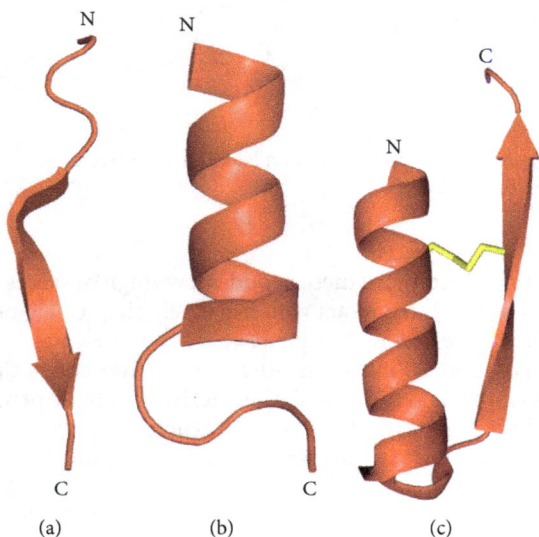

FIGURE 2: The structural comparison of peptides (a) LF1-11, (b) lactoferrampin, and (c) lactoferricin in the native structure of human lactoferrin (PDB: 1LFG).

TABLE 2: Comparison of amino acid sequences of LF1-11 from lactoferrin from six species.

Human	GRRRSVQWCAV	11
Bovine	APRKNVRWCTI	11
Buffalo	APRKNVRWCTI	11
Equine	APRKSVRWCTI	11
Caprine	APRKNVRWCAI	11
Camel	ASKKSVRWCTT	11

It was speculated that the mechanism of antimicrobial action of LF1-11 is mitochondrial damage, with the extracellular ATP being essential but not sufficient for LF1-11 to exert its candidacidal activity [69]. In later studies, it was found that uptake of calcium by mitochondria is vital for killing of *Candida albicans* by the LF1-11 [73].

In another study, it was found that LF1-11 is responsible for directing the GM-CSF-driven monocyte differentiation toward macrophages that produces both pro- and anti-inflammatory cytokines. It was speculated that the peptide could be used as agent to empower the innate immune response of the host for infections. These results demonstrated the importance of the further development of LF1-11 as a promising drug against microbial infections in patients who may have compromised immune systems [74].

The cellular target for the immunomodulatory activity of LF1-11 was found to be myeloperoxidase, to which LF1-11 binds and inhibits after entering the monocytes. A molecular modeling study by the same group demonstrated that LF1-11 bound at the active site of the enzyme. The importance of

the first two arginines and the cysteine at the tenth position was further substantiated by the fact that peptides which did not possess these necessary residues were not as effective in binding with myeloperoxidase [75].

The sequence comparison of LF1-11 among the six species (Table 2) shows that unlike human LF1-11, which contains three arginines in the positions 2–5 (R2–R4), the peptide from other species contains only one arginine (R3). Yet, it is noteworthy that the R4 has been replaced by lysine, which is also a basic residue in all the other species, thereby maintaining the highly cationic nature of the peptide throughout the species. Also, in all the cases except human, arginine occurs at the seventh position also. The most significant change is seen in the second residue which is proline in all cases except in human and camel. Notably, the hydrophobic residues, V6, and W8 are conserved in all the cases.

3. Lactoferrampin

Lactoferrampin, comprised of residues 268–284 in the N1 domain of lactoferrin, has been identified as an antimicrobial peptide and plays a key role in membrane-mediated activities of lactoferrin [76, 77]. It exhibits broad antimicrobial action against several gram-positive and gram-negative bacteria, notably, *Bacillus subtilis*, *Escherichia coli*, *Pseudomonas aeruginosa*, and *Staphylococcus aureus*, as well as candidacidal activity [76].

The antimicrobial action of this peptide was also found to be more potent than the native lactoferrin. This peptide was found to be located in close proximity to lactoferricin. The structure of lactoferrampin revealed an amphipathic alpha-helix which begins with the N-terminus and ends at the 11th residue, followed by a C-terminus tail [78].

It is reported that the cleavage of this peptide at both the termini resulted in considerable decrease of the candidacidal activity. The C-terminal residues of lactoferrampin are most critical for its antimicrobial action, possibly because the C-terminus consists of several residues with positive charges which are clustered together. But truncation of C-terminal side did not alter the ability of this peptide to adopt helical conformations. Also, substitution of the basic residues at the C-terminus led to decrease in potency of this peptide [77, 79]. The N-terminal residues, truncated up to the sequence 270–284, are essential for maintaining the structure of this peptide in a helical conformation [77].

The helical conformation of this peptide was found to be critical for the potency against gram-positive bacteria as established when the bactericidal activities of two lactoferrampin peptides, lactoferrampin 265–284 and lactoferrampin 268–284, were compared [80]. Lactoferrampin 265–284, which consists of additional three residues, Asp-Leu-Ile, showed a broader specificity since the Asp-Leu-Ile sequence increases the tendency of this peptide to assume an alpha-helical conformation. Both the peptides possessed bactericidal activity against certain species of gram-positive and gram-negative bacteria. Compared to lactoferrampin 268–284, higher concentrations of lactoferrampin 265–284 were required to kill the gram-negative bacteria, E. coli and P. aeruginosa. The killing activity expressed as LC_{50} value (the concentration that produced 50% reduction in viable counts of the microorganisms) was found to be about 5.8 μmol/L for lactoferrampin 268–284 which is about 4 times higher than lactoferrampin 265–284 [80].

The mode of action of this peptide on bacteria is by bacterial membrane binding and membrane disruption. It is established that lactoferrampin is internalized within few minutes with the bacterial membrane permeabilization followed by cellular damage [81, 82].

Distinct vesicle-like structures by the lactoferrampin peptide were also observed by freeze-fracture transmission electron microscopy in the membrane of C. albicans [81]. It is speculated that this peptide exerted detergent-like activity, disturbing the hydrophobic interphase of the lipid bilayer.

Several studies have revealed that the determinants for antimicrobial action are the orientation and structure of bovine lactoferrampin in bacterial membranes [78, 83–85]. The solution structure of bovine lactoferrampin suggests that it adopts an amphipathic alpha-helical conformation across the first 11 residues of the peptide but remains comparatively random at the C-terminus [78, 85]. The interaction between the N-terminal tryptophan residue and model membranes of varying composition was evaluated suggesting that W1 is inserted into the membrane at the lipid/water interface [78]. Along with this, the orientation of the phenyl side chain of F11 found to be in same direction as the indole ring of W1 also suggested that the amphipathic N-terminal helix anchors the peptide to membrane with these two residues that facilitates peptide folding [78, 86]. The same group has suggested that the hydrophobic patch in between the two residues as well as Leu, Ile, and Ala side chains are responsible for interaction between the peptide and the hydrophobic core of a phospholipid bilayer [83]. In addition, the

TABLE 3: Comparison of amino acid sequences of lactoferrampin from lactoferrin from six species.

Human	WNLLRQAQEKFGKDKSP	17
Bovine	WKLLSKAQEKFGKNKSR	17
Buffalo	WKLLSKAQEKFGKNKSG	17
Equine	WKLLHRAQEEFGRNKSS	17
Caprine	WELLRKAQEKFGKNKSQ	17
Camel	WKLLVKAQEKFGRGKPS	17

helix capping residues Asp-Leu-Ile in the N-terminus of the peptide has been found to mediate the depth of membrane insertion by enhancing the affinity for negatively charged vesicles [84]. Bovine lactoferrampin had been shown to have greater affinity to acidic phospholipids than that to neutral phospholipids [85]. Haney et al. have speculated a two-step model of antimicrobial action by this peptide where the C-terminus positive charge cluster helps in the primary attraction of lactoferrampin to the membrane followed by the helix formation at the N-terminus that interacts to the surface of the bacterial lipid bilayer [78].

The sequence comparison of lactoferrampin from six different species shows uniform preponderance of cationic amino acid residues among hydrophobic residues (Table 3). The hydrophobic domain contains W1 in all the species that is involved in membrane insertion [87]. Bovine lactoferrampin 268–284 has a net positive charge of 5+ at neutral pH with hydrophobic domain. The hydrophobic moment (μ) of the peptide which is a measure of lipophilicity was found to be 5.42. In contrast, the human lactoferrampin has a net charge of 2+ resulting in reduced antimicrobial activity [76]. However, by increasing the net positive charge near the C-terminal end of human lactoferrampin, a significant increase in its antibacterial and candidacidal activity was obtained [83]. The basic amino acid residues crucial for the antimicrobial action were found to be conserved among all the six species.

4. Lactoferricin

Lactoferricin is a multifunctional, 25-residue peptide that is generated upon cleavage of native lactoferrin by pepsin and represents amino acid residues number 17–41 in lactoferrin. The lactoferricin peptide is different from the other peptides described so far as it contains a disulfide bond between residues Cys 20 and Cys 37 in human lactoferrin and Cys 19 and Cys 36 in bovine lactoferrin. The peptide has an abundance of basic amino acids like lysine and arginine as well as hydrophobic residues like tryptophan and phenylalanine.

The first report on lactoferricin in 1992 described this peptide to be more potent as an antibacterial agent in comparison with the intact lactoferrin and it was demonstrated to cause a rapid loss of colony-forming capacity in most of its targets. However, some strains like Pseudomonas fluorescens, Enterococcus faecalis, and Bifidobacterium bifidum strains were found to be resistant to lactoferricin [88].

The antibacterial activity of this peptide was attributed to its action of releasing lipopolysaccharide from bacterial strains and, hence, disruption of cytoplasmic membrane

TABLE 4: Amino acid sequences of lactoferricin from lactoferrin from six species.

Human	TKCFQWQRNMRKVRGPPVSCIKRDS	25
Bovine	FKCRRWQWRMKKLGAPSITCVRRAF	25
Buffalo	LKCHRWQWRMKKLGAPSITCVRRAF	25
Equine	AKCAKFQRNMKKVRGPSVSCIRKTS	25
Caprine	SKCYQWQRRMRKLGAPSITCVRRTS	25
Camel	KKCAQWQRRMKKVRGPSVTCVKKTS	25

permeability after cell binding [89–91]. Apart from having a broad antibacterial spectrum, lactoferricin was found to be highly potent against *Candida albicans* [89, 92, 93]. Recently, it has also been shown to have antiviral [94, 95] and antiprotozoal activities [96]. It also displayed other activities like inhibition of tumor metastasis [97] and induction of apoptosis in human leukemic cells [98].

The mechanism of action of lactoferricin was attributed to 11-amino-acid amphipathic alpha-helical region which is positioned on the outer surface of the N-lobe of lactoferrin. The proline at the 26th position (P26) was found to be essential for the antibacterial activity, and it was speculated to be responsible for disruption of the helical region, and hence the helicity of the peptide was predicted to be an essential aspect of the antibacterial action of this peptide [99]. Lactoferricin was found to be produced in the human stomach, indicating that this peptide is definitely generated in vivo for host defense [100].

The comparison of the antimicrobial activities of lactoferricin from human, bovine, murine, and caprine showed that bovine lactoferricin was the most potent [101]. The minimal inhibitory concentration (MIC) of lactoferricin 3 differs according to their source [102]. A comparison of the MIC values of lactoferricin shows that bovine lactoferricin is the most potent. The MIC of bovine lactoferricin against certain *E. coli* strains has been found to be around 30 μg/mL while that derived from human is more than 100 μg/mL. The efficiency of antibacterial activity of bovine lactoferricin is due to the presence of high amount of net positive charge (+8) and hydrophobic residues (primarily W6, W8, and M10) [51, 88]. The action of this peptide is also dependent on the pH [89, 103].

It was shown that only six central residues (4–9) among the twenty five residues of the peptide are required for its antimicrobial activity [104]. It may be noted that the tetrapeptide KRDS is present only in human lactoferrin while it shows variations in the sequence of others. It has been reported that KRDS inhibits platelet aggregation [105, 106].

The mode of action of lactoferrin peptides is best studied in bovine lactoferricin. The bovine lactoferricin has been demonstrated to interact with the negatively charged elements in the membrane of susceptible bacteria and disrupt the cell membrane. A synthetic peptide derived from human lactoferricin has been found to be effective in depolarizing the bacterial cytoplasmic membrane with a loss of pH gradient [107].

The permeabilizing effect of bovine lactoferricin causes membrane disruption resulting in inhibition of macromolecular biosynthesis and ultimately cell death [108]. The mode of action is however different in gram-positive and gram-negative bacteria. In gram-negative bacteria antimicrobial peptides act on lipopolysaccharides and in gram-positive bacteria they act on lipoteichoic and teichoic acids. In addition to antimicrobial properties, lactoferricin derived from human and bovine origin has also been found to be effective in inhibiting the classical complement pathway. This implicates a role of these peptides in suppression of inflammatory effects caused by bacteria [109].

The sequence analysis of the lactoferricin from various species indicates that unlike bovine lactoferricin there is only one tryptophan at position 6 (Table 4). A further exception to this is equine which has no tryptophan residue in either of the positions. This shows that the two tryptophans in lactoferricin are important for its optimal activity against microbes [110].

The solution structures of lactoferricin have been determined from bovine [111] and human [112] sources. The bovine lactoferricin adopted a distorted antiparallel beta sheet, in complete contrast with its conformation in the intact lactoferrin, as observed in the structures obtained by X-ray crystallography. However, the solution structure of human lactoferricin was closer to its structure in native lactoferrin since the amphipathic helix was preserved from Gln14 to Lys29. However, the beta-sheet character was not observed in the solution structure of human lactoferricin either.

5. Lipopolysaccharide Neutralization Activity of Lactoferrin-Derived Peptides

Lipopolysaccharide (LPS), the outer membrane component of gram-negative bacteria, is one of the major causes of endotoxin-induced production of inflammatory cytokines [113] and septic shock [114]. Lactoferrin has been shown to neutralize the effect of LPS-induced toxicity by binding to LPS [115, 116]. The cationic peptide derived from lactoferrin which is responsible for this interaction and release of LPS is first identified to be lactoferricin [90]. The residues from 28 to 34 of lactoferrin corresponding to the region in human lactoferrin have been identified to have a high affinity for binding to LPS [117]. Soluble LPS can interact with bovine lactoferrin. The initial binding of the peptide with *E. coli* has been found to be due to interaction with bacterial LPS [118]. Further studies have shown that bovine lactoferricin can arrest the LPS-induced cytokine release by suppressing the IL-6 response in human monocytic cells stimulated by LPS [119]. A synthetic peptide corresponding to the antibacterial region of human lactoferricin was also found to facilitate depolarization of the bacterial cytoplasmic membrane, loss of the pH gradient, and a bactericidal effect in *E. coli* [60]. Modelling studies using synthetic peptides derived from human and bovine lactoferricin have shown that these cationic peptides with their positively charged residues first interact with LPS carrying negative charges. This is followed by hydrophobic interactions between the tryptophan residues of the peptides and the lipid A molecule of LPS to promote structural disorganization [120]. Similarly, a synthetic peptide corresponding to 11 residues of human lactoferricin near its N-terminus has been found to bind to LPS and neutralize the LPS-induced adverse effects in vitro and in monocytes [121, 122]. In yet another study using NMR, it has been observed that this peptide folds into a "T-shaped" conformation formed by its hydrophobic core and the two clusters of hydrophilic residues of the peptide targets the two phosphate moieties of lipid A in LPS [123].

6. Comparison of Lactoferrin Antimicrobial Peptides with Other Antimicrobial Peptides Found in Nature

The antimicrobial peptides found in nature are classified into four groups according to a combination of their sequence homologies, functional similarities, and common three-dimensional structures [124].

The four groups include Group 1, which consists of linear, cationic, and amphipathic-helical peptides, for example, cecropins, magainins, bombinins, and temporins; Group 2, which consists of β-strands connected by intramolecular disulfide bridges, for example, human β-defensin-2, tachyplesins, and protegrins; Group 3, which consists of linear peptides with an extended structure, characterized by overrepresentation of one or more amino acids, for example, tritrpticin and indolicidin; and Group 4, which consists of peptides containing a looped structure, for example, bactenecin, brevinins, and esculentin.

In the light of the above classification, human LF1-11 (GRRRSVQWCAV) consists of a highly variable loop region and a short β-strand and is arginine rich, and hence can be classified in Group 4. However, the same cannot be said about LF1-11 from other species, since their conformation may be similar to the human LF1-11 in the structure, but they are not rich in arginines. The arginine-rich fragment of this peptide is similar to other cationic arginine-rich peptides found in nature which have cell-penetrating activity and hence can traverse the plasma membrane of eukaryotic cells [125]. A significant example of arginine-rich peptide that has cell-penetrating property is arginine-rich HIV Tat peptide (GRKKR-RQRRRPPQ) [126].

On the other hand, lactoferrampin belongs to Group 1 which consists of linear, cationic, and amphipathic-helical peptides. The alpha-helical amphipathic character of lactoferrampin has been compared with other Group I peptides like magainins, bombinins, cecropin A, and temporins and are depicted by the helical wheel representation of the peptides in which the charged and polar residues are found aligned along one side and most of the amino acids with nonpolar side chains occupy the opposite side of the helical cylinder (Figure 3). The spatial segregation of the hydrophobic and hydrophilic residues designates the amphipathic nature of the peptides [127, 128]. These peptides upon interaction with target membranes fold into an amphipathic α-helix with one face of the helix predominantly containing the hydrophobic amino acids and the opposite face the charged amino acids [129]. The presence of a prominent hydrophobic face is

TABLE 5: Comparison of amino acid sequences of lactoferricin and other tryptophan and arginine containing antimicrobial peptides. The active hexapeptide of lactoferricin and its corresponding matching residues in other antimicrobial peptides are indicated in red.

Lactoferricin	FKCRRWQWRMKKLGAPSITVCVRRAF
Tripticin	VRRFPWWWPWPFLRR
Lf1-11	GRRRSVQWCAV
Combi-1	RRWWRF
Indolicidin	ILPWKWPWWPWRR
Puroindoline	FPVTWRWWKWWKG
Lys H	RAWVAWR

observed in the helical wheel representations of magainin 2, bombinin, and temporin (Figures 3(a), 3(b), and 3(c)) whereas a pronounced cationic domain is present on the hydrophilic surface of the helical wheel diagram of bovine lactoferrampin like that in cecropin A (Figures 3(d) and 3(e)). The positively charged domain is more distinct in bovine than human lactoferrampin (data not shown). The analysis suggests that there is very little similarity in the amino acid sequence within the group; however there is a distinct trend in the distribution of different types of residue, that is, hydrophobic and charged, polar, and so forth, within the secondary structure of the helix.

Lactoferricin has been shown to display a similarity with an antimicrobial peptide, magainin. Both peptides are able to traverse the bacterial cytoplasmic membrane [130]. Sequence similarities between lactoferricin and dermaseptin and magainins suggest that lactoferricin may act as an amphipathic alpha-helix [131]. The active hexapeptide fragment within bovine lactoferricin peptide showed distinct similarities with LF1-11 and other amphipathic tryptophan and arginine-rich antimicrobial peptides found in other sources apart from lactoferrin (Table 5). Bovine lactoferricin contains two tryptophan residues at positions 6 and 8 and two arginines at positions 4 and 5. These two amino acids have chemical properties which are one of the critical components of antimicrobial peptides [132].

7. Conclusion

Enzymatic digestion of lactoferrin results in the generation of antimicrobial peptides which display antimicrobial properties, in some cases, with greater potency than the native lactoferrin, possibly for the protection of neonates against the invading pathogens. These peptides, all from the N-lobe of lactoferrin, show a remarkable similarity to cationic antimicrobial peptides found in other invertebrate and vertebrate species. These peptides are conserved in lactoferrin, structurally and functionally in most species. Though there may be minor variations in the sequence and the conformational features among these lactoferrin peptides from various species, the basic framework tends to be similar and conserved. This indicates that these peptides play a significant role in the antimicrobial function of this protein. The antimicrobial effect of cationic peptides of different origins is due to cytoplasmic membrane disruption of the target cell as well as

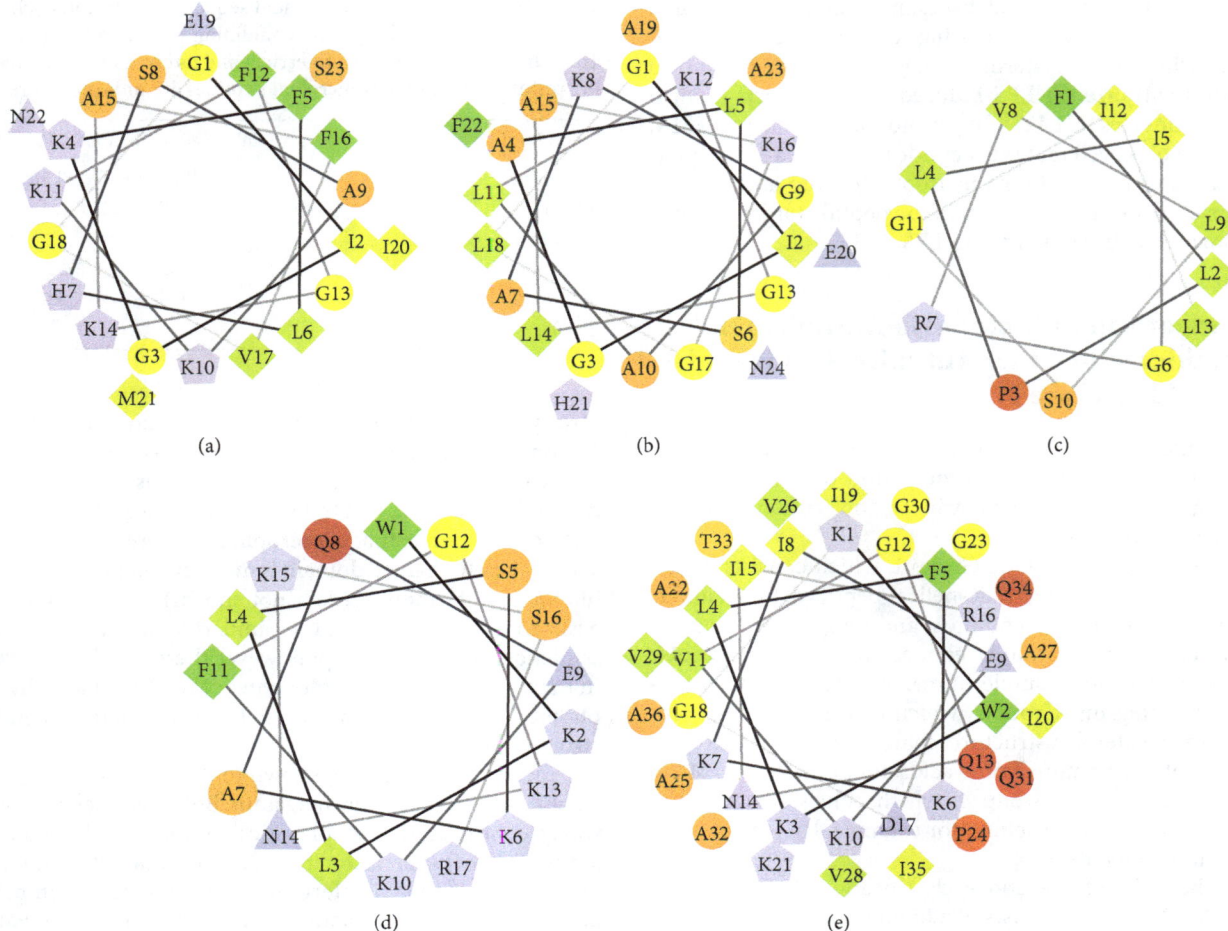

FIGURE 3: Helical wheel representation of (a) magainin 2, (b) bombinin, (c) temporin, (d) lactoferrampin, and (e) cecropin A. The hydrophilic residues are shown as circles, hydrophobic residues as diamonds, potentially negatively charged as triangles, and potentially positively charged as pentagons. Hydrophobicity is color coded as well: the most hydrophobic residue is green, and the amount of green is decreasing proportionally to low hydrophobicity, coded as yellow. Hydrophilic residues are coded red with pure red being the uncharged residues, and the amount of red decreasing proportionally to the hydrophilicity. The potentially charged residues are light blue. (The plots were made using the software created by Don Armstrong and Raphael Zidovetzki. Version: 0.10 p06 12/14/2001 DLA modified by Jim Hu.)

immunomodulation. The difference in their functional properties is due to the difference in their amino acid composition inspite of sharing amphipathic and cationic characteristics. The presence of all these antimicrobial peptides in a single domain of lactoferrin suggests that the protein acts on the membrane interface and disturbs the membrane integrity resulting in its antimicrobial activity.

Since lactoferrin is found in the milk and is ingested throughout the life of all neonates and most adults, it may be an excellent agent for administration to humans. In the future, these lactoferrin peptides could serve as leads for drug development for antimicrobial therapy.

Conflict of Interests

The authors declare that they have no conflict of interests.

Acknowledgments

T. P. Singh thanks the Department of Biotechnology (DBT) for the Distinguished Biotechnology Research Professor scheme. M. Sinha thanks DST for financial grants under Fast-Track program in Life Sciences.

References

[1] J. A. Hoffmann and J. M. Reichhart, "Drosophila innate immunity: an evolutionary perspective," *Nature Immunology*, vol. 3, no. 2, pp. 121–126, 2002.

[2] H. Kaufmann, R. Medzhitov, and S. Gordon, *The Innate Immune Response to Infection*, ASM Press, Washington, DC, USA, 2004.

[3] A. Iwasaki and R. Medzhitov, "Regulation of adaptive immunity by the innate immune system," *Science*, vol. 327, no. 5963, pp. 291–295, 2010.

[4] B. Beutle, "Innate immunity: an overview," *Molecular Immunology*, vol. 40, no. 12, pp. 845–859, 2004.

[5] R. E. W. Hancock and A. Patrzykat, "Clinical development of cationic antimicrobial peptides: from natural to novel antibiotics," *Current Drug Targets*, vol. 2, no. 1, pp. 79–83, 2002.

[6] K. A. Brogden, M. Ackermann, P. B. McCray Jr., and B. F. Tack, "Antimicrobial peptides in animals and their role in host defences," *International Journal of Antimicrobial Agents*, vol. 22, no. 5, pp. 465–478, 2003.

[7] M. Pasupuleti, A. Schmidtchen, and M. Malmsten, "Antimicrobial peptides: key components of the innate immune system," *Critical Reviews in Biotechnology*, vol. 32, no. 2, pp. 143–171, 2012.

[8] M. Maes, A. Loyter, and A. Friedler, "Peptides that inhibit HIV-1 integrase by blocking its protein-protein interactions," *FEBS Journal*, vol. 279, no. 16, pp. 2795–2809, 2012.

[9] W. van't Hof, E. C. Veerman, E. J. Helmerhorst, and A. V. Amerongen, "Antimicrobial peptides: properties and applicability," *Biological Chemistry*, vol. 382, no. 4, pp. 597–619, 2001.

[10] Y. Shai, "Mode of action of membrane active antimicrobial peptides," *Biopolymers*, vol. 66, no. 4, pp. 236–248, 2002.

[11] S. L. Sand, J. Nissen-Meyer, O. Sand, T. M. Haug, and A. Plantaricin, "A cationic peptide produced by *Lactobacillus plantarum*, permeabilizes eukaryotic cell membranes by a mechanism dependent on negative surface charge linked to glycosylated membrane proteins," *Biochimica et Biophysica Acta*, vol. 1828, no. 2, pp. 249–259, 2012.

[12] M. E. Quiñones-Mateu, M. M. Lederman, Z. Feng et al., "Human epithelial beta-defensins 2 and 3 inhibit HIV-1 replication," *AIDS*, vol. 17, no. 16, pp. 39–48, 2003.

[13] S. Sinha, N. Cheshenko, R. I. Lehrer, and B. C. Herold, "NP-1, a rabbit α-defensin, prevents the entry and intercellular spread of herpes simplex virus type 2," *Antimicrobial Agents and Chemotherapy*, vol. 47, no. 2, pp. 494–500, 2003.

[14] W. Wang, S. M. Owen, D. L. Rudolph et al., "Activity of alpha- and theta-defensins against primary isolates of HIV-1," *Journal of Immunology*, vol. 173, no. 1, pp. 515–520, 2004.

[15] B. Yasin, W. Wang, M. Pang et al., "Theta defensins protect cells from infection by herpes simplex virus by inhibiting viral adhesion and entry," *Journal of Virology*, vol. 78, no. 10, pp. 5147–5156, 2004.

[16] Y. J. Gordon, E. G. Romanowski, and A. M. McDermott, "A review of antimicrobial peptides and their therapeutic potential as anti-infective drugs," *Current Eye Research*, vol. 30, no. 7, pp. 505–515, 2005.

[17] J. Davies, "Inactivation of antibiotics and the dissemination of resistance genes," *Science*, vol. 264, no. 5157, pp. 375–382, 1994.

[18] J. Verhoef, "Antibiotic resistance: the pandemic," *Advances in Experimental Medicine and Biology*, vol. 531, pp. 301–313, 2003.

[19] K. M. Shea, "Antibiotic resistance: what is the impact of agricultural uses of antibiotics on children's health," *Pediatrics*, vol. 112, no. 1, pp. 253–258, 2003.

[20] M. Zasloff, "Innate immunity, antimicrobial peptides, and protection of the oral cavity," *The Lancet*, vol. 360, no. 9340, pp. 1116–1117, 2002.

[21] A. R. Koczulla and R. Bals, "Antimicrobial peptides: current status and therapeutic potential," *Drugs*, vol. 63, no. 4, pp. 389–406, 2003.

[22] M. Zasloff, "Antimicrobial peptides, innate immunity, and the normally sterile urinary tract," *Journal of the American Society of Nephrology*, vol. 18, no. 11, pp. 2810–2816, 2007.

[23] D. Yang, A. Biragyn, D. M. Hoover, J. Lubkowski, and J. J. Oppenheim, "Multiple roles of antimicrobial defensins, cathelicidins, and eosinophil-derived neurotoxin in host defense," *Annual Review of Immunology*, vol. 22, pp. 181–215, 2004.

[24] R. Capparelli, F. De Chiara, N. Nocerino et al., "New perspectives for natural antimicrobial peptides: application as antinflammatory drugs in a murine model," *BMC Immunology*, vol. 13, article 61, 2012.

[25] R. E. Hancock, "The therapeutic potential of cationic peptides," *Expert Opinion on Investigational Drugs*, vol. 7, no. 2, pp. 167–174, 1998.

[26] R. E. Hancock, "Cationic antimicrobial peptides: towards clinical applications," *Expert Opinion on Investigational Drugs*, vol. 9, no. 8, pp. 1723–1729, 2000.

[27] M. G. Scott, H. Yan, and R. E. W. Hancock, "Biological properties of structurally related α-helical cationic antimicrobial peptides," *Infection and Immunity*, vol. 67, no. 4, pp. 2005–2009, 1999.

[28] M. G. Scott and R. E. Hancock, "Cationic antimicrobial peptides and their multifunctional role in the immune system," *Critical Reviews in Immunology*, vol. 20, no. 5, pp. 407–431, 2000.

[29] K. Y. Choi, L. N. Chow, and N. Mookherjee, "Cationic host defence peptides: multifaceted role in immune modulation and inflammation," *Journal of Innate Immunity*, vol. 4, no. 4, pp. 361–370, 2012.

[30] M. E. Selsted, M. J. Novotny, W. L. Morris, Y. Q. Tang, W. Smith, and J. S. Cullor, "Indolicidin, a novel bactericidal tridecapeptide amide from neutrophils," *The Journal of Biological Chemistry*, vol. 267, no. 7, pp. 4292–4295, 1992.

[31] C. Lawyer, S. Pai, M. Watabe et al., "Antimicrobial activity of a 13 ammo acid tryptophan-rich peptide derived from a putative porcine precursor protein of a novel family of antibacterial peptides," *FEBS Letters*, vol. 390, no. 1, pp. 95–98, 1996.

[32] J. E. Blochet, C. Chevalier, E. Forest et al., "Complete amino acid sequence of puroindoline, a new basic and cystine-rich protein with a unique tryptophan-rich domain, isolated from wheat endosperm by Triton x-114 phase partitioning," *FEBS Letters*, vol. 329, no. 3, pp. 336–340, 1993.

[33] S. E. Blondelle, E. Takahashi, K. T. Dinh, and R. A. Houghten, "The antimicrobial activity of hexapeptides derived from synthetic combinatorial libraries," *Journal of Applied Bacteriology*, vol. 78, no. 1, pp. 39–46, 1995.

[34] A. Pellegrini, U. Thomas, N. Bramaz, S. Klauser, P. Hunziker, and R. Von Fellenberg, "Identification and isolation of a bactericidal domain chicken egg white lysozyme," *Journal of Applied Microbiology*, vol. 82, no. 3, pp. 372–378, 1997.

[35] P. Bulet, R. Sticklin, and L. Menin, "Anti-microbial peptides: from invertebrates to vertebrates," *Immunological Reviews*, vol. 198, pp. 169–184, 2004.

[36] A. Giangaspero, L. Sandri, and A. Tossi, "Amphipathic alpha helical antimicrobial peptides," *European Journal of Biochemistry*, vol. 268, no. 3, pp. 5589–5600, 2001.

[37] A. Tossi, L. Sandri, and A. Giangaspero, "Amphipathic, alpha-helical antimicrobial peptides," *Biopolymers*, vol. 55, no. 1, pp. 4–30, 2000.

[38] J. He, R. Eckert, T. Pharm et al., "Novel synthetic antimicrobial peptides against *Streptococcus* mutants," *Antimicrobial Agents of Chemotherapy*, vol. 51, no. 4, pp. 1351–1358, 2007.

[39] P. L. Masson and J. F. Heremans, "Metal-combining properties of human lactoferrin (red milk protein). 1. The involvement of bicarbonate in the reaction," *European Journal of Biochemistry*, vol. 6, no. 4, pp. 579–584, 1968.

[40] P. L. Masson, J. F. Heremans, and J. Ferin, "Presence of an Iron-binding protein (lactoferrin) in the genital tract of the human female. I. Its immunohistochemical localization in the endometrium," *Fertility and Sterility*, vol. 19, no. 5, pp. 679–689, 1968.

[41] P. L. Masson, J. F. Heremans, and E. Schonne, "Lactoferrin, an iron-binding protein in neutrophilic leukocytes," *The Journal of Experimental Medicine*, vol. 130, no. 3, pp. 643–658, 1969.

[42] P. Masson, J. F. Heremans, and J. Prignot, "Immunohistochemical localization of the iron-binding protein lactoferrin in human bronchial glands," *Experientia*, vol. 21, no. 10, pp. 604–605, 1965.

[43] L. H. Vorland, "Lactoferrin: a multifunctional glycoprotein," *Acta Pathologica, Microbiologica et Immunologica Scandinavica*, vol. 107, no. 11, pp. 971–981, 1999.

[44] R. Chierici, "Antimicrobial actions of lactoferrin," *Advances in Nutrition Research*, vol. 10, pp. 247–269, 2001.

[45] P. Valenti, M. Marchetti, F. Superti et al., "Antiviral activity of lactoferrin," *Advances in Experimental Medicine and Biology*, vol. 443, pp. 199–203, 1998.

[46] H. Nikawa, L. P. Samaranayake, J. Tenovuo, K. M. Pang, and T. Hamada, "The fungicidal effect of human lactoferrin on *Candida albicans* and *Candida krusei*," *Archives of Oral Biology*, vol. 38, no. 12, pp. 1057–1063, 1993.

[47] C. C. Yen, C. J. Shen, W. H. Hsu et al., "Lactoferrin: an iron-binding antimicrobial protein against *Escherichia coli* infection," *Biometals*, vol. 24, no. 4, pp. 585–594, 2011.

[48] S. Farnaud and R. W. Evans, "Lactoferrin—a multifunctional protein with antimicrobial properties," *Molecular Immunology*, vol. 40, no. 7, pp. 395–405, 2003.

[49] A. M. Cole, H. I. Liao, O. Stuchlik et al., "Cationic polypeptides are required for antibacterial activity of human airway fluid," *Journal of Immunology*, vol. 169, no. 12, pp. 6985–6991, 2002.

[50] B. E. Haug, M. L. Skar, and J. S. Svendsen, "Bulky aromatic amino acids increase the antibacterial activity of 15-residue bovine lactoferricin derivatives," *Journal of Peptide Science*, vol. 7, no. 8, pp. 425–432, 2001.

[51] M. B. Strøm, O. Rekdal, and J. S. Svendsen, "Antibacterial activity of 15-residue lactoferricin derivatives," *The Journal of Peptide Research*, vol. 56, no. 5, pp. 265–274, 2000.

[52] A. R. Lizzi, V. Carnicelli, M. M. Clarkson, A. Di Giulio, and A. Oratore, "Lactoferrin derived peptides: mechanisms of action and their perspectives as antimicrobial and antitumoral agents," *Mini-Reviews in Medicinal Chemistry*, vol. 9, no. 6, pp. 687–695, 2009.

[53] M. Tomita, M. Takase, W. Bellamy, and S. Shimakura, "A review: the active peptide of lactoferrin," *Acta Paediatrica Japonica*, vol. 36, no. 5, pp. 585–591, 1994.

[54] J. L. Gifford, H. N. Hunter, and H. J. Vogel, "Lactoferricin: a lactoferrin-derived peptide with antimicrobial, antiviral, antitumor and immunological properties," *Cellular and Molecular Life Sciences*, vol. 62, no. 22, pp. 2588–2598, 2005.

[55] M. Tomita, H. Wakabayashi, K. Shin, K. Yamauchi, T. Yaeshima, and K. Iwatsuki, "Twenty-five years of research on bovine lactoferrin applications," *Biochimie*, vol. 91, no. 1, pp. 52–57, 2009.

[56] N. Orsi, "The antimicrobial activity of lactoferrin: current status and perspectives," *BioMetals*, vol. 17, no. 3, pp. 189–196, 2004.

[57] H. M. Baker and E. N. Baker, "Lactoferrin and iron: structural and dynamic aspects of binding and release," *Biometals*, vol. 17, no. 3, pp. 209–216, 2004.

[58] P. P. Ward and O. M. Conneely, "Lactoferrin: role in iron homeostasis and host defense against microbial infection," *Biometals*, vol. 17, no. 3, pp. 203–208, 2004.

[59] H. Jenssen and R. E. W. Hancock, "Antimicrobial properties of lactoferrin," *Biochimie*, vol. 91, no. 1, pp. 19–29, 2009.

[60] O. Aguilera, H. Ostolaza, L. M. Quirós, and J. F. Fierro, "Permeabilizing action of an antimicrobial lactoferricin-derived peptide on bacterial and artificial membranes," *FEBS Letters*, vol. 462, no. 3, pp. 273–277, 1999.

[61] S. Sánchez-Gómez, M. Lamata, J. Leiva et al., "Comparative analysis of selected methods for the assessment of antimicrobial and membrane-permeabilizing activity: a case study for lactoferricin derived peptides," *BMC Microbiology*, vol. 8, article 196, 2008.

[62] S. Farnaud, C. Spiller, L. C. Moriarty et al., "Interactions of lactoferricin-derived peptides with LPS and antimicrobial activity," *FEMS Microbiology Letters*, vol. 233, no. 2, pp. 193–199, 2004.

[63] Y. Shai, "Mechanism of the binding, insertion and destabilization of phospholipid bilayer membranes by α-helical antimicrobial and cell non-selective membrane-lytic peptides," *Biochimica et Biophysica Acta*, vol. 1462, no. 1-2, pp. 55–70, 1999.

[64] H. J. Vogel, D. J. Schibli, W. Jing, E. M. Lohmeier-Vogel, R. F. Epand, and R. M. Epand, "Towards a structure-function analysis of bovine lactoferricin and related tryptophan- and arginine-containing peptides," *Biochemistry and Cell Biology*, vol. 80, no. 1, pp. 49–63, 2002.

[65] L. Dijkshoorn, C. P. J. M. Brouwer, S. J. P. Bogaards, A. Nemec, P. J. Van Den Broek, and P. H. Nibbering, "The synthetic n-terminal peptide of human lactoferrin, hLF(1-11), is highly effective against experimental infection caused by multidrug-resistant *Acinetobacter baumannii*," *Antimicrobial Agents and Chemotherapy*, vol. 48, no. 12, pp. 4919–4921, 2004.

[66] H. P. Stallmann, C. Faber, A. L. Bronckers et al., "Histatin and lactoferrin derived peptides: antimicrobial properties and effects on mammalian cells," *Peptides*, vol. 26, no. 12, pp. 2355–2359, 2005.

[67] A. Lupetti, A. Paulusma-Annema, M. M. Welling et al., "Synergistic activity of the N-terminal peptide of human lactoferrin and fluconazole against *Candida* species," *Antimicrobial Agents and Chemotherapy*, vol. 47, no. 1, pp. 262–267, 2003.

[68] A. Lupetti, C. P. J. M. Brouwer, S. J. P. Bogaards et al., "Human lactoferrin-derived peptide's antifungal activities against disseminated *Candida albicans* infection," *Journal of Infectious Diseases*, vol. 196, no. 9, pp. 1416–1424, 2007.

[69] A. Lupetti, A. Paulusma-Annema, M. M. Welling, S. Senesi, J. T. Van Dissel, and P. H. Nibbering, "Candidacidal activities of human lactoferrin peptides derived from the N terminus," *Antimicrobial Agents and Chemotherapy*, vol. 44, no. 12, pp. 3257–3263, 2000.

[70] P. H. Nibbering, E. Ravensbergen, M. M. Welling et al., "Human lactoferrin and peptides derived from its N terminus are highly effective against infections with antibiotic-resistant bacteria," *Infection and Immunity*, vol. 69, no. 3, pp. 1469–1476, 2001.

[71] P. H. C. Van Berkel, M. E. J. Geerts, H. A. van Veen, M. Mericskay, H. A. De Boer, and J. H. Nuijens, "N-terminal stretch Arg2, Arg3, Arg4 and Arg5 of human lactoferrin is essential for binding to heparin, bacterial lipopolysaccharide, human lysozyme and DNA," *Biochemical Journal*, vol. 328, no. 1, pp. 145–151, 1997.

[72] H. P. Stallmann, C. Faber, A. L. J. J. Bronckers, A. V. Nieuw Amerongen, and P. I. J. M. Wuismann, "Osteomyelitis prevention in rabbits using antimicrobial peptide hLF1-11- or gentamicin-containing calcium phosphate cement," *Journal of Antimicrobial Chemotherapy*, vol. 54, no. 2, pp. 472–476, 2004.

[73] A. Lupetti, C. P. J. M. Brouwer, H. E. C. Dogterom-Ballering et al., "Release of calcium from intracellular stores and subsequent uptake by mitochondria are essential for the candidacidal activity of an N-terminal peptide of human lactoferrin," *Journal of Antimicrobial Chemotherapy*, vol. 54, no. 3, pp. 603–608, 2004.

[74] A. M. van der Does, S. J. P. Bogaards, B. Ravensbergen, H. Beekhuizen, J. T. Van Dissel, and P. H. Nibbering, "Antimicrobial peptide hLF1-11 directs granulocyte-macrophage colony-stimulating factor-driven monocyte differentiation toward macrophages with enhanced recognition and clearance of pathogens," *Antimicrobial Agents and Chemotherapy*, vol. 54, no. 2, pp. 811–816, 2010.

[75] A. M. van der Does, P. J. Hensbergen, S. J. Bogaards et al., "The human lactoferrin-derived peptide hLF1-11 exerts immunomodulatory effects by specific inhibition of myeloperoxidase activity," *Journal of Immunology*, vol. 188, no. 10, pp. 5012–5019, 2012.

[76] M. I. A. van der Kraan, J. Groenink, K. Nazmi, E. C. I. Veerman, J. G. M. Bolscher, and A. V. Nieuw Amerongen, "Lactoferrampin: a novel antimicrobial peptide in the N1-domain of bovine lactoferrin," *Peptides*, vol. 25, no. 2, pp. 177–183, 2004.

[77] M. I. A. van der Kraan, K. Nazmi, A. Teeken et al., "Lactoferrampin, an antimicrobial peptide of bovine lactoferrin, exerts its candidacidal activity by a cluster of positively charged residues at the C-terminus in combination with a helix-facilitating N-terminal part," *Biological Chemistry*, vol. 386, no. 2, pp. 137–142, 2005.

[78] E. F. Haney, F. Lau, and H. J. Vogel, "Solution structures and model membrane interactions of lactoferrampin, an antimicrobial peptide derived from bovine lactoferrin," *Biochimica et Biophysica Acta*, vol. 1768, no. 10, pp. 2355–2364, 2007.

[79] M. I. A. van der Kraan, C. van der Made, K. Nazmi et al., "Effect of amino acid substitutions on the candidacidal activity of LFampin 265-284," *Peptides*, vol. 26, no. 11, pp. 2093–2097, 2005.

[80] M. I. A. van der Kraan, K. Nazmi, W. van't Hof, A. V. Nieuw Amerongen, E. C. I. Veerman, and J. G. M. Bolscher, "Distinct bactericidal activities of bovine lactoferrin peptides LFampin 268-284 and LFampin 265-284: Asp-Leu-Ile makes a difference," *Biochemistry and Cell Biology*, vol. 84, no. 3, pp. 358–362, 2006.

[81] M. I. A. van der Kraan, J. van Marle, K. Nazmi et al., "Ultrastructural effects of antimicrobial peptides from bovine lactoferrin on the membranes of *Candida albicans* and *Escherichia coli*," *Peptides*, vol. 26, no. 9, pp. 1537–1542, 2005.

[82] H. Flores Villaseñor, A. Canizalez Román, M. Reyes Lopez et al., "Bactericidal effect of bovine lactoferrin, LFcin, LFampin and LFchimera on antibiotic-resistant *Staphylococcus aureus* and *Escherichia coli*," *Biometals*, vol. 23, no. 3, pp. 569–578, 2010.

[83] E. F. Haney, H. N. Hunter, K. Matsuzaki, and H. J. Vogel, "Solution NMR studies of amphibian antimicrobial peptides: linking structure to function?" *Biochimica et Biophysica Acta*, vol. 1788, no. 8, pp. 1639–1655, 2009.

[84] E. F. Haney, S. Nathoo, H. J. Vogel, and E. J. Prenner, "Induction of non-lamellar lipid phases by antimicrobial peptides: a potential link to mode of action," *Chemistry and Physics of Lipids*, vol. 163, no. 1, pp. 82–93, 2010.

[85] A. Tsutsumi, N. Javkhlantugs, A. Kira et al., "Structure and orientation of bovine lactoferrampin in the mimetic bacterial membrane as revealed by solid-state NMR and molecular dynamics simulation," *Biophysical Journal*, vol. 103, no. 8, pp. 1735–1743, 2012.

[86] E. F. Haney, K. Nazmi, J. G. Bolscher, and H. J. Vogel, "Structural and biophysical characterization of an antimicrobial peptide chimera comprised of lactoferricin and lactoferrampin," *Biochimica et Biophysica Acta*, vol. 1818, no. 3, pp. 762–775, 2012.

[87] M. Schiffer, C. H. Chang, and F. J. Stevens, "The functions of tryptophan residues in membrane proteins," *Protein Engineering*, vol. 5, no. 3, pp. 213–214, 1992.

[88] W. Bellamy, M. Takase, H. Wakabayashi, K. Kawase, and M. Tomita, "Antibacterial spectrum of lactoferricin B, a potent bactericidal peptide derived from the N-terminal region of bovine lactoferrin," *Journal of Applied Bacteriology*, vol. 73, no. 6, pp. 472–479, 1992.

[89] W. Bellamy, H. Wakabayashi, M. Takase, K. Kawase, S. Shimamura, and M. Tomita, "Killing of *Candida albicans* by lactoferricin B, a potent antimicrobial peptide derived from the N-terminal region of bovine lactoferrin," *Medical Microbiology and Immunology*, vol. 182, no. 2, pp. 97–105, 1993.

[90] K. Yamauchi, M. Tomita, T. J. Giehl, and R. T. Ellison, "Antibacterial activity of lactoferrin and a pepsin-derived lactoferrin peptide fragment," *Infection and Immunity*, vol. 61, no. 2, pp. 719–728, 1993.

[91] J. H. Kang, M. K. Lee, K. L. Kim, and K. S. Hahm, "Structure-biological activity relationships of 11-residue highly basic peptide segment of bovine lactoferrin," *International Journal of Peptide and Protein Research*, vol. 48, no. 4, pp. 357–363, 1996.

[92] H. Wakabayashi, S. Abe, S. Teraguchi, H. Hayasawa, and H. Yamaguchi, "Inhibition of hyphal growth of azole-resistant strains of *Candida albicans* by triazole antifungal agents in the presence of lactoferrin-related compounds," *Antimicrobial Agents and Chemotherapy*, vol. 42, no. 7, pp. 1587–1591, 1998.

[93] H. Wakabayashi, T. Okutomi, S. Abe, H. Hayasawa, M. Tomita, and H. Yamaguchi, "Enhanced anti-Candida activity of neutrophils and azole antifungal agents in the presence of lactoferrin-related compounds," *Advances in Experimental Medicine and Biology*, vol. 443, pp. 229–237, 1998.

[94] H. Jenssen, "Anti herpes simplex virus activity of lactoferrin/lactoferricin—an example of antiviral activity of antimicrobial protein/peptide," *Cellular and Molecular Life Sciences*, vol. 62, no. 24, pp. 3002–3013, 2005.

[95] M. Ikeda, A. Nozaki, K. Sugiyama et al., "Characterization of antiviral activity of lactoferrin against hepatitis C virus infection in human cultured cells," *Virus Research*, vol. 66, no. 1, pp. 51–63, 2000.

[96] Y. Omata, M. Satake, R. Maeda et al., "Reduction of the infectivity of *Toxoplasma gondii* and *Eimeria stiedai* sporozoites by treatment with bovine lactoferricin," *The Journal of Veterinary Medical Science*, vol. 63, no. 2, pp. 187–190, 2001.

[97] Y. C. Yoo, S. Watanabe, R. Watanabe, K. Hata, K. I. Shimazaki, and I. Azuma, "Bovine lactoferrin and lactoferricin, a peptide derived from bovine lactoferrin, inhibit tumor metastasis in mice," *Japanese Journal of Cancer Research*, vol. 88, no. 2, pp. 184–190, 1997.

[98] Y. C. Yoo, R. Watanabe, Y. Koike et al., "Apoptosis in human leukemic cells induced by lactoferricin, a bovine milk protein-devived peptide: involvement of reactive oxygen species," *Biochemical and Biophysical Research Communications*, vol. 237, no. 3, pp. 624–628, 1997.

[99] D. S. Chapple, D. J. Mason, C. L. Joannou, E. W. Odell, V. Gant, and R. W. Evans, "Structure-function relationship of antibacterial synthetic peptides homologous to a helical surface region on human lactoferrin against *Escherichia coli* serotype O111," *Infection and Immunity*, vol. 66, no. 6, pp. 2434–2440, 1998.

[100] H. Kuwata, T. T. Yip, M. Tomita, and T. W. Hutchens, "Direct evidence of the generation in human stomach of an antimicrobial peptide domain (lactoferricin) from ingested lactoferrin," *Biochimica et Biophysica Acta*, vol. 1429, no. 1, pp. 129–141, 1998.

[101] L. H. Vorland, H. Ulvatne, J. Andersen et al., "Lactoferricin of bovine origin is more active than lactoferricins of human, murine and caprine origin," *Scandinavian Journal of Infectious Diseases*, vol. 30, no. 5, pp. 513–517, 1998.

[102] K. Shin, K. Yamauchi, S. Teraguchi et al., "Antibacterial activity of bovine lactoferrin and its peptides against enterohaemorrhagic *Escherichia coli* O157:H7," *Letters in Applied Microbiology*, vol. 26, no. 6, pp. 407–411, 1998.

[103] E. M. Jones, A. Smart, G. Bloomberg et al., "Lactoferricin, a new antimicrobial peptide," *Journal of Applied Bacteriology*, vol. 77, no. 2, pp. 208–214, 1994.

[104] M. Tomita, M. Takase, H. Wakabayashi, and W. Bellamy, "Antimicrobial peptides of lactoferrin," *Advances in Experimental Medicine and Biology*, vol. 357, pp. 209–218, 1994.

[105] E. Mazoyer, S. Lévy-Toledano, F. Rendu et al., "KRDS, a new peptide derived from human lactotransferrin, inhibits platelet aggregation and release reaction," *European Journal of Biochemistry*, vol. 194, no. 1, pp. 43–49, 1990.

[106] G. Wu, C. Ruan, L. Drouet, and J. Caen, "Inhibition effects of KRDS, a peptide derived from lactotransferrin, on platelet function and arterial thrombus formation in dogs," *Haemostasis*, vol. 22, no. 1, pp. 1–6, 1992.

[107] O. Aguilera, H. Ostolaza, L. M. Quirós, and J. F. Fierro, "Permeabilizing action of an antimicrobial lactoferricin-derived peptide on bacterial and artificial membranes," *FEBS Letters*, vol. 462, no. 3, pp. 273–277, 1999.

[108] H. Ulvatne, Ø. Samuelsen, H. H. Haukland, M. Krämer, and L. H. Vorland, "Lactoferricin B inhibits bacterial macromolecular synthesis in *Escherichia coli* and *Bacillus subtilis*," *FEMS Microbiology Letters*, vol. 237, no. 2, pp. 377–384, 2004.

[109] Ø. Samuelsen, H. H. Haukland, H. Ulvatne, and L. H. Vorland, "Anti-complement effects of lactoferrin-derived peptides," *FEMS Immunology and Medical Microbiology*, vol. 41, no. 2, pp. 141–148, 2004.

[110] B. E. Haug and J. S. Svendsen, "The role of tryptophan in the antibacterial activity of a 15-residue bovine lactoferricin peptide," *Journal of Peptide Science*, vol. 7, no. 4, pp. 190–196, 2001.

[111] P. M. Hwang, N. Zhou, X. Shan et al., "Three-dimensional solution structure of lactoferricin B, an antimicrobial peptide derived from bovine lactoferrin," *Biochemistry*, vol. 37, no. 12, pp. 4288–4298, 1998.

[112] H. N. Hunter, A. R. Demcoe, H. Jenssen, T. J. Gutteberg, and H. J. Vogel, "Human lactoferricin is partially folded in aqueous solution and is better stabilized in a membrane mimetic solvent," *Antimicrobial Agents and Chemotherapy*, vol. 49, no. 8, pp. 3387–3395, 2005.

[113] D. L. Rosenstreich and S. N. Vogel, "Central role of macrophages in the host response to endotoxin," in *Microbiology*, D. Schlessinger, Ed., pp. 11–15, American Society for Microbiology, Washington, DC, USA, 1980.

[114] D. C. Morrison and J. L. Ryan, "Endotoxins and disease mechanisms," *Annual Review of Medicine*, vol. 38, pp. 417–432, 1987.

[115] K. Miyazawa, C. Mantel, L. Lu, D. C. Morrison, and H. E. Broxmeyer, "Lactoferrin-lipopolysaccharide interactions. Effect on lactoferrin binding to monocyte/macrophage-differentiated HL-60 cells," *Journal of Immunology*, vol. 146, no. 2, pp. 723–729, 1991.

[116] B. J. Appelmelk, Y. Q. An, M. Geerts et al., "Lactoferrin is a lipid A-binding protein," *Infection and Immunity*, vol. 62, no. 6, pp. 2628–2632, 1994.

[117] E. Elass-Rochard, A. Roseanu, D. Legrand et al., "Lactoferrin-lipopolysaccharide interaction: involvement of the 28-34 loop region of human lactoferrin in the high-affinity binding to *Escherichia coli* 055B5 lipopolysaccharide," *Biochemical Journal*, vol. 312, part 3, pp. 839–845, 1995.

[118] L. H. Vorland, H. Ulvatne, Ø. Rekdal, and J. S. Svendsen, "Initial binding sites of antimicrobial peptides in *Staphylococcus aureus* and *Escherichia coli*," *Scandinavian Journal of Infectious Diseases*, vol. 31, no. 5, pp. 467–473, 1999.

[119] I. Mattsby-Baltzer, A. Roseanu, C. Motas, J. Elverfors, I. Engberg, and L. Å. Hanson, "Lactoferrin or a fragment thereof inhibits the endotoxin-induced interleukin-6 response in human monocytic cells," *Pediatric Research*, vol. 40, no. 2, pp. 257–262, 1996.

[120] S. Farnaud, C. Spiller, L. C. Moriarty et al., "Interactions of lactoferrin-derived peptides with LPS and antimicrobial activity," *FEMS Microbiology Letters*, vol. 233, no. 2, pp. 193–199, 2004.

[121] G. H. Zhang, D. M. Mann, and C. M. Tsai, "Neutralization of endotoxin in vitro and in vivo by a human lactoferrin-derived peptide," *Infection and Immunity*, vol. 67, no. 3, pp. 1353–1358, 1999.

[122] A. Majerle, J. Kidrič, and R. Jerala, "Enhancement of antibacterial and lipopolysaccharide binding activities of a human lactoferrin peptide fragment by the addition of acyl chain," *Journal of Antimicrobial Chemotherapy*, vol. 51, no. 5, pp. 1159–1165, 2003.

[123] B. Japelj, P. Pristovšek, A. Majerle, and R. Jerala, "Structural origin of endotoxin neutralization and antimicrobial activity of a lactoferrin-based peptide," *The Journal of Biological Chemistry*, vol. 280, no. 17, pp. 16955–16961, 2005.

[124] H. G. Boman, "Peptide antibiotics and their role in innate immunity," *Annual Review of Immunology*, vol. 13, pp. 61–92, 1995.

[125] N. Schmidt, A. Mishra, G. H. Lai, and G. C. Wong, "Arginine-rich cell-penetrating peptides," *FEBS Letters*, vol. 584, no. 9, pp. 1806–1813, 2010.

[126] S. Futaki, T. Suzuki, W. Ohashi et al., "Arginine-rich peptides. An abundant source of membrane-permeable peptides having potential as carriers for intracellular protein delivery," *The Journal of Biological Chemistry*, vol. 276, no. 8, pp. 5836–5840, 2001.

[127] M. L. Mangoni, N. Grovale, A. Giorgi, G. Mignogna, M. Simmaco, and D. Barra, "Structure-function relationships in bombinins H, antimicrobial peptides from *Bombina* skin secretions," *Peptides*, vol. 21, no. 11, pp. 1673–1679, 2000.

[128] R. Lai, Y. T. Zheng, J. H. Shen et al., "Antimicrobial peptides from skin secretions of Chinese red belly toad *Bombina maxima*," *Peptides*, vol. 23, no. 3, pp. 427–435, 2002.

[129] J. B. McPhee and R. E. W. Hancock, "Function and therapeutic potential of host defense peptides," *Journal of Peptide Science*, vol. 11, no. 11, pp. 677–687, 2005.

[130] H. H. Haukland, H. Ulvatne, K. Sandvik, and L. H. Vorland, "The antimicrobial peptides lactoferricin B and magainin 2 cross over the bacterial cytoplasmic membrane and reside in the cytoplasm," *FEBS Letters*, vol. 508, no. 3, pp. 389–393, 2001.

[131] E. W. Odell, R. Sarra, M. Foxworthy, D. S. Chapple, and R. W. Evans, "Antibacterial activity of peptides homologous to a loop region in human lactoferrin," *FEBS Letters*, vol. 382, no. 1-2, pp. 175–178, 1996.

[132] D. I. Chan, E. J. Prenner, and H. J. Vogel, "Tryptophan- and arginine-rich antimicrobial peptides: structures and mechanisms of action," *Biochimica et Biophysica Acta*, vol. 1758, no. 9, pp. 1184–1202, 2006.

16

Diet-Induced Obesity in Mice Overexpressing Neuropeptide Y in Noradrenergic Neurons

Suvi T. Ruohonen,[1] Laura H. Vähätalo,[1,2] and Eriika Savontaus[1,3]

[1] Department of Pharmacology, Drug Development, and Therapeutics and Turku Center for Disease Modeling, University of Turku, Itäinen Pitkäkatu 4B, 20520 Turku, Finland
[2] Fin Pharma Doctorate Program Drug Discovery Section, University of Turku, Itäinen Pitkäkatu 4B, 20520 Turku, Finland
[3] Unit of Clinical Pharmacology, Turku University Hospital, Itäinen Pitkäkatu 4B, 20520 Turku, Finland

Correspondence should be addressed to Suvi T. Ruohonen, suvi.ruohonen@utu.fi

Academic Editor: Hubert Vaudry

Neuropeptide Y (NPY) is a neurotransmitter associated with feeding and obesity. We have constructed an NPY transgenic mouse model (OE-NPY^{DBH} mouse), where targeted overexpression leads to increased levels of NPY in noradrenergic and adrenergic neurons. We previously showed that these mice become obese on a normal chow. Now we aimed to study the effect of a Western-type diet in OE-NPY^{DBH} and wildtype (WT) mice, and to compare the genotype differences in the development of obesity, insulin resistance, and diabetes. Weight gain, glucose, and insulin tolerance tests, fasted plasma insulin, and cholesterol levels were assayed. We found that female OE-NPY^{DBH} mice gained significantly more weight without hyperphagia or decreased activity, and showed larger white and brown fat depots with no difference in UCP-1 levels. They also displayed impaired glucose tolerance and decreased insulin sensitivity. OE-NPY^{DBH} and WT males gained weight robustly, but no difference in the degree of adiposity was observed. However, 40% of OE-NPY^{DBH} but none of the WT males developed hyperglycaemia while on the diet. The present study shows that female OE-NPY^{DBH} mice were not protected from the obesogenic effect of the diet suggesting that increased NPY release may predispose females to a greater risk of weight gain under high caloric conditions.

1. Introduction

Neuropeptide Y (NPY) is one of the most common peptides in the brain and an abundant neurotransmitter in the peripheral sympathetic nervous system (SNS). NPY has been linked to several disorders associated with metabolic syndrome. It plays a well-established role in the hypothalamic control of body energy balance by promoting feeding and lipid storage in white adipose tissue (WAT) [1]. However, in pair-fed rats, central NPY administration still leads to increased fat accumulation, which suggests that NPY has an important role in promoting adiposity independent of food intake [2]. NPY outside the hypothalamus in the regulation of energy homeostasis has not been widely studied, although NPY and its receptors are located in key peripheral tissues, such as WAT, liver, and pancreas. NPY inhibits lipolysis in adipocytes via Y_1 receptors [3], and could modulate adipose

tissue expansion by regulating angiogenesis [4]. NPY also inhibits insulin release via pancreatic Y_1 receptors [5]. On the other hand, NPY in the brainstem could modulate sympathetic tone, which is known to have multiple effects on energy homeostasis. To address the role of NPY colocalized with noradrenaline in SNS and brain noradrenergic neurons, we created a transgenic mouse model (OE-NPY^{DBH} mouse), where NPY is overexpressed under the dopamine-betahydroxylase (DBH) promoter. This resulted in moderately increased levels of NPY protein (1.3–1.8-fold) peripherally in adrenal glands and centrally in noradrenergic neurons of the brainstem, but not in the hypothalamus [6]. The OE-NPY^{DBH} mice displayed increased adiposity and liver triglyceride accumulation without changes in body weight, food consumption, or physical activity on a normal chow diet [6]. Hyperinsulinemia and impaired glucose tolerance develop with age and increased adiposity in these mice [6].

The increased adiposity was observed in both genders at all ages studied, but the major metabolic complications such as impaired glucose tolerance, hyperinsulinemia, and hepatosteatosis were only present in male OE-NPY^{DBH} mice. Furthermore, the OE-NPY^{DBH} mice showed elevated sympathetic tone and increased responses to stress [7], and increased susceptibility to arterial thickening after vascular injury [8]. Others have shown that stress-induced activation of the NPY system combined with a high-calorie diet results in augmented obesity [4]. The primary finding of Kuo et al. was increased fat mass without change in food intake or body weight after two weeks of daily stress and high fat diet. Prolonged stress and diet led to liver steatosis, impaired glucose tolerance and obesity, which were attenuated by local Y_2R antagonist administration and fat-targeted Y_2-gene knockdown procedure. These data combined show that extrahypothalamic NPY regulates adiposity independent of food intake and that NPY together with stress are potent factors in several pathways leading to a complex disorder called the metabolic syndrome.

In humans, association studies of a polymorphism in the *NPY* gene have linked NPY to metabolic disturbances. The *NPY*1228T > C polymorphism (rs16139), which causes an amino acid substitution of Leucine 7 to Proline 7 (L7P) in the signal peptide [9] is functional, enhancing the secretion of NPY [10, 11]. It is associated with an earlier onset of type 2 diabetes in carriers of L7P [12], as well as with other traits of the metabolic syndrome [13] especially in obese subjects [14]. Recently, it was reported that there is a gender difference in the NPY-mediated effects of the Proline 7 allele [15]. The associations of the L7P seem to be more pronounced in men than women, which is in agreement with our findings in OE-NPY^{DBH} mice indicating that males develop more severe metabolic changes.

Human obesity and metabolic syndrome are often due to lifestyle changes including increased sedentary work and decreased physical activity, and increased accessibility and affordability of dense, high-calorie foods. An increase in dietary fat and sucrose content has been shown to produce hyperglycaemia and obesity in various strains of mice and rats. In the present study, we exposed male and female OE-NPY^{DBH} and wildtype (WT) control mice to an energy-dense Western-type diet for seven weeks, and compared the genotype and sex differences in the impacts of the diet on the development of obesity, insulin resistance, and impaired glucose tolerance. Based on the findings in these mice, the possible reasons, that is, food intake and physical activity, were studied further in the OE-NPY^{DBH} female mice.

2. Materials and Methods

2.1. Experiment 1. Eight-week-old male and female OE-NPY^{DBH} mice heterozygous for the NPY transgene [6] and their WT littermates on a C57Bl/6N genetic background were used. Animal care was in accordance with the guidelines of the European Convention for the Protection of Vertebrate Animals used for Experimental and other Scientific purposes (Council of Europe no. 123, Strasbourg 1985), and all experimental procedures were approved by the institutional animal care and use committee. OE-NPY^{DBH} mice are stress-sensitive [7], and therefore the mice in this study were housed with their same-sex siblings instead of single cages and maintained on a 12-h light/dark cycle (lights on at 6 am) with free access to food and water. The number of animals in the male groups was 9-10, and in the female groups 15-16. The mice were placed on a Western-type diet (42% kcal fat from milk fat, 43% carbohydrates from sucrose and corn starch, 15% protein, and 0.15% supplementary cholesterol, code 829100, Special Diets Services, Essex, UK) and fed *ad libitum* for seven weeks. The weight gain was monitored weekly.

After five weeks on the diet, the mice were fasted from 6:00 h to 10:00 h, and administered intraperitoneally (ip) with glucose (10% w/v, $1 \, g \, kg^{-1}$ body weight) for glucose tolerance test (GTT). Tail vein blood glucose was measured immediately before the injection and at 20, 40, 60, and 90 min with a glucose analyzer (Precision Xtra, Abbott Diabetes Care, Abbott Park, IL, USA). Areas under the resultant curves (AUC) were calculated with the trapezoidal method in GraphPad Prism 5.01 software. The induction of hyperglycemia was defined as a blood glucose level over $13.8 \, mmol \, L^{-1}$ ($250 \, mg \, dL^{-1}$) in mice following a fast started on the morning of the experiment as guided by mouse metabolic phenotyping centers (http://www.mmpc.org/) established by the National Institutions of Health (NIH) and by the Animal Models of Diabetic Complications Consortium (AMDCC, http://www.amdcc.org/). The same level of fasted blood glucose in determination of hyperglycaemia in mice has been used by others as well [16, 17]. After six weeks on the diet, the mice were fasted for 1 h. Recombinant human insulin (Protaphane, Novo Nordisk) at a dose of 0.5 or $1.0 \, IU \, kg^{-1}$ was injected ip for insulin tolerance test (ITT). Blood glucose was measured immediately before the injection and at 20, 40, and 60 min after the injection.

At sacrifice, the mice were fasted from 6:00 h to 10:00 h, the rectal temperature was measured (Ellab, Roedovre, Denmark), and the animals were anesthetized with ketamine (Ketalar $75 \, mg \, kg^{-1}$) and medetomidine (Domitor $1 \, mg \, kg^{-1}$). Terminal blood samples were obtained from inferior vena cava with 1 mL syringe and 23 G needle into heparinized tubes. Plasma was centrifuged (4000 rpm, 10 min) and stored at $-70°C$ until analyzed. WAT weight was determined by collecting the subcutaneous, epididymal/gonadal, and retroperitoneal fat pads. The amount of fat was calculated as a sum of these different WAT depots. In addition, brown fat and liver weights were determined, and the tissues were snap frozen in liquid nitrogen for further analyses. Liver triglyceride content was measured as described previously [6]. Brown fat samples were cryosectioned on microscopic slides and stained with hematoxylin and eosin (H&E) for standard morphology. Plasma concentrations of insulin (Mercodia Ultrasensitive Mouse Insulin ELISA, Mercodia AB, Uppsala, Sweden) and total cholesterol (BioVision Cholesterol Quantitation kit, BioVision Inc., Mountain View, CA, USA) were determined according to the manufacturers' instructions.

2.2. Brown Adipose Tissue (BAT) Thermogenic Capacity Analysis with Real-Time Quantitative PCR (qPCR). Total RNA in BAT was extracted with Trizol Reagent (Invitrogen, Carlsbad, CA, USA) combined with DNase treatment (TURBO DNA-free Kit, Ambion Inc., Austin, TX, USA). RNA was converted to cDNA with High Capacity RNA-to-cDNA Kit (Applied Biosystems) according to the manufacturer's instructions. Predesigned TaqMan Gene Expression assay (Applied Biosystems, assay ID Mm01244861_m1) for uncoupling protein-1 (UCP-1) was used to analyze the mRNA levels in BAT. UCP-1 levels were quantitated relatively to the housekeeping gene β-actin, (Mouse ACTB, VIC/MGB Probe, Primer Limited with 7300 Real-Time PCR System Applied Biosystems). Relative C_T method and the formula $2^{-\Delta\Delta CT}$ were used as the quantitation method.

2.3. Experiment 2. In order to study the genotype differences in the female gender in more detail, another set of female mice age-matched to experiment 1 were studied. This time we used OE-NPY^{DBH} mice homozygous for the transgene and C57Bl/6N WT female mice. The production of the homozygous line was done as follows. Heterozygous OE-NPY^{DBH} mice were bred together in order to produce WT, heterozygous and homozygous pups, which were genotyped by qPCR. Genomic DNA was isolated from tail biopsies with a commercial kit (Puregene DNA Purification Kit, Gentra, Minneapolis, MN, USA). Forward (5'-TGGCTGGAGTGCGATCTTC-3') and reverse (5'-GAGTTTGACCGTCTACGTGC-3') primers, and the TaqMan-MGB probe (6FAM-CCGATACTGTCCTCGTC-MGB) were designed for the LacZ reporter gene in the transgene construct with the Primer Express 3.0 software (Applied Biosystems). LacZ levels were quantitated relatively to the housekeeping gene β-actin, (Mouse ACTB, VIC/MGB Probe, Primer Limited). Homozygous OE-NPY^{DBH} and littermate WT mice were selected for breeding (OE-NPY^{DBH} × OE-NPY^{DBH} and WT × WT), thus creating purely transgenic or WT lines housed separately. The first litters from the homozygous and WT mouse lines were genotyped as described above to verify all pups had the same genotype for the transgene.

The female mice ($n = 7$–12) were placed on the Western-type diet for five weeks and their weight gain and food consumption were measured on a weekly basis. In addition, after two weeks on the diet, a 24-h spontaneous physical activity along with food intake was monitored with a photobeam recording system (San Diego Instruments, San Diego, CA, USA) in 10 minute intervals. Mice were placed in single cages and had a 23-h adjustment period before monitoring the 24-h activity. Whole body fat mass was measured *in vivo* after the five-week period with an EchoMRI-700 (Echo Medical Systems, Houston, TX, USA).

2.4. Reproduction Analyses. The reproduction history of all three genotype lines was followed and the number of total pups as well as female versus male pups was analyzed. The number of WT, heterozygous OE-NPY^{DBH}, and homozygous OE-NPY^{DBH} dames used in breeding was 45, 26, and 34, respectively.

2.5. Statistical Analyses. The results were analyzed with a Student's parametric t-test or with a Mann-Whitney's nonparametric U-test to compare the groups of OE-NPY^{DBH} and WT mice. Weight gain, GTT and ITT were analyzed with two-way ANOVA for repeated measurements and Bonferroni posthoc tests. χ^2 test was used to calculate the statistical difference in the number of male mice with hyperglycaemia as 0 = fasting glucose <13.8 mmol L^{-1} or 1 = fasting glucose >13.8 mmol L^{-1}. Statistical analyses were carried out using GraphPad Prism 5.01 (GraphPad Software, San Diego, CA, USA). Data are presented as means ± SEM for the indicated number of observations. Means were considered significant when $P < 0.05$.

3. Results

3.1. Experiment 1: Body Weight Gain during the Diet. Female OE-NPY^{DBH} mice gained more weight compared with their WT littermates on the Western diet (Figures 1(a) and 1(c)). In males, prominent weight gain was observed in both genotypes, but no difference in the weight gain pattern was observed between the OE-NPY^{DBH} and WT mice (Figures 1(b) and 1(d)). The female transgenic mice also had larger WAT and BAT mass as measured by individual fat pads or the sum of WAT pads compared with WT female mice (Figures 1(e) and 1(g)). In males, a significant difference between the genotypes was only observed in the epididymal WAT weight (Figures 1(f) and 1(h)).

3.2. Experiment 1: GTT. Glucose tolerance was significantly impaired in female OE-NPY^{DBH} mice compared with WT mice, as shown by the much greater rise in blood glucose in OE-NPY^{DBH} mice over the studied 90 min time period following administration of glucose (Figure 2(a)). In addition, the AUC value in OE-NPY^{DBH} mice was significantly higher than in WT controls (Table 1). In male mice, the glucose tolerance was similar between the genotypes and thus no statistical significance between the groups was reached (Figure 2(b)). However, four out of ten OE-NPY^{DBH} male mice, but none of the nine WT controls, had a 4-h fasting glucose value over 13.8 mmol L^{-1}. Thus, they were considered to show severe hyperglycaemia [16, 17]. This difference reached statistical difference of $P < 0.05$ with the χ^2 test. There was no difference in GTT AUC values between the genotypes in males (Table 1).

3.3. Experiment 1: ITT. Insulin sensitivity was assessed by investigating the action of exogenous insulin on blood glucose levels. Blood glucose levels were higher in female OE-NPY^{DBH} mice compared with their WT controls at all investigated time points after intraperitoneal injection of insulin (Figure 2(c)). Insulin was first administered with 0.5 IU/kg, which was sufficient in females but had no effect on circulating glucose levels in males. Hence, the experiment was repeated with a higher dose of 1.0 IU kg^{-1}. This dose decreased blood glucose on average by 35% in eight out of nine WT and seven out of ten OE-NPY^{DBH} male mice (Figure 2(d)), whereas the rest of the mice (1/9 of WT and 3/10 of OE-NPY^{DBH}) were resistant to the effect of insulin,

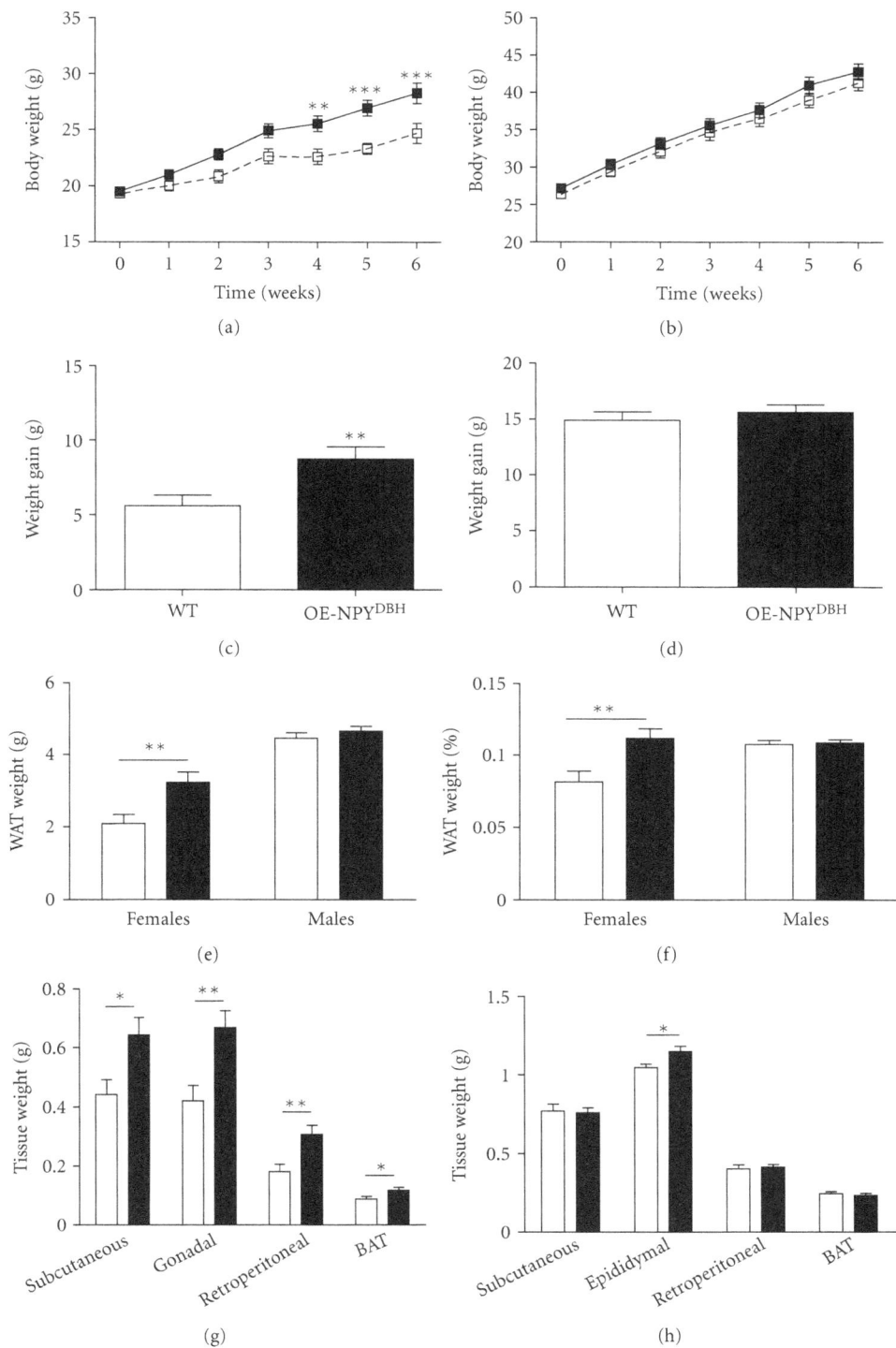

FIGURE 1: The effect of the Western-type diet on body weight gain and adiposity in wildtype (WT) and heterozygous OE-NPY^{DBH} mice. Body weight gain curves (a, b), total weight gain (c, d), and fat tissue weights (e–h) were measured as described in *Experiment* 1 in Section 2, and presented here with females in the left panel (a, c, e, g) and males on the right (b, d, f, h). (a, c) Heterozygous female OE-NPY^{DBH} mice gained significantly more weight during the diet in comparison with their WT littermates. (b, d) The male mice were equally susceptible to weight gain in both genotypes. (e) Mean total WAT weights, (f) total WAT per body weight in percentages, and (g-h) different WAT subclass and BAT weights in female (e, g) and male (f, h) wildtype and OE-NPY^{DBH} mice. Values are expressed as means ± SEM. $n = 15$, WT females; $n = 16$, OE-NPY^{DBH} females; $n = 9$, WT males; $n = 10$, OE-NPY^{DBH} males. White squares and bars = WT; Black squares and bars = OE-NPY^{DBH}. $*P < 0.05$ with a Student's t-test; $**P < 0.01$ with a Bonferroni posthoc test in repeated measures two-way ANOVA (a) or with a Student's t-test (c–g); $***P < 0.001$ with a Bonferroni posthoc test in repeated measures two-way ANOVA.

TABLE 1: Metabolic and endocrinological parameters in wildtype (WT) and heterozygous OE-NPY^{DBH} mice after seven weeks on an energy-dense Western-type diet (42% kcal fat, 43% carbohydrates, 15% protein, and 0.15% supplementary cholesterol).

	Males			Females		
	WT ($n = 9$)	OE-NPY^{DBH} ($n = 10$)	P value	WT ($n = 15$)	OE-NPY^{DBH} ($n = 16$)	P value
Plasma insulin (μg L^{-1})	1.4 ± 0.19	1.8 ± 0.21	0.21	0.6 ± 0.19[a]	0.5 ± 0.07[b]	0.74
Plasma cholesterol (μg μL^{-1})	4.3 ± 0.3	4.4 ± 0.2	0.81	1.0 ± 0.2[c]	1.3 ± 0.3[b]	0.32
Liver triglycerides (mg g^{-1} tissue)	3.0 ± 0.3	3.1 ± 0.1	0.87	4.6 ± 0.3	4.6 ± 0.3	0.96
GTT (AUC) (mmol L^{-1} × min)	1446 ± 83.0	1668 ± 113.3	0.13	961.3 ± 50.2	1115 ± 38.8	0.02

[a] $n = 8$; [b] $n = 9$; [c] $n = 7$; ND: not determined. P values were obtained with a Student's t-test.

that is, the blood glucose levels were the same or even higher than at baseline. No significant difference between the genotypes in blood glucose values after insulin dosing in males was observed.

3.4. Experiment 1: Plasma Insulin and Cholesterol Concentrations. Circulating insulin and total cholesterol levels were determined after a 4-h fast. No difference in either parameter was observed between the genotypes in males or females (Table 1).

3.5. Experiment 1: Liver Triglyceride Content. Liver weights were measured at sacrifice. The female OE-NPY^{DBH} mice had significantly heavier livers *per se* (WT: 1.1 ± 0.07; OE-NPY^{DBH}: 1.3 ± 0.06 grams; P < 0.05), but when the weights were corrected with body weights, the statistical difference was lost (data not shown). In males, the actual (WT: 2.5 ± 0.21; OE-NPY^{DBH}: 2.8 ± 0.22 grams; P = NS) and normalized liver weights were identical between the genotypes. Hepatic lipids were determined by assessing their triglyceride contents. Gross visual examination of the livers suggested steatosis as observed by a pale abnormal color with naked eyes. No difference in triglyceride levels between the genotypes in males or females was observed (Table 1).

3.6. Experiment 1: Body Temperatures. Rectal body temperatures did not differ between the genotypes in males (WT: 35.4 ± 0.24; OE-NPY^{DBH}: 35.7 ± 0.29 degrees centigrade) or females (WT: 35.9 ± 0.36; OE-NPY^{DBH}: 35.7 ± 0.37 degrees centigrade).

3.7. Experiment 1: Brown Fat UCP-1 Expression Levels and Tissue Morphology. UCP-1 expression levels in BAT were measured with qPCR by using the average mRNA expression of the WT group as a calibrator. No difference in UCP-1 levels in males (WT: 1.13 ± 0.18; OE-NPY^{DBH}: 1.16 ± 0.15 fold mRNA expression; P = NS) or females (WT: 1.09 ± 0.13; OE-NPY^{DBH}: 1.24 ± 0.16 fold mRNA expression; P = NS) were observed. The tissue morphology showed atypical BAT tissue with lipid deposition in both male groups and in the OE-NPY^{DBH} female group whereas the WT females showed normal BAT morphology (Figure 3).

3.8. Experiment 2: Body Weight Gain, Feeding Behavior and Locomotor Activity during the Diet. In order to verify and to find the reason for the findings in females, the early steps of the weight gain and adiposity were further studied in a homozygous OE-NPY^{DBH} line. A similar weight gain pattern was observed as in the heterozygous mice in comparison with WT controls, that is, the weights start to differ after four weeks on the diet and are significantly higher at five weeks from the beginning of the diet (Figure 4(a)). Food intake measured weekly in group-housed (food consumed divided by the number of animals) (Figure 4(b)) or in 47-h individually housed mice (WT: 6.7 ± 0.7; OE-NPY^{DBH}: 6.3 ± 1.2 grams; P = NS) was not different between the genotypes. A whole body EchoMRI analysis was performed at week five to measure the amount of fat *in vivo*. The OE-NPY^{DBH} females showed increased adiposity compared with WT mice (Figure 4(c)), which explains the difference in body weights. 24-h Locomotor activity of female WT and OE-NPY^{DBH} mice measured in photo-beam cages after two weeks on the diet revealed no differences between the genotypes either in horizontal or vertical movements (data not shown). Total activity per hour calculated as the sum of these two parameters is presented in Figure 5.

3.9. Reproduction Analyses. Based on 82 litters from 45 WT dames, WT female mice produced on average 3.5 female (total 284) and 3.6 male (total 294) pups. Heterozygous OE-NPY^{DBH} female mice (35 litters from 26 dames) produced 4.0 female (total 141) and 3.8 male (total 134) pups and homozygous OE-NPY^{DBH} mice (55 litters from 34 dames) 3.3 female (total 176) and 3.7 male (total 199) pups.

4. Discussion

In our previous work, we observed that both the male and female transgenic mice overexpressing NPY in noradrenergic and adrenergic neurons showed increased adiposity without significant increase in body weight or food intake on a normal chow diet [6]. In the current study, we hypothesized that NPY overexpression would render the transgenic mice even more susceptible to obesity induced by a Western-type diet. In line with the hypothesis, the female OE-NPY^{DBH} mice gained significantly more weight compared to their WT littermate controls (40% versus 26% of initial weight) and showed an increased adipose tissue mass. In contrast, the male OE-NPY^{DBH} mice gained weight similarly to their WT controls (55% of initial weight). Thus, NPY overexpression

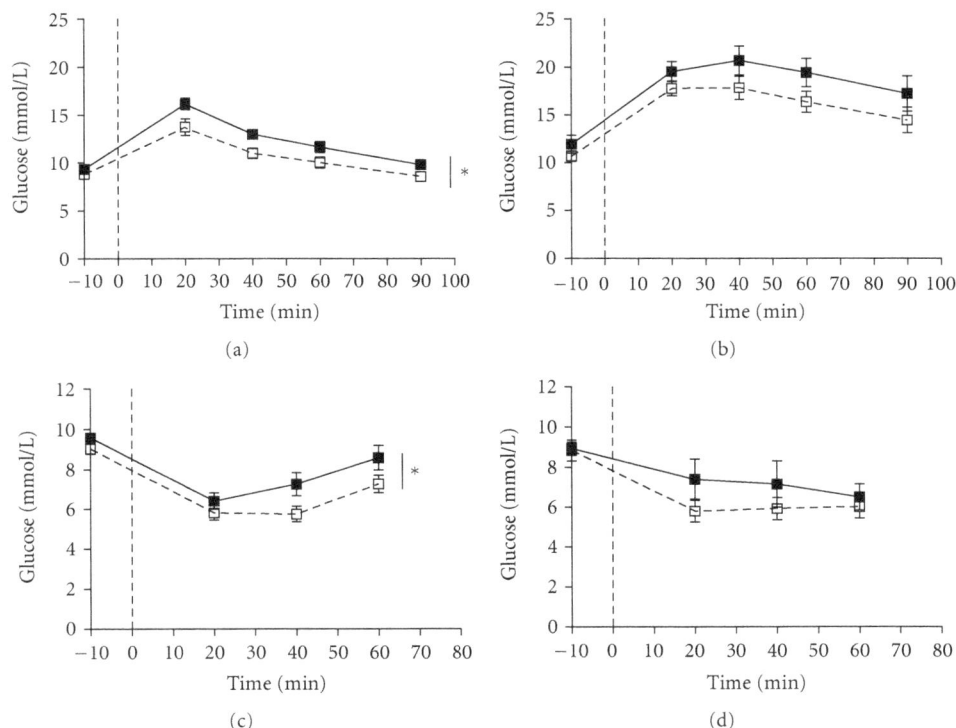

FIGURE 2: The effect of the Western-type diet on glucose homeostasis in wildtype (WT) and OE-NPY^{DBH} mice. Intraperitoneal glucose (a, b) and insulin (c, d) tolerance tests (GTT and ITT, resp.). Females are presented in the left panel (a, c) and males on the right (b, d). Mean ± SEM blood glucose values are shown at each studied time point in GTT (baseline, 20, 40, 60, and 90 min) and ITT (baseline, 20, 40, and 60 min) in female ($n = 15$, WT; $n = 16$, OE-NPY^{DBH}) and male ($n = 9$, WT; $n = 10$, OE-NPY^{DBH}) mice. The administration of 1.0 g kg^{-1} glucose (a, b) and insulin 0.5 IU kg^{-1} (c) or 1.0 IU kg^{-1} (d) is marked at 0 min. White squares = WT; black squares = OE-NPY^{DBH}. $*P < 0.05$ with repeated measures two-way ANOVA.

overrode the resistance of the C57Bl/6 female mice [18] to the diet-induced obesity, but susceptibility of the male C57Bl/6 mice to the diet induced obesity [18] overrode the NPY's adiposity inducing effect.

The mechanisms of increased weight gain in the female OE-NPY^{DBH} mice were studied in more detail in a separate group of mice. This time the mice were homozygous for the transgene and thus housed with same-sex, same-genotype siblings, which made it possible to study the feeding behavior between the genotypes in unstressed conditions (group housing). However, the results were also verified in single-housed mice. The results showed that enhanced weight gain is not due to increased food consumption during the time preceding the difference in body weights. The mechanism seems not to be impaired thermogenesis or decreased physical activity either. Rodents resistant to the diet-induced obesity increase BAT thermogenesis via increased UCP-1 activity to avoid weight gain [19]. Similar to the OE-NPY^{DBH} and WT male mice, the female OE-NPY^{DBH} mice showed atypical BAT morphology with white fat-like appearance and large lipid vacuoles in BAT. This implied the onset of brown adipocyte degeneration often associated with obesity and impaired thermogenesis in mice [20, 21]. However, the BAT thermogenic capacity, as measured by

UCP-1 expression levels in BAT, did not differ between the genotypes.

We show in this study that there are no differences in the number of litters or litter sizes between the OE-NPY^{DBH} or WT mice. Thus, prenatal and suckling conditions in this regard are similar for the pups. Therefore, different challenges in the developmental and prepubertal environment are an unlikely explanation for the weight gain in the diet-induced obesity in the OE-NPY^{DBH} mice.

Sexually dimorphic responses to NPY-associated metabolic changes have been reported in NPY Y$_1$ receptor knock-out mice, which develop late-onset obesity without hyperphagia that is more pronounced in female than male mice [22]. The difference was attributed to decreased skeletal muscle mitochondrial oxidative capacity in female Y$_1$ knock-out mice [23]. Gonadal steroid hormones seem to play a role in the fat accumulation in the Y$_1$ knock-out mice as weight gain does not start until puberty [24, 25]. Estrogens seem to protect female mice from obesity as evidenced for instance by weight gain after gonadectomy, and their effects are at least in part mediated by hypothalamic NPY [26, 27]. Estrogens have been shown to inhibit feeding also in the brainstem [28]. Our results may imply that the overexpression of NPY in adrenergic and noradrenergic

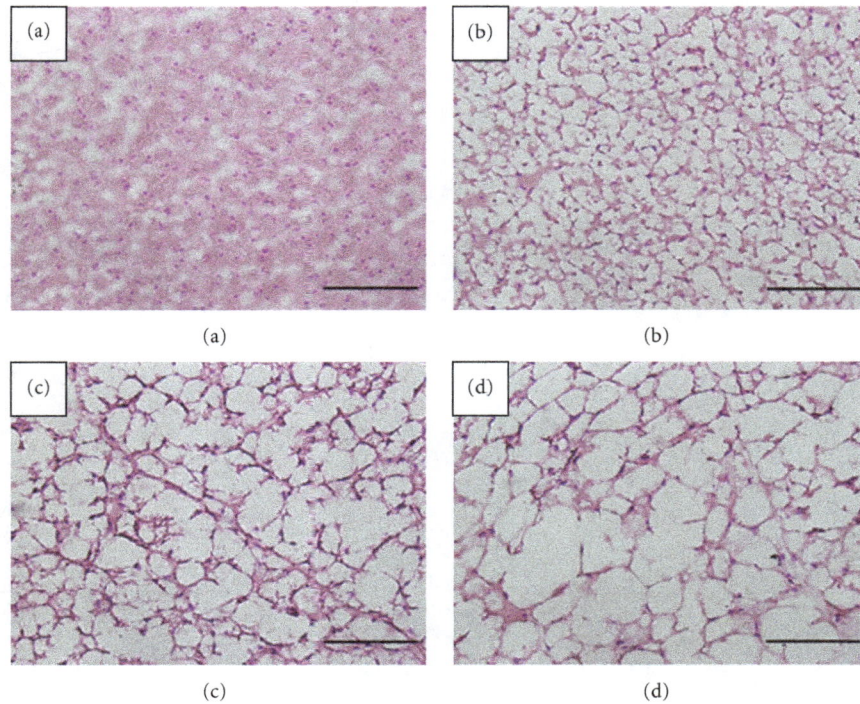

FIGURE 3: Brown adipose tissue (BAT) morphology after seven weeks on the Western-type diet. (a) WT female, (b) OE-NPY^{DBH} female, (c) WT male, and (d) OE-NPY^{DBH} male. Scale bar is 100 μm.

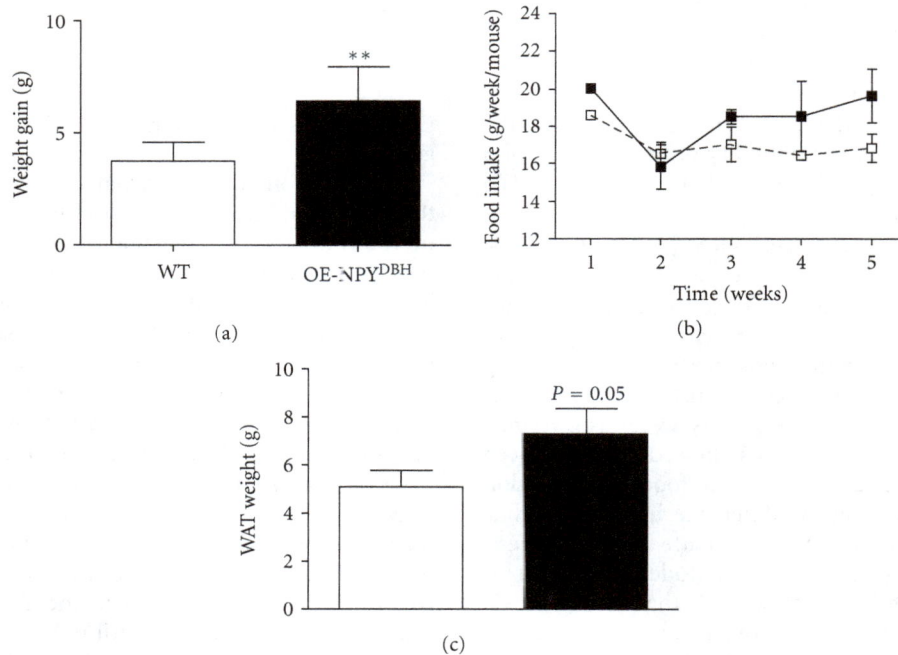

FIGURE 4: The effect of the Western-type diet on body weight gain and adiposity in wildtype (WT) and homozygous OE-NPY^{DBH} mice. Total weight gain (a), food intake (b), and total amount of fat tissue (c) were measured as deascribed in *Experiment* 2 in Section 2. Values are expressed as means ± SEM. $n = 12$, WT females; $n = 7$, OE-NPY^{DBH} females. White squares and bars = WT; black squares and bars = OE-NPY^{DBH}. **$P < 0.01$ with a Mann-Whitney test; $P = 0.05$ with a Student's t-test.

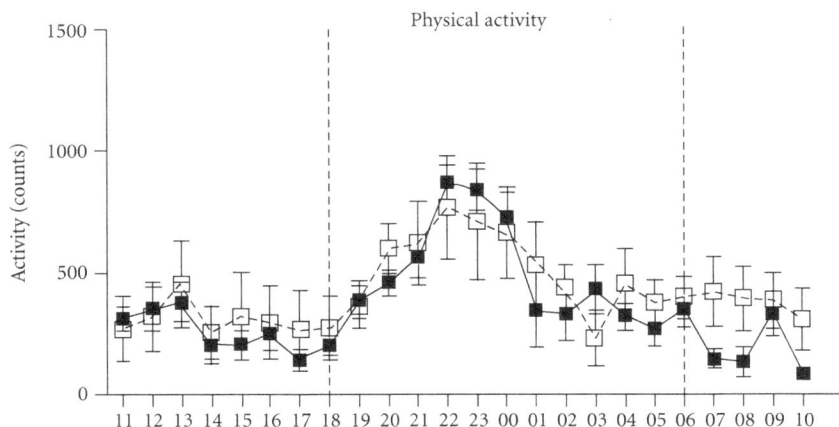

FIGURE 5: Physical activity during the Western-type diet in wildtype (WT) and homozygous OE-NPY^{DBH} female mice. The total activity was calculated as the sum of horizontal and vertical movement counts over the 24-h period starting at 11:00 o'clock a.m. The dark period (18:00–06:00 h) is marked with vertical lines. Mean ± SEM count values are shown for each hour ($n = 12$, WT; $n = 7$, OE-NPY^{DBH}). White squares = WT; Black squares = OE-NPY^{DBH}.

neurons overrides the protective effects of estrogens in female mice, suggesting that NPY in the brainstem and in the SNS may also play a role in mediating estrogen action on body composition.

Along with the increased adiposity, the OE-NPY^{DBH} female mice displayed impaired glucose tolerance and altered insulin sensitivity or counterregulatory effect of glucose 40 min after the ip administration of glucose. It is generally acknowledged that female mice are less likely to develop disturbances in glucose metabolism, which may be due to the antiobesity effects of estrogens in females [29, 30]. Previously, chow-fed female OE-NPY^{DBH} mice were shown to possess normal glucose tolerance despite of increased adiposity and sensitivity to stress [6, 7], and impairment in glucose tolerance in the current study occurred with increasing levels of obesity. Thus, disturbances in glucose metabolism in the female OE-NPY^{DBH} mice seem to be caused by increased adiposity rather than by direct effects of NPY or sympathoadrenal system on glucose metabolism, which is supported by a positive correlation with the area under the curve value for GTT and body weight in both female genotypes (WT, $r = 0.54$, $P < 0.05$; OE-NPY^{DBH}, $r = 0.45$, $P = 0.08$). In contrast, the diet-induced hepatosteatosis that similarly affected the WT and OE-NPY^{DBH} mice is not likely to explain the impaired glucose tolerance in the female OE-NPY^{DBH} mice. However, increased glycogenolysis in the liver in response to the insulin-induced decline may be responsible for the altered insulin sensitivity presented in Figure 3(c), which remains to be studied further.

Although GTT showed very high levels of blood glucose with no genotype differences in the male mice, the OE-NPY^{DBH} mice had a tendency towards reduced insulin-induced decline in glucose levels as evidenced by the more numerous individuals that did not respond to insulin in ITT. In addition, forty per cent of the OE-NPY^{DBH} males but none of the WT littermates developed severe hyperglycaemia defined as a blood glucose level over 13.8 mmol L^{-1} following

a 4-h fast. Interestingly, these tendencies occurred without major differences in WAT depot weights or at the level of hepatosteatosis in the OE-NPY^{DBH} male mice. This suggests that increased NPY may have a diabetogenic effect in the context of obesity and supports the association of the L7P polymorphism with an earlier onset of diabetes in the obese human population [14].

5. Conclusions

OE-NPY^{DBH} female mice showed a more pronounced diet-induced obesity, glucose intolerant, and insulin resistant phenotype on a Western-type energy-dense diet than their WT littermates that could not be explained by increased feeding, decreased activity, or impaired thermogenesis. This suggests that increased NPY release may predispose women and females in general, to a greater risk of weight gain under high caloric conditions. In contrast, there was no difference between the male OE-NPY^{DBH} and WT mice in the degree of obesity, although the OE-NPY^{DBH} mice seemed to be more susceptible to developing diabetes while on the diet. Interestingly, we have shown that the OE-NPY^{DBH} males fed with chow and the OE-NPY^{DBH} females with a Western diet display similar traits with the humans who carry the Proline 7 allele in their NPY sequence, which support the associations between the rs16139 polymorphism and high NPY levels and various metabolic risks. Furthermore, these results strengthen the hypothesis that NPY has an important role in promoting adiposity via extrahypothalamic pathways.

Acknowledgments

The authors would like to thank Ms. Raija Kaartosalmi and Ms. Anna-Maija Penttinen for technical assistance and Ms. Pirkko Huuskonen for the reviewing of the language. This study was supported by the Academy of Finland, the Finnish Cultural Foundation (Varsinais-Suomi fund), the European

Foundation for the Study of Diabetes, and the Novo Nordisk Foundation.

References

[1] B. Beck, "Neuropeptide Y in normal eating and in genetic and dietary-induced obesity," *Philosophical Transactions of the Royal Society B*, vol. 361, no. 1471, pp. 1159–1185, 2006.

[2] N. Zarjevski, I. Cusin, R. Vettor, F. Rohner-Jeanrenaud, and B. Jeanrenaud, "Chronic intracerebroventricular neuropeptide-Y administration to normal rats mimics hormonal and metabolic changes of obesity," *Endocrinology*, vol. 133, no. 4, pp. 1753–1758, 1993.

[3] R. L. Bradley, J. P. R. Mansfield, and E. Maratos-Flier, "Neuropeptides, including neuropeptide y and melanocortins, mediate lipolysis in murine adipocytes," *Obesity Research*, vol. 13, no. 4, pp. 653–661, 2005.

[4] L. E. Kuo, J. B. Kitlinska, J. U. Tilan et al., "Neuropeptide Y acts directly in the periphery on fat tissue and mediates stress-induced obesity and metabolic syndrome," *Nature Medicine*, vol. 13, no. 7, pp. 803–811, 2007.

[5] H. R. Patel, Y. Qi, E. J. Hawkins et al., "Neuropeptide Y deficiency attenuates responses to fasting and high-fat diet in obesity-prone mice," *Diabetes*, vol. 55, no. 11, pp. 3091–3098, 2006.

[6] S. T. Ruohonen, U. Pesonen, N. Moritz et al., "Transgenic mice overexpressing neuropeptide y in noradrenergic neurons: a novel model of increased adiposity and impaired glucose tolerance," *Diabetes*, vol. 57, no. 6, pp. 1517–1525, 2008.

[7] S. T. Ruohonen, E. Savontaus, P. Rinne et al., "Stress-induced hypertension and increased sympathetic activity in mice overexpressing neuropeptide y in noradrenergic neurons," *Neuroendocrinology*, vol. 89, no. 3, pp. 351–360, 2009.

[8] S. T. Ruohonen, K. Abe, M. Kero et al., "Sympathetic nervous system-targeted neuropeptide Y overexpression in mice enhances neointimal formation in response to vascular injury," *Peptides*, vol. 30, no. 4, pp. 715–720, 2009.

[9] M. K. Karvonen, U. Pesonen, M. Koulu et al., "Association of a leucine(7)-to-proline(7) polymorphism in the signal peptide of neuropeptide Y with high serum cholesterol and LDL cholesterol levels," *Nature Medicine*, vol. 4, no. 12, pp. 1434–1437, 1998.

[10] J. Kallio, U. Pesonen, K. Kaipio et al., "Altered intracellular processing and release of neuropeptide Y due to leucine 7 to proline 7 polymorphism in the signal peptide of preproneuropeptide Y in humans," *The FASEB Journal*, vol. 15, no. 7, pp. 1242–1244, 2001.

[11] G. C. Mitchell, Q. Wang, P. Ramamoorthy, and M. D. Whim, "A common single nucleotide polymorphism alters the synthesis and secretion of neuropeptide Y," *Journal of Neuroscience*, vol. 28, no. 53, pp. 14428–14434, 2008.

[12] U. Jaakkola, U. Pesonen, E. Vainio-Jylhä, M. Koulu, M. Pöllönen, and J. Kallio, "The Leu7Pro polymorphism of neuropeptide Y is associated with younger age of onset of type 2 diabetes mellitus and increased risk for nephropathy in subjects with diabetic retinopathy," *Experimental and Clinical Endocrinology and Diabetes*, vol. 114, no. 4, pp. 147–152, 2006.

[13] U. Pesonen, "NPY L7P polymorphism and metabolic diseases," *Regulatory Peptides*, vol. 149, no. 1–3, pp. 51–55, 2008.

[14] U. Jaakkola, J. Kallio, R. J. Heine et al., "Neuropeptide Y polymorphism significantly magnifies diabetes and cardiovascular disease risk in obesity: the Hoorn Study," *European Journal of Clinical Nutrition*, vol. 63, no. 1, pp. 150–152, 2009.

[15] U. Jaakkola, T. Kakko, H. Seppälä et al., "The Leu7Pro polymorphism of the signal peptide of neuropeptide Y (NPY) gene is associated with increased levels of inflammatory markers preceding vascular complications in patients with type 2 diabetes," *Microvascular Research*, vol. 80, no. 3, pp. 433–439, 2010.

[16] Z. Qi, H. Fujita, J. Jin et al., "Characterization of susceptibility of inbred mouse strains to diabetic nephropathy," *Diabetes*, vol. 54, no. 9, pp. 2628–2637, 2005.

[17] B. G. Han, C. M. Hao, E. E. Tchekneva et al., "Markers of glycemic control in the mouse: comparisons of 6-h and over-night-fasted blood glucoses to Hb A1c," *American Journal of Physiology*, vol. 295, no. 4, pp. E981–E986, 2008.

[18] C. M. Novak, C. M. Kotz, and J. A. Levine, "Central orexin sensitivity, physical activity, and obesity in diet-induced obese and diet-resistant rats," *American Journal of Physiology*, vol. 290, no. 2, pp. E396–E403, 2006.

[19] V. Kus, T. Prazak, P. Brauner et al., "Induction of muscle thermogenesis by high-fat diet in mice: association with obesity-resistance," *American Journal of Physiology*, vol. 295, no. 2, pp. E356–E367, 2008.

[20] S. Enerbäck, A. Jacobsson, E. M. Simpson et al., "Mice lacking mitochondrial uncoupling protein are cold-sensitive but not obese," *Nature*, vol. 387, no. 6628, pp. 90–94, 1997.

[21] A. Hamann, J. S. Flier, and B. B. Lowell, "Decreased brown fat markedly enhances susceptibility to diet-induced obesity, diabetes, and hyperlipidemia," *Endocrinology*, vol. 137, no. 1, pp. 21–29, 1996.

[22] A. Kushi, H. Sasai, H. Koizumi, N. Takeda, M. Yokoyama, and M. Nakamura, "Obesity and mild hyperinsulinemia found in neuropeptide Y-Y1 receptor-deficient mice," *Proceedings of the National Academy of Sciences of the United States of America*, vol. 95, no. 26, pp. 15659–15664, 1998.

[23] L. Zhang, L. MacIa, N. Turner et al., "Peripheral neuropeptide Y Y1 receptors regulate lipid oxidation and fat accretion," *International Journal of Obesity*, vol. 34, no. 2, pp. 357–373, 2010.

[24] F. P. Pralong, C. Gonzales, M. J. Voirol et al., "The neuropeptide Y Y1 receptor regulates leptin-mediated control of energy homeostasis and reproductive functions," *The FASEB Journal*, vol. 16, no. 7, pp. 712–714, 2002.

[25] C. Gonzales, M. J. Voirol, M. Giacomini, R. C. Gaillard, T. Pedrazzini, and F. P. Pralong, "The neuropeptide Y Y1 receptor mediates NPY-induced inhibition of the gonadotrope axis under poor metabolic conditions," *The FASEB Journal*, vol. 18, no. 1, pp. 137–139, 2004.

[26] J. J. Bonavera, M. G. Dube, P. S. Kalra, and S. P. Kalra, "Anorectic effects of estrogen may be mediated by decreased neuropeptide-Y release in the hypothalamic paraventricular nucleus," *Endocrinology*, vol. 134, no. 6, pp. 2367–2370, 1994.

[27] D. G. Baskin, B. J. Norwood, M. W. Schwartz, and D. J. Koerker, "Estradiol inhibits the increase of hypothalamic neuropeptide Y messenger ribonucleic acid expression induced by weight loss in ovariectomized rats," *Endocrinology*, vol. 136, no. 12, pp. 5547–5554, 1995.

[28] S. Thammacharoen, T. A. Lutz, N. Geary, and L. Asarian, "Hindbrain administration of estradiol inhibits feeding and activates estrogen receptor-α-expressing cells in the nucleus tractus solitarius of ovariectomized rats," *Endocrinology*, vol. 149, no. 4, pp. 1609–1617, 2008.

[29] M. L. Klebig, J. E. Wilkinson, J. G. Geisler, and R. P. Woychik, "Ectopic expression of the agouti gene in transgenic mice

causes obesity, features of type II diabetes, and yellow fur," *Proceedings of the National Academy of Sciences of the United States of America*, vol. 92, no. 11, pp. 4728–4732, 1995.

[30] E. H. Leiter and H. D. Chapman, "Obesity-induced diabetes (diabesity) in C57BL/KsJ mice produces aberrant trans-regulation of sex steroid sulfotransferase genes," *The Journal of Clinical Investigation*, vol. 93, no. 5, pp. 2007–2013, 1994.

17

Intracellular Loop 2 Peptides of the Human 5HT1a Receptor are Differential Activators of Gi

Brian Hall, Carley Squires, and Keith K. Parker

Department of Biomedical and Pharmaceutical Sciences (MPH I02), Center for Structural and Functional Neuroscience, Skaggs School of Pharmacy, The University of Montana, 32 Campus Drive No. 1552, Missoula, MT 59812-1552, USA

Correspondence should be addressed to Keith K. Parker, keith.parker@umontana.edu

Academic Editor: Piero Andrea Temussi

Peptide mimics of intracellular loop 2 (ic2) of the human 5HT1a receptor have been studied with respect to their ability to inhibit agonist binding via interference with receptor-G-protein coupling. These peptides give shallow concentration-effect relationships. Additionally, these peptides have been studied with respect to their ability to trigger the signal transduction system of this Gi-coupled receptor. Two signaling parameters have been quantified: concentration of intracellular cAMP and changes in incorporation into the G protein of a stable analog of GTP. In both cases, peptide mimics near midloop of ic2 actually show agonist activity with efficacy falling off toward both loop termini near TM 3 and TM 4. Previous results have suggested that the loop region near the TM3/ic2 interface is primarily responsible for receptor-G-protein coupling, while the current result emphasizes the mid-ic2 loop region's ability to activate the G protein following initial coupling. A limited number of peptides from the receptor's TM5/ic3 loop vicinity were also studied regarding agonist inhibition and G-protein activation. These peptides provide additional evidence that the human 5HT1a receptor, TM5/ic3 loop region, is involved in both coupling and activation actions. Overall, these results provide further information about potential pharmacological intervention and drug development with respect to the human 5HT1a receptor/G-protein system. Finally, the structural evidence generated here provides testable models pending crystallization and X-ray analysis of the receptor.

1. Introduction

Regulation of serotonergic (5-hydroxytryptamine; 5HT) function in animals impacts numerous physiological and pathological processes [1]. 5HT is broadly represented in biological systems as a regulator and modulator via nervous, hormonal, and autacoidal means [2–5]. For example, serotonin [6] has been implicated in the pathophysiology of migraine. This association with migraine is shared with many other factors including adipokines such as leptin; hypothalamic hormones, Orexin A and B (also known appetite regulators as is 5HT); numerous neurotransmitters [7]; autacoids; hormones, and ions like calcium, and magnesium. The range of biological molecules that interact with serotonergic processes suggests that various signaling pathways may be shared, and that the potential for dynamic, collaborative regulation exists. Better understanding of the molecular basis underlying these signaling processes is not only critical to

greater fundamental knowledge but to therapeutic development.

Various receptors (R), including the 5HT3R's that are ligand-controlled ion channels, are crucial to these regulatory processes [4]. All other known 5HTR's are structurally different than these ion channels, being serpentine membrane R's [8], coupled (C) to the cells interior by G (GTP binding) proteins (P), which in turn regulate key effectors such as adenylyl cyclase (AC, [9]). These GPCR's share the structural characteristic of 7-transmembrane (7TM) helical segments [10–13]. For many years, the only crystal structure was of rhodopsin, the prototype GPCR, in its interaction with the G-protein transducin [14]. Recently a breakthrough has occurred, with crystallization of the beta-adrenergic receptor (BAR) and publication of X-ray structures [15–18]. This long-awaited event has set the stage for other GPCR. Progressive developments have been demonstrated by crystal structures for the adenosine A2R [19], the CXDR4 chemokine R [20], and the dopamine D3R [21]. Crystal structures for other GPCR, including that those 5HTR's that are GPCR's should soon follow [22].

Of the GPCR recognized as 5HTR, the 5HT1aR (a relative of BAR) is one of the most highly studied [23–25], and it has been associated with physiological and pathological processes as diverse as thermoregulation, cognitive flexibility, and control of mood [26–33]. Depression, underlying anxiety disorders, and related psychopathologies are a particular theme [34–39]. Multiple strategies have been used to dissect the complex pathways underscoring these physiological and pathological processes [40–47]. One approach centers around analysis of allosteric sites of action on receptors. Peptide mimics of intracellular loop regions of 5HT1aR have been used as probes of the receptor-G-protein interface in this context [48–54]. The current communication continues our analysis with these peptide probes particularly emphasizing intracellular loop 2 (ic2) with some, limited comparative data from intracellular loop 3 (ic3). The results with ic2 and ic3 are suggestive of potential sites for regulation and therapeutic drug development.

2. Materials and Methods

2.1. Cell Culture.
Chinese Hamster ovary (CHO) cells expressing the H5-HT1aR [55, 56] were cultured in Ham's F-12 medium fortified with 10% fetal calf serum and 200 ug/mL geneticin. Cultures were maintained at 37°C in a humidified atmosphere of 5% CO_2. Cells were subcultured or assayed upon confluency (5–8 days). Cloned H5-HT1aR was kindly provided by Dr. John Raymond (Medical U. of South Carolina; [41]). The cell line has been tested for mycoplasma with a PCR kit (ATCC) and is free of contamination.

2.2. Receptor Preparation.
Cells were trypsinized and centrifuged at low speed in ice-cold medium [53]. The pellet was resuspended in ice-cold Earle's Balanced Salt Solution followed by centrifugation. Cells were resuspended in 10 mL of ice-cold binding buffer (50 mM Tris, 4 mM $CaCl_2$, 10 μM pargyline, and pH 7.4), homogenized with Teflon-glass, and centrifuged for 450,000 g-min. at 4°C. For a crude membrane preparation, the pellet was resuspended in 30 mL of ice-cold binding buffer, homogenized on Teflon-glass and then by Polytron (setting 4) for 5 seconds, and stored on ice and assayed within the next 1.5 hours [54].

2.3. Assay of Receptor Activity.
Binding of the agonist [3H]8-OH-DPAT ([3H]8-hydroxy-2-(di-n-propylamino)tetralin) to H5-HT1aR followed well-characterized protocols [49, 50, 53]. Radioligands were purchased from New England Nuclear (NEN), Boston, MA, and 1 mL reaction mixtures, in triplicate, were incubated for 30 min. in a 30°C shaker. The 1 mL mixture was 700 μL of receptor preparation; 100 μL of binding buffer (for total binding) or 10 μM 5-HT (for nonspecific binding), 100 μL of the tritiated agent (concentration of 0.5 nM [3H] 8-OH-DPAT), and 100 μL of peptide or binding buffer in the case of controls.

Reactions were stopped by addition of 4 mL of ice-cold 50 mM Tris buffer, pH 7.4, and vacuum filtration on glass fiber filters (Whatman GF/B). Filters were rinsed twice in 5 mL of ice-cold Tris buffer, dried, and counted in 5 mL of Ecoscint (National Diagnostics) liquid scintillation fluid in a Beckman LS 6500. Homogenates were assayed for protein to maintain a nominal value of 50 μg protein per filter [57]. All tubes were run in triplicate.

2.4. cAMP Assay.
CHO cells were cultured to confluency in 12- or 24-well plates. Medium was aspirated, and the cells were rinsed twice in warm, serum-free F-12 medium. Cells were incubated for 20 min. at 37°C in 0.5 mls of serum-free F-12 medium containing 100 uM isobutylmethylxanthine (IBMX) and the following substances (final concs.) alone or in combination (see Figures 3 and 5): 30 μM forskolin (FSK; for all treatments); 1 μM 5-HT; peptide concentrations as noted in figure legends. Reactions were stopped by aspiration of medium and addition of 0.5 mL of 100 mM HCl. After 10 min., well contents were removed and centrifuged at 4000 rpm. Supernatants were diluted in 100 mM HCl, and cAMP was quantified [53] directly in a microplate format by enzyme immunoassay (EIA) with a kit from Assay Designs (Ann Arbor). Triplicate-independent samples were assayed.

2.5. [35S]GTPγS Assay.
H5-HT1aR membranes from transfected CHO cells were incubated with 5-HT (0.1 μM) and/or peptide concentrations as noted in figure legends (see Figures 2 and 4) and the following incubation mixture: 20 mM HEPES buffer, pH 7.4, 5 mM $MgCl_2$, 1 mM EDTA, 1 mM DTT, 100 mM NaCl, 100 uM GDP, 10 μM pargyline, 0.2 mM ascorbate, and 0.1 nM [35S]GTPγS [53, 58]. Mixtures were incubated for 30 min. at 30°C, and were terminated by dilution in cold buffer. The mixture was filtered on GF/C filters, rinsed twice in buffer, dried and counted by liquid scin-tillation. All values reported in are for specific binding (total nonspecific) of triplicates. Nonspecific binding was determined in the presence of cold γ-S-GTP (10 uM). Negative control is the above mixture minus test drug or 5HT. Positive control contains 5HT.

2.6. Data Analysis.
All statistics (means, standard errors of the mean (SEM), t tests and ANOVA, Pearson correlation coefficients (r), and graphical procedures (including drug-receptor-binding analysis) were conducted with PSI-Plot (Version 8) software (Poly Software International), Prism (version 4.0c), or using a Hewlett-Packard Graphing Calculator, HP48. The apriori was $\alpha = 0.05$ for all experiments. Experiments were conducted with a minimum of $n = 3$, in triplicate. Most experiments were $n = 3$–5. In some cases (indicated in figure legends), different n's and multiplicates were used.

2.7. Peptide Preparation.
These highly purified (greater than 95%) peptides were purchased from New England Peptide LLC. The peptides are segments of ic2 and ic3 of the cloned H5HT1aR. Peptides stored at −20°C were initially dissolved in deionized water. Subsequent dilutions were in binding

TABLE 1: ic2 and ic3 peptide mimics. The primary amino acid sequences for the H5HT1aR ic2 loop peptide mimics P11 and P's 21–27, and for ic3 (P1, P12, and P13). The receptor's amino terminal is to the left. Sequences for H5HT1aR from Kobilka et al., 1987 [56]. P11 is from a previous study by Thiagaraj et al., 2007 [53], and P1 from Hayataka et al., 1988 [49] (both included for comparative purposes).

P11	IALDRYWAITD		
P21	LDRYWAITD**P**		
P22	RYWAITDP**ID**		
P23	WAITDPID**YV**		
P24	ITDPIDYV**NK**		
P25	DPIDYVNK**RT**		
P26	IDYVNKRT**PR**		
P27	YVNKRTPR**PR**		
P1	IFRAARFRIRKTVKK		
P12		KTVKKVEKTG	
P13		VKKVEKTGAD	

buffer. The peptides examined in this study are listed in Table 1.

3. Results

3.1. Intracellular Loop 2 (ic2). The size of H5HT1a's ic2 (about 20 amino acids) makes it a tempting target for analyzing the loop's coupling to and activation of Gi [23, 24]. Our previous work with ic2 emphasized the N-terminal region of the loop with a peptide we call P11 (Table 1). Results with this peptide suggest that the loop residues near TM 3 are vital for coupling of the loop to Gi but are not involved in G-protein activation [53]. Results from the Varrault group [48] looked at the entire loop without distinguishing subregions; their conclusions were that the entire loop is responsible for activation (they did not differentiate between coupling and activation). The following question arises: can coupling and activation characteristics be identified for the loop on a subregional basis? Our preliminary work at the N-terminal aspect of H5HT1a'a ic3 suggested to us that the techniques we use could be productive in addressing such a question [49, 50, 53]. Thus, we synthesized peptides of 10 residues each that progress from the N-terminus of ic2 to the C-terminus two amino acids at a time (Table 1). Beyond the parent peptide, P11, this results in seven additional peptides (P21–P27). Agonist inhibition [59, 60] was used as a measure of coupling efficacy. Any agent or process that uncouples a receptor from its G-protein partner increases the probability that the receptor will be in a lower affinity state for agonist binding. This results in concentration-dependent agonist dissociation relationships that reflect affinity of the uncoupler for the G protein (and potentially by analogy the affinity of the cognate receptor loop region for the G protein). Two determinants of G-protein activation (stable GTP binding to Gi and changes in intracellular cAMP concentration) were used to monitor a peptide's ability to

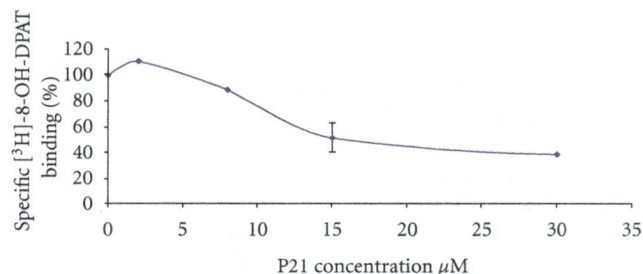

FIGURE 1: P21 noncentration-dependent displacement of bound 8-OH-DPAT. This curve represents the change in specific binding of [^{3}H]-8-OH-DPAT, a 5HT1aR agonist, to the receptor in the presence of various concentrations of the ic2 peptide mimic P21. Nominal binding of agonist at control levels was 400 fmoles/mg protein.

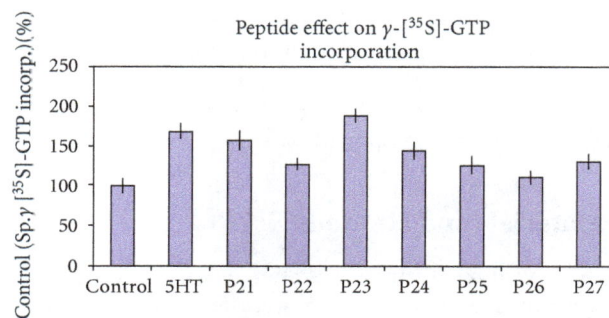

FIGURE 2: ic2 peptide effect on γ-[^{35}S]-GTP incorporated into Gi, a measure of G-protein activation. Control is the basal amount of γ-[^{35}S]-GTP incorporated into Gi in CHO cells expressing the human 5HT1aR, set as 100%. The Y-axis is the percent of specifically bound (total minus nonspecific) γ-[^{35}S]-GTP. All other treatments are percents of the control value. All peptides are 30 uM concentration and 5HT 10^{-7} M concentration. Error is expressed as SEM.

perturb G protein following coupling. The overall results for these eight peptides are in Table 2.

As shown in Figure 1 with results from peptide P21 as an example, these peptides give shallow concentration-effect curves for the measure of coupling and agonist inhibition. Similar experiments with all peptides form the basis for the summarized coupling results found in Table 2. Note that limited peptide solubility and lack of efficacy prevented complete IC50 determination for all peptides (P24–27). The uM concentration ranges for activity of these peptides, and, shallow concentration-effect relationships in, are typical for other peptides we and others have analyzed [48, 53].

Figure 2 gives results for the eight peptides' ability to foster incorporation of GTP into Gi using a radioactively labeled, reasonably stable form of GTP ([35S] gamma-S-GTP). Relative to control (buffer alone; no agonist nor peptide) midloop residues as represented by peptide P23 are most effective in directing incorporation of GTP into Gi. Efficacy declines in both N- and C-terminal directions from P23 although the results for P21 are anomalous in this regard. It is not clear whether this result for P21 is meaningful or due to experimental error although the results for intracellular

TABLE 2: ic2 Peptide mimic effect on [3H]8-OH-DPAT binding. All binding inhibition values are percent of control agonist (ag.) bound. The upper portion of the table is for peptides nearer the C-terminus, including P11 from Thiagaraj et al., 2007 [53]. These peptides decreased the specific high affinity binding of 5HT1aR agonist [^{3}H]-8-OH-DPAT by 50% at the given concentration. The lower portion of the table (P24 on) is the ic2 peptides toward the C terminus. These peptides were less effective at decreasing specific high affinity binding of [^{3}H]-8-OH-DPAT, and values given are percent of control at the given concentration. Values for intracellular cAMP are relative to FSK-stimulated control. All values for incorporation of γ-[^{35}S]-GTP into Gi are percent of control. Nominal values for control binding were 400 fmoles/mg protein.

Peptide	Conc. (uM)	% cont. ag, bound, (SEM)	[cAMP] (SEM)	GTP Incorp. (SEM)
Control			100 (6)	100 (7)
5HT			21 (4)	168 (12)
P11	7	50 (1)	87 (8)	100 (3)
P21	15	52 (4)	122 (8)	158 (11)
P22	16	51 (2)	71 (2)	128 (9)
P23	30	50 (22)	42 (4)	188 (10)
P24	10	94 (9)	45 (7)	146 (17)
P25	30	87 (12)	64 (3)	126 (10)
P26	30	75 (19)	100 (5)	111 (9)
P27	30	95 (5)	132 (6)	130 (7)

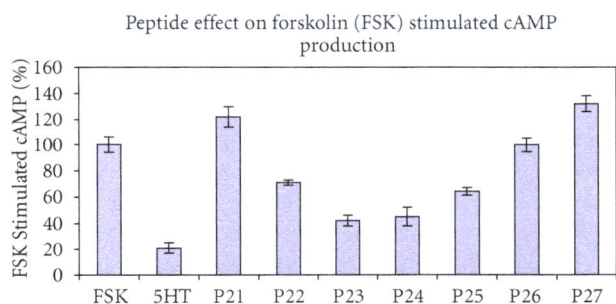

FIGURE 3: ic2 peptide effect on forskolin-stimulated cAMP production, a measure of activated-G-protein regulation of adenylyl cyclase. Forskolin (FSK) stimulated cAMP production by adenylyl cyclase (AC) is in CHO cells expressing the human 5HT1aR. FSK (30 uM) is the control, which is set to 100%. All other treatments are expressed as a percent of the control value. Peptide concentrations are 30 uM. All treatments include isobutylmethylxanthine (IBMX) an inhibitor of the metabolism of cAMP by phosphodiesterase. Error is expressed as SEM.

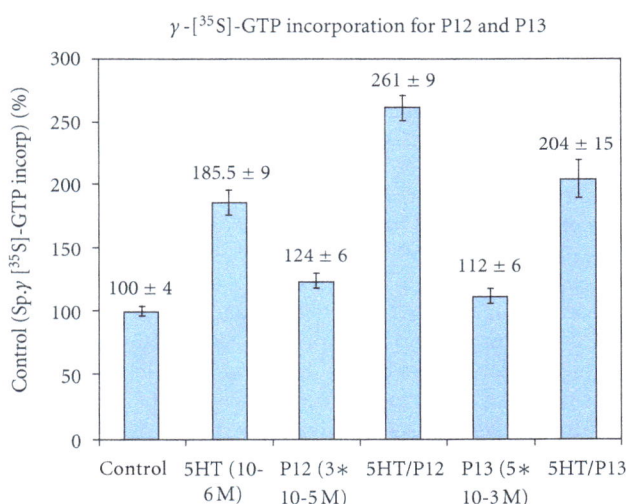

FIGURE 4: P12 and P13-stimulated incorporation of γ-[^{35}S]-GTP control is the basal amount of γ-[^{35}S]-GTP incorporated into Gi in CHO cells expressing the human 5HT1aR set as 100%. The Y axis is the percent of specifically bound γ-[^{35}S]-GTP. All other treatments are percents of the control value. Peptide concentrations are 30 uM. *$P < 0.01$ P12 versus control; €$P < 0.01$ 5HT versus 5HT/P12. *P13 versus control $P < 0.01$; €5HT versus 5HT/P13 $P < 0.01$.

cAMP (Figure 3) may shed some light on this situation. Note that GTP binding by Gi is an agonist-dependent process (see 5HT in the Figure); thus, when peptides increase GTP incorporation above control level, the implication is that the peptides are representing native loop regions under the influence of agonist.

Figure 3 shows a parallel set of results whereby the peptides' ability to change intracellular cAMP concentrations following coupling to Gi is determined (control is the FSK stimulated level; agonist; e.g., 5HT activates Gi and lowers cAMP levels below the control reading). Again, peptide P23, representing mid-loop residues, is most effective. In contrast to the results for GTP incorporation in Figure 2, the cAMP results have a smooth drop off in efficacy on both sides of P23. Overall, the trends peaking at P23 and declining

on both sides are parallel for GTP incorporation (Figure 2) and cAMP concentrations (Figure 3). Note that basal levels of intracellular cAMP are quite low, and the experimental protocol for these experiments involves artificially raising cAMP concentrations via stimulation of adenylyl cyclase with forskolin (control) and comparison of peptide results to that produced by the agonist serotonin.

3.2. Intracellular Loop 3 (ic3). H5HT1aR's ic3 is much larger than ic2 (about 130 amino acids); nevertheless, we did a very

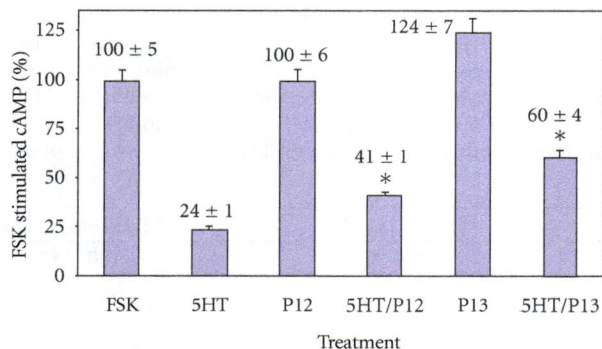

FIGURE 5: P12 and P13 effect of forskolin-stimulated cAMP production forskolin (FSK) stimulated cAMP production by adenylyl cyclase (AC) in CHO cells expressing the human 5HT1aR. These experiments were a measure of second messenger regulation by G protein. FSK is the control, which is set to 100% All other treatments are expressed as a percent of the control value. Peptide concentrations are 30 uM. All treatments include isobutylmethyl xanthine (IBMX), an inhibitor of the metabolism of cAMP. 5HT versus 5HT/P12 and 5HT versus 5HT/P13 *$P < 0.05$.

TABLE 3: ic3 Peptide mimic coupling and signal transduction data. Summary of data generated for all ic3 experiments with P12 and P13. P1 is included as a reference, from Hayataka et al., 1998 [49]. Nominal values for control agonist binding were 400 fmoles/mg protein.

Peptide	Agonist (%) inhibition	$[35S]$-γ-S-GTP incorporation% above conro	% Inhibition of FSK-stimulated cAMP
P1*	50 (3 uM)	30 (1 uM)	10 (10 uM)
P12	28 (30 uM)	24 (30 uM)	0 (30 uM)
P13	50 (15 uM)	12 (30 uM)	−24 (30 uM)

limited number of comparisons at the N-terminal (TM5) region of ic3, continuing preliminary work [50–53] and using the same approach as with ic2 by synthetically building 10-MER's two amino acids at a time from the parent (P1; Table 1). Table 3 gives coupling and activation summaries for the two peptides, P12 and P13 (Table 1). As with the ic2 peptides, the ic3 peptides, P12 and P13, give shallow, uM concentration-effect relationships (data not shown in graphical form as in Figure 1). For coupling, if 50% is listed, then that is the IC 50; if another value is listed, that is the maximum inhibition possible with the highest soluble concentration. Both P12 and P13 produce small but significant incorporation increases of GTP based upon the amount of $[^{35}S]$-GTP incorporated into Gi (Figure 4), while the outcomes for changes in intracellular cAMP concentrations are more complex (Figure 5): for peptide P12, intracellular cAMP concentration is not altered; unusually, peptide P13 actually increases intracellular cAMP concentration. A possible explanation for the combined results for P12 and P13 is given in the Discussion section.

4. Discussion

H5HT1aR is linked to numerous important physiological and pathological processes. Additionally, the receptor is a close relative, not only of other 5HT1 type receptors, but also the beta adrenergic receptors and other GPCR's [13, 42, 55, 56]. Because of these characteristics, structural determinations of the receptor are crucial matters. Despite recent critical structural advances with the beta adrenergic receptors [10, 15, 16, 18], the 5HT1aR is uncrystallized and its structure awaits X-ray analysis [22].

In previous work [49, 50, 53, 54], we have demonstrated the utility of an agonist-based inhibition system associated with signal transduction parameters to study interactions of the receptor with its cognate G protein, Gi. In the

current investigation we have presented further information about the H5HT1aR/Gi interface that should provide testable hypotheses anticipating the ultimate structural analysis of the receptor.

Data collected in previous and current experiments have implicated a role for ic2 in receptor coupling and G-protein activation. The N-terminus end of ic2, involving the sequence IALDRYWAITDPIDYV and including peptides P11 (previous work) and P21–P23 (current work), is important for coupling to the G protein. Evidence for this includes presence of the highly conserved DRY sequence for GPCR's [51] and from the present study, IC50's for the peptides' coupling capacity, with ranges from 7 to 30 uM. Decay of G-protein coupling activity was observed as the peptides progress towards the C-terminus of ic2 (P24–27). As the amino acids seem to wane in importance for receptor coupling, they increase for part of the distance in importance for G-protein activation with its peak at the P23 amino acid stretch WAITDPIDYV. This is clearly shown by the bell-shaped progression of the data bars for the incorporation of γ-$[^{35}S]$-GTP into Gi_{α} (Figure 2). This can be superimposed over the inverted bell-shaped depression for intracellular levels of FSK-stimulated cAMP production (Figure 3) following peptide treatment.

The C-terminal end of ic2 consisting of the amino acids RTPRPR may serve as an anchor, helping to hold the amino terminal of ic2 in a favorable orientation for coupling to the G protein [48]. Also interesting about the carboxy terminal end of ic2 is the presence of the 2 proline (P) residues separated by only 1 amino acid. These proline residues in close proximity to each other introduce a kink in the receptor structure constraining its range of motion. These data demonstrates the clear role for H5HT1aR's ic2 in coupling receptor to G protein, and toward the loop's middle, G-protein activation.

For ic3, the inhibition of AC by Gi is an important regulator of intracellular signal transduction. The current peptides tested, P12 and P13, differed in their ability to regulate this step in the cascade. P12 was unable to decrease the FSK-stimulated levels of cAMP (Figure 5). This is in contrast to the action of 5HT which was able to significantly decrease the FSK-stimulated levels of cAMP. P13 had the opposite effect; it increased cAMP concentrations (Figure 5)! This suggests that the two new amino acids (AD added to form P13) from ic3 are potentially at the beginning of a region of the loop which has negative regulatory properties on Gi blunting its normal ability to regulate AC. It is interesting to speculate about the differences in data from the γ-[^{35}S]-GTP (Figure 4) incorporation assays and cAMP assays (Figure 5). P12 slightly increases GTP incorporation while P13 statistically does not. Thus, P12 activates Gi but cAMP changes do not ensue. P13 does not activate Gi, but a cAMP change occurs in the atypical direction. With the relatively small changes produced by these two peptides in both signaling measures, one possibility is experimental error that has not been accounted for. It is possible that the peptides are acting at some sites other than the proposed receptor-G-protein interface or that the process at the interface is more complex.

The most tantalizing possibility is that the newly explored region represented by P12 and P13 is the beginning of a region of ic3 involved in coupling of receptor to G protein still capable of regulating Gi. Additionally, the perturbation of Gi in this case involves different conformational changes that activate Gi but in a novel way. This would produce the opposite effect on cAMP concentration and would be equivalent to the downstream actions of an inverse agonist at the ligand binding site. Since 5HT1aR is capable of constitutive activity [25], inverse agonism is possible, and it will be fascinating to see if the P12/P13 region is involved in this activity once the crystal structure is available. In this context then, P12 would represent a transitional region between "normal" and "atypical" Gi regulation while P13 is in the atypical subregion.

While the data support this region's (P12/P13) role in receptor-G-protein coupling, the peptides' ability to uncouple declines relative to previously studied peptides whose structures represent segments closer to ic3's N-terminus. P12 and 13 are beyond (toward the C-terminus) the key RFRI region of P1 previously identified as key to that part of 5HT1aR's ic3-N-terminus responsible for G-protein activation [50].

Varrault et al. [48] demonstrated that the C-terminal section of i3 is involved in G protein coupling and regulation. So, if our work can be interpreted to mean that peak coupling and activating properties are associated with ic3's N-terminal residues and Varrault's work can be interpreted to mean that peak coupling and activating properties are associated with ic3's C-terminal residues, then what role will hold for the vast internal region of the loop in 5HT1aR? GPCR ic3's are variable in size in rhodopsin versus 5HT1aR and BAR's, which have larger ic3 loops (at least twice the size of rhodopsin's ic3). It would be meaningful to extend this peptide approach into the midloop region of H5HT1a's ic3, and then as a crystal structure becomes available the comparisons of 5HT1aR loop function with BAR and rhodopsin will be fascinating.

Neither of the peptides (P12 and P13) are as potent as 5HT at incorporating GTP into Gi$_\alpha$. It is possible that multiple regions are responsible for G-protein activation, and the individual peptides mimic only part of this structure [61], thereby producing a diminished effect relative to 5HT. Also, a given peptide region, even one that is absolutely critical in the native structure, may not have the most

efficacious tertiary structure without the full loop being present. It is crucial to point out that the parent ic3 peptide (P1) contains the full TVKK sequence at its N-terminus. This sequence is part of the so-called Ric-8 [62, 63] region that has been shown in other GPCR as crucial to G-protein regulation. Significantly, the P1 relatives (P12 and P13) under discussion in this communication are at a transition point for this sequence; P12 contains the full TVKK stretch while P13 has lost the T! One additional observation may be pertinent. For GTP incorporation, for both peptides, the combination of peptide plus 5HT produces markedly greater incorporation than that produced by 5HT alone. This may suggest that 5HT and the peptides may be perturbing separate sites on the receptor and/or G protein.

In summary, this peptide mimic study for intracellular loops 2 and 3 of the H5HT1aR was designed to examine which segments were involved in coupling and activation of Gi. The results reported here in combination with previously reported work conclude that the amino terminal ends of ic2 and ic3 are important for coupling the receptor and G protein. The activation of G-protein peaks at P23 (WAITDPIDYV) in ic2 (mid-loop). The activity is decreased as the structures move in either direction away from this core sequence. The curious results of increased cAMP concentrations caused by P13 suggests that the two new amino acids (AD) in P13 are the beginning of a new region of ic3 which has negative regulatory properties on Gi. That is, the new region may be one that is not normally activated by agonists; however, in the presence of inverse agonists and the different conformational changes they produce, the new region may couple to and activate Gi in a way that regulates AC in a way we define as inverse agonism. The combined results with H5HT1aR ic2 and ic3 peptides should lead to testable crystallographic hypotheses with drugs having differential intrinsic activities. Beyond the final judgment of these peptide probes in the structural sense, the information produced may be useful as independent pharmacological observations. Pragmatic implications of the work may be relevant in a framework where the multiple, differential activities of the peptides can be used by medicinal chemists to build unique pharmacological agents targeting unutilized sites at the receptor-G-protein interface.

Acknowledgments

The authors would like to express deepest appreciation to the Department of Biomedical and Pharmaceutical Sciences, the Skaggs School of Pharmacy, the College of Health Professions and Biomedical Sciences, all of The University of Montana (UM), Missoula. Without the resources and human support of these units and The University, the project could not have occurred. Special thanks to Dr. David Freeman for his editorial assistance. Research was conducted under the generous and essential sponsorship of the following NIH Grants: UM-CHPBS Endowment Fund Program NIH S21-MD000236, RR10169 and GM/OD 54302-01/02, and P20 RR 15583 to the UM COBRE Center for Structural and Functional Neurosciences from NCRR.

References

[1] E. C. Azmitia, "Serotonin and brain: evolution, neuroplasticity, and homeostasis," *International Review of Neurobiology*, vol. 77, pp. 31–56, 2006.

[2] M. Filip and M. Bader, "Overview on 5-HT receptors and their role in physiology and pathology of the central nervous system," *Pharmacological Reports*, vol. 61, no. 5, pp. 761–777, 2009.

[3] D. E. Nichols and C. D. Nichols, "Serotonin receptors," *Chemical Reviews*, vol. 108, no. 5, pp. 1614–1641, 2008.

[4] N. M. Barnes and T. Sharp, "A review of central 5-HT receptors and their function," *Neuropharmacology*, vol. 38, no. 8, pp. 1083–1152, 1999.

[5] D. Hoyer, D. E. Clarke, J. R. Fozard et al., "International Union of Pharmacology classification of receptors for 5-hydroxytryptamine (serotonin)," *Pharmacological Reviews*, vol. 46, no. 2, pp. 157–203, 1994.

[6] P. J. Goadsby, A. R. Charbit, A. P. Andreou, S. Akerman, and P. R. Holland, "Neurobiology of migraine," *Neuroscience*, vol. 161, no. 2, pp. 327–341, 2009.

[7] K. K. Parker, "Involvement of adipokines in migraine headache," in *Extracellular & Intracellular Signaling*, J. D. Adams and K. K. Parker, Eds., Royal Society of Chemistry, Cambridge, UK, 2011.

[8] H. R. Bourne, "G-proteins and GPCRs: from the beginning," *Ernst Schering Foundation symposium proceedings*, no. 2, pp. 1–21, 2006.

[9] S. C. Sinha and S. R. Sprang, "Structures, mechanism, regulation and evolution of class III nucleotidyl cyclases," *Reviews of Physiology, Biochemistry and Pharmacology*, vol. 157, pp. 105–140, 2006.

[10] B. K. Kobilka, "G protein coupled receptor structure and activation," *Biochimica et Biophysica Acta*, vol. 1768, no. 4, pp. 794–807, 2007.

[11] B. K. Kobilka and X. Deupi, "Conformational complexity of G-protein-coupled receptors," *Trends in Pharmacological Sciences*, vol. 28, no. 8, pp. 397–406, 2007.

[12] R. J. Lefkowitz, J. P. Sun, and A. K. Shukla, "A crystal clear view of the β_2-adrenergic receptor," *Nature Biotechnology*, vol. 26, no. 2, pp. 189–191, 2008.

[13] D. M. Rosenbaum, S. G. F. Rasmussen, and B. K. Kobilka, "The structure and function of G-protein-coupled receptors," *Nature*, vol. 459, no. 7245, pp. 356–363, 2009.

[14] J. M. Baldwin, "Structure and function of receptors coupled to G proteins," *Current Opinion in Cell Biology*, vol. 6, no. 2, pp. 180–190, 1994.

[15] V. Cherezov, D. M. Rosenbaum, M. A. Hanson et al., "High-resolution crystal structure of an engineered human β_2-adrenergic G protein-coupled receptor," *Science*, vol. 318, no. 5854, pp. 1258–1265, 2007.

[16] S. G. F. Rasmussen, H. J. Choi, D. M. Rosenbaum et al., "Crystal structure of the human β_2 adrenergic G-protein-coupled receptor," *Nature*, vol. 450, no. 7168, pp. 383–387, 2007.

[17] D. M. Rosenbaum, V. Cherezov, M. A. Hanson et al., "GPCR engineering yields high-resolution structural insights into β_2-adrenergic receptor function," *Science*, vol. 318, no. 5854, pp. 1266–1273, 2007.

[18] D. M. Rosenbaum, C. Zhang, J. A. Lyons et al., "Structure and function of an irreversible agonist-β_2 adrenoceptor complex," *Nature*, vol. 469, no. 7329, pp. 236–240, 2011.

[19] V. P. Jaakola, M. T. Griffith, M. A. Hanson et al., "The 2.6 angstrom crystal structure of a human A2A adenosine receptor bound to an antagonist," *Science*, vol. 322, no. 5905, pp. 1211–1217, 2008.

[20] B. Wu, E. Y. T. Chien, C. D. Mol et al., "Structures of the CXCR4 chemokine GPCR with small-molecule and cyclic peptide antagonists," *Science*, vol. 330, no. 6007, pp. 1066–1071, 2010.

[21] E. Y. T. Chien, W. Liu, Q. Zhao et al., "Structure of the human dopamine D3 receptor in complex with a D2/D3 selective antagonist," *Science*, vol. 330, no. 6007, pp. 1091–1095, 2010.

[22] S. Topiol and M. Sabio, "X-ray structure breakthroughs in the GPCR transmembrane region," *Biochemical Pharmacology*, vol. 78, no. 1, pp. 11–20, 2009.

[23] J. R. Raymond, Y. V. Mukhin, T. W. Gettys, and M. N. Garnovskaya, "The recombinant 5-HT(1A) receptor: G protein coupling and signalling pathways," *British Journal of Pharmacology*, vol. 127, no. 8, pp. 1751–1764, 1999.

[24] J. R. Raymond, Y. V. Mukhin, A. Gelasco et al., "Multiplicity of mechanisms of serotonin receptor signal transduction," *Pharmacology and Therapeutics*, vol. 92, no. 2-3, pp. 179–212, 2001.

[25] J. C. Martel, A. M. Ormiere, N. Leduc, M. B. Assie, D. Cussac, and A. Newman-Tancredi, "Native rat hippocampal 5-HT$_{1A}$ receptors show constitutive activity," *Molecular Pharmacology*, vol. 71, no. 3, pp. 638–643, 2007.

[26] J. Guptarak, A. Selvamani, and L. Uphouse, "GABAA-5-HT$_{1A}$ receptor interaction in the mediobasal hypothalamus," *Brain Research*, vol. 1027, no. 1-2, pp. 144–150, 2004.

[27] E. B. Russo, A. Burnett, B. Hall, and K. K. Parker, "Agonistic properties of cannabidiol at 5-HT$_{1A}$ receptors," *Neurochemical Research*, vol. 30, no. 8, pp. 1037–1043, 2005.

[28] C. P. Muller, R. J. Carey, J. P. Huston, and M. A. de Souza Silva, "Serotonin and psychostimulant addiction: focus on 5-HT$_{1A}$-receptors," *Progress in Neurobiology*, vol. 81, no. 3, pp. 133–178, 2007.

[29] C. Jonnakuty and C. Gragnoli, "What do we know about serotonin?" *Journal of Cellular Physiology*, vol. 217, no. 2, pp. 301–306, 2008.

[30] D. Kozaric-Kovacic, "Psychopharmacotherapy of posttraumatic stress disorder," *Croatian Medical Journal*, vol. 49, no. 4, pp. 459–475, 2008.

[31] J. L. Rausch, M. E. Johnson, K. E. Kasik, and S. M. Stahl, "Temperature regulation in depression: functional 5HT1A receptor adaptation differentiates antidepressant response," *Neuropsychopharmacology*, vol. 31, no. 10, pp. 2274–2280, 2006.

[32] E. Akimova, R. Lanzenberger, and S. Kasper, "The serotonin-1A receptor in anxiety disorders," *Biological Psychiatry*, vol. 66, no. 7, pp. 627–635, 2009.

[33] G. S. Kranz, S. Kasper, and R. Lanzenberger, "Reward and the serotonergic system," *Neuroscience*, vol. 166, no. 4, pp. 1023–1035, 2010.

[34] L. A. Catapano and H. K. Manji, "G protein-coupled receptors in major psychiatric disorders," *Biochimica et Biophysica Acta*, vol. 1768, no. 4, pp. 976–993, 2007.

[35] B. le Francois, M. Czesak, D. Steubl, and P. R. Albert, "Transcriptional regulation at a HTR1A polymorphism associated with mental illness," *Neuropharmacology*, vol. 55, no. 6, pp. 977–985, 2008.

[36] L. B. Resstel, R. F. Tavares, S. F. Lisboa, S. R. Joca, F. M. Correa, and F. S. Guimaraes, "5-HT$_{1A}$ receptors are involved in the cannabidiol-induced attenuation of behavioral & cardiovascular responses to acute restraint stress in rats," *British Journal of Pharmacology*, vol. 156, no. 1, pp. 181–188, 2009.

[37] J. Savitz, I. Lucki, and W. C. Drevets, "5-HT$_{1A}$ receptor function in major depressive disorder," *Progress in Neurobiology*, vol. 88, no. 1, pp. 17–31, 2009.

[38] G. V. Carr and I. Lucki, "The role of serotonin receptor subtypes in treating depression: a review of animal studies," *Psychopharmacology*, vol. 213, no. 2-3, pp. 265–287, 2011.

[39] P. N. Yadav, A. I. Abbas, M. S. Farrell et al., "The presynaptic component of the serotonergic system is required for clozapine's efficacy," *Neuropsychopharmacology*, vol. 36, no. 3, pp. 638–651, 2011.

[40] J. P. Changeux and S. J. Edelstein, "Allosteric mechanisms of signal transduction," *Science*, vol. 308, no. 5727, pp. 1424–1428, 2005.

[41] J. H. Turner, M. N. Garnovskaya, and J. R. Raymond, "Serotonin 5-HT$_{1A}$ receptor stimulates c-Jun N-terminal kinase and induces apoptosis in Chinese hamster ovary fibroblasts," *Biochimica et Biophysica Acta*, vol. 1773, no. 3, pp. 391–399, 2007.

[42] M. C. Lagerstrom and H. B. Schioth, "Structural diversity of G protein-coupled receptors and significance for drug discovery," *Nature reviews. Drug Discovery*, vol. 7, no. 4, pp. 339–357, 2008.

[43] M. J. Millan, P. Marin, J. Bockaert, and C. Mannoury la Cour, "Signaling at G-protein-coupled serotonin receptors: recent advances and future research directions," *Trends in Pharmacological Sciences*, vol. 29, no. 9, pp. 454–464, 2008.

[44] B. Sjgren, L. L. Blazer, and R. R. Neubig, "Regulators of G protein signaling proteins as targets for drug discovery," *Progress in Molecular Biology and Translational Science*, vol. 91, no. C, pp. 81–119, 2010.

[45] J. A. Allen and B. L. Roth, "Strategies to discover unexpected targets for drugs active at G protein-coupled receptors," *Annual Review of Pharmacology and Toxicology*, vol. 51, pp. 117–144, 2011.

[46] S. Ganguly, A. H. A. Clayton, and A. Chattopadhyay, "Organization of higher-order oligomers of the serotonin$_{1A}$ receptor explored utilizing homo-FRET in live cells," *Biophysical Journal*, vol. 100, no. 2, pp. 361–368, 2011.

[47] A. Ivetac and J. Andrew McCammon, "Mapping the druggable allosteric space of g-protein coupled receptors: a fragment-based molecular dynamics approach," *Chemical Biology and Drug Design*, vol. 76, no. 3, pp. 201–217, 2010.

[48] A. Varrault, Dung Le Nguyen, S. McClue, B. Harris, P. Jouin, and J. Bockaert, "5-Hydroxytryptamine$_{1A}$ receptor synthetic peptides: mechanisms of adenylyl cyclase inhibition," *The Journal of Biological Chemistry*, vol. 269, no. 24, pp. 16720–16725, 1994.

[49] K. Hayataka, M. F. O'Connor, N. Kinzler, J. T. Weber, and K. K. Parker, "A bioactive peptide from the transmembrane 5-intracellular loop 3 region of the human 5HT$_{1A}$ receptor," *Biochemistry and Cell Biology*, vol. 76, no. 4, pp. 657–660, 1998.

[50] T. C. Ortiz, M. C. Devereaux, and K. K. Parker, "Structural variants of a human 5-HT$_{1A}$ receptor intracellular loop 3 peptide," *Pharmacology*, vol. 60, no. 4, pp. 195–202, 2000.

[51] N. Kushwaha, S. C. Harwood, A. M. Wilson et al., "Molecular determinants in the second intracellular loop of the 5-hydroxytryptamine-1A receptor for G-protein coupling," *Molecular Pharmacology*, vol. 69, no. 5, pp. 1518–1526, 2006.

[52] A. O. Shpakov and M. N. Pertseva, "Molecular mechanisms for the effect of mastoparan on G proteins in tissues of vertebrates and invertebrates," *Bulletin of Experimental Biology and Medicine*, vol. 141, no. 3, pp. 302–306, 2006.

[53] H. V. Thiagaraj, T. C. Ortiz, M. C. Devereaux Jr., B. Seaver, B. Hall, and K. K. Parker, "Regulation of G proteins by human 5-HT$_{1A}$ receptor TM3/i2 and TM5/i3 loop peptides," *Neurochemistry International*, vol. 50, no. 1, pp. 109–118, 2007.

[54] B. Hall, A. Burnett, A. Christians et al., "Thermodynamics of peptide and non-peptide interactions with the human 5HT$_{1A}$ receptor," *Pharmacology*, vol. 86, no. 1, pp. 6–14, 2010.

[55] A. Fargin, J. R. Raymond, M. J. Lohse, B. K. Kobilka, M. G. Caron, and R. J. Lefkowitz, "The genomic clone G-21 which resembles a β-adrenergic receptor sequence encodes the 5-HT$_{1A}$ receptor," *Nature*, vol. 335, no. 6188, pp. 358–360, 1988.

[56] B. K. Kobilka, T. Frielle, S. Collins et al., "An intronless gene encoding a potential member of the family of receptors coupled to guanine nucleotide regulatory proteins," *Nature*, vol. 329, no. 6134, pp. 75–79, 1987.

[57] M. M. Bradford, "A rapid and sensitive method for the quantitation of microgram quantities of protein utilizing the principle of protein dye binding," *Analytical Biochemistry*, vol. 72, no. 1-2, pp. 248–254, 1976.

[58] T. Wieland and K. H. Jakobs, "[1] Measurement of receptor-stimulated guanosine 5'-O-(γ-thio)triphosphate binding by G proteins," *Methods in Enzymology*, vol. 237, pp. 3–13, 1994.

[59] M. E. Maguire, P. M. van Arsdale, and A. G. Gilman, "An agonist specific effect of guanine nucleotides on binding to the β adrenergic receptor," *Molecular Pharmacology*, vol. 12, no. 2, pp. 335–339, 1976.

[60] S. J. Peroutka, R. M. Lebovitz, and S. H. Snyder, "Serotonin receptor binding sites affected differentially by guanine nucleotides," *Molecular Pharmacology*, vol. 16, no. 3, pp. 700–708, 1979.

[61] S. Kalipatnapu and A. Chattopadhyay, "Membrane organization and function of the serotonin$_{1A}$ receptor," *Cellular and Molecular Neurobiology*, vol. 27, no. 8, pp. 1097–1116, 2007.

[62] G. G. Tall, A. M. Krumins, and A. G. Gilman, "Mammalian Ric-8A (synembryn) is a heterotrimeric Gα protein guanine nucleotide exchange factor," *The Journal of Biological Chemistry*, vol. 278, no. 10, pp. 8356–8362, 2003.

[63] C. J. Thomas, G. G. Tall, A. Adhikari, and S. R. Sprang, "Ric-8A catalyzes guanine nucleotide exchange on Gαi1 bound to the GPR/GoLoco exchange inhibitor AGS3," *The Journal of Biological Chemistry*, vol. 283, no. 34, pp. 23150–23160, 2008.

18

Platelet-Rich Plasma Peptides: Key for Regeneration

Dolores Javier Sánchez-González,[1, 2] **Enrique Méndez-Bolaina,**[3, 4]
and Nayeli Isabel Trejo-Bahena[2, 5]

[1] Subsección de Biología Celular y Tisular, Escuela Médico Militar, Universidad del Ejército y Fuerza Aérea,
   11200 México City, MEX, Mexico
[2] Sociedad Internacional para la Terapia Celular con Células Madre, Medicina Regenerativa y Antienvejecimiento S.C. (SITECEM),
   53840 Naucalpan, MEX, Mexico
[3] Facultad de Ciencias Químicas, Universidad Veracruzana, 94340 Orizaba, VER, Mexico
[4] Centro de Investigaciones Biomédicas-Doctorado en Ciencias Biomédicas, Universidad Veracruzana, 91000 Xalapa, VER, Mexico
[5] Área de Medicina Física y Rehabilitación, Hospital Central Militar, 11200 México City, MEX, Mexico

Correspondence should be addressed to Dolores Javier Sánchez-González, javiersglez@yahoo.com

Academic Editor: Frédéric Ducancel

Platelet-derived Growth Factors (GFs) are biologically active peptides that enhance tissue repair mechanisms such as angiogenesis, extracellular matrix remodeling, and cellular effects as stem cells recruitment, chemotaxis, cell proliferation, and differentiation. Platelet-rich plasma (PRP) is used in a variety of clinical applications, based on the premise that higher GF content should promote better healing. Platelet derivatives represent a promising therapeutic modality, offering opportunities for treatment of wounds, ulcers, soft-tissue injuries, and various other applications in cell therapy. PRP can be combined with cell-based therapies such as adipose-derived stem cells, regenerative cell therapy, and transfer factors therapy. This paper describes the biological background of the platelet-derived substances and their potential use in regenerative medicine.

1. Introduction

Platelets are nonnuclear cellular fragments derived from megakaryocytes in the bone marrow; they are specialized secretory elements that release the contents of their intracellular granules in response to activation. They were discovered by Bizzozero in the 19th century [1] and after Wright observed that megakaryocytes are platelet precursors [2]. Actually we know that platelets synthesize proteins and that pattern of peptides synthesis changes in response to cellular activation [3].

Platelets contain a great variety of proteins molecules, among which are the high presence of signaling, membrane proteins, protein processing, cytoskeleton regulatory proteins, cytokines, and other bioactive peptides that initiate and regulate basic aspects of wound healing [3]. It is known, through efforts such as the platelet proteome project, that more than 300 proteins are released by human platelets in response to thrombin activation [4]. Proteome platelet includes 190 membrane-associated and 262 phosphorylated

proteins, which were identified via independent proteomic and phospho proteomic profiling [5].

When platelets fall precipitously below critical levels (usually under 10,000 to 20,000 per cubic millimeter), molecular disassembly opens the zippers formed by adjacent intercellular endothelial junctions, causing extravasation of erythrocytes into the surrounding tissues. In addition to their well-known function in hemostasis, platelets also release substances that promote tissue repair, angiogenesis, and inflammation [6]. Furthermore, they induce the migration and adherence of bone-marrow-derived cells to sites of angiogenesis; platelets also induce differentiation of endothelial-cell progenitors into mature endothelial cells [7].

At the site of the injury, platelets release an arsenal of potent regenerative and mitogenic substances that are involved in all aspects of the wound-healing process including a potential point-of-care biologic treatment following myocardial injury [8]. Based on this, platelet called Platelet-rich Plasma (PRP) has been extensively used for orthopaedic

TABLE 1: Peptidic growth factors present in platelet-rich plasma (PRP).

Name	Cytogenetic location	Biologic activities
Transforming growth factor, beta-I; TGFB1	19q13.2	Controls proliferation, differentiation, and other functions in many cell types
Platelet-derived growth factor, alpha polypeptide; PDGFA	7p22.3	Potent mitogen for connective tissue cells and exerts its function by interacting with related receptor tyrosine kinases
Platelet-derived growth factor, beta polypeptide; PDGFB	22q13.1	Promotes cellular proliferation and inhibits apoptosis
Platelet-derived growth factor C; PDGFC	4q32.1	Increases motility in mesenchymal cells, fibroblasts, smooth muscle cells, capillary endothelial cells, and neurons
Platelet-derived growth factor D; PDGFD	11q22.3	Involved in developmental and physiologic processes, as well as in cancer, fibrotic diseases, and arteriosclerosis
Insulin-like growth factor I; IGF1	12q23.2	Mediates many of the growth-promoting effects of growth hormone
Fibroblast growth factor I; FGF1	5q31.3	Induces liver gene expression, angiogenesis and fibroblast proliferation
Epidermal growth factor; EGF	4q25	Induces differentiation of specific cells, is a potent mitogenic factor for a variety of cultured cells of both ectodermal and mesodermal origin
Vascular endothelial growth factor A; VEGFA	6p21.1	Is a mitogen primarily for vascular endothelial cells, induces angiogenesis
Vascular endothelial growth factor B; VEGFB	11q13.1	Is a regulator of blood vessel physiology, with a role in endothelial targeting of lipids to peripheral tissues
Vascular endothelial growth factor C; VEGFC	4q34.3	Angiogenesis and endothelial cell growth, and can also affect the permeability of blood vessels

Includes, name, cytogenetic location, and biologic activities of platelet growth factors. Furthermore, PRP content other proteins like interleukin-8, macrophage inflammatory protein-1 alpha, and platelet factor-4.

applications; for topical therapy of various clinical conditions, including wounds and soft tissue injuries; and suitable alternative to fetal calf serum for the expansion of mesenchymal stem cells from adipose tissue (see Table 1) [9–12].

In this paper we are going to talk about the platelets, platelet-derived particles, and their biological effects in regenerative medicine.

2. Platelets

Platelets are the first element to arrive at the site of tissue injury and are particularly active in the early inflammatory phases of the healing process [6]. They play a role in aggregation, clot formation, homeostasis through cell membrane adherence, and release of substances that promote tissue repair and that influence the reactivity of blood vessels and blood cell types involved in angiogenesis, regeneration, and inflammation [13]. Platelet secretory granules contain growth factors (GFs), signaling molecules, cytokines, integrins, coagulation proteins, adhesion molecules, and some other molecules, which are synthesized in megakaryocytes and packaged into the granules through vesicle trafficking processes [14]. Three major storage compartments in platelets are alpha granules, dense granules, and lysosomes [14].

Platelets mediate these effects through degranulation, in which platelet-derived GF (PDGF), insulin-like GF (IGF1),

transforming GF-beta 1 (TGF-β1), vascular endothelial GF (VEGF), basic fibroblastic GF (bFGF), and epidermal GF (EGF) are released from alpha granules [15]. In fact, the majority of the platelet substances are contained in alpha granules (see Table 2) [15]. When platelets are activated, they exocytose the granules; this process is mediated by molecular mechanisms homologous to other secretory cells, uniquely coupled to cell activation by intracellular signaling events [16].

Among bioactive molecules stored and released from platelets dense granules are catecholamines, histamine, serotonin, ADP, ATP, calcium ions, and dopamine, which are active in vasoconstriction, increased capillary permeability, attract and activate macrophages, tissue modulation and regeneration. These non-GF molecules have fundamental effects on the biologic aspects of wound healing [5].

For their numerous functions, platelets have developed a set of platelet receptors that are the contact between platelets and their surroundings; they determine the reactivity of platelets with a wide range of agonists and adhesive proteins. Some of these receptors are expressed only on activated platelets [6]. Certain biological mechanisms present in the platelets are shared with other cells, and therefore they contain some common cytoplasmic enzymes, signal transduction molecules, and cytoskeletal components [14].

TABLE 2: Some bioactive peptides present in the alpha granules of platelets.

General activity categories	Specific molecules	Cytogenetic location	Biologic activities
Clotting factors and related proteins	Tissue factor pathway inhibitor; TFPI	2q32.1	Regulates the tissue factor-(TF-) dependent pathway of blood coagulation
	Kininogen; KNG	3q27.3	Plays an important role in assembly of the plasma kallikrein
	Growth arrest-specific 6; GAS6	13q34	Stimulates cell proliferation
	Multimerin; MMRN	4q22	Carrier protein for platelet factor V
	Antithrombin; AT	1q25.1	Is the most important inhibitor of thrombin
	Protein S; PROS1	3q11.1	Inhibits blood clotting
	Coagulation factor V; F5	1q24.2	Acts as a cofactor for the conversion of prothrombin to thrombin by factor Xa
	Coagulation factor XI; F11	4q35.2	It participates in blood coagulation as a catalyst in the conversion of factor IX to factor IXa in the presence of calcium ions
Fibrinolytic factors and related proteins	Plasminogen; PLG	6q26	Induces plasmin production (leads to fibrinolysis)
	Plasminogen activator inhibitor 1; PAI1	7q22.1	Regulation of plasmin production
	Alpha-2-plasmin inhibitor	17p13.3	Inactivation of plasmin
	Osteonectin; ON	5q33.1	Inhibits cell-cycle progression and influences the synthesis of extracellular matrix (ECM)
	Histidine-rich glycoprotein; HRG	3q27.3	Interacts with heparin and thrombospondin
	Thrombin-activatable fibrinolysis inhibitor; TAFI	13q14.13	Attenuates fibrinolysis
	Alpha-2-Macroglobulin; A2M	12p13.31	Carrier of specific growth factors and induces cell signaling
Proteases and antiproteases	Tissue inhibitor of metalloproteinase 4; TIMP4	3p25.2	Inhibits matrix metalloproteinases (MMPs), a group of peptidases involved in degradation of the extracellular matrix
	Complement component 1 inhibitor; C1NH	11q12.1	Inhibits serine proteinases including plasmin, kallikrein, and coagulation factors XIa and XIIa
	Alpha-1-antitrypsin (serpin peptidase inhibitor)	14q32.13	Acute phase protein, inhibits a wide variety of proteases and enzymes
	Nexin 2; SNX2	5q23.2	Modulates intracellular trafficking of proteins to various organelles
Basic proteins	Platelet factor 4; PF4	4q13.3	Inhibition of angiogenesis
	β-thromboglobulin (Pro-platelet basic protein; PPBP)	4q13.3	Platelet activation, inhibition of angiogenesis
	Endostatin (Collagen, type XVIII, Alpha-1; COL18A1)	21q22.3	Inhibitors of endothelial cell migration and angiogenesis
Adhesive proteins	Fibrinogen; FG	4q31.3	Blood clotting cascade (fibrin clot formation)
	Fibronectin; FN	2q35	Binds to cell-surface integrins, affecting cell adhesion, cell growth, migration, and differentiation
	Vitronectin; VTN	17q11.2	Induces cell adhesion, chemotaxis
	Thrombospondin I; THBS1	15q14	Inhibition of angiogenesis
	Laminin-8	18p11.31-p11.23	Modulates cell contact interactions

It is described general activity categories, specific molecules, cytogenetic location and biologic activities. Furthermore, alpha granules include growth factors of Table 1, membrane glycoproteins, and others proteins like albumin and immunoglobulins.

The platelet lifespan is approximately 7 to 9 days, which they spend circulating in the blood in their resting form. When adhered to exposed endothelium or activated by agonists, they change their shape and secrete the contents of the granules (including ADP, fibrinogen, and serotonin), which is followed by platelet aggregation [7]. Initiation of the signaling event within the platelet leads to the reorganization of the platelet cytoskeleton, which is visible as an extremely rapid shape change [17].

3. Platelet-Rich Plasma (PRP)

Platelets are activated either by adhesion to the molecules that are exposed on an injured endothelium, such as von Willebrand Factor (vWF), collagen, fibronectin, and laminin, or by physiologic agonists such as thrombin, ADP, collagen, thromboxane A2, epinephrine, and platelet-activating factors [18].

PRP has been used clinically in humans since the 1970s for its healing properties attributed of autologous GF and secretory proteins that may enhance the healing process on a cellular level [19]. Furthermore, PRP enhances the recruitment, proliferation, and differentiation of cells involved in tissue regeneration [20]. PRP-related products, also known as platelet-rich concentrate, platelet gel, preparation rich in growth factors (PRGF), and platelet releasate, have been studied with in vitro and in vivo experiments in the fields of surgical sciences mainly [21].

Depending on the device and technique used, PRP can contain variable amounts of plasma, erythrocytes, white blood cells, and platelets. The platelet concentration should be increased above baseline or whole blood concentration. It is generally agreed upon that PRP should have a minimum of 5 times the number of platelets compared to baseline values for whole blood to be considered "platelet rich" [22].

This conclusion is supported by in vitro work showing a positive dose-response relationship between platelet concentration and proliferation of human mesenchymal stem cells, proliferation of fibroblasts, and production of type I collagen [23]. This suggests that the application of autologous PRP can enhance wound healing, as has been demonstrated in controlled animal studies for both soft and hard tissues [24, 25].

Autologous PRP represents an efficacious treatment for its use in wound healing like chronic diabetic foot ulceration due to multiple growth factors, is safe for its autologous nature, and is produced as needed from patient blood. Like we said, key for self-regeneration [26].

Upon activation, platelets release their granular contents into the surrounding environment. The platelet alpha granules are abundant and contain many of the GFs responsible for the initiation and maintenance of the healing response [14]. These GFs have been shown to play an important role in all phases of healing. The active secretion of these proteins by platelets begins within 10 minutes after clotting, with more than 95% of the presynthesized GFs secreted within 1 hour. After this initial burst, the platelets synthesize and secrete additional proteins for the balance of their life (5–10 days) [10].

The fibrin matrix formed following platelet activation also has a stimulatory effect on wound healing. The fibrin matrix forms by polymerization of plasma fibrinogen following either external activation with calcium or thrombin or internal activation with endogenous tissue thromboplastin [23]. This matrix traps platelets allowing a slow release of a natural combination of GF while providing a provisional matrix that provides a physical framework for wound stem cells and fibroblast migration and presentation of other biological mediators such as adhesive glycoproteins [27, 28].

PRP with a platelet concentration of at least 1 000 000 platelets/μL in 5 mL of plasma is associated with the enhancement of healing [29]. PRP can potentially enhance healing by the delivery of various GF and cytokines from the alpha granules contained in platelets and has an 8-fold increase in GF concentrations compared with that of whole blood [30].

The use of PRP to enhance bone regeneration and soft tissue maturation has increased dramatically in the fields of orthopedics, periodontics, maxillofacial surgery, urology, and plastic surgery over the last years. However, controversies exist in the literature regarding the added benefit of this procedure. While some authors have reported significant increases in bone formation and maturation rates [21], others did not observe any improvement [31].

The wound-healing process is a complex mechanism characterized by four distinct, but overlapping, phases: hemostasis, inflammation, proliferation, and remodeling [8]. The proliferative phase includes blood vessel formation by endothelial cells and bone synthesis by osteoblasts. All these events are coordinated by cell-cell interactions and by soluble GF released by various cell types. Recent reviews have emphasized the need for additional research aiming to characterize PRP in terms of GF content and their physiological roles in wound healing [32–34].

Thrombin represents a strong inducer of platelet activation leading to GF release [35]. It is also known that particulate grafts, when combined with calcium and thrombin treated PRP, possess better handling characteristics and higher GF content [31]. Typically, thrombin concentrations used in clinical applications vary between 100 and 200 units per mL [21], while platelet aggregation is maximum in the range of 0.5 to 4 units per mL [36].

The basic cytokines identified in platelets play important roles in cell proliferation, chemotaxis, cell differentiation, regeneration, and angiogenesis [28]. A particular value of PRP is that these native cytokines are all present in "normal" biologic ratios. The platelets in PRP are delivered in a clot, which contains several cell adhesion molecules including fibronectin, fibrin, and vitronectin. These cell adhesion molecules play a role in cell migration and thus also add to the potential biologic activity of PRP. The clot itself can also play a role in wound healing by acting as conductive matrix or "scaffold" upon which cells can adhere and begin the wound-healing process [28].

PRP can only be made from anticoagulated blood. Preparation of PRP begins by addition of citrate to whole blood to bind the ionized calcium and inhibit the clotting cascade [15]. This is followed by one or two centrifugation steps. The first centrifugation step separates the red and white blood cells from plasma and platelets. The second centrifugation step further concentrates the platelets, producing the PRP separate from platelet-poor plasma [19].

An important point is that clotting leads to platelet activation, resulting in release of the GF from the alpha granules, otherwise known as degranulation. Approximately 70% of the stored GFs are released within 10 minutes, and nearly 100% of the GFs are released within 1 hour. Small amounts of GF may continue to be produced by the platelet during the rest of its lifespan (1 week) [21].

A method to delay the release of GF is possible by addition of calcium chloride ($CaCl_2$) to initiate the formation of autogenous thrombin from prothrombin. The $CaCl_2$ is added during the second centrifugation step and results in formation of a dense fibrin matrix. Intact platelets are subsequently trapped in the fibrin matrix and release GF slowly over a 7-day period. The fibrin matrix itself may also contribute to healing by providing a conductive scaffold for cell migration and new matrix formation [27].

4. Growth Factors (GFs)

Platelets are known to contain high concentrations of different GF and are extremely important in regenerative process; activation of the platelet by endothelial injury initiates the wound-healing process [30]. When platelets are activated, their alpha granules are released, resulting in an increased concentration of GF in the wound milieu [14].

There is increasing evidence that the platelet cell membranes themselves also play a crucial role in wound healing through their GF receptor sites [28]. GFs are found in a wide array of cells and in platelet alpha granules [37]. Table 1 gives an overview of some of the more extensively studied GFs and their involvement in wound healing. There are many more, both discovered and undiscovered, GFs. The platelet is an extremely important cell in wound healing because it initiates and plays a major role in the wound regenerative process [38].

The first discovered GF was EGF in 1962 by Cohen [39]. It was not until 1989 before clinical trials with EGF were attempted to demonstrate enhanced wound healing. Studies did demonstrate that EGF can accelerate epidermal regeneration and enhance healing of chronic wounds [40].

PDGF was discovered in 1974 and is ubiquitous in the body. It is known to be released by platelet alpha granules during wound healing and stimulate the proliferation of many cells, including connective tissue cells. In fact, thus far, high-affinity cell-surface receptors specific for PDGF have only been demonstrated on connective tissue cells. When released, PDGF is chemotactic for monocytes, neutrophils, and fibroblasts. These cells release their own PDGF, thus creating a positive autocrine feedback loop [41]. Other functions of PDGF include effects on cell growth, cellular migration, metabolic effects, and modulation of cell membrane receptors [42].

PDGFs were first identified as products of platelets which stimulated the proliferation in vitro of connective tissue cell types such as fibroblasts [43].

The PDGF system, comprising four isoforms (PDGF-A, -B, -C, and -D) and two receptor chains (PDGFR-alpha and -beta), plays important roles in wound healing, atherosclerosis, fibrosis, and malignancy. Components of the system are expressed constitutively or inducibly in most renal cells [42]. They regulate a multitude of pathophysiologic events, ranging from cell proliferation and migration to extracellular matrix accumulation, production of pro- and anti-inflammatory mediators, tissue permeability, and regulation of hemodynamics [43].

Inactivation of PDGF-B and PDGF beta receptor (PDGFRb) genes by homologous recombination in embryonic stem cells shows cardiovascular, hematological, and renal defects. The latter is particularly interesting since it consists of a specific cellular defect: the complete loss of kidney glomerular mesangial cells and the absence of urine collection in the urinary bladder [43].

PDGF-C and PDGFR-alpha contribute to the formation of the renal cortical interstitium. Almost all experimental and human renal diseases are characterized by altered expression of components of the PDGF system. Infusion or systemic overexpression of PDGF-B or -D induces prominent mesangioproliferative changes and renal fibrosis. Intervention studies identified PDGF-C as a mediator of renal interstitial fibrosis and PDGF-B and -D as key factors involved in mesangioproliferative disease and renal interstitial fibrosis [43–45].

Fréchette et al., demonstrated that the release of PDGF-B, TGF-beta1, bFGF, and VEGF is significantly regulated by the amount of calcium and thrombin added to the PRP and that PRP supernatants are more mitogenic for endothelial cells than whole-blood supernatants [11]. Other GFs such as epidermal growth factor (EGF), transforming growth factor-alpha (TGF-alpha), insulin-like growth factor-1 (IGF-1), angiopoietin-2 (Ang-2), and interleukin-1beta (IL-1beta) are also known to play important roles in the wound-healing process [28].

In 2008, Wahlström et al., demonstrated that growth factors released from platelets had potent effects on fracture and wound healing. The acidic tide of wound healing, that is, the pH within wounds and fractures, changes from acidic pH to neutral and alkaline pH as the healing process progresses [44]. They investigated the influence of pH on lysed platelet concentrates regarding the release of growth factors. The platelet concentrates free of leukocyte components were lysed and incubated in buffers with pH between 4.3 and 8.6. Bone morphogenetic protein-2 (BMP-2), platelet-derived growth factor (PDGF), transforming growth factor-beta (TGF-beta), and vascular endothelial growth factor (VEGF) were measured by quantitative enzyme-linked immunosorbent assays. BMP-2 was only detected in the most acidic preparation (pH 4.3), which is interesting since BMP-2 has been reported to be an endogenous mediator of fracture repair and to be responsible for the initiation of fracture healing. These findings indicate that platelets release substantial amounts of BMP-2 only under conditions of low pH, the milieu associated with the critical initial stage of fracture healing [44].

Recently, Bir et al., demonstrated stromal cell-derived factor 1-α (SDF-1α) PRP from diabetic mice. The concentration (pg/mL) of different growth factors was significantly higher in the PRP group than in the platelet-poor plasma (PPP) group. The concentrations (pg/mL) of SDF-1α (10,790 $\pm$ 196 versus 810 $\pm$ 39), PDGF-BB (45,352 $\pm$ 2,698 versus 958 $\pm$ 251), VEGF (53 $\pm$ 6 versus 30 $\pm$ 2), bFGF (29 $\pm$ 5 versus 9 $\pm$ 5), and IGF-1 (20,628 $\pm$ 1,180 versus 1,214 $\pm$ 36) were significantly higher in the PRP group than in the PPP group, respectively [46].

5. Platelet GF as Treatment

5.1. In Vitro Studies. Knowledge of GF and their function is far from complete. Many of the known functions were learned through in vitro study. Although many GFs are associated with wound healing, PDGF and TGF-β1 appear to be two of the more integral modulators [46]. PDGF has activity in early wound healing (during the acid tide). In vitro studies have shown that at lower pH (5.0), platelet concentrate lysate has increased concentrations of PDGF, with an increased capacity to stimulate fibroblast proliferation [23]. TGF-β increases the production of collagen from fibroblasts [47]. Its release in vitro is enhanced by neutral or alkaline pH, which correspond to the later phases of healing [10]. Through modulation of interleukin-1 production by macrophages, PRP may inhibit excessive early inflammation that could lead to dense scar tissue formation [48]. Insulin-like GF-I (IGF-1) has also been extensively studied for its ability to induce proliferation, differentiation, and hypertrophy of multiple cell lines. Separate analyses of GF in PRP have shown significant increases in PDGF, VEGF, TGF-β1, and EGF, compared with their concentrations in whole blood [15, 49].

IGF-1 has two important functions: chemotaxis for vascular endothelial cells into the wound which results in angiogenesis and promoting differentiation of several cell lines including chondroblasts, myoblasts, osteoblasts, and hematopoietic cells [50].

TGF-β is a member of the newest family of proteins discovered. Two major sources of this protein are the platelet and macrophage. TGF-β causes chemotactic attraction and activation of monocytes, macrophages, and fibroblasts. The activated fibroblasts enhance the formation of extracellular matrix and collagen and also stimulate the cells ability to contract the provisional wound matrix [51]. Macrophages infiltration promotes TGF-β that induces extracellular matrix such as collagen and fibronectin; however alpha-mangostin prevents the increase in this molecule in rats with Cisplatin-induced nephrotoxicity [52].

5.2. In Vivo Studies. In vivo study is much more complex due to the inability to control the environment. A further complexing matter is the fact that the same GF, depending on the presence or absence of other peptides, may display either stimulatory or inhibitory activity within the same cell. Also, a particular GF can alter the binding affinity of another GF receptor [53].

Release of PDGF can have a chemotactic effect on monocytes, neutrophils, fibroblasts, stem cells, and osteoblasts. This peptide is a potent mitogen for mesenchymal cells including fibroblasts, smooth muscle cells and glial cells [54] and is involved in all three phases of wound healing, including angiogenesis, formation of fibrous tissue, and reepithelialization [41].

TGF beta released from platelet alpha granules is a mitogen for fibroblasts, smooth muscle cells, and osteoblasts. In addition, it promotes angiogenesis and extracellular matrix production [41]. VEGF promotes angiogenesis and can promote healing of chronic wounds and aid in endochondral ossification. EGF, another platelet-contained GF, is a mitogen for fibroblasts, endothelial cells, and keratinocytes and also is useful in healing chronic wounds [55].

IGF, another platelet-contained GF regulates bone maintenance and is also an important modulator of cell apoptosis, and, in combination with PDGF, can promote bone regeneration [56].

However, there are conflicting results with regard to IGF-1, where the majority of studies reported no increase in IGF-1 in PRP, compared with whole blood. There are also conflicting results regarding the correlation between the GF content and platelet counts in PRP [57]. The basis of these contradictions is not fully understood and may be related to variability in patient age, health status, or platelet count. Alternatively, differences in GF content and platelet count may be due to the various methods of processing, handling, and storing of samples, in addition to the type of assay performed. The diversity of PRP products should be taken into account when interpreting and comparing results and methods for generating PRP [10].

VEGF, discovered 25 years ago, was initially referred to as vascular permeability factor [58]. In mammals, there are at least four members of the VEGF family: VEGF-A, VEGF-B, and the VEGF-C/VEGF-D pair, which has a common receptor, VEGF receptor 3 (VEGF-R3) [59]. VEGF-A is a proangiogenic cytokine during embryogenesis and contributes to vascular integrity: selective knockout of VEGF-A in endothelial cells increases apoptosis, which compromises the integrity of the junctions between endothelial cells [60, 61]. VEGF-B, which can form heterodimers with VEGF-A, occurs predominantly in brown fat, myocardium, and skeletal muscle [62]. VEGF-C and VEGF-D seem to regulate lymphangiogenesis. The expression of VEGF-R3 in adults is restricted to the lymphatics and fenestrated endothelium [63]. Neuropilin 1 and neuropilin 2 are receptors that bind specific VEGF family members and are important in neuronal development and embryonic vasculogenesis [64].

Megakaryocytes and platelets contain the three major isoforms of VEGF-A; after exposure to thrombin in vitro, they release VEGF-A [65–67]. VEGF-A alters the endothelial-cell phenotype by markedly increasing vascular permeability, upregulating expression of urokinase, tissue plasminogen activator, connexin, osteopontin, and the vascular-cell adhesion molecule [68].

TABLE 3: Platelet-plasma-derived peptides are current in clinical use and clinical trials.

Year	Researchers	Health problems	Clinical protocols	Level of evidence	Results
2005	Carreon et al.	Bone healing in instrumented posterolateral spinal fusions	Retrospective cohort study to evaluate rates of nonunionin patients ($n = 76$) with autologous iliac bone graft augmented with platelet gel	Level 4, case control group of 76 randomly selected patients who were matched and grafted with autogenous iliac bone graft with no platelet gel	Nonunion rate in platelet gel group was 25%; 17% in control group ($P = .18$)
2006	Mishra and Pavelko	Chronic elbow tendinitis	Cohort, 15 patients injected with PRP	Level 2, 5 controls	Decreased pain at 2 years (measured by visual analog pain score)
	Savarino et al.	Bone healing in varus HTOs for genu varus	Randomized case control, 5 patients with bone grafted with bone chips and PRP	Level 4, 5 controls bone grafted without PRP	No functional or clinical difference; histology shows increased amounts of osteoid and osteoblasts in PRP group
2007	Sánchez et al.	Achilles tear healing	Case control, 6 repairs with PRP	Level 3, 6 matched retrospective controls	Improved ROM and early return to activity with PRP by $\pm$ 4–7 weeks
	Dallari et al.	Bone healing in varus HTOs for genu varus	Prospective randomized control: group A, bone chips with platelet gel ($n = 11$); group B, bone chips, BMC, and platelet gel ($n = 12$)	Level 1, 10 controls treated with bone chips only	Biopsies at 6 weeks after surgery showed increased osteoid and osteoblasts in groups A and B; radiographic differences decreased with time; no clinical difference at 1 year among groups
	Kitoh et al.	Bone healing in distraction osteogenesis for limb lengthening and short stature	Retrospective, comparison case control; at 3 weeks, patients injected with expanded BMC with or without PRP ($n = 32$ bones)	Level 3, 60 bones in retrospective control group (high % of congenital etiologies versus PRP group)	Average healing in BMC + PRP was 34 ± 4 d/cm; control group average was 73.4 ± 27 d/cm ($P = .003$)
2009	Sánchez et al.	Bone healing in nonunions	Retrospective, case series; 16 nonhypertrophic nonunions treated with either surgery and PRGF or percutaneous injections of PRGF to stimulate ($n = 3$) without surgery	Level 4, no control group	84% healed after surgical treatment; unclear if PRGF made a difference

Some published human clinical orthopaedic PRP studies. PRP: platelet-rich plasma; ROM: range of motion; HTO: high tibial osteotomy; BMC: bone marrow cells; PRGF: preparation rich in growth factors [27]. Taken from Foster et al. [27].

6. PRP in Cell Therapy and Regenerative Medicine

PRP can be combined with cell-based therapies such as adipose-derived stem cells, regenerative cell therapy, and transfer factors therapy [69]. While this is a relatively new concept, the strategy is appealing as the regenerative matrix graft delivers a potent trilogy of regenerative cells, fibrin matrix, and GF [15]. The applications are similar to those for PRP alone with the added benefit of regenerative cell enrichment.

Verrier et al., demonstrated that cultures of human mesenchymal stem cell (MSC) supplemented with platelet-released supernatant (PRS) had differentiation towards an osteoblastic phenotype in vitro possibly mediated by bone morphogenetic protein-2 (BMP-2). PRS showed an osteoinductive effect on MSC, as shown by an increased expression of typical osteoblastic marker genes such as collagen I, bone

sialoprotein II, BMP-2, and matrix metalloproteinase-13 (MMP-13), as well as by increased Ca^{++} incorporation [70].

Furthermore, the role of platelets in hemostasis may be influenced by alteration of the platelet redox state, the presence of endogenous or exogenous antioxidants, and the formation of reactive oxygen and nitrogen species [71]. As discussed by Sobotková et al., [71], trolox and resveratrol inhibit aggregation of washed platelets and PRP activated by ADP, collagen, and thrombin receptor-activating peptide. Antioxidants, apart from nonspecific redox or radical-quenching mechanisms, inhibit platelet activation also by specific interaction with target proteins. In this context, powerful natural antioxidants, like nordihydroguaiaretic acid (NDGA) extracted from *Larrea tridentata* [72], S-allylcysteine (SAC), the most abundant organosulfur compound in aged garlic extract (AG) [73], sulforaphane (SFN), an isothiocyanate produced by the enzymatic action of myrosinase on glucoraphanin, a glucosinolate contained in cruciferous vegetables [74], and acetonic and methanolic extracts of *Heterotheca inuloides*, can be administered with ample safety margin in patients treated with PRP [75].

7. Advantages, Limitations, and Precautions

PRP as autologous procedure eliminates secondary effects and unnecessary risks for chemically processed strange molecules, is a natural reserver of various growth factors that can be collected autologously, and is costeffective [76–81] (see Table 3). Thus for clinical use, no special considerations concerning antibody formation and infection risk are needed. The key of our health and our own regeneration resides in our own body. Nevertheless this treatment is not the panacea, it is only the beginning in this new age of the regenerative medicine. Some clinical devices to automatically prepare PRPs are available at present. PRP are consistently being used clinically in the department of orthopedics and plastic surgery (oral, maxillary facial) for a long time. On the basis of research evidence, some publications have reported positive results in either bone or soft tissue healing. It is recommended to avoid the abuse and use generalized of this procedure before any disease. Until now, their clinical applications still are limited. However, some research concludes that there is no or little benefit from PRP. This is likely due to faster degradation of growth factors in PRP since some authors suggest using sustained release form of PRP to deliver optimal effect of PRP. Gelatin hydrogel is also being used clinically as a slow, sustained release of carrier for growth factors in our center recently [47].

8. Conclusions

PRP as therapeutic option is a powerful tool nowadays for the localized delivery of great variety of biologically active GF to the site of injury and is supported by its simplicity, potential cost effectiveness, safety, and permanent availability [11].

Platelet concentrates are potentially useful in wound-healing applications because they function as both a tissue sealant and a drug delivery system that contains a host of powerful mitogenic and chemotactic GFs. However, the method of PRP preparation has a potentially significant impact on the different levels of platelet recovery and activation [10]. Platelet activation during preparation of the platelet concentrate can result in early alpha granule release and loss of the GF during the collection process. It is therefore critical to recognize that each PRP preparation method may differ in regard to platelet number, platelet activation rates, and GF profiles [15].

In this regard, therefore, it is critical to define the extent of platelet activation that occurs during graft preparation. If platelets become activated and release the contents of the alpha granules during the centrifugation process, the GF will be diluted and lost into the plasma. To ensure that platelets are intact until the PRP fraction has been collected, platelet surface marker for platelet activation P selectin can be measured [30]. However, in all methods, the application of PRP is really providing a sufficient dose of these useful bioactive peptides for wound healing and regenerative process [82].

Acknowledgments

This work was supported by award SNI-33834, to D. J. Sánchez-González and SNI-38264 to E. Méndez-Bolaina from CONACYT Investigators National System from México and International Society for Cellular Therapy with Stem Cells, Regenerative Medicine and Anti-aging (SITECEM, México).

References

[1] G. Bizzozero, "Su di un nuovo elemento morfologico del sangue dei mammiferi e della sua importanza nella trombosi e nella coagulazione," *L'Osservatore*, vol. 17, pp. 785–787, 1881.

[2] J. H. Wright, "The origin and nature of blood plates," *Boston Medical Surgical Journal*, vol. 154, pp. 643–645, 1906.

[3] A. S. Weyrich, H. Schwertz, L. W. Kraiss, and G. A. Zimmerman, "Protein synthesis by platelets: historical and new perspectives," *Journal of Thrombosis and Haemostasis*, vol. 7, no. 2, pp. 241–246, 2009.

[4] J. A. Coppinger, G. Cagney, S. Toomey et al., "Characterization of the proteins released from activated platelets leads to localization of novel platelet proteins in human atherosclerotic lesions," *Blood*, vol. 103, no. 6, pp. 2096–2104, 2004.

[5] A. H. Qureshi, V. Chaoji, D. Maiguel et al., "Proteomic and phospho-proteomic profile of human platelets in basal, resting state: insights into integrin signaling," *PLoS One*, vol. 4, no. 10, Article ID e7627, 2009.

[6] R. L. Nachman and S. Rafii, "Platelets, petechiae, and preservation of the vascular wall," *The New England Journal of Medicine*, vol. 359, no. 12, pp. 1261–1270, 2008.

[7] K. Jurk and B. E. Kehrel, "Platelets: physiology and biochemistry," *Semin Thromb Hemost*, vol. 31, pp. 381–392, 2005.

[8] A. Mishra, J. Velotta, T. J. Brinton et al., "RevaTen platelet-rich plasma improves cardiac function after myocardial injury," *Cardiovascular Revascularization Medicine*, vol. 12, no. 3, pp. 158–163, 2011.

[9] P. Borzini and L. Mazzucco, "Platelet-rich plasma (PRP) and platelet derivatives for topical therapy. What is true from the biological view point?" *ISBT Science Series*, vol. 2, pp. 272–281, 2007.

[10] B. Cole and S. Seroyer, "Platelet-rich plasma: where are we now and where are we going?" *Sports Health*, vol. 2, no. 3, pp. 203–210, 2010.

[11] J. P. Fréchette, I. Martineau, and G. Gagnon, "Platelet-rich plasmas: growth factor content and roles in wound healing," *Journal of Dental Research*, vol. 84, no. 5, pp. 434–439, 2005.

[12] A. Kocaoemer, S. Kern, H. Klüter, and K. Bieback, "Human AB serum and thrombin-activated platelet-rich plasma are suitable alternatives to fetal calf serum for the expansion of mesenchymal stem cells from adipose tissue," *Stem Cells*, vol. 25, no. 5, pp. 1270–1278, 2007.

[13] N. Borregaard and J. B. Cowland, "Granules of the human neutrophilic polymorphonuclear leukocyte," *Blood*, vol. 89, no. 10, pp. 3503–3521, 1997.

[14] F. Rendu and B. Brohard-Bohn, "The platelet release reaction: granules' constituents, secretion and functions," *Platelets*, vol. 12, no. 5, pp. 261–273, 2001.

[15] E. Anitua, I. Andia, B. Ardanza, P. Nurden, and A. T. Nurden, "Autologous platelets as a source of proteins for healing and tissue regeneration," *Thrombosis and Haemostasis*, vol. 91, no. 1, pp. 4–15, 2004.

[16] A. Garcia, N. Zitzmann, and S. P. Watson, "Analyzing the platelet proteome," *Seminars in Thrombosis and Hemostasis*, vol. 30, no. 4, pp. 485–489, 2004.

[17] A. I. Mininkova, "Platelet structure and functions (a review of literature). Part 1," *Klinichescheskaya Laboratornaya Diagnostika*, no. 11, pp. 21–26, 2010.

[18] A. I. Mininkova, "Investigation of platelets by the flow cytofluorometric technique (a review of literature). Part 2," *Klinichescheskaya Laboratornaya Diagnostika*, no. 4, pp. 25–30, 2011.

[19] O. Mei-Dan, L. Laver, M. Nyska, and G. Mann, "Platelet rich plasma—a new biotechnology for treatment of sports injuries," *Harefuah*, vol. 150, no. 5, pp. 453–457, 2011.

[20] A. T. Nurden, "Platelets, inflammation and tissue regeneration," *Thrombosis and Haemostasis*, vol. 105 suppl 1, pp. S13–S33, 2011.

[21] R. E. Marx, E. R. Carlson, R. M. Eichstaedt, S. R. Schimmele, J. E. Strauss, and K. R. Georgeff, "Platelet-rich plasma: growth factor enhancement for bone grafts," *Oral Surgery, Oral Medicine, Oral Pathology, Oral Radiology, and Endodontics*, vol. 85, no. 6, pp. 638–646, 1998.

[22] L. Brass, "Understanding and evaluating platelet function," *Hematology / the Education Program of the American Society of Hematology Education Program*, vol. 2010, pp. 387–396, 2010.

[23] Y. Liu, A. Kalen, O. Risto, and O. Wahlström, "Fibroblast proliferation due to exposure to a platelet concentrate in vitro is pH dependent," *Wound Repair and Regeneration*, vol. 10, no. 5, pp. 336–340, 2002.

[24] C. A. Carter, D. G. Jolly, C. E. Worden, D. G. Hendren, and C. J. M. Kane, "Platelet-rich plasma gel promotes differentiation and regeneration during equine wound healing," *Experimental and Molecular Pathology*, vol. 74, no. 3, pp. 244–255, 2003.

[25] B. L. Eppley, W. S. Pietrzak, and M. Blanton, "Platelet-rich plasma: a review of biology and applications in plastic surgery," *Plastic and Reconstructive Surgery*, vol. 118, no. 6, pp. 147e–159e, 2006.

[26] K. M. Lacci and A. Dardik, "Platelet-rich plasma: support for its use in wound healing," *Yale Journal of Biology and Medicine*, vol. 83, no. 1, pp. 1–9, 2010.

[27] T. E. Foster, B. L. Puskas, B. R. Mandelbaum, M. B. Gerhardt, and S. A. Rodeo, "Platelet-rich plasma: from basic science to clinical applications," *American Journal of Sports Medicine*, vol. 37, no. 11, pp. 2259–2272, 2009.

[28] S. Werner and R. Grose, "Regulation of wound healing by growth factors and cytokines," *Physiological Reviews*, vol. 83, no. 3, pp. 835–870, 2003.

[29] R. E. Marx, "Platelet-rich plasma (PRP): what is PRP and what is not PRP?" *Implant Dentistry*, vol. 10, no. 4, pp. 225–228, 2001.

[30] B. L. Eppley, J. E. Woodell, and J. Higgins, "Platelet quantification and growth factor analysis from platelet-rich plasma: implications for wound healing," *Plastic and Reconstructive Surgery*, vol. 114, no. 6, pp. 1502–1508, 2004.

[31] S. J. Froum, S. S. Wallace, D. P. Tarnow, and S. C. Cho, "Effect of platelet-rich plasma on bone growth and osseointegration in human maxillary sinus grafts: three bilateral case reports," *International Journal of Periodontics and Restorative Dentistry*, vol. 22, no. 1, pp. 45–53, 2002.

[32] A. R. Sanchez, P. J. Sheridan, and L. I. Kupp, "Is platelet-rich plasma the perfect enhancement factor? A current review," *International Journal of Oral and Maxillofacial Implants*, vol. 18, no. 1, pp. 93–103, 2003.

[33] T. F. Tozum and B. Demiralp, "Platelet-rich plasma: a promising innovation in dentistry," *Journal Canadian Dental Association*, vol. 69, no. 10, p. 664, 2003.

[34] E. G. Freymiller and T. L. Aghaloo, "Platelet-rich plasma: ready or not?" *Journal of Oral and Maxillofacial Surgery*, vol. 62, no. 4, pp. 484–488, 2004.

[35] M. I. Furman, L. Liu, S. E. Benoit, R. C. Becker, M. R. Barnard, and A. D. Michelson, "The cleaved peptide of the thrombin receptor is a strong platelet agonist," *Proceedings of the National Academy of Sciences of the United States of America*, vol. 95, no. 6, pp. 3082–3087, 1998.

[36] J. P. Maloney, C. C. Silliman, D. R. Ambruso, J. Wang, R. M. Tuder, and N. F. Voelkel, "In vitro release of vascular endothelial growth factor during platelet aggregation," *American Journal of Physiology*, vol. 275, no. 3, pp. H1054–H1061, 1998.

[37] N. T. Bennett and G. S. Schultz, "Growth factors and wound healing: biochemical properties of growth factors and their receptors," *American Journal of Surgery*, vol. 165, no. 6, pp. 728–737, 1993.

[38] D. R. Knighton, T. K. Hunt, K. K. Thakral, and W. H. Goodson, "Role of platelets and fibrin in the healing sequence: an in vivo study of angiogenesis and collagen synthesis," *Annals of Surgery*, vol. 196, no. 4, pp. 379–388, 1982.

[39] S. Cohen, "Isolation of a mouse submaxillary gland protein accelerating incisor eruption and eyelid opening in the new born animal," *Journal of Biological Chemistry*, vol. 237, pp. 1555–1562, 1962.

[40] G. L. Brown, L. B. Nancy, J. Griffen et al., "Enhancement of wound healing by topical treatment with epidermal growth factor," *The New England Journal of Medicine*, vol. 321, no. 2, pp. 76–79, 1989.

[41] G. Hosgood, "Wound healing: the role of platelet-derived growth factor and transforming growth factor beta," *Veterinary Surgery*, vol. 22, no. 6, pp. 490–495, 1993.

[42] H. N. Antoniades and L. T. Williams, "Human platelet-derived growth factor: structure and function," *Federation Proceedings*, vol. 42, no. 9, pp. 2630–2634, 1983.

[43] J. Floege, F. Eitner, and C. E. Alpers, "A new look at platelet-derived growth factor in renal disease," *Journal of the American Society of Nephrology*, vol. 19, no. 1, pp. 12–23, 2008.

[44] O. Wahlström, C. Linder, A. Kalén, and P. Magnusson, "Acidic preparations of platelet concentrates release bone morphogenetic protein-2," *Acta Orthopaedica*, vol. 79, no. 3, pp. 433–437, 2008.

[45] C. Betsholtz, "Role of platelet-derived growth factors in mouse development," *International Journal of Development Biology*, vol. 39, no. 5, pp. 817–825, 1995.

[46] S. C. Bir, J. Esaki, A. Marui et al., "Therapeutic treatment with sustained-release platelet-rich plasma restores blood perfusion by augmenting ischemia-induced angiogenesis and arteriogenesis in diabetic mice," *Journal of Vascular Research*, vol. 48, no. 3, pp. 195–205, 2011.

[47] S. C. Bir, J. Esaki, A. Marui et al., "Angiogenic properties of sustained release platelet-rich plasma: characterization in vitro and in the ischemic hind limb of the mouse," *Journal of Vascular Surgery*, vol. 50, no. 4, pp. 870–879, 2009.

[48] A. Mishra and T. Pavelko, "Treatment of chronic elbow tendinosis with buffered platelet-rich plasma," *American Journal of Sports Medicine*, vol. 34, no. 11, pp. 1774–1778, 2006.

[49] G. Weibrich, W. K. Kleis, G. Hafner, and W. E. Hitzler, "Growth factor levels in platelet-rich plasma and correlations with donor age, sex, and platelet count," *Journal of Craniomaxillofacial Surgery*, vol. 30, no. 2, pp. 97–102, 2002.

[50] M. M. Rechler and S. P. Nissley, "Insulin-like growth factors," in *Handbook of Experimental Pharm: Peptide Growth Factors and Their Receptors*, M. B. Sporn and A. B. Roberts, Eds., vol. 96, pp. 263–367, Springer, Berlin, Germany, 1990.

[51] G. F. Pierce, T. A. Mustoe, J. Lingelbach, V. R. Masakowski, P. Gramates, and T. F. Deuel, "Transforming growth factor β reverses the glucocorticoid-induced wound healing deficit in rats. Possible regulation in macrophages by platelet-derived growth factor," *Proceedings of the National Academy of Sciences of the United States of America*, vol. 86, no. 7, pp. 2229–2233, 1989.

[52] J. M. Pérez-Rojas, C. Cruz, P. García-López et al., "Renoprotection byα-mangostin is related to the attenuation in renal oxidative/nitrosative stress induced by cisplatin nephrotoxicity," *Free Radical Research*, vol. 43, no. 11, pp. 1122–1132, 2009.

[53] M. G. Goldner, "The fate of the second leg in the diabetic amputee," *Diabetes*, vol. 9, pp. 100–103, 1960.

[54] J. Yu, C. Ustach, and H. R. Kim, "Platelet-derived growth factor signaling and human cancer," *Journal of Biochemistry and Molecular Biology*, vol. 36, no. 1, pp. 49–59, 2003.

[55] S. P. Bennett, G. D. Griffiths, A. M. Schor, G. P. Leese, and S. L. Schor, "Growth factors in the treatment of diabetic foot ulcers," *British Journal of Surgery*, vol. 90, no. 2, pp. 133–146, 2003.

[56] E. M. Spencer, A. Tokunaga, and T. K. Hunt, "Insulin-like growth factor binding protein-3 is present in the α-granules of platelets," *Endocrinology*, vol. 132, no. 3, pp. 996–1001, 1993.

[57] T. McCarrel and L. Fortier, "Temporal growth factor release from platelet-rich plasma, trehalose lyophilized platelets, and bone marrow aspirate and their effect on tendon and ligament gene expression," *Journal of Orthopaedic Research*, vol. 27, no. 8, pp. 1033–1042, 2009.

[58] H. F. Dvorak, "Discovery of vascular permeability factor (VPF)," *Experimental Cell Research*, vol. 312, no. 5, pp. 522–526, 2006.

[59] T. Tammela, B. Enholm, K. Alitalo, and K. Paavonen, "The biology of vascular endothelial growth factors," *Cardiovascular Research*, vol. 65, no. 3, pp. 550–563, 2005.

[60] S. Lee, T. T. Chen, C. L. Barber et al., "Autocrine VEGF signaling is required for vascular homeostasis," *Cell*, vol. 130, no. 4, pp. 691–703, 2007.

[61] P. Carmeliet, V. Ferreira, G. Breier et al., "Abnormal blood vessel development and lethality in embryos lacking a single VEGF allele," *Nature*, vol. 380, no. 6573, pp. 435–439, 1996.

[62] B. Olofsson, M. Jeltsch, U. Eriksson, and K. Alitalo, "Current biology of VEGF-B and VEGF-C," *Current Opinion in Biotechnology*, vol. 10, no. 6, pp. 528–535, 1999.

[63] T. A. Partanen, J. Arola, A. Saaristo et al., "VEGF-C and VEGF-D expression in neuroendocrine cells and their receptor, VEGFR-3, in fenestrated blood vessels in human tissues," *FASEB Journal*, vol. 14, no. 13, pp. 2087–2096, 2000.

[64] M. Klagsbrun, S. Takashima, and R. Mamluk, "The role of neuropilin in vascular and tumor biology," *Advances in Experimental Medicine and Biology*, vol. 515, pp. 33–48, 2002.

[65] R. J. Levine, S. E. Maynard, C. Qian et al., "Circulating angiogenic factors and the risk of preeclampsia," *The New England Journal of Medicine*, vol. 350, no. 7, pp. 672–683, 2004.

[66] J. Folkman, "Angiogenesis: an organizing principle for drug discovery?" *Nature Reviews Drug Discovery*, vol. 6, no. 4, pp. 273–286, 2007.

[67] R. Möhle, D. Green, M. A. Moore, R. L. Nachman, and S. Rafii, "Constitutive production and thrombin-induced release of vascular endothelial growth factor by human megakaryocytes and platelets," *Proceedings of the National Academy of Sciences of the United States of America*, vol. 94, no. 2, pp. 663–668, 1997.

[68] M. Lucerna, A. Zernecke, R. de Nooijer et al., "Vascular endothelial growth factor-A induces plaque expansion in ApoE

knock-out mice by promoting de novo leukocyte recruitment," *Blood*, vol. 109, no. 1, pp. 122–129, 2007.

[69] D. J. Sánchez-González, C. A. Sosa-Luna, and I. Vásquez-Moctezuma, "Transfer factors in medical therapy," *Medicina Clinica*, vol. 137, no. 6, pp. 273–277, 2011.

[70] S. Verrier, T. R. Meury, L. Kupcsik, P. Heini, T. Stoll, and M. Alini, "Platelet-released supernatant induces osteoblastic differentiation of human mesenchymal stem cells: potential role of BMP-2," *European Cells & Materials*, vol. 20, pp. 403–414, 2010.

[71] A. Sobotková, L. Másová-Chrastinová, J. Suttnar et al., "Antioxidants change platelet responses to various stimulating events," *Free Radical Biology and Medicine*, vol. 47, no. 12, pp. 1707–1714, 2009.

[72] E. Floriano-Sánchez, C. Villanueva, O. N. Medina-Campos et al., "Nordihydroguaiaretic acid is a potent in vitro scavenger of peroxynitrite, singlet oxygen, hydroxyl radical, superoxide anion and hypochlorous acid and prevents in vivo ozone-induced tyrosine nitration in lungs," *Free Radical Research*, vol. 40, no. 5, pp. 523–533, 2006.

[73] C. Cruz, R. Correa-Rotter, D. J. Sánchez-González et al., "Renoprotective and antihypertensive effects of S-allylcysteine in 5/6 nephrectomized rats," *American Journal of Physiology*, vol. 293, no. 5, pp. F1691–F1698, 2007.

[74] C. E. Guerrero-Beltrán, M. Calderón-Oliver, E. Tapia et al., "Sulforaphane protects against cisplatin-induced nephrotoxicity," *Toxicology Letters*, vol. 192, no. 3, pp. 278–285, 2010.

[75] E. Coballase-Urrutia, J. Pedraza-Chaverri, N. Cárdenas-Rodríguez et al., "Hepatoprotective effect of acetonic and methanolic extracts of Heterotheca inuloides against CCl(4)-induced toxicity in rats," *Experimental and Toxicologic Pathology*, vol. 63, no. 4, pp. 363–370, 2011.

[76] M. Sánchez, E. Anitua, J. Azofra, I. Andía, S. Padilla, and I. Mujika, "Comparison of surgically repaired achilles tendon tears using platelet-rich fibrin matrices," *American Journal of Sports Medicine*, vol. 35, no. 2, pp. 245–251, 2007.

[77] L. Savarino, E. Cenni, C. Tarabusi et al., "Evaluation of bone healing enhancement by lyophilized bone grafts supplemented with platelet gel: a standardized methodology in patients with tibial osteotomy for genu varus," *Journal of Biomedical Materials Research*, vol. 76, no. 2, pp. 364–372, 2006.

[78] D. Dallari, L. Savarino, C. Stagni et al., "Enhanced tibial osteotomy healing with use of bone grafts supplemented with platelet gel or platelet gel and bone marrow stromal cells," *Journal of Bone and Joint Surgery*, vol. 89, no. 11, pp. 2413–2420, 2007.

[79] H. Kitoh, T. Kitakoji, H. Tsuchiya, M. Katoh, and N. Ishiguro, "Transplantation of culture expanded bone marrow cells and platelet rich plasma in distraction osteogenesis of the long bones," *Bone*, vol. 40, no. 2, pp. 522–528, 2007.

[80] L. Y. Carreon, S. D. Glassman, Y. Anekstein, and R. M. Puno, "Platelet gel (AGF) fails to increase fusion rates in instrumented posterolateral fusions," *Spine*, vol. 30, no. 9, pp. E243–E247, 2005.

[81] M. Sanchez, E. Anitua, R. Cugat et al., "Nonunions treated with autologous preparation rich in growth factors," *Journal of Orthopaedic Trauma*, vol. 23, no. 1, pp. 52–59, 2009.

[82] P. R. Siljander, "Platelet-derived microparticles - an updated perspective," *Thrombosis Research*, vol. 127, no. 2, suppl 2, pp. S30–S33, 2011.

α-RgIB: A Novel Antagonist Peptide of Neuronal Acetylcholine Receptor Isolated from *Conus regius* Venom

Maria Cristina Vianna Braga,[1,2] **Arthur Andrade Nery,**[3] **Henning Ulrich,**[3]
Katsuhiro Konno,[4] **Juliana Mozer Sciani,**[5] **and Daniel Carvalho Pimenta**[5]

[1] *CAT/CEPID, Instituto Butantan, Avenida Vital Brasil 1500, 05503-900 São Paulo, SP, Brazil*
[2] *Ministério da Ciência, Tecnologia e Inovação, Esplanada dos Ministérios, Bloco E, 70067-900 Brasília, DF, Brazil*
[3] *Departamento de Bioquímica, Instituto de Química, Universidade de São Paulo, Av. Lineu Prestes 748,
  05508-900 São Paulo, SP, Brazil*
[4] *Institute of Natural Medicine, University of Toyama, 2630 Sugitani, Toyama 930-0194, Japan*
[5] *Laboratório de Bioquímica e Biofísica, Instituto Butantan, Avenida Vital Brasil 1500, 05503-900 São Paulo, SP, Brazil*

Correspondence should be addressed to Daniel Carvalho Pimenta; dcpimenta@butantan.gov.br

Academic Editor: Ayman El-Faham

Conus venoms are rich sources of biologically active peptides that act specifically on ionic channels and metabotropic receptors present at the neuromuscular junction, efficiently paralyzing the prey. Each species of *Conus* may have 50 to 200 uncharacterized bioactive peptides with pharmacological interest. *Conus regius* is a vermivorous species that inhabits Northeastern Brazilian tropical waters. In this work, we characterized one peptide with activity on neuronal acetylcholine receptor (nAChR). Crude venom was purified by reverse-phase HPLC and selected fractions were screened and sequenced by mass spectrometry, MALDI-ToF, and ESI-Q-ToF, respectively. A new peptide was identified, bearing two disulfide bridges. The novel 2,701 Da peptide belongs to the cysteine framework I, corresponding to the cysteine pattern CC-C-C. The biological activity of the purified peptide was tested by intracranial injection in mice, and it was observed that high concentrations induced hyperactivity in the animals, whereas lower doses caused breathing difficulty. The activity of this peptide was assayed in patch-clamp experiments, on nAChR-rich cells, in whole-cell configuration. The peptide blocked slow rise-time neuronal receptors, probably α3β4 and/or α3β4α5 subtype. According to the nomenclature, the new peptide was designated as α-RgIB.

1. Introduction

Marine mollusks from *Conus* genus may produce from 50 up to 200 biologically active molecules that can be injected in the prey to capture or be employed as defense and/or escape mechanisms to deter competitors. The peptide toxins, called conopeptides, are composed of 10–40 amino acids (including nonnatural amino acids) and are abundant in the venom. Peptides presenting a rigid structure due to more than one disulfide bridges are common, being called conotoxins. These peptides act specifically on ionic channels and/or neuromuscular receptors [1, 2].

Conotoxins are classified according to three schemes: the similarities between the endoplasmatic reticulum signal sequence of the conotoxin precursors (gene superfamilies), the cysteine patterns of conotoxin mature peptide regions (cysteine frameworks), and the specificities to pharmacological targets (pharmacological families) [3, 4].

Conopeptides of the pharmacological family α, which acts on neuronal acetylcholine receptor, have been found in the A, D, L, M, and S gene superfamilies [5, 6].

Typically, α-conotoxins are peptides with 12 to 16 amino acid residues and two disulfide bridges, presenting the pattern CC-C-C. These peptides are competitive antagonists of the nicotinic acetylcholine receptors (nAChR) and display high selectivity by subtypes of this receptor [5, 7–9]. After the blockage of the muscular acetylcholine receptor, the α-conotoxins significantly decrease the amplitude of the motor end plate postsynaptic potentials in vertebrates, paralyzing the prey [10].

In the Brazilian tropical coast, there are approximately 18 species of cone snails [11]. *Conus regius* (Gmelin, 1791) is a vermivorous species that inhabits rock and coral deep waters of Florida (USA), Central America, and the Northeast and East coast of Brazil, including Fernando de Noronha archipelago [12].

In this work we described a novel peptide from *Conus regius* venom, belonging to the α-conotoxins family. This peptide blocks the neuronal acetylcholine receptors on PC12 cells, which comprise $\alpha 3\beta 4$ and/or $\alpha 3\beta 4\alpha 5$ subtypes receptors, probably target of the peptide.

2. Material and Methods

2.1. Reagents. All the employed reagents were of analytical grade and were purchased from Sigma Co (St Louis, MO, USA), unless otherwise stated.

2.2. Animals and Venom. Specimens of *C. regius* were collected at Fernando de Noronha Archipelago, Pernambuco, Brazil. The Brazilian Environmental Agency (IBAMA—Instituto Brasileiro do Meio Ambiente e dos Recursos Naturais Renováveis) license numbers were 030/2000 and 087/2001, and the process number was 02001, 000775/00-00. Venom was extracted from the specimens as previously described [13]. The crude venom was obtained by dissection of the venom duct gland and then freeze-dried and stored at −80°C.

Voucher material is deposited in the malacological collection of Zoology Museum of University of São Paulo, São Paulo, Brazil.

2.3. Peptide Fractionation and Purification. A reversed-phase binary HPLC system (LC-8A, Shimadzu Co., Japan) was used for sample fractionation. The lyophilized crude venom powder was solubilized into 0.1% trifluoroacetic acid (TFA) and aliquots were loaded in a Shim-pack Prep-ODS C18 column (Shimadzu, 3 μm, C18, 300 Å, 250 × 20 mm) in a two-solvent system: (A) trifluoroacetic acid (TFA)/H_2O (1:1000) and (B) TFA/Acetonitrile (ACN)/H_2O (1:900:100). The sample was eluted at a constant flow rate of 8 mL·min^{-1} with a 0 to 60% gradient of solvent B over 60 min. The HPLC column eluates were monitored by a Shimadzu SPD-10A detector scanning 220 nm.

For α-RgIB purification, the interest peak was fractionated in a Merck C18 column (300 × 4.6 mm), in a 19 to 21% B gradient over 20 min, at a constant flow rate of 8 mL·min^{-1}. A subsequent purification step was still necessary to obtain the peptide. This purification was conducted in a Merck C18 column (300 × 4.6 mm), in an isocratic elution at 35% B (TFA/methanol/H_2O 1:900:100) at a constant flow rate of 1 mL·min^{-1}.

2.4. Mass Spectrometry Analysis. Molecular mass analyses of the peaks and the peptides were performed on a micro-LC-MS Ettan (Amersham Biosciences, Sweden) coupled in

a Q-ToF Ultima API (Micromass, Manchester, UK) and/or by MALDI-TOF mass spectrometry on a Ettan MALDI-ToF/Pro System (Amersham Biosciences, Sweden).

The analysis in the micro-LC-MS Ettan (Amersham Biosciences, Sweden) was performed in a μRPC C2/C18 ST 1.0/150 column (Amersham Biosciences, Sweden), with two solvents: (A) formic acid (FA)/H_2O (1:1000) and (B) FA/ACN/H_2O (1:900:100). The sample was eluted at a constant flow rate of 50 μL·min^{-1} with a 5 to 65% gradient of solvent B over 60 min. Q-Tof operated under positive ionization mode. For MALDI-TOF analyses, a-cyano-4-hydroxycinnamic acid was used as matrix.

2.5. "De Novo" Peptide Sequencing. Mass spectrometric "de novo" peptide sequencing was carried out in positive ionization mode on a Q-TOF Ultima API fitted with an electrospray ion source (Micromass, Manchester, UK). Briefly, the amounts of previously lyophilized peptide were dissolved in 50 mM ammonium acetate, reduced with 50 mM DTT, alkylated by 150 mM iodoacetamide, and hydrolyzed by 25 nM trypsin, according to slight modifications of Westermeier and Naven [14]. The reaction products were then lyophilized and dissolved in 50% ACN, containing 0.1% FA and injected into the source at 5 μL·min^{-1} by a Hamilton infusion pump, or directly injected using a Rheodyne 7010 sample loop coupled to a LC-10A VP Shimadzu pump operating at 20 μL·min^{-1} constant flow rate. The instrument control and data acquisition were conducted by MassLynx 4.0 data system (Micromass, Manchester, UK) and experiments were performed by scanning a mass-to-charge ratio (m/z) of 50–1800 using a scan time of 2 s applied during the whole chromatographic process. The mass spectra corresponding to each signal from the total ion current (TIC) chromatogram were averaged, allowing an accurate molecular mass determination. External calibration of the mass scale was performed with NaI. For the MS/MS analysis, collision energy ranged from 18 to 45 and the precursor ions were selected under a 1-m/z window.

2.6. Biological Activity. The biological activity of α-RgIB was determined in Swiss Webster mice (5.5 to 7 g body weight) by observation of the behavioral disorders after intracranial injection [15] of the peptide diluted in NaCl 0.9%, in concentration of 0.1, 0.5 and 1 nmol. Alterations were compared to animals injected with NaCl 0.9% (control). All animals were observed by 60 min.

2.7. Patch Clamp. BC$_3$H1 cells, mouse myocytes which express nicotinic acetylcholine receptors, were acquired by ATCC (CRL-1443) and maintained in culture according to Sine and Taylor [16] to electrophysiological experiments. In order to verify the subtype of neuronal nicotinic receptor that the peptide acts, PC12 cells were employed and maintained in culture according to Greene et al. [17].

Individual cells were subjected to a patch-clamp, at a whole cell configuration, according to Hamill et al. [18] and Urlich et al. [19]. Cells were maintained in an extracellular solution containing 25 mM HEPES, 5.3 mM KCl, 144.8 mM

NaCl, 1.2 mM $MgCl_2$, 2.38 mM $CaCl_2$, and 10 mM glucose (pH 7.4). A recording electrode was filled with intracellular solution containing 25 mM HEPES, 141 mM KCl, 10 mM NaCl, 2 mM $MgCl_2$, and 1 mM EGTA (pH 7.4). Experiments were carried out at room temperature (20–24°C).

Throughout the experiment, the membrane potential was clamped at a −60 mV for BC_3H_1 cells and −70 mV for PC12 cells, holding potential using an Axon Axopatch amplifier (Molecular Devices, California, USA). Data were recorded and digitized by Clampex 8.2 software (Molecular Devices, California, USA) and plots were made using Origin 7.0 software (OriginLab Corp., Northampton, MA).

Control currents were performed with 1.5 mM carbamylcholine, using the cell-flow technique [20]. After the carbamylcholine administration, the peptide (10 μM) was incubated on cells, and then another dose of the agonist was incubated [21].

2.8. Data Fitting, Statistical Analyses, and Sequence Alignment. When data fitting was performed, results were presented as the calculated value ± standard deviation (SD). Otherwise, data correspond to the mean of three individual experiments. Peptide sequence alignment was performed using ClustalW software [22]. The 3D model of the peptide α-RgIB, as well as three PDB deposited 3D solution structures, was created by I-TASSER [23, 24].

3. Results

3.1. Purification. The crude venom from *C. regius* was fractionated by RP-HPLC, as shown in Figure 1(a). Some peaks could be detected along the profile, and the arrow indicates the peak of interest. Two subsequent chromatographic steps were necessary to purify the peptide (Figures 1(b) and 1(c)), under the conditions described material and methods section. After the third step of purification, the purity and the molecular mass of the peptide were assessed by MALDI-TOF/MS (Figure 1(d)).

3.2. "De Novo" Peptide Sequencing. After cysteine bridge reduction and alkylation, the reaction product was digested with trypsin. The obtained peptides were submitted to MS/MS analyses (Figure 2) and ions were selected and fragmented by collision with argon (CIF), yielding daughter ion spectra (Figure 3) that was processed with BioLynx and manually checked for accuracy of interpretation. Since the digestion allowed peptides with missed cleavage sites, it was possible to assemble the fragments without the aid of another digestion with a different enzyme. The sequenced peptides, their charge states, and theoretical molecular mass are presented in Table 1.

The peptide sequence was determined to be TWEECCKNPGCRNNHVDRCRGQV. This sequence has 4 cysteine residues with pattern CC-C-C, typical from conotoxins of framework I [25]. This peptide was named α-RgIB, according to the guidelines for conotoxins nomenclature ConoServer and has been assigned the following UNIPROT accession number: C0HJA8 [3, 6, 26].

3.3. Sequence Features. A sequence alignment was performed with all α-conotoxins available at UniProt (supplementary Table 1 of the supplementary material available online at http://dx.doi.org/10.1155/2013/543028). Based on this large alignment, a phylogeny was constructed (supplementary Figure 1) and the peptide sequences present at the branch containing α-RgIB were realigned (Figure 4). ClustalW standard annotations consider, as expected, the Cys residues as consensus (*), and the Glu at the 8th aligned position, as being highly conserved (:). Moreover, according to the algorithm standard notation, the Pro residue at the 13th aligned position is also conserved (.). Among the UNIPROT database, the SwissModel tool could not identify any suitable template for structure prediction of α-RgIB, therefore an external application was used. Figure 5 presents a 3D model of α-RgIB; created by I-TASSER [23, 24], as well as three PDB deposited 3D solution structures of α-RgIA (P0C1D0) mutants, a conotoxin that specifically and potently blocks the $\alpha 9\alpha 10$ nAChR [27]. In spite of α-RgIB N- and C-terminal extensions and longer interbridge peptide sequence, the model and the structures are tridimensionally related, for example, a C-shaped structure, held by the Cys-bridges.

3.4. Biological Activity. The *in vivo* biological activity of the peptide was assessed by means of intracranial injection in Swiss Webster mice. Following 1 nmol injection, the animals displayed a hyperactive behavior, defecating and urinating all the time, which was not observed for the control group that received saline solution. Auditory stimuli, for example, a hand-clap or hitting the cage, also triggered the hyperactive behavior. Interestingly, the lower doses (0.1 and 0.5 nmol), caused the animals to have difficulty in breathing. Although the peptide promoted behavioral disorders, it was not lethal to the animals.

3.5. Patch Clamp. Whole-cell voltage clamp measurement was used to verify the ion currents on acetylcholine receptors. BC_3H1 cells, which express the acetylcholine muscle type receptors on the surface, and PC12, which terminally differentiate in neurons and express nicotinic neuronal receptors [21] were selected for the experiments. Carbamylcholine, a stable and well-characterized analogue of acetylcholine, was used as an agonist [28], for it elicits a fast activating current that rapidly desensitizes during the application.

10 μM α-RgIB was not able to induce any change in the ion currents on BC_3H_1 cells (data not shown), as well as a higher dose (30 μM) of the peptide. d-tubocurarine (a classic nicotinic receptor antagonist) was used as a positive control and successfully to block this channel (data not shown).

After the incubation of the peptide with PC12 cells, fast and slow desensitization of the receptor was observed. Figure 6(b) shows that on neuronal slow rise-time receptors, α-RgIB is able to block the ion current by 40%, compared to cells stimulated with carbamylcholine (Figure 6(a)). The blockage was irreversible and persistent, once the current

(a)

(b)

(c)

(d)

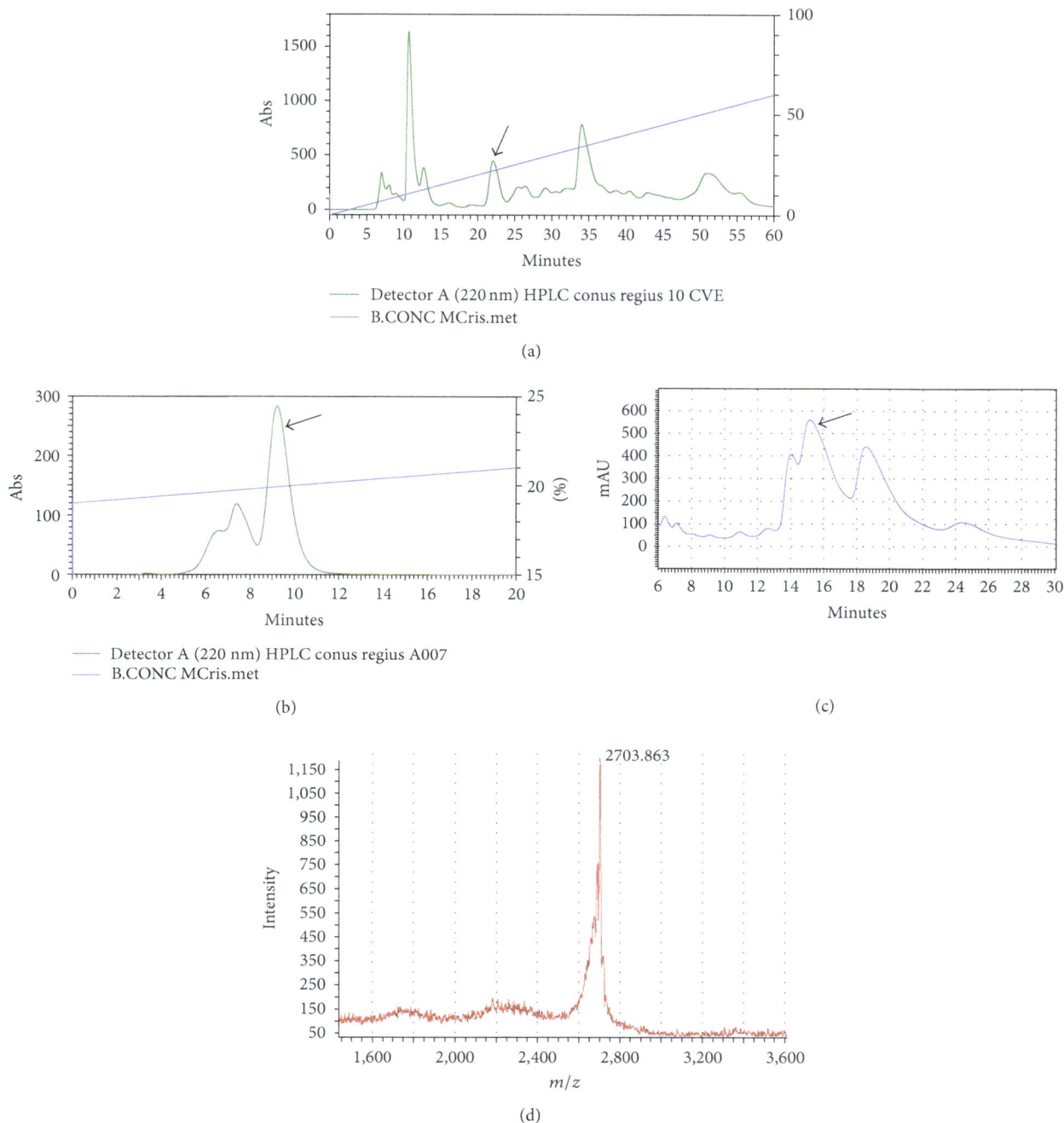

FIGURE 1: (a) Representative RP-HPLC of the crude *C. regius* venom. The arrow indicates the peak of interest. (b) Representative RP-HPLC of the selected peak (arrow), indicating the presence of impurities. (c) Isocratic elution of the isolated the selected fraction from chromatogram B. The arrow indicates the peak of interest. (d) MALDI-TOF/MS profile of the purified peptide.

does not recover after a new application of the agonist, carbamylcholine (Figure 6(c)).

4. Discussion

Conotoxins are classified according to the similarities between the signal sequence of the conotoxin precursors (gene superfamilies), the cysteine patterns of conotoxin mature peptide regions (cysteine frameworks), and the specificities to pharmacological targets [3, 4].

This new peptide was termed α-Rg-IB because the peptide acts on neuronal acetylcholine receptors ("α"), was extracted from a *Conus regius* specimen ("Rg"), displays a cysteine framework I—CC-C-C ("I"), and was the second

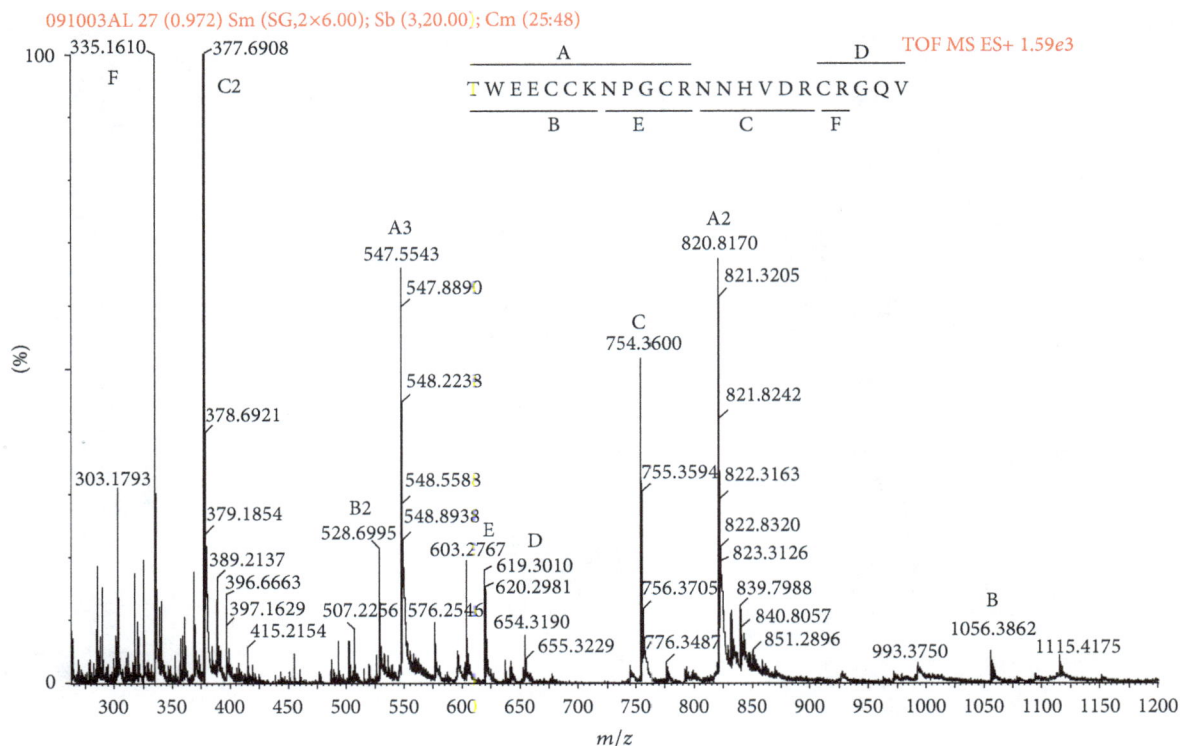

FIGURE 2: Representative ESI-Q-TOF/MS profile of the trypsin digested purified peptide. The deduced sequence is printed above the spectrum, together with the tryptic peptides (A–F). The MS profile indicates the tryptic peptides and the charge states.

TABLE 1: Theoretical and experimental m/z values for the tryptic peptides obtained after the enzymatic digestion of α-RgIB.

Ions	[M + H]+		[M + 2H]2+		[M + 3H]3+		Sequence
	Exp.	Theor.	Exp.	Theor.	Exp.	Theor.	
A	1640.63[1,2]	1639.839[1,2]	820.82[1,2]	820.423[1,2]	547.55[1,2]	547.282[1,2]	TWEECCKNPGCR
B	1056.39[1,2]	1055.180[1,2]	528.70[1,2]	528.094[1,2]	—	x	TWEECCK
C	754.37	754.359	377.69	377.896	—	(252.125)	NNHVDR
D	619.30[2]	619.298[2]	—[3]	(310.366)[2,4]	—	x	CRGQV
E	603.28[2]	603.267[2]	—	(302.137)	—	x	NPGCR
F	335.16[2]	335.147[2]	—	x[5]	—	x	CR

[1] Acetylation (N-terminal, variable modification).
[2] Carbamidomethyl cysteine (fixed modification).
[3] Not detected.
[4] Not observed.
[5] Not expected.

peptide discovered with both being from *C. regius* with a cysteine framework I ("B") [25].

α-RgIA was the first α-conotoxin described from *C. regius*, acting on neuronal nicotinic receptors. This peptide has been thoroughly characterized in terms of its primary and three-dimensional structures [29], as well as regarding its biological effect, for example, the blockage of the $\alpha9\alpha10$ nAChR [30, 31]. α-RgIA and α-RgIB come from the same animal, belong to the same toxin family, and possess similar biological effects; however, their amino acid sequences differ. Figure 4 shows the ClustalW alignment of α-RgIB and

its closest phylogenetic relatives (supplementary material), besides α-RgIA, which was not considered to be similar (according to MEGA5), but was manually inserted in the figure for the benefit of sequence comparison. α-RgIA is shorter, both in the N- and C-terminal flanking regions, as well as in the inter-Cys-bridge region. Nevertheless, in a considerably small universe of possibilities (8 out 12, since 4 amino acids are necessarily Cys), α-RgIA and α-RgIB bare considerable similarities: the Pro, at the 13th aligned position, and the charged residues at the 17th and 18th aligned positions. It is noteworthy to mention that,

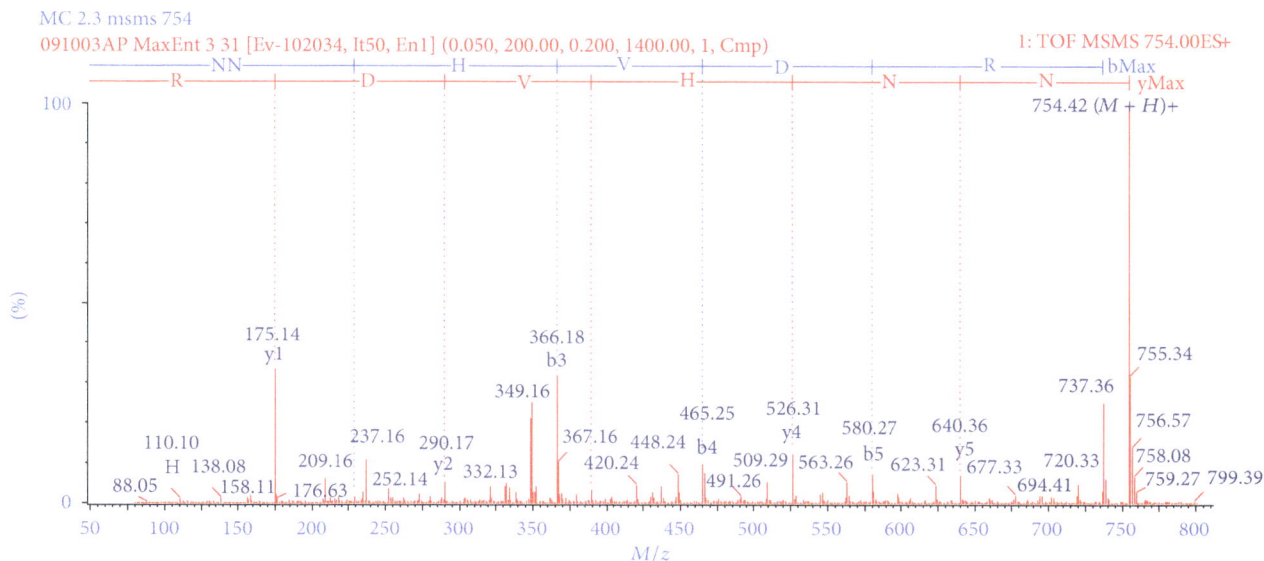

FIGURE 3: MaxEnt3 deconvoluted annotated representative MS/MS profile of the CID spectrum of tryptic peptide C, from Figure 2. y and b series are annotated above the spectrum, as well as other fragments.

Clustal 2.1 multiple sequence alignment

```
α-RgIB     ---- TWEECCKNPGCRNNHVDRCRGQV ----------------   23
A6M938     ------ NDCCHNAPCRNNHPGIC --------------------    17
A1X8C3     --- GMWDECCDDPPCRQNNMEHCPAS ----------------     23
Q2I2R6     ------- ECCDDPPCRQNNMEHCPAS ----------------     19
A1X8C2     --- GVWDECCKDPQCRQNHMQHCPAR ----------------     23
P0C8U6     ------ DDCCPDPACRQNHPELCSTR ----------------     20
P0C8U7     ------ DDCCPDPACRQNHPEICPSR ----------------     20
P0C8U9     NAWLTPEECCAAPACREMILEFCLAGEAFAAALDGFRRLPYR       42
P0C8U8     NAWFTPEECCAAPACRGMILEFCLAGEAFAAALDGFRRLPYR       42
P0C8V0     ISEMTWEECCTNPVCRQHYMHYC --------------------    23
P01519     -------ECC- NPACGRHYS--C --------------------   13
                  .**    .   *       *

P0C1D0     GCCSDPRCRYR       C                             12
```

FIGURE 4: ClustalW alignment of α-RgIB and the closest phylogenetic α-conotoxins relatives (calculated according to supplemental Figure 3).(*) Consensus; (:) highly conserved; (.) conserved. The bold underlined amino acid residues of α-RgIB were also considered to be conserved. A6M938: α-conotoxin-like Lp1.10 C. *leopardus*/homology; A1X8C3: α-conotoxin-like Lp1.7 C. *leopardus*/transcript; Q2I2R6: α-conotoxin-like Ltl.3 C. *litteratus*/transcript; A1X8C2: α-conotoxin-like Lp1.8 C. *leopardus*/transcript; P0C8U6: α-conotoxin-like PuSG1.1 C. pulicarius/transcript; P0C8U9: α-conotoxin-like Pu1.5 C. *pulicarius*/transcript; P0C8U8: α-conotoxin-like Pu1.4 C. *pulicarius*/transcript; P0C8U7/α-conotoxin-like PuSG1.2 C. *pulicarius*/trasncritpt; P0C8V0: α-conotoxin-like Pu1.6 C. *pulicarius*/transcript; P01519: α-conotoxin GIA C. *geographus*/protein; P0C1D0: α-conotoxin RgIA, C. *regius*/protein. (Key: UniProt Accesion code: toxin/Conus species/evidence level).

in spite of the phylogenetic analyses, all conotoxins listed in Figure 4 (except α-RgIA and α-RgIB) come from other *Conus* species: *C. leopardus* (A6 M938, A1X8C2, A1X8C3), *C. litteratus* (Q2I2R6), *C. pulicarius* (P0C8U6, P0C8U7, P0C8U8, P0C8U9, and P0C8V0) and *C. geographus* (P01519). Moreover, only P01519 has been detected at the protein level and has been characterized as active on the muscular nicotinic receptors [32].

Besides α-RgIA, the following toxins have been isolated from *C. regius*: P85009; P85010; P85011; P85012; and P85013, all α-conotoxin-like peptides belonging to superfamily A; P85016; P85017; P85018; P85019; P85020; P85021 and P85022, all belonging to the M-superfamily of conotoxins [33].

Moreover, our group has also identified two conotoxins, as well: Rg11a, belonging to the I_1 superfamily (P84197, [34]); and Rg9.1, belonging to the P-superfamily (Q8I6V7; direct submission).

There is no high level of homology between α-RgIB and the conopeptides described until the present moment; therefore, the identification of a proper 3D structure to serve as a template for homology modeling is deprecated. Instead, a structure was predicted by using I-TASSER server [23, 24]. Figure 5 shows that, in spite of the low homology with α-RgIA, α-RgIB model assumed the same basic shape as the NMR determined structures of the α-RgIA mutants, available at the PDB database [27].

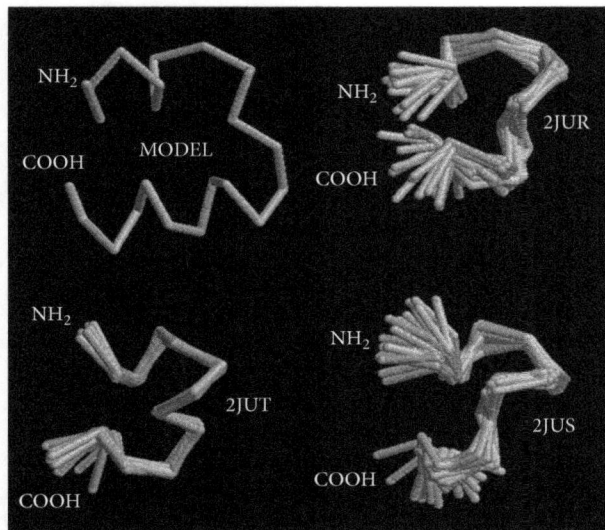

FIGURE 5: I-TASSER model of α-RgIB and PDB NMR superimposed structures of α-RgI-A.

Regarding the rather unique amino acid sequence of α-RgIB and thoroughly analyzing our data, we could not rule out the possibility that one of the glutamic acid (Glu) residues of this novel conotoxin would be a gamma-carboxyglutamic acid residue (Gla). Our suspicions arouse from the slightly higher deviation between the theoretical and calculated molecular mass values for A and B ions (Table 1), that could reflect that a side chain carboxylation and not an N-terminal acetylation would be present. Moreover, conotoxins are known for presenting posttranslation modifications, Gla included [35–38] and, even though the MALDI data of the crude peptide support the proposed peptide sequence, MALDI ionization is also a source of facile decarboxylation for Gla residues [39]. Our future experiments with *C. regius* conotoxins (α-RgIB included) will clarify this matter.

α-conotoxins bind to nicotinic acetylcholine receptors. The subgroup α3/5 of α-conotoxins, from piscivorous *Conus*, has the motif CCX_3CX_5C and can cause paralysis of the prey by the binding on muscle nicotinic receptors. Another subgroup, α4/3, that present the motif CCX_4CX_3C, bind on neuronal nicotinic receptors. The main subgroup of α-conotoxins is α4/7, with motif CCX_4CX_7C. These peptides bind in all classes of nicotinic receptors: muscular (e.g., α-conotoxin EI), homomeric neuronal (e.g., α-conotoxins PnIB), and heteromeric neuronal (α-conotoxins MII and AuIB) [9].

Neuronal nicotinic acetylcholine receptors (nAChRs) belong to the pentameric superfamily of Cys-loop ligand gated ionic channels. They are composed of either homomeric α or heteromeric α and β subunits assembled from a family of 12 distinct neuronal nicotinic subunits (α2–α10; β2–β4) [5]. The combination of subunits α2, α3, and α4 with β2 and β4 results in a functional receptor, as well α7, α8, and α9 homomeric receptors [40, 41]. In our experiments, it was verified by RT-PCR (supplemental Figure 3) that the pool of PC12 cells expressed α3, α5, α7, β2, and β4 subunits of neuronal nicotinic receptors, the same pattern found by Sargent [40] in PC12 cells. However, in spite of α-Rg-IB affinity by the PC12 nicotinic receptors, there are still other neuronal nicotinic receptors that may be higher affinity targets for these toxins that were not explored in the present work.

AuIB, from *Conus aulicus*, which is also an α-conotoxin, blocks the α3β4 receptors; however, the currents can be recovered after the toxin washing [42]. In our experiments, the no-recovery of α-RgIB is probably due to the irreversible action of the peptide on the receptor. Successive applications of the agonist (carbamolycholine), in control experiments, did not cause recovery of the ion currents on slow rise-time receptors (data not shown). Besides the irreversible action, the peptide may also be able to prolong the desensitization time of the receptor, since the repeated CBC administration on α-RgIB-treated PC12 cells was not able to recover the initial current, which is either caused by the irreversible binding of a low affinity toxin or the prolonging of the desensitization time of the receptor (or both).

Sudweeks and Yakel [43] showed that α3, α7, and β2 subunits of nAChR are correlated to fast rise-time receptors. The slow desensitization is a characteristic of α3β4 receptor, while α3β2 receptor is from fast desensitization [44]. The fast desensitization receptors, on PC12 cells, contain α3β2, α3β2α5, and α7 subunits, while slow desensitization receptors are formed by subunits α3β4 and α3β4α5. α-RgIB was able to inhibit the currents elicited by carbamolycholine on PC12 cells, mainly on the slow desensitization component, which comprise, in our model, α3β4 and α3β4α5 receptors.

The intracranial injection assay was performed to investigate whether there would be any direct activity of the toxin in the central nervous system (CNS), once peptides can promote behavioral alterations by acting on receptors and ionic channels on CNS. These alterations can indicate activities on specific ionic channels. For example, ω-conotoxin GVIA causes trembling on the mice, which indicates an action on calcium ionic channels [45]. The α-nicotinic acetylcholine receptor (nAChR) is associated to attention-deficit/hyperactivity disorder [46] which corroborates our observations of α-RgIB-treated hyperactive mice.

In conclusion, we have isolated a novel conotoxin from *Conus regius* and, by means of a combination of biochemical, structural and pharmacological assays were able to classify this peptide in the α-family and named it α-RgIB. There are still several peptides to be explored in the *C. regius* venom, as our previous qualitative investigations have shown [34] and the current study has focused on the biochemical characterization of one such novel peptide. Further studies are still necessary to better characterize the structural and pharmacological properties of α-RgIB.

Acknowledgments

This work was supported by Grants from the Brazilian funding agencies FAPESP and CNPq, including the INCTTOX PROGRAM.

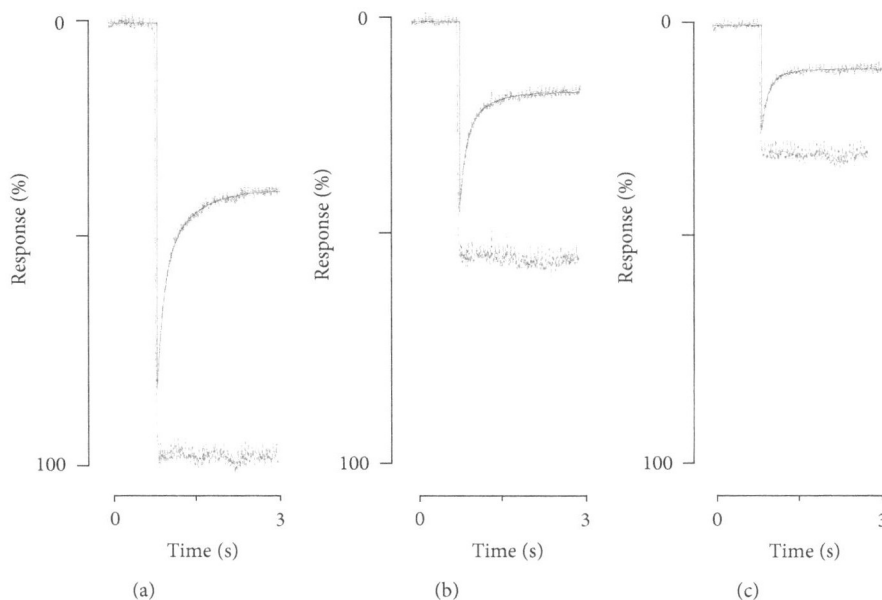

FIGURE 6: PC12 whole-cell characteristic patch clamp currents (expressed as a percentage of response to 1.5 mM carbamoylcholine (CBC)) (a), 1.5 mM CBC + 10 μM α-RgIB (b), and 1.5 mM CBC (c). Cells were kept at −70 mV.

References

[1] B. M. Olivera and L. J. Cruz, "Conotoxins, in retrospect," *Toxicon*, vol. 39, no. 1, pp. 7–14, 2001.

[2] S. R. Woodward, L. J. Cruz, B. M. Olivera, and D. R. Hillyard, "Constant and hypervariable regions in conotoxin propeptides," *The EMBO Journal*, vol. 9, no. 4, pp. 1015–1020, 1990.

[3] Q. Kaas, R. Yu, A. H. Jin, S. Dutertre, and D. J. Craik, "ConoServer: updated content, knowledge, and discovery tools in the conopeptide database," *Nucleic Acids Research*, vol. 40, pp. D325–D330, 2012.

[4] Q. Kaas, J. C. Westermann, R. Halai, C. K. L. Wang, and D. J. Craik, "ConoServer, a database for conopeptide sequences and structures," *Bioinformatics*, vol. 24, no. 3, pp. 445–446, 2008.

[5] R. M. Jones, G. E. Cartier, J. M. McIntosh, G. Bulaj, V. E. Farrar, and B. M. Olivera, "Composition and therapeutic utility of conotoxins from genus *Conus*: patent status 1996–2000," *Expert Opinion on Therapeutic Patents*, vol. 11, no. 4, pp. 603–623, 2001.

[6] Q. Kaas, J. C. Westerman, and D. J. Craik, "Conopeptide characterization and classifications: an analysis using ConoServer," *Toxicon*, vol. 55, no. 8, pp. 1491–1509, 2010.

[7] O. B. McManus, J. R. Musick, and C. Gonzalez, "Peptide isolated from the venom of *Conus geographus* block neuromuscular transmission," *Neuroscience Letters*, vol. 25, no. 1, pp. 57–62, 1981.

[8] B. M. Olivera, W. R. Gray, and L. J. Cruz, "Marine snail venoms," in *Marine Toxins and Venoms: Handbook of Natural Toxins*, A. T. Tu, Ed., Marcel Dekker, New York, NY, USA, 1989.

[9] H. Terlau and B. M. Olivera, "*Conus* venoms: a rich source of novel ion channel-targeted peptides," *Physiological Reviews*, vol. 84, no. 1, pp. 41–68, 2004.

[10] H. R. Arias and M. P. Blanton, "α-conotoxins," *International Journal of Biochemistry and Cell Biology*, vol. 32, no. 10, pp. 1017–1028, 2000.

[11] E. C. Rios, *Brazilian Marine Mollusks Iconography*, Fundação Universidade do Rio Grande, Rio Grande do Sul, Brazil, 1975.

[12] V. R. Eston, A. E. Migotto, E. C. Oliveira Filho, S. A. Rodrigues, and J. C. Freitas, "Vertical distribution of benthic marine organisms on rocky coasts of the Fernando de Noronha archipelago (Brazil)," *Boletim do Instituto Paulista de Oceanografia*, vol. 34, pp. 37–53, 1986.

[13] L. J. Cruz, G. Corpuz, and B. M. Olivera, "A preliminary study of *Conus* venom protein," *The Veliger*, vol. 18, pp. 302–308, 1976.

[14] R. Westermeier and T. Naven, *Proteomics in Practice: Laboratory Manual of Proteome Analysis*, Wiley-VCH, Weinheim, Germany, 2002.

[15] C. Clark, B. M. Olivera, and L. J. Cruz, "A toxin from the venom of the marine snail *Conus geographus* which acts on the vertebrate central nervous system," *Toxicon*, vol. 19, no. 5, pp. 691–699, 1981.

[16] S. M. Sine and P. Taylor, "Functional consequences of agonist-mediated state transitions in the cholinergic receptor. Studies in cultured muscle cells," *The Journal of Biological Chemistry*, vol. 254, no. 9, pp. 3315–3325, 1979.

[17] L. A. Greene, J. M. Aletta, A. Rukenstein, and S. H. Green, "PC12 pheochromocytoma cells: culture, nerve growth factor treatment, and experimental exploitation," *Methods in Enzymology*, vol. 147, pp. 207–216, 1987.

[18] O. P. Hamill, A. Marty, E. Neher, B. Sakmann, and F. J. Sigworth, "Improved patch-clamp techniques for high-resolution current recording from cells and cell-free membrane patches," *Pflugers Archiv European Journal of Physiology*, vol. 391, no. 2, pp. 85–100, 1981.

[19] H. Ulrich, J. E. Ippolito, O. R. Pagán, V. A. Eterović, R. M. Hann, H. Shi et al., "*In vitro* selection of RNA molecules that displace cocaine from the membrane-bound nicotinic acetylcholine receptor," *Proceedings of the National Academy of Sciences of the United States of America*, vol. 95, pp. 14051–14056, 1998.

[20] J. B. Udgaonkar and G. P. Hess, "Chemical kinetic measurements of a mammalian acetylcholine receptor by a fast-reaction

technique," *Proceedings of the National Academy of Sciences of the United States of America*, vol. 84, no. 24, pp. 8758–8762, 1987.

[21] A. A. Nery, R. R. Resende, A. H. Martins, C. A. Trujillo, V. A. Eterovic, and H. Ulrich, "Alpha7 nicotinic acetylcholine receptor expression and activity during neuronal differentiation of PC12 pheochromocytoma cells," *Journal of Molecular Neuroscience*, vol. 41, no. 3, pp. 329–339, 2010.

[22] M. A. Larkin, G. Blackshields, N. P. Brown, R. Chenna, P. A. McGettigan, and H. McWilliam, "ClustalW and ClustalX version 2," *Bioinformatics*, vol. 23, pp. 2947–2948, 2007.

[23] A. Roy, A. Kucukural, and Y. Zhang, "I-TASSER: a unified platform for automated protein structure and function prediction," *Nature protocols*, vol. 5, no. 4, pp. 725–738, 2010.

[24] Y. Zhang, "Template-based modeling and free modeling by I-TASSER in CASP7," *Proteins*, vol. 69, no. S8, pp. 108–117, 2007.

[25] W. R. Gray, A. Luque, B. M. Olivera, J. Barrett, and L. J. Cruz, "Peptide toxins from *Conus geographus* venom," *The Journal of Biological Chemistry*, vol. 256, no. 10, pp. 4734–4740, 1981.

[26] B. M. Olivera, G. Bulaj, J. Garrett, H. Terlau, and J. Imperial, "Peptide toxins from the venoms of cone snails and other toxoglossan gastropods," in *Animal Toxins: State of the Art—Perspectives in Health and Biotechnology*, M. E. Lima, Ed., Editora UFMG, Belo Horizonte, Brazil, 2009.

[27] M. Ellison, C. Haberlandt, M. E. Gomez-Casati et al., "α-RgIA: a novel conotoxin that specifically and potently blocks the α9α10 nAChR," *Biochemistry*, vol. 45, no. 5, pp. 1511–1517, 2006.

[28] C. Grewer and G. P. Hess, "On the mechanism of inhibition of the nicotinic acetylcholine receptor by the anticonvulsant MK-801 investigated by laser-pulse photolysis in the microsecond-to-millisecond time region," *Biochemistry*, vol. 38, no. 24, pp. 7837–7846, 1999.

[29] R. J. Clark, N. L. Daly, R. Halai, S. T. Nevin, D. J. Adams, and D. J. Craik, "The three-dimensional structure of the analgesic α-conotoxin, RgIA," *FEBS Letters*, vol. 582, no. 5, pp. 597–602, 2008.

[30] B. Callaghan, A. Haythornthwaite, G. Berecki, R. J. Clark, D. J. Craik, and D. J. Adams, "Analgesic α-conotoxins Vc1.1 and RgIA inhibit N-type calcium channels in rat sensory neurons via GABAB receptor activation," *Journal of Neuroscience*, vol. 28, no. 43, pp. 10943–10951, 2008.

[31] M. Ellison, Z. P. Feng, A. J. Park et al., "α-RgIA, a novel conotoxin that blocks the α9α10 nAChR: structure and identification of key receptor-binding residues," *Journal of Molecular Biology*, vol. 377, no. 4, pp. 1216–1227, 2008.

[32] D. R. Groebe, W. R. Gray, and S. N. Abramson, "Determinants involved in the affinity of α-conotoxins GI and SI for the muscle subtype of nicotinic acetylcholine receptors," *Biochemistry*, vol. 36, no. 21, pp. 6469–6474, 1997.

[33] A. Franco, K. Pisarewicz, C. Moller, D. Mora, G. B. Fields, and F. Marì, "Hyperhydroxylation: a new strategy for neuronal targeting by venomous marine molluscs," *Progress in molecular and subcellular biology*, vol. 43, pp. 83–103, 2006.

[34] M. C. V. Braga, K. Konno, F. C. V. Portaro et al., "Mass spectrometric and high performance liquid chromatography profiling of the venom of the Brazilian vermivorous mollusk *Conus regius*: feeding behavior and identification of one novel conotoxin," *Toxicon*, vol. 45, no. 1, pp. 113–122, 2005.

[35] E. Czerwiec, D. E. Kalume, P. Roepstorff et al., "Novel γ-carboxyglutamic acid-containing peptides from the venom of *Conus textile*," *The FEBS Journal*, vol. 273, no. 12, pp. 2779–2788, 2006.

[36] Q. Dai, Z. Sheng, J. H. Geiger, F. J. Castellino, and M. Prorok, "Helix-helix interactions between homo- and heterodimeric γ-carboxyglutamate-containing conantokin peptides and their derivatives," *The Journal of Biological Chemistry*, vol. 282, no. 17, pp. 12641–12649, 2007.

[37] K. H. Gowd, V. Twede, M. Watkins et al., "Conantokin-P, an unusual conantokin with a long disulfide loop," *Toxicon*, vol. 52, no. 2, pp. 203–213, 2008.

[38] K. Hansson, B. Furie, B. C. Furie, and J. Stenflo, "Isolation and characterization of three novel Gla-containing *Conus marmoreus* venom peptides, one with a novel cysteine pattern," *Biochemical and Biophysical Research Communications*, vol. 319, no. 4, pp. 1081–1087, 2004.

[39] T. Nakamura, Z. Yu, M. Fainzilber, and A. L. Burlingame, "Mass spectrometric-based revision of the structure of a cysteine-rich peptide toxin with γ-carboxyglutamic acid, TxVIIA, from the sea snail, *Conus textile*," *Protein Science*, vol. 5, no. 3, pp. 524–530, 1996.

[40] P. B. Sargent, "The diversity of neuronal nicotinic acetylcholine receptors," *Annual Review of Neuroscience*, vol. 16, pp. 403–443, 1993.

[41] F. Wang, V. Gerzanich, G. B. Wellst et al., "Assembly of human neuronal nicotinic receptor α5 subunits with α3, β2, and β4 subunits," *The Journal of Biological Chemistry*, vol. 271, no. 30, pp. 17656–17665, 1996.

[42] S. Luo, J. M. Kulak, G. E. Cartier et al., "α-conotoxin AuIB selectively blocks α3β4 nicotinic acetylcholine receptors and nicotine-evoked norepinephrine release," *Journal of Neuroscience*, vol. 18, no. 21, pp. 8571–8579, 1998.

[43] S. N. Sudweeks and J. L. Yakel, "Functional and molecular characterization of neuronal nicotinic ACh receptors in rat CA1 hippocampal neurons," *Journal of Physiology*, vol. 527, no. 3, pp. 515–528, 2000.

[44] S. Bohler, S. Gay, S. Bertrand et al., "Desensitization of neuronal nicotinic acetylcholine receptors conferred by N-terminal segments of the β2 subunit," *Biochemistry*, vol. 40, no. 7, pp. 2066–2074, 2001.

[45] B. M. Olivera, L. J. Cruz, and D. Yashikami, "Effects of *Conus* peptides on the behavior of mice," *Current Opinion in Neurobiology*, vol. 9, no. 6, pp. 772–777, 1999.

[46] T. Dinklo, H. Shaban, J. W. Thuring et al., "Characterization of 2-[[4-fluoro-3-(trifluoromethyl)phenyl]amino]-4-(4-pyridinyl)-5-thiazolemethanol (JNJ-1930942), a novel positive allosteric modulator of the α7 nicotinic acetylcholine receptor," *Journal of Pharmacology and Experimental Therapeutics*, vol. 336, no. 2, pp. 560–574, 2011.

20

Nociceptin Signaling Involves a Calcium-Based Depolarization in *Tetrahymena thermophila*

Thomas Lampert,[1] **Cheryl Nugent,**[2] **John Weston,**[2] **Nathanael Braun,**[2] **and Heather Kuruvilla**[2]

[1] *Department of Biological Sciences, State University of New York at Buffalo, 109 Cooke Hall, Buffalo, NY 14260, USA*
[2] *Department of Science and Mathematics, Cedarville University, 251 North Main Street, Cedarville, OH 45314, USA*

Correspondence should be addressed to Heather Kuruvilla; heatherkuruvilla@cedarville.edu

Academic Editor: Hubert Vaudry

Tetrahymena thermophila are free-living, ciliated eukaryotes. Their behavioral response to stimuli is well characterized and easily observable, since cells swim toward chemoattractants and avoid chemorepellents. Chemoattractant responses involve increased swim speed or a decreased change in swim direction, while chemorepellent signaling involves ciliary reversal, which causes the organism to jerk back and forth, swim in small circles, or spin in an attempt to get away from the repellent. Many food sources, such as proteins, are chemoattractants for these organisms, while a variety of compounds are repellents. Repellents in nature are thought to come from the secretions of predators or from ruptured organisms, which may serve as "danger" signals. Interestingly, several peptides involved in vertebrate pain signaling are chemorepellents in *Tetrahymena*, including substances P, ACTH, PACAP, VIP, and nociceptin. Here, we characterize the response of *Tetrahymena thermophila* to three different isoforms of nociceptin. We find that G-protein inhibitors and tyrosine kinase inhibitors do not affect nociceptin avoidance. However, the calcium chelator, EGTA, and the SERCA calcium ATPase inhibitor, thapsigargin, both inhibit nociceptin avoidance, implicating calcium in avoidance. This result is confirmed by electrophysiology studies which show that 50 μM nociceptin-NH2 causes a sustained depolarization of approximately 40 mV, which is eliminated by the addition of extracellular EGTA.

1. Introduction

Nociceptin/orphanin FQ (hereafter referred to as nociceptin) is a peptide involved in vertebrate pain signaling. The endogenous receptor for this ligand is ORL-1/NCR [1, 2]. A number of signaling pathways have been implicated in vertebrate nociceptin signaling. A partial listing of molecules involved in this signaling cascade would include $G_{i/o}$ proteins [1], neuronal nitric oxide synthase (nNOS) [3], and Erk-dependent signaling [4]. In addition, signaling through the nociceptin receptor induces a reduction in calcium influx via P/Q-type calcium channels in rat brain [5].

Tetrahymena thermophila are free-living, unicellular eukaryotes. While *T. thermophila* do not feel pain, they are capable of sensing chemoattractants and chemorepellents in

their environment. This allows them to find food and possibly to escape predation [6]. A recent review by Csaba [7] details the response of *T. thermophila* to a number of chemoattractants and chemorepellents, including their response to many vertebrate hormones. Indeed, *T. thermophila* appear to synthesize and respond to a number of vertebrate hormones, including serotonin, melatonin, adrenocorticotropic hormone, and insulin [7].

A number of chemorepellents which have been characterized in *T. thermophila* are polycationic peptides, including lysozyme [8], the lysozyme fragment CB2 [9], PACAP [10], and nociceptive peptides including bradykinin and substance P [11]. Lysozyme signaling involves a calcium-based depolarization [12]. Lysozyme and PACAP appear to share a signaling pathway [9], which involves cAMP and phospholipase C

[13], as well as NOS and cGMP [14]. A related peptide, VIP, also uses these signaling pathways and cross-adapts with lysozyme and PACAP, suggesting that *Tetrahymena* are signaling through a generalized polycation receptor [15].

Nociceptin is a polycationic peptide that is commercially available in three different isoforms. Nociceptin carries a charge of +4 at pH 7.0, while nociceptin-NH2 carries a charge of +5 at pH 7.0. Nociceptin-$Arg_{14}Lys_{15}$ carries a charge of +6 at pH 7.0. Our hypothesis was that all three of the nociceptin analogues would be chemorepellents in *T. thermophila* and that more highly charged nociceptin isoforms will have a lower EC_{100} in behavioral assays than isoforms which carry a lesser charge.

2. Materials and Methods

2.1. Cell Cultures. *Tetrahymena thermophila*, strain B2086, a generous gift from Hennessey [6] (SUNY Buffalo), was used for all of the experiments. Cells were grown at 25°C in the axenic medium of Dentler [16], without shaking or addition of antibiotics. Two-day old cell cultures were used for all behavioral assays described below.

2.2. Chemicals and Solutions. Behavioral assays were carried out in a buffer of pH 7.0 containing 10 mM Trizma base, 0.5 mM MOPS, and 50 μM $CaCl_2$. All repellents and inhibitors used were dissolved in this buffer.

All nociceptin isoforms, thapsigargin, J-113397, and EGTA, were purchased from Tocris Biosciences, Bristol, UK.

2.3. Behavioral Assays. Behavioral assays were carried out as previously described [8, 10, 17]. Ten milliliters of *T. thermophila* culture was washed by centrifugation in a clinical centrifuge at high speed, and the pellet was reconstituted in 10 mL buffer. This wash step was repeated twice, and cells were reconstituted in 5 mL of buffer for use in behavioral assays. To perform the behavioral assays, 300 μL of cell suspension was transferred to the first well of a microtiter plate. Cells were then transferred individually using a micropipette into the second well of the microtiter plate, which contained 300 μL of buffer as a control. Cells were then transferred to a third well containing 300 μL of nociceptin. Behavior of the cells was observed for the first 5 seconds after transfer to the third well, and the percentage of cells exhibiting avoidance behavior was noted. Varying concentrations of each peptide were used until we determined the minimum concentration at which 90% of the cells exhibited avoidance behavior (EC_{100}). Each trial represents 10 cells. A minimum of 6 trials was performed for each data point.

Pharmacological inhibition assays were performed similar to the behavioral assays described previously. After being washed in buffer, cells were exposed to pharmacological agents known to block specific signaling pathways and incubated for 15 minutes to 2 hours. Cells were then transferred to a solution containing nociceptin at EC_{100} and then monitored for avoidance behavior. Each trial represents 10 cells. A minimum of 6 trials was performed for each data point.

Cross-adaptation assays were performed as previously described [10, 17]. Briefly, 300 μL of cells were placed into the first well of a 3-well microtiter plate. Cells were then individually transferred to the second well of the 3-well microtiter plate, which contained a repellent. The cells were allowed to adapt to this repellent for 10–15 minutes or until cells showed baseline avoidance (an avoidance of no more than 20%). Cells were then individually transferred to the third well of the 3-well microtiter plate, which contained the repellent to be tested for cross-adaptation and monitored for avoidance behavior. Data which showed 20% or fewer cells exhibiting avoidance was considered "baseline avoidance." Baseline avoidance is the number of cells in our assay which show avoidance behavior when being transferred from one well containing buffer to another well containing the same buffer and usually ranges from 5 to 20%. Cells exhibiting baseline avoidance in response to this assay were considered to be cross-adapted. Each trial represents 10 cells. A minimum of 6 trials was performed for each data point.

2.4. Electrophysiology. Standard one-electrode whole-cell membrane potential recordings were recorded as the previously reported procedures in *Tetrahymena thermophila* [9, 12]. The recording buffer contained were carried out in a buffer of pH 7.0 containing 10 mM Trizma base, 0.5 mM MOPS, and 1 mM $CaCl_2$. Membrane potentials were displayed on a digital oscilloscope and retained on a chart recorder during continuous bath perfusion at a rate of approximately 20.0 mL/min. The recording bath had a volume of approximately 1 mL. Solutions were changed by switching valves connected either to buffer or to the experimental solution without changing the flow rate of the perfusion system.

3. Results

All isoforms of nociceptin were chemorepellents in *T. thermophila* (Figure 1). Nociceptin, which has a charge of +4 at our assay pH of 7.0, had an EC_{100} of 100 μM in our behavioral assay. Nociceptin-NH2, which has a charge of +5 under assay conditions, had an EC_{100} of 50 μM, while nociceptin-Arg_{14}-Lys_{15} which has a charge of +6 under assay conditions had an EC_{100} of 25 μM. Avoidance was observed for 1–5 seconds but was seen for as long as 10–15 minutes (not shown). After cells acclimated to the nociceptin, they returned to forward swimming.

Cross-adaptation assays (Table 1) show that all three isoforms of nociceptin cross-adapt with one another. However, nociceptin-adapted cells did not cross-adapt to PACAP-38, and PACAP-adapted cells did not cross-adapt to nociceptin. Since all three nociceptin isoforms cross-adapted to one another, implying a common signaling pathway, we used 50 μM nociceptin-NH2 in all subsequent pharmacological and behavioral assays.

Studies with pharmacological agents known to block G-protein signaling, tyrosine kinase signaling, and broad spectrum kinase activity had no effect on avoidance behavior

TABLE 1: Nociceptin cross-adaptation studies. Cells were adapted to a given ligand by incubating them in that ligand for 10–15 minutes or until avoidance behavior ceased. Cells were then moved into another ligand and were scored positively or negatively for avoidance. Cross-adaptation with various analogues of nociceptin all show avoidance values that are at or below baseline ($\leq 20\%$; [10]). However, cross-adaptation with the polycationic peptide, PACAP, does not cross-adapt with nociception, implying that nociception is using a pathway that is distinct from the previously described lysozyme/PACAP receptor [10, 15]. N represents the number of trials conducted. Each trial consisted of 10 cells, which were individually scored as positive or negative for avoidance.

	Nociceptin	Nociceptin-NH$_2$	Nociceptin Arg$_{14}$Lys$_{15}$	PACAP 1-38
Nociceptin	9.2 ± 8.2	5 ± 8.3	0 ± 0	96.6 ± 5.8
	$N = 13$	$N = 6$	$N = 6$	$N = 6$
Nociceptin-NH$_2$	5 ± 7.5	16.9 ± 12.2	14.5 ± 12.4	100 ± 0
	$N = 9$	$N = 8$	$N = 12$	$N = 6$
Nociceptin Arg$_{14}$Lys$_{15}$	3.3 ± 5.8	13.3 ± 12.1	16.6 ± 16.3	91.25 ± 9.9
	$N = 6$	$N = 6$	$N = 6$	$N = 8$
PACAP 1-38	97.5 ± 4.6	100 ± 0	100 ± 0	13.3 ± 5.8
	$N = 8$	$N = 10$	$N = 10$	$N = 6$

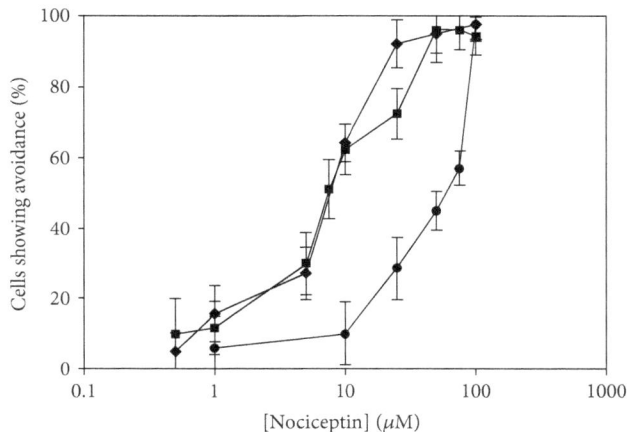

FIGURE 1: Nociceptin is a chemorepellent in *Tetrahymena thermophila*. Nociceptin (closed circles), nociceptin-NH2 (closed squares), and nociceptin-Arg$_{14}$Lys$_{15}$ (closed diamonds) all caused avoidance in *Tetrahymena thermophila*. The EC$_{100}$ of each compound was correlated with its charge. Nociceptin, which has a net charge of +4, had an EC$_{100}$ of 100 μM. Nociceptin-NH2, which has a net charge of +5, had an EC$_{100}$ of 50 μM. Finally, nociceptin-Arg$_{14}$Lys$_{15}$, which had a net charge of +6, had an EC$_{100}$ of 25 μM. $N \geq 6$. N represents the number of trials conducted. Each trial consisted of 10 cells, which were individually scored as positive or negative for avoidance.

FIGURE 2: Calcium chelators inhibit the behavioral response to 50 μM nociceptin-NH2 in *Tetrahymena thermophila*. EGTA (closed circles) reduces avoidance to 20% (near baseline) at a concentration of 50 μM. The IC$_{50}$ of EGTA is approximately 7.5 μM. Thapsigargin (open triangles) reduced avoidance by 50% at a concentration of 100 μM; however, increasing the concentration to 300 μM did not cause a significant decrease in avoidance beyond that seen with 100 μM thapsigargin. $N \geq 6$. N represents the number of trials conducted. Each trial consisted of 10 cells, which were individually scored as positive or negative for avoidance.

in *T. thermophila* (Table 2). However, studies with the calcium chelator, EGTA, and the SERCA ATPase inhibitor, thapsigargin, both affected nociceptin avoidance (Figure 2). A concentration of 50 μM EGTA was sufficient to reduce avoidance to a baseline avoidance of 20%. Thapsigargin, however, never reduced avoidance to baseline under the conditions of our assay. The highest concentration of thapsigargin we were able to achieve in our assay was 300 μM. Thapsigargin reduced avoidance by 50% at a concentration of 100 μM; however, increasing the concentration to 300 μM did not decrease avoidance beyond that seen with 100 μM thapsigargin.

Whole-cell electrophysiology studies indicate that nociceptin-NH2 is a depolarizing signal in *T. thermophila*

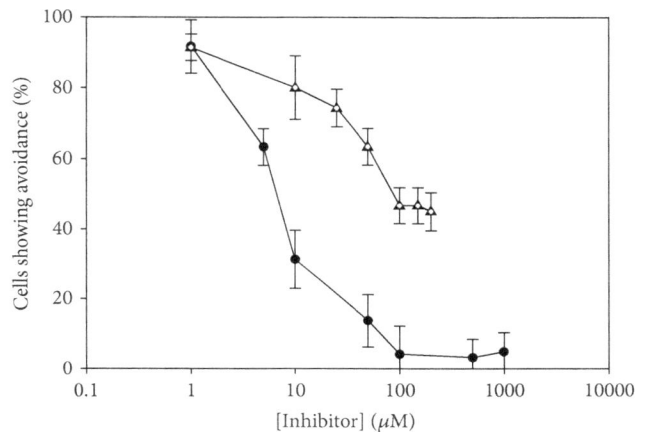

(Figure 3). A nociceptin-NH2 concentration of just 5 μM was sufficient to elicit a depolarization of approximately 20 mV, though this concentration does not cause behavioral avoidance above baseline levels in *Tetrahymena* (Figures 3(a) and 1). Fifty μM nociceptin-NH2, which is the EC$_{100}$ for behavioral avoidance in *Tetrahymena*, elicited a depolarization of approximately 40 mV (Figure 3(b)). The depolarization produced by 50 μM nociceptin-NH2 was eliminated by the addition of 1 mM EGTA to the external medium (Figure 3(c)).

J-113397, a competitive inhibitor of the human nociceptin receptor, inhibited the behavioral response to 50 μM nociceptin-NH2 in *Tetrahymena thermophila* when applied

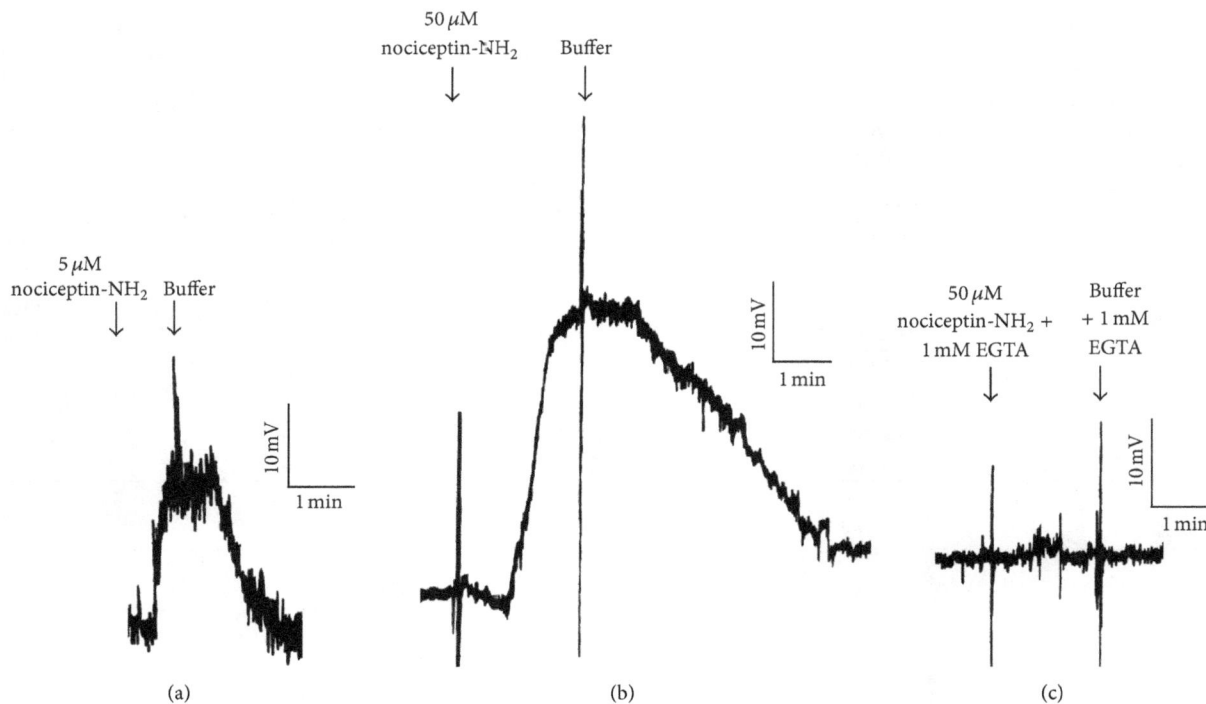

FIGURE 3: Nociceptin-NH2 is a depolarizing signal in *Tetrahymena thermophila*. (a) $5\,\mu M$ nociceptin-NH2 causes a depolarization of approximately 20 mV, though this concentration does not often provoke a behavioral response in *Tetrahymena*. (b) $50\,\mu M$ nociceptin-NH2 causes a depolarization of approximately 40 mV. This concentration is the EC_{100} for behavioral avoidance in *Tetrahymena*. (c) The depolarization produced by $50\,\mu M$ nociceptin-NH2 is eliminated by the addition of 1 mM EGTA to the external medium, implying that calcium is involved in the depolarization.

TABLE 2: Pharmacological inhibitors which act on G-protein mediated receptor pathways and tyrosine kinase pathways do not significantly impact nociception avoidance. N represents the number of trials conducted. Each trial consisted of 10 cells, which were individually scored as positive or negative for avoidance.

Pharmacological inhibitor	Pathway inhibited	Percentage of cells avoiding nociceptin	N
Control	None	94.28 ± 5.34	8
$50\,\mu M$ RpcAMPs	Adenylyl cyclase	96.67 ± 5.16	6
1 mM GDP-β-S	G-proteins	100.0 ± 0.0	6
$1\,\mu M$ U-73122	Phospholipase C	97.78 ± 4.40	9
$1\,\mu M$ U-73345	Inactive analogue of U-73122	96.67 ± 5.16	6
$10\,\mu M$ calphostin C	Protein kinase C	96.0 ± 5.0	10
1 mM 1400 W	NOS	96.0 ± 5.0	10
$100\,\mu M$ tyrphostin 47	Receptor tyrosine kinases	92.5 ± 9.57	6
$100\,\mu M$ AG126	Map kinase pathway	96.0 ± 6.99	10
$150\,\mu M$ SU 6668	Receptor tyrosine kinases	91.66 ± 4.08	6
$300\,\mu M$ apigenin	Protein kinases	91.2 ± 6.4	8
3 mM H-9	Protein kinases	96.6 ± 5.1	6

extracellularly (Figure 4). Baseline avoidance to nociceptin was achieved by the addition of $50\,\mu M$ of J-113397.

4. Discussion

Our results confirmed our hypothesis that all three nociceptin isoforms tested would serve as chemorepellents in *T. thermophila* (Figure 1). In addition, the EC_{100} of each compound was correlated with the charge, with the most highly charged isoform having the lowest EC_{100}, although all of the EC_{100} values were in a similar range. The correlation of lower EC_{100} values with a higher charge is consistent with what we have seen using other charged peptides in *T. thermophila*. For example, when we have used various peptides derived from ACTH, the more highly charged peptides caused avoidance at lower concentrations than did the less

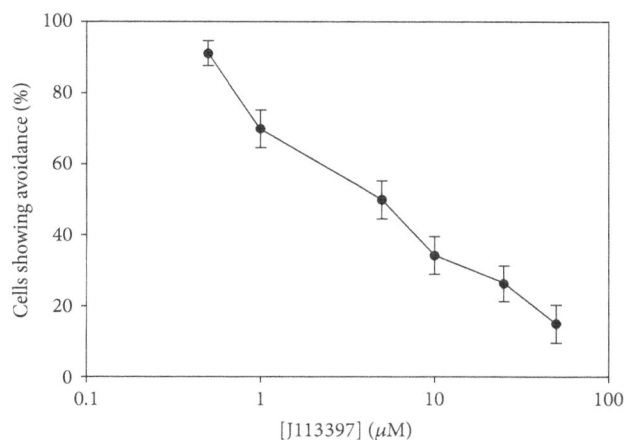

FIGURE 4: J-113397, a competitive inhibitor of the human nociceptin receptor, inhibits the behavioral response to 50 μM nociceptin-NH2 in *Tetrahymena thermophila* when applied extracellularly. The IC$_{50}$ of this compound is approximately 5 μM. $N \geq 6$. N represents the number of trials conducted. Each trial consisted of 10 cells, which were individually scored as positive or negative for avoidance.

highly charged peptides [11]. In addition, our previous studies with PACAP and VIP [15] show that PACAP is effective at causing avoidance at a 1000-fold lower concentration than VIP, though presumably acting through the same receptor and/or signaling pathway. The isoform of PACAP that we used in the 2003 study, PACAP-38-NH2, has a net charge of +11 at pH 7.0, while VIP has a net charge of just +4 at the same pH. While factors other than charge are certainly involved in the interaction between these peptides and their putative receptor, it is highly probable that charge is playing a role in these interactions, possibly by increasing the affinity of ligand for its receptor. In the case of nociceptin, the charge differences were relatively small as were the differences in EC$_{100}$.

Cells acclimated to nociception within 10–15 minutes of first being exposed to it (not shown). All isoforms of nociceptin were cross-adapted to one another, indicating that all forms of nociceptin were using the same receptor and/or signaling pathway. This is similar to what has previously been shown for lysozyme [8] and PACAP/VIP [15]. Since PACAP, lysozyme, and VIP appear to share a common receptor [10, 15], we cross-adapted cells to nociceptin and PACAP to determine whether nociceptin was using the same receptor/signaling pathway as the three previously studied polycationic ligands. As Table 1 shows, PACAP-adapted cells did not cross-adapt to nociceptin and nociceptin-adapted cells did not cross-adapt to PACAP. This indicates that nociceptin signals through a pathway that does not involve the previously described polycation receptor.

The previously studied PACAP response appears to be mediated through a G-protein-coupled receptor which uses adenylyl cyclase, phospholipase C, and nitric oxide synthase [10, 13–15]. In order to further ascertain whether nociceptin was using a separate signaling pathway, we used pharmacological inhibitors to block G-protein-linked receptors and

associated pathways. None of these inhibitors blocked avoidance to nociceptin (Table 2), giving further evidence that the previously described polycation receptor is not being used in nociceptin signaling. This also differs from the vertebrate nociceptin receptor, which signals through G$_{i/o}$ proteins [1].

Since a tyrosine kinase has been implicated in GTP signaling in *T. thermophila* [18] as well as insulin signaling [19], we also tested a battery of protein kinase and tyrosine kinase inhibitors to determine whether nociceptin signaling would be inhibited. None of these inhibitors affected nociceptin signaling (Table 2). Interestingly, genomic studies of *Tetrahymena* [20] show no evidence of the presence of a tyrosine kinase in this organism.

Since a calcium-based depolarization is elicited by the addition of lysozyme [12] as well as the lysozyme fragment, CB$_2$ [8], to *T. thermophila,* we wished to determine whether calcium was involved in nociceptin signaling in this organism. Studies with the external calcium chelator, EGTA (Figure 2) indicated that extracellular calcium was necessary for behavioral avoidance to nociceptin, since concentrations of EGTA above 50 μM reduced avoidance down to baseline. Baseline avoidance in this organism is determined by counting the number of cells that show avoidance behavior when transferred from one well of buffer to another well of the same buffer [10]. The SERCA ATPase inhibitor, thapsigargin, was used to determine whether internal calcium stores were required in order for avoidance to occur. As seen in Figure 2, exposure of cells to 100 μM thapsigargin reduced avoidance by approximately 50%. However, the avoidance response was not completely inhibited, indicating that while intracellular calcium may play a role in avoidance, lack of intracellular calcium stores depleted by thapsigargin may be partially compensated for by allowing extracellular calcium into the cytosol. Notably, the thapsigargin concentration used in this study was much higher than what we used in a previous study [18], in which only 1 nM thapsigargin was necessary in order to block the behavioral response to GTP. This is further evidence that extracellular calcium is primarily responsible for nociceptin avoidance. Calcium is not necessary for avoidance to all peptides, however, since avoidance of netrin-1, semaphorin 3C, and fragments of ACTH is unaffected by addition of either EGTA or thapsigargin [11].

Whole-cell electrophysiology studies indicate that nociceptin causes a depolarization in *T. thermophila* (Figure 3), even at concentrations that normally do not cause a behavioral response in this organism (Figures 3(a) and 1). When the EC$_{100}$ of nociceptin-NH2 was used, the amplitude of the depolarization increased (Figure 3(b)). Finally, we were able to remove the depolarization by the addition of EGTA to the external medium (Figure 3(c)), implying that calcium is involved in the depolarization. This is similar to the previously described responses to lysozyme [12] and the lysozyme fragment, CB2 [9].

The involvement of calcium in nociceptin avoidance in *T. thermophila* is rather different from the human response to nociceptin, which involves closing calcium channels [5]. However, we did use J-113397, which is a competitive inhibitor of the human nociceptin receptor [21], in order to determine if it could also block *T. thermophila* avoidance to nociceptin. As shown in Figure 4, 50 μM J-113397 was effective in reducing avoidance to baseline. This drug had no effect on avoidance to ACTH fragments (data not shown), suggesting that the response was specific to nociceptin. While we have not identified the receptor or signaling pathway that nociceptin is using in *T. thermophila*, these data suggest that there may be commonalities between the human nociceptin receptor and a possible nociceptin-binding protein in *T. thermophila*.

In summary, we have shown that nociceptin is a chemorepellent in *Tetrahymena* which elicits a depolarization. It does not act through the previously described polycation receptor nor does it signal through a G-protein-mediated receptor like the vertebrate nociceptin receptor. However, the J113397 studies imply that *Tetrahymena* may possess some type of receptor that shares binding characteristics with the human nociceptin receptor. Further studies may help elucidate the signaling mechanisms used in nociceptin avoidance in *T. thermophila*. If the receptor is identified, comparisons between the human nociceptin receptor and the unknown nociceptin-sensing mechanism in *T. thermophila* would be instructive.

5. Conclusions

The vertebrate signaling peptide, nociceptin, is a chemorepellent in *Tetrahymena thermophila*. The effectiveness of signaling is impacted by the charge of the nociceptin isoform, with more highly charged forms of nociceptin requiring lower concentrations to signal effectively. Nociceptin does not signal through the previously described polycation receptor of *Tetrahymena thermophila* nor does it signal through a G-protein-linked receptor, as it does in humans. However, nociceptin avoidance in *Tetrahymena thermophila* is blocked by addition of J-113397, a competitive inhibitor of the vertebrate nociceptin receptor. This suggests that the vertebrate nociceptin receptor and its analog in *Tetrahymena* may share common binding characteristics. Finally, nociceptin signaling provokes a depolarization, which pharmacological studies suggest may be caused by an influx of calcium.

Abbreviations

ACTH: Adrenocorticotropic hormone
EC_{100}: Concentration of chemorepellent that causes avoidance in 100% of individuals
EGTA: Ethylene glycol tetraacetic acid
IC_{50}: Concentration of inhibitor that blocks 50% of avoidance
PACAP: Pituitary adenylyl cyclase activating polypeptide
VIP: Vasoactive intestinal peptide.

Conflict of Interests

None of the authors of this paper has any financial relationship with the vendors mentioned in this paper. There is no conflict of interests to declare.

References

[1] B. E. Hawes, M. P. Graziano, and D. G. Lambert, "Cellular actions of nociceptin: transduction mechanisms," *Peptides*, vol. 21, no. 7, pp. 961–967, 2000.

[2] E. Hashiba, C. Harrison, G. Calo' et al., "Characterisation and comparison of novel ligands for the nociceptin/orphanin FQ receptor," *Naunyn-Schmiedeberg's Archives of Pharmacology*, vol. 363, no. 1, pp. 28–33, 2001.

[3] L. Xu, E. Okuda-Ashitaka, S. Matsumura et al., "Signal pathways coupled to activation of neuronal nitric oxide synthase in the spinal cord by nociceptin/orphanin FQ," *Neuropharmacology*, vol. 52, no. 5, pp. 1318–1325, 2007.

[4] C. Goeldner, D. Reiss, J. Wichmann, H. Meziane, B. L. Kieffer, and A. M. Ouagazzal, "Nociceptin receptor impairs recognition memory via interaction with NMDA receptor-dependent mitogen-activated protein kinase/extracellular signal-regulated kinase signaling in the hippocampus," *Journal of Neuroscience*, vol. 28, no. 9, pp. 2190–2198, 2008.

[5] H. S. Gompf, M. G. Moldavan, R. P. Irwin, and C. N. Allen, "Nociceptin/orphanin FQ (N/OFQ) inhibits excitatory and inhibitory synaptic signaling in the suprachiasmatic nucleus (SCN)," *Neuroscience*, vol. 132, no. 4, pp. 955–965, 2005.

[6] T. M. Hennessey, "Responses of the ciliates *Tetrahymena* and *Paramecium* to external ATP and GTP," *Purinergic Signalling*, vol. 1, no. 2, pp. 101–110, 2005.

[7] G. Csaba, "The hormonal system of the unicellular *Tetrahymena*: a review with evolutionary aspects," *Acta Microbiologica et Immunologica Hungarica*, vol. 59, no. 2, pp. 131–156.

[8] H. G. Kuruvilla, M. Y. Kim, and T. M. Hennessey, "Chemosensory adaptation to lysozyme and GTP involves independently regulated receptors in *Tetrahymena thermophila*," *Journal of Eukaryotic Microbiology*, vol. 44, no. 3, pp. 263–268, 1997.

[9] H. G. Kuruvilla and T. M. Hennessey, "Chemosensory responses of *Tetrahymena thermophila* to CB2, a 24-amino-acid fragment of lysozyme," *Journal of Comparative Physiology A*, vol. 184, no. 5, pp. 529–534, 1999.

[10] S. R. Mace, J. G. Dean, J. R. Murphy, J. L. Rhodes, and H. G. Kuruvilla, "PACAP-38 is a chemorepellent and an agonist for the lysozyme receptor in *Tetrahymena thermophila*," *Journal of Comparative Physiology A*, vol. 186, no. 1, pp. 39–43, 2000.

[11] S. Ort, T. Warren, J. Morrow, E. Infante, A. Yager, and H. Kuruvilla, "Polycationic peptides are chemorepellents in Tetrahymena thermophila," manuscript in preparation.

[12] H. G. Kuruvilla and T. M. Hennessey, "Purification and characterization of a novel chemorepellent receptor from *Tetrahymena thermophila*," *Journal of Membrane Biology*, vol. 162, no. 1, pp. 51–57, 1998.

[13] D. L. Hassenzahl, N. K. Yorgey, M. D. Keedy et al., "Chemorepellent signaling through the PACAP/lysozyme receptor is mediated through cAMP and PKC in *Tetrahymena thermophila*," *Journal of Comparative Physiology A*, vol. 187, no. 3, pp. 171–176, 2001.

[14] J. Lucas, M. Riddle, J. Bartholomew et al., "PACAP-38 signaling in *Tetrahymena thermophila* involves NO and cGMP," *Acta Protozoologica*, vol. 43, no. 1, pp. 15–20, 2004.

[15] M. Keedy, N. Yorgey, J. Hilty, A. Price, D. Hassenzahl, and H. Kuruvilla, "Pharmacological evidence suggests that the lysozyme/PACAP receptor of *Tetrahymena thermophila* is a polycation receptor," *Acta Protozoologica*, vol. 42, no. 1, pp. 11–17, 2003.

[16] W. L. Dentler, "Fractionation of *Tetrahymena* ciliary membranes with Triton X-114 and the identification of a ciliary membrane ATPase," *Journal of Cell Biology*, vol. 107, no. 6, pp. 2679–2688, 1988.

[17] E. D. Robinette, K. T. Gulley, K. J. Cassity et al., "A comparison of the polycation receptors of *Paramecium tetraurelia* and *Tetrahymena thermophila*," *Journal of Eukaryotic Microbiology*, vol. 55, no. 2, pp. 86–90, 2008.

[18] J. Bartholomew, J. Reichart, R. Mundy et al., "GTP avoidance in *Tetrahymena thermophila* requires tyrosine kinase activity, intracellular calcium, NOS, and guanylyl cyclase," *Purinergic Signalling*, vol. 4, no. 2, pp. 171–181, 2008.

[19] P. Kovacs and G. Csaba, "Effect of inhibitors and activators of tyrosine kinase on insulin imprinting in *Tetrahymena*," *Cell Biochemistry and Function*, vol. 10, no. 4, pp. 267–271, 1992.

[20] J. A. Eisen, R. S. Coyne, M. Wu et al., "Macronuclear genome sequence of the ciliate *Tetrahymena thermophila*, a model eukaryote," *PLoS Biology*, vol. 4, no. 9, article e286, 2006.

[21] C. Trapella, R. Guerrini, L. Piccagli et al., "Identification of an achiral analogue of J-113397 as potent nociceptin/orphanin FQ receptor antagonist," *Bioorganic and Medicinal Chemistry*, vol. 14, no. 3, pp. 692–704, 2006.

21

Characterization of Selective Antibacterial Peptides by Polarity Index

C. Polanco,[1,2] J. L. Samaniego,[3] T. Buhse,[1] F. G. Mosqueira,[4] A. Negron-Mendoza,[5]
S. Ramos-Bernal,[5] and J. A. Castanon-Gonzalez[2]

[1] Centro de Investigaciones Químicas, Universidad Autónoma del Estado de Morelos, Avenida Universidad 1001,
62209 Cuernavaca, MOR, Mexico
[2] Subdireccion de Epidemiologia Hospitalaria y Control de Calidad de la Atencion Medica, Instituto Nacional de Ciencias Medicas y
Nutricion Salvador Zubiran, Vasco de Quiroga 15, Piso 4, Colonia Seccion XVI, 14000 Mexico City, DF, Mexico
[3] Departamento de Matematicas, Facultad de Ciencias, Universidad Nacional Autonoma de México, Ciudad Universitaria,
04510 Mexico City, DF, Mexico
[4] Direccion General de Divulgacion de la Ciencia, Universidad Nacional Autonoma de Mexico, Ciudad Universitaria, P.O. Box 70487,
04510 Mexico City, DF, Mexico
[5] Instituto de Ciencias Nucleares, Universidad Nacional Autonoma de México, Ciudad Universitaria, P.O. Box 70543,
04510 Mexico City, DF, Mexico

Correspondence should be addressed to C. Polanco, polanco@unam.mx

Academic Editor: Jean-Marie Zajac

In the recent decades, antibacterial peptides have occupied a strategic position for pharmaceutical drug applications and became subject of intense research activities since they are used to strengthen the immune system of all living organisms by protecting them from pathogenic bacteria. This work proposes a simple and easy statistical/computational method through a peptide polarity index measure by which an antibacterial peptide subgroup can be efficiently identified, that is, characterized by a high toxicity to bacterial membranes but presents a low toxicity to mammal cells. These peptides also have the feature not to adopt to an alpha-helicoidal structure in aqueous solution. The double-blind test carried out to the whole Antimicrobial Peptide Database (November 2011) showed an accuracy of 90% applying the polarity index method for the identification of such antibacterial peptide groups.

1. Introduction

The increasing resistance of pathogen agents towards multiple drugs has oriented parts of the investigation in bioinformatics to fast and efficient techniques that can predict the remarkable impact of antibacterial peptide action. These techniques can help to enhance the sometimes cumbersome chemical synthetic approach as well as the subsequent trial and error experiments to identify the peptide performance.

Among the proposed various classifications of peptides, one of it refers to the alpha-helicoidal versus beta-sheet conformation that the peptides can adopt in aqueous solution. This classification refers to the predominance of certain amino acids in the linear sequence of the peptides such as proline-arginine, cathelicidin, or cysteine. It is important

to note that such classification appears to be without any influence on the toxicity or selectivity of the peptide once it got in contact with the target membrane [1, 2].

Although nature was used as the main source of peptides with antibacterial properties in the past [3], parts of the research efforts are now more directed towards synthetic strategies. One of these synthetic approaches generate the peptides by replacing and/or removing constitutive amino acids from a natural peptide known for its antibacterial action [4], thus trying to reduce its size while keeping or increasing its toxicity [5]. Another technique consists of joining two peptides that individually do not exhibit antibacterial properties but combined turn out to be highly toxic [6].

To obtain efficient antibacterial peptides by measuring the potential action of each altered peptide with the-above described methods would result in a possibility combination that exceeds by far the capacity of the known verification methods in the laboratory. For instance, the number of possible peptides to be formed from one peptide with 8 amino acids in length would be $20^8 = 25,600,000,000$ peptides. This is the reason why contemporary technique profiles to construct antibacterial peptides are the result of joint computational and/or mathematical methods to simulate peptide variations and then to evaluate and qualify these variations to eventually determine if the peptide complies with the required purposes. However, these methods with the aim to simulate the properties of the peptides as well as to evaluate their performance respecting all possible combinatorics are highly complex in their mathematical/computational model design.

In this paper, we present a statistical method that can be attributed to a single physical-chemical property, which is easy to computerize and that efficiently identifies antibacterial peptide subgroups for its highly selective toxicity to bacteria, hereinafter referred to as "Selective Cationic Amphipathic Antibacterial Peptides" (SCAAPs). A SCAAP is characterized by being less than 60 amino acids in length, not adopting an alpha-helicoidal structure in neutral aqueous solution, and showing a therapeutic index higher than 75 [7]. The therapeutic index of a peptide is defined as the ratio between the minimum inhibitory concentration observed against mammalian and bacterial cells [7, 8]; that is, the higher the value, the more specific the peptide for bacterial-like membranes. Hence SCAAPs display strong lytic activity against bacteria but exhibit no toxicity against normal eukaryotic cells such as erythrocytes [9].

Our method determines an index that we call polarity index that uses the existent 20 proteic amino acid classification differentiated by its side chain R that divides them in four types and three categories [10]. The three general categories of side chains are nonpolar, polar but uncharged, and charged polar. The nonpolar residues include those with aliphatic hydrocarbon side chains: Gly, Ala, Val, Leu, Ilu, Pro, one aromatic group, Phe, and one "pseudo-hydrocarbon," Met. The polar but neutral category contains two hydroxyl-containing residues, Ser and Thr; two amides, Asn and Gln; two with aromatic rings, Tyr and Trp; one with a sulfhydryl group, Cys. In the charged polar class there are two amino acids with acidic groups, Asp and Glu, and three bases, His, Lys, and Arg (Table 1). The polarity index only makes use of that classification to get the SCAAP characteristic blueprint that in a double-blind test applied to all known peptides registered in the APD database (November 2011) [11] showed a very high efficiency.

2. Methods

2.1. Physicochemical Properties. Peptides can be expressed linearly as an amino acid sequence [12]. Such representation gives the peptide a unique blueprint. From this sequence, mathematical/computational algorithms have been designed with different complexity levels that measure a variety of physicochemical properties [13]. Among the properties on which the linear peptide representation focuses are two that

TABLE 1: 20 proteinogenic amino acid classification differentiated by their side chain according to their polarity [10].

Symbol	Category	1-letter code
P−	polar	D, E, Y
N	neutral	C, G, N, Q, S, T
P+	basic hydrophilic	H, K, R
NP	non polar residues	A, F, I, L, M, P, V, W

define if a peptide falls into the category of SCAAP [7]; that is, when its measure meets simultaneously the parameters established for the following physicochemical properties:

(i) isoelectric point [14] (IP) from 9.65 to 11.80,

(ii) hydrophobic moment [14] (HM) from 0.16 to 0.57.

Note that the original parameter values [7] have been extended. For this work, it was decided to take these two properties at a maximum range without considering the so-called AGADIR property, which is the tendency for not adopting an alpha-helicoidal structure in aqueous solution. As we have already verified [13], this property is not of significance for peptides with a length smaller than 22 amino acids.

A statistical-computational method was designed based only on one physicochemical property: polarity, which quickly and efficiently discerns if a peptide falls into the category of SCAAP or not. The verification was carried out by evaluating the IP and HM physicochemical properties.

2.2. Polarity Index Method. The polarity index method uses the 20 amino acid classification differentiated by their side chains that fall into four polarity groups: [P+] polar, [N] neutral, [P+] basic hydrophilic, and [NP] nonpolar residues (Table 1).

From these four groups, a polarity $\mathbf{P}[i, j]$ matrix is built with 16 elements that have as rows and columns the four different polarity groups set in the order P+, P−, N, NP and where $\mathbf{P}[i, j]$ matrix elements $[i, j]$ represent the 16 possible interactions of the groups.

The method consists of the following steps.

(i) Creating a $\mathbf{P}[i, j]$ incidence matrix from the subject peptide.

(ii) Generating a $\mathbf{Q}[i, j]$ incidence matrix from the SCAAP set.

(iii) Comparing the incidences from both $\mathbf{P}[i, j]$ and $\mathbf{Q}[i, j]$ matrices.

2.3. $\mathbf{P}[i, j]$ Incidence Matrix from a Subject Peptide. The $\mathbf{P}[i, j]$ incidence matrix is built by adding to each of its elements the matches that occurred in the peptide subject sequence from the left to the right with two amino acids in length and by moving one amino acid to the right at the time until it arrives at the peptide side end. Each amino acid pair is related to its polarity group. From that association, we identify row i and column j. To the $\mathbf{P}[i, j]$ matrix element

TABLE 2: SCAAP subjects [7]. IP: estimated isoelectric point. HM: hydrophobic moment. TI: calculated therapeutic index. Peptides Cecropin-A and CA(1-8)M(1-18)NH2 were used to determine the $P[i, j]$ incidence matrix.

Entry	Peptide	Sequence	IP	HM	TI
1	(KIAKKIA)2NH2	KIAKKIAKIAKKIA	11.5	0.48	86.2
2	(KLGKKLG)3NH2	KLGKKLGKLGKKLGKLGKKLG	11.7	0.49	98.3
3	**Cecropin-A**	**KWKLFKKIEKVGQNIRDGIIKAGPAVAVVGQATQIAK**	**11.2**	**0.44**	**1000.0**
4	Melittin	GIGAVLKVLTTGLPALISWIKRKRQQ	12.6	0.46	500.0
5	Magainin 2	GIGKFLHSAKKFGKAFVGEIMNS	10.8	0.56	75.0
6	CA(1-13)M(1-13)NH2	KWKLFKKIEKVGQGIGAVLKVLTTGL	11.1	0.53	400.0
7	**CA(1-8)M(1-18)NH2**	**KWKLFKKIGIGAVLKVLTTGLPALIS**	**10.4**	**0.43**	**2000.0**

TABLE 3: Number of matches in a typical SCAAP sequence in each peptide database with single or multiple action on fungi, viruses, mammalian cells, Gram+/Gram− bacteria, cancer cells, insects, parasites, and sperms (see also Section 2.6) [7].

Total	Action	Fungi	Viruses	Mammalian cells	Bacteria	Cancer cells	Insects	Parasites	Sperms
879	Unique	0/77	0/22	0/10	51/743	1/16	0/2	0/9	0/0
2644	Multiple	62/638	7/122	20/205	76/1489	21/121	3/20	5/40	1/9

will be added 1, resulting thus in $P[i, j] = P[i, j] + 1$. Finally, the $P[i, j]$ incidence matrix relative frequency distribution is normalized and weighted over a 0.30 factor. This last step helps to enhance the peptide distinctive characteristics by increasing the effect of the relative frequency position of the amino acids pairs in the incidence matrix $P[i, j]$.

2.4. $Q[i, j]$ Incidence Matrix from a SCAAP Set.

The $Q[i, j]$ incidence matrix is determined following the same procedure as for the $P[i, j]$ incidence matrix. The peptide used here is the set of peptide sequences described in Table 2. The peptides used here as SCAAP templates were reported as SCAAP subjects by Del Rio et al. [7]. From the 7 peptides submitted, only those with a therapeutic index higher or equal to 1000 were chosen (Table 2, entries 3 and 7).

2.5. $P[i, j]$ and $Q[i, j]$ Matrices Comparison.

In both the $P[i, j]$ and the $Q[i, j]$ matrices five stated positions M_1, M_2, M_3, M_4, and M_{16} were identified, where the subscript numeral stands for the element position in the matrix. The first row in the matrix represents the first four positions, the second row the next four positions, and so forth until allocating the last four positions to the last row. The position of the four elements with higher incidence would be M_1, M_2, M_3, and M_4 while M_{16} being the one with the lowest incidence. If the sequences $\{M_1, M_2, M_3, M_4, M_{16}\}$ for both matrices coincide, the peptide is classified as SCAAP

2.6. Trial Data Preparation.

1894 peptides registered in the Antimicrobial Peptide Database (APD) [11] (November 2011) were analyzed and classified by their single and multiple action against fungi, virus, mammalian cells, Gram+/Gram− bacteria, cancer cells, insects, parasites, and sperms. Peptides with more than one action were not included. The single action database only includes peptides with confirmed experimental action on a single pathogen agent, in contrast to multiple-action databases that contain peptides with action on two or more pathogen agents. On

this basis, the figures in multiple action databases are over-represented.

The verification of peptides found in the single-action database on Gram+/Gram− bacteria was carried out by validating both the isoelectric point (IP) and hydrophobic moment (HM) in the ranges stated (see Section 2.1). The integrity of the APD database information was verified by checking identified peptides by their action in the whole extent of the database itself.

3. Results

Due to the importance of detecting possible peptide pathogenic action, the use of computer programs that evaluate peptic sequences to predict their action on different pathogen agents such as fungi, virus, mammalian cells, and Gram+/Gram− bacteria has become a standard practice among different research groups. The polarity index method is one of these computer programs, but it differs in measuring exclusively one physicochemical property to identify a SCAAP.

The $P[i, j]$ Incidence matrix delivered by the polarity index method to identify a SCAAP used two peptides known by their toxic activity on Gram+/Gram+ bacteria (Table 2, entries 3 and 7) that turned out to be $\{M_1, M_2, M_3, M_4, M_{16}\} = \{16, 4, 13, 15, 10\}$. SCAAP subjects identified from the provided single pathogenic action peptide database were fungi (0/77), viruses (0/22), mammalian cells (0/10), Gram+/Gram+ bacteria (51/743), cancer cells (1/16), insects (0/2), parasites (0/9), and sperms (0/0) (Table 3).

Note that the polarity index method only identified SCAAP subjects basically in the bacterial group. Whereas SCAAP subjects identified from the multiple pathogenic action peptide database were fungi (62/638), viruses (7/122), mammalian cells (20/205), Gram+/Gram+ bacteria (76/1489), cancer cells (21/121), insects (3/20), parasites (5/40), and sperms (1/9) (Table 3). Among the 743 peptides with a single action on Gram+/Gram− bacteria, the polarity index method identified 51 SCAAP subjects (Table 4), their

TABLE 4: SCAAP subjects identified by the polarity index method in APD Gram+/Gram− bacteria database [11] where peptides have action only on bacteria. IP: estimated isoelectric point. HM: hydrophobic moment. Status: (X) not accepted for its IP and HM parameters, because the corresponding calculations were out of the ranges [7] (see Section 2.1).

No.	Peptide	Sequence	IP	HM	Status	Reference
1	Clavanin D (sea squirt, tunicate, invertebrates, animals)	AFKLLGRIIHHVGNFVYGFSHVF	10.85	0.54		[21]
2	Palustrin-1b (frog, amphibians, animals; XXU)	ALFSILRGLKKLGNMGQAFVNCKIYKKC	10.80	0.49		[22]
3	Palustrin-1d (frog, amphibians, animals; XXU)	ALSILKGLEKLAKMGIALTNCKATKKC	10.50	0.35		[22]
4	Palustrin-1c (frog, amphibians, animals; XXU)	ALSILRGLEKLAKMGIALTNCKATKKC	10.60	0.35		[22]
5	Brevinin-1PRc (frog, amphibians, animals; XXU)	FFPMLAGVAARVVPKVICLITKKC	10.50	0.38		[23]
6	Brevinin-1Be (frog, amphibians, animals; XXU)	FLPAIVGAAAKFLPKIFCVISKKC	10.30	0.43		[24]
7	Brevinin-1HSa (frog, amphibians, animals; XXU)	FLPAVLRVAAKIVPTVFCAISKKC	10.50	0.40		[25]
8	Brevinin-1Ba (frog, amphibians, animals; XXU)	FLPFIAGMAAKFLPKIFCAISKKC	10.30	0.50		[24]
9	Brevinin-1Bc (frog, amphibians, animals; XXU)	FLPFIAGVAAKFLPKIFCAISKKC	10.30	0.49		[24]
10	RANATUERIN 4 (ranatuerin-4, frog, amphibians, animals; XXU)	FLPFIARLAAKVFPSIICSVTKKC	10.50	0.46		[26]
11	Phylloseptin-H11 (PLS-H11, Phylloseptin-13, PS-13; frog, amphibians, animals; XXA)	FLSLIPHAINAVGVHAKHF	9.65	0.36		[27]
12	Phylloseptin-H5 (phylloseptin-7, PLS-H5, PS-7, XXA, frog, amphibians, animals)	FLSLIPHAINAVSAIAKHF	9.65	0.45		[28]
13	Phylloseptin-H2 (PLS-H2, Phylloseptin-2, PS-2) (XXA, frog, amphibians, animals)	FLSLIPHAINAVSTLVHHF	7.80	0.46	X	[29]
14	Phylloseptin-B1 (PLS-B1, PBN1; frog, amphibians, animals; XXA)	FLSLIPHIVSGVAALAKHL	9.65	0.46		[30]
15	Papilosin (tunicate, ascidian, invertebrates, sea animals)	GFWKKVGSAAWGGVKAAAKGAAVGGLNALAKHIQ	11.40	0.32		[31]
16	SMAP-34 (sheep myeloid antimicrobial peptide-34; OaMAP34, ovine cathelicidin, sheep, ruminant, animals)	GLFGRLRDSLQRGGQKILEKAERIWCKIKDIFR	10.43	0.48		[32]
17	Caerin 1.17 (frog, amphibians, animals; XXA)	GLFSVLGSVAKHLLPHVAPIIAEKL	9.50	0.49		[33]
18	Caerin 1.18 (frog, amphibians, animals; XXA)	GLFSVLGSVAKHLLPHVVPVIAEKL	9.50	0.50		[33]
19	Fallaxidin 3.2 (XXA, frog, amphibians, animals)	GLLDFAKHVIGIASKL	9.50	0.49		[34]
20	Fallaxidin 3.1 (XXA, frog, amphibians, animals)	GLLDLAKHVIGIASKL	9.50	0.48		[34]
21	Dahlein 5.2 (frog, amphibians, animals)	GLLGSIGNAIGAFIANKLKPK	11.10	0.52		[35]
22	Caerin 1.2 (XXA, frog, amphibians, animals)	GLLGVLGSVAKHVLPHVVPVIAEHL	7.02	0.49	X	[36]

TABLE 4: Continued.

No.	Peptide	Sequence	IP	HM	Status	Reference
23	Caerin 1.4 (XXA, frog, amphibians, animals)	GLLSSLSSVAKHVLPHVVPVIAEHL	7.02	0.48	X	[36]
24	Palustrin-2SIb (frog, amphibians, animals; XXU)	GLWNSIKIAGKKLFVNVLDKIRCKVAGGCKTSPDVE	10.10	0.36		[37]
25	XPF (the xenopsin precursor fragment, African clawed frog, amphibians, animals)	GWASKIGQTLGKIAKVGLKELIQPK	11.00	0.40		[38]
26	Pleurocidin (fish, animals)	GWGSFFKKAAHVGKHVGKAALTHYL	11.00	0.34		[39]
27	Cecropin (insects, invertebrates, animals)	GWLKKIGKKIERVGQNTRDATVKGLEVAQQAANVAATVR	11.30	0.36		[40]
28	Pm_mastoparan PMM (insects, invertebrates, animals; XXA; derivatives)	INWKKIASIGKEVLKAL	10.80	0.37		[41]
29	Hinnavin II (Hin II, insects, invertebrates, animals; JJsn)	KWKIFKKIEHMGQNIRDGLIKAGPAVQVVGQAATIYKG	10.12	0.45		[42]
30	Ostrich AvBD2 (Ostrich avian beta defensin 2, ostricacin-1, OSP-1, birds, animals; BBL)	LFCRKGTCHFGGCPAHLVKVGSCFGFRACCKWPWDV	8.94	0.33	X	[43]
31	Clavanin D (sea squirt, tunicate, invertebrates, animals)	LFKLLGKIIHHVGNFVHGFSHVF	10.80	0.56		[44]
32	Enterocin Q (EntQ, class 2d bacteriocins; leaderless, that is, no signal peptide, bacteria)	MNFLKNGIAKWMTGAELQAYKKKYGCLPWEKISC	10.00	0.39		[45]
33	Temporin-1Lb (Temporin 1Lb, frog, amphibians, animals)	NFLGTLINLAKKIM	10.80	0.41		[24]
34	Bovine beta-defensin 6 (bBD-6,cow, ruminant, animals)	QGVRNHVTCRIYGGFCVPIRCPGRTRQIGTCFGRPVKCC-RRW	11.30	0.37		[46]
35	mBD-4 (mBD4, mouse beta-defensin 4, or Defb4, animals; 3S=S)	QIINNPITCMTNGAICWGPCPTAFRQIGNCGHFKVRCCKIR	8.91	0.33	X	[47]
36	ChBac5 (Pro-rich; Arg-rich, goat cathelicidin, ruminant, animals)	RFRPPIRRPFIRPPFNPPFRPPVRPPFRPPFRPPFRPPIGPFP	13.45	0.50		[48]
37	Cyclic dodecapeptide (OaDode, ovine cathelicidin, sheep, ruminant, animals)	RICRIIFLRVCR	12.00	0.36		[49]
38	Bactenecin (cyclic dodecapeptide, bovine cathelicidin, cow, cattle, ruminant, animals; BBMm; JJsn; derivatives: Bac2A)	RLCRIVVIRVCR	12.00	0.33		[50]
39	RL-37 (RL37, cathelicidin, Old World monkey, primates, animals)	RLGNFFRKVKEKIGGGLKKVGQKIKDFLGNLVPRTAS	11.90	0.59		[51]
40	BACTENECIN 7 (bac-7, bac 7; bac7; Pro-rich; cow cathelicidin, ruminant, animals; BBL, SeqAR, BBPP)	RRIRPRPPRLPRPRPRPLPFPRPGPRPIPRPLPFPRPGPRPI-PRPLPFPRPGPRPIPRPL	13.20	0.25		[52]
41	Bac4 (Pro-rich, Arg-rich; cow cathelicidin, ruminant, animals)	RRLHPQHQRFPRERPWPKPLSLPLPRPGPRPWPKPL	12.90	0.23		[53]
42	Hyphancin IIIE (insects, invertebrates, animals)	RWKFFKKIEFVGQNVRDGLIKAGPAIQVLGAAKAL	11.80	0.41		[54]
43	Cecropin B (insects, invertebrates, animals)	RWKIFKKIEKMGRNIRDGIVKAGPAIEVLGSAKAI	11.40	0.42		[55]
44	Hyphancin IIID (insects, invertebrates, animals)	RWKIFKKIERVGQNVRDGIIKAGPAIQVLGTAKAL	11.80	0.41		[54]
45	Hyphancin IIIG (insects, invertebrates, animals)	RWKVFKKIEKVGRHIRDGVIKAGPAITVVGQATAL	11.80	0.38		[54]

No.	Peptide	Sequence	IP	HM	Status	Reference
46	Hyphancin IIIF (insects, invertebrates, animals)	RWKVFKKIEKVGRNIRDGVIKAGPAIAVVGQAKAL	11.80	0.39		[54]
47	Phylloseptin-H4 (Phylloseptin-6, PLS-H4, PS-6, XXA, frog, amphibians, animals)	SLIPHAINAVSAIAKHF	9.65	0.47		[29]
48	Pep5 (Lantibiotic, type 1, class 1 bacteriocin, Gram-positive bacteria; XXT3; XXW3)	TAGPAIRASVKQCQKTLKATRLFTVSCKGKNGCK	11.10	0.27		[56]
49	Clavanin C (sea squirt, tunicate, invertebrates, animals)	VFHLLGKIIHHVGNFVYGFSHVF	9.55	0.47		[21]
50	Andropin (insects, invertebrates, animals)	VFIDILDKVENAIHNAAQVGIGFAKPFEKLINPK	7.50	0.45	X	[57]
51	Clavanin A (urochordates, sea squirts, and sea pork, tunicate, invertebrates, animals)	VFQFLGKIIHHVGNFVHGFSHVF	9.71	0.61		[21]

IP and HM parameters were calculated and 46 of them are in the ranges previously mentioned in Section 2.1; that is, IP = 9.65–11.80 and HM = 0.16–0.57.

The APD database information integrity verification [11] showed 14 peptides not classified yet. When their activity as SCAAP was double checked by the polarity index method, there was a mismatch. The APD database margin of error did not exceed 8%.

4. Discussion

All different peptide classifications achieved over the decades seem to be directed to validate the peptide action and toxicity. However, it appears that these two characteristics are intrinsically related to the space where the peptide interacts as well as to the structural form of the subject membrane. Missing peptide specificity in the studied isolated peptides indicates that nature avoids peptide specificity in order not to favor certain pathogen agents in their blocking action.

Most peptides found experimentally show multiple actions on pathogen agents. Thus it appears that the detection and prediction of antibacterial peptides—in our case SCAAP—is more related to general, nonspecific peptide profiles that are well known for their antibacterial action. For that reason and as given in the present case, more efficient algorithms should rather evaluate fundamental characteristics of such peptides and search for small differences among them.

The design of bioinformatical algorithms to detect antimicrobial peptides is basically of two types.

(i) Based on a system of differential equations [15] that characterizes the peptide properties with an exponentially growing complexity.

(ii) The inclusion of multiple peptide characteristics without affecting its complexity [16] where the efficiency greatly depends on a skillful peptide set selection.

Our polarity index method falls in the latter category and is characterized by the following.

(i) Effectively excluding multiple action peptides, with a margin of error less than 10% and single-action peptides with a margin of error less than 6%.

(ii) Its efficiency to identify SCAAP subjects which is higher than 90%.

(iii) The simplicity of the computational method which is easy to implement for massive parallel processing in GPUs [17].

(iv) Its straightforwardness by measuring the peptide polarity exclusively and from this information effectively classifying its pathogenic action.

The algorithm involved in this method allows simple modifications to identify in a general level peptide groups by their pathogenic action and in a more specific level to refine the peptide search and identification as in the group used here.

The polarity index method uses the amino acid polarity classification; however there are other types of classifications [18, 19] that use the amino acid side chain chemical properties such as the neutral pH charge, their type of chemical structure, the reactivity, the elements present, or the ability to form hydrogen bonds. These classifications can be used to generate a more specific peptide blueprint when searched, with features that would not be considered otherwise.

As this method is a simple mathematical and computational algorithm, it does not demand heavy computational resources as processing memory or speed; therefore it can be used to explore peptide regions. These peptide regions can be worked out by evaluating massively all possible peptide combinations with the same length [20], thus taking advantage of the polarity index method simplicity to determine their activity.

5. Conclusion

The statistical/computacional polarity index method is an effective algorithm to find potential antibacterial peptides from a public domain database. These peptides have been denominated "Selective Cationic Amphipathic Antibacterial Peptides" (SCAAP). The method features a high efficiency to exclude peptides that exhibit single pathogenic action on other pathogens than bacteria, and it is equally efficient to exclude multiple-action peptides. In summary, the polarity index method is an adaptable and efficient method to detect and predict SCAAPs and it is a useful analysis and modeling tool for biological sequences using a single physicochemical property.

6. Availability

The polarity index computational implementation is listed in the Appendix section.

Appendix

Source Program for the Detection of SCAAP by the Polarity Index Method

```
c    Author Carlos Polanco 2011.

c

c    Program Detection of SCAAP by Polarity-index
     method.
C

c    Operating System: GNU Linux Fedora 14

c

c    Compilation: gfortran program. f

c

c    Execution: ./a.out AEVAPAPAAAAPAKAPKKKA-
     AAKPKKAGPS

c

c

     implicit none
     character * 1 arreglo(100), arreglo3(500)
     character * 500 backup
     character * 1 convert
     integer convertN, tipo2
     integer base(16), candidato(16), aciertos2, acier-
     tos0
     integer aciertost, aciertos3, aciertos4, aciertos14,
     aciertos24
     integer aciertos34, aciertos44, aciertos04, acier-
     tos1, aciertos5
     integer x1, x2, x3, x4, n, j, i, k
     real tipo1
     real comodin
     double precision matriz(4, 4)
     double precision total, peso(4, 4)
     equivalence (arreglo3, backup)
     open (2, file = "candidate0.dat")
34   format (f8.4, 1x, I2)
52   format (A3)
c    Relative frequency position of pairs of amino acid
     in the
c    candidate SCAAP
c

     peso (4, 4) = 0.272727281/0.272727281
     peso (1, 4) = 0.209790215/0.272727281
     peso (4, 1) = 0.164335668/0.272727281
     peso (1, 1) = 0.087412588/0.272727281
     peso (4, 3) = 0.083916083/0.272727281
     peso (3, 3) = 0.062937066/0.272727281
     peso (3, 4) = 0.059440561/0.272727281
     peso (3, 1) = 0.024475524/0.272727281
     peso (2, 1) = 0.006993007/0.272727281
     peso (1, 3) = 0.006993007/0.272727281
     peso (4, 2) = 0.006993007/0.272727281
     peso (2,4) = 0.003496503/0.272727281
     peso (2, 3) = 0.003496503/0.272727281
     peso (1, 2) = 0.003496503/0.272727281
     peso (3, 2) = 0.003496503/0.272727281
     peso (2, 2) = 0.000000000/0.272727281
c    Position of pairs of amino acid in the candidate
     SCAAP
c

     base(1) = 16
     base(2) = 4
     base(3) = 13
     base(4) = 15
     base(5) = 12
     base(6) = 1
     base(7) = 11
     base(8) = 9
     base(9) = 3
     base(10)= 14
     base(11)= 6
     base(12)= 8
     base(13)= 2
     base(14)= 7
```

```
      base(15)= 5
      base(16)= 10
      do i = 1, 4
         do j = 1, 4
         matriz (i, j) = 0
         enddo
      enddo
      x1 = 0
      x2 = 0
      x3 = 0
      x4 = 0
      total = 0
      k = 0
      n = 0
c     Command to gets the peptide (sequence of amino
      acid in letter-code)
c
      call getarg (1, backup)
      do i =1,500
         if (arreglo3(i). ne. " ") n = n + 1
      enddo
      do i = 1, n
         arreglo (i) = convert(arreglo3(i))
      enddo
c     Procedure to determine the relative frequency
c     distribution of amino acid in the sequence
c
      do i = 1, (n − 1)
         if (arreglo(i).eq. "1") x1 = x1 + 1
         if (arreglo(i).eq. "2") x2 = x2 + 1
         if (arreglo(i).eq. "3") x3 = x3 + 1
         if (arreglo(i).eq. "4") x4 = x4 + 1
         if (arreglo(i).eq. "0") goto 100
         if (arreglo(i).ne. "0") total = total +1
         matriz (convertN (arreglo (i)), convertN
         (arreglo (i + 1))) =
      &  matriz (convertN (arreglo (i)), convertN
         (arreglo (i + 1))) + 1
      enddo
100   do i = 1, 4
         do j = 1, 4
            k = k + 1
            write (2, 34) (matriz(i, j) * peso(i, j) *
      *1.3)/total, k
         enddo
```

```
      enddo
      close(1)
      close(2)
      call system ("sort −r candidate0.dat > candi-
      date1.dat")
      open (3, file = "candidate1.dat")
      open (4, file = "candidate0.dat")
      do i = 1, 16
         read (3, *) tipo1, tipo2
         write (4, *) tipo2
      enddo
      close(3)
      close(4)
      open (2, file = "candidate0.dat")
c     Procedure to evaluate if the sequence of peptide is
      or
c     not candidate SCAAP
c
      do i = 1, 16
      read (2, *, END = 101) candidato(i)
      enddo
      call parte04 (base, candidato, aciertos0)
      call parte14 (base, candidato, aciertos1)
      call parte54 (base, candidato, aciertos5)
      if  ((aciertos0.eq.1).  and.(aciertos1.eq.3).and.
      (aciertos5.eq.1))then
         write (6, 52) "Yes"
      else
         write (6, 52) "No"
        endif
      call system ("rm candidate0.dat")
      call system ("rm candidate1.dat")
101   stop
      end
c
c     Subroutines and functions
c
c     Verification of position 1
c
      subroutine parte04(base, candidato, aciertos0)
      integer base(16), candidato(16), aciertos0
      aciertos0 = 0
      if (candidato(1).eq. base(1)) aciertos0 = aciertos0
      + 1
      return
```

```
      end
c     Verification of positions 2, 3 and 4
c
      subroutine parte14(base, candidato, aciertos1)
      integer base(16), candidato(16), aciertos1
      aciertos1 = 0
      do i = 2, 4
      if (candidato(i).eq. base(i)) aciertos1= aciertos1 + 1
      enddo
      return
      end
c     Verification of position 16
c
      subroutine parte54 (base, candidato, aciertos5)
      integer base(16), candidato(16), aciertos5
      aciertos5 = 0
      if (candidato(16).eq. base(16)) aciertos5 = acier-
      tos5 + 1
      return
      end
c     Conversion letters to the corresponding groups of
      polarity (in numbers)
c
      character function convert(tipo)
      character * 1 tipo
      if (tipo.eq. "A") convert = "4"
      if (tipo.eq. "C") convert = "3"
      if (tipo.eq. "D") convert = "2"
      if (tipo.eq. "E") convert = "2"
      if (tipo.eq. "F") convert = "4"
      if (tipo.eq. "G") convert = "3"
      if (tipo.eq. "H") convert = "1"
      if (tipo.eq. "I") convert = "4"
      if (tipo.eq. "K") convert = "1"
      if (tipo.eq. "L") convert = "4"
      if (tipo.eq. "M") convert = "4"
      if (tipo.eq. "N") convert = "3"
      if (tipo.eq. "P") convert = "4"
      if (tipo.eq. "Q") convert = "3"
      if (tipo.eq. "R") convert = "1"
      if (tipo.eq. "S") convert = "3"
      if (tipo.eq. "T") convert = "3"
      if (tipo.eq. "V") convert = "4"
      if (tipo.eq. "W") convert = "4"
      if (tipo.eq. "Y") convert = "2"
      if (tipo.eq. "X") convert = "0"
      return
      end
c     Conversion number in code-letters to numbers in
      code-numbers
c
      integer function convertN(tipo)
      character * 1 tipo
      if (tipo.eq. "1") convertN = 1
      if (tipo.eq. "2") convertN = 2
      if (tipo.eq. "3") convertN = 3
      if (tipo.eq. "4") convertN = 4
      return
      end
```

Conflicts of Interest

We declare that we do not have any financial and personal relationship with other people or organizations that could inappropriately influence (bias) our work.

Authors Contribution

Experiments conception and design were done by C. Polanco and J. L. Samaniego. Experimental performance was made by C. Polanco. Data analysis was made by T. Buhse. Results discussion was made by: T. Buhse, F. G. Mosqueira, A. Negron-Mendoza, S. Ramos-Bernal, and J. A. Castanon-Gonzalez.

Acknowledgments

The authors acknowledge the support given by the Departamento de Computo and the Instituto de Ciencias Nucleares, Universidad Nacional Autonoma de México, and by Concepcion Celis Juarez for proofreading the paper.

References

[1] T. Unger, Z. Oren, and Y. Shai, "The effect of cyclization of magainin 2 and melittin analogues on structure, function, and model membrane interactions: implication to their mode of action," *Biochemistry*, vol. 40, no. 21, pp. 6388–6397, 2001.

[2] M. R. Yeaman and N. Y. Yount, "Mechanisms of antimicrobial peptide action and resistance," *Pharmacological Reviews*, vol. 55, no. 1, pp. 27–55, 2003.

[3] H. G. Boman, "Peptide antibiotics and their role in innate immunity," *Annual Review of Immunology*, vol. 13, pp. 61–92, 1995.

[4] J. M. Saugar, M. J. Rodríguez-Hernández, B. G. De La Torre et al., "Activity of cecropin A-melittin hybrid peptides against colistin-resistant clinical strains of Acinetobacter baumannii: molecular basis for the differential mechanisms of action," *Antimicrobial Agents and Chemotherapy*, vol. 50, no. 4, pp. 1251–1256, 2006.

[5] Y. Cao, R. Q. Yu, Y. Liu et al., "Design, recombinant expression, and antibacterial activity of the cecropins-melittin hybrid antimicrobial peptides," *Current Microbiology*, vol. 61, no. 3, pp. 169–175, 2010.

[6] L. Z. Wan, Y. Park, I. S. Park et al., "Improvement of bacterial cell selectivity of melittin by a single Trp mutation with a peptoid residue," *Protein and Peptide Letters*, vol. 13, no. 7, pp. 719–725, 2006.

[7] G. Del Rio, S. Castro-Obregon, R. Rao, H. M. Ellerby, and D. E. Bredesen, "APAP, a sequence-pattern recognition approach identifies substance P as a potential apoptotic peptide," *FEBS Letters*, vol. 494, no. 3, pp. 213–219, 2001.

[8] H. M. Ellerby, W. Arap, L. M. Ellerby et al., "Anti-cancer activity of targeted pro-apoptotic peptides," *Nature Medicine*, vol. 5, no. 9, pp. 1032–1038, 1999.

[9] S. Y. Shin, J. H. Kang, S. Y. Jang, Y. Kim, K. L. Kim, and K. S. Hahm, "Effects of the hinge region of cecropin A(1-8)-magainin 2(1-12), a synthetic antimicrobial peptide, on liposomes, bacterial and tumor cells," *Biochimica et Biophysica Acta*, vol. 1463, no. 2, pp. 209–218, 2000.

[10] R. E. Hausman and G. M. Cooper, *The Cell: A Molecular Approach*, ASM Press, Washington, DC, USA, 2004.

[11] G. Wang, X. Li, and Z. Wang, "APD2: the updated antimicrobial peptide database and its application in peptide design," *Nucleic Acids Research*, vol. 37, no. 1, pp. D933–D937, 2009.

[12] K. Iwai and T. Ando, "N → O acyl rearrangement," *Methods in Enzymology*, vol. 11, pp. 263–282, 1967.

[13] C. Polanco and J. L. Samaniego, "Detection of selective cationic amphipatic antibacterial peptides by Hidden Markov models," *Acta Biochimica Polonica*, vol. 56, no. 1, pp. 167–176, 2009.

[14] D. Eisenberg, R. M. Weiss, T. C. Terwilliger, and W. Wilcox, "Hydrophobic moments and protein structure," *Faraday Symposia of the Chemical Society*, vol. 17, pp. 109–120, 1982.

[15] E. L. Ince, *Ordinary Differential Equations*, Dover, New York, NY, USA, 1956.

[16] B. Resch, "Hidden Markov Models. A tutorial for the course computational intelligence. Signal processing and speech communication laboratory," 2004, http://speech.tifr.res.in/tutorials/hmmTutExamplesAlgo.pdf.

[17] W. M. W. Hwu, *GPU Computing Gems Emerald Edition*, Elsevier, New York, NY, USA, 2011.

[18] J. Davies and B. Shaffer Littlewood, *Elementary Biochemistry—An Introduction to the Chemistry of Living Cells*, Prentice Hall, Upper Saddle River, NJ, USA, 1979.

[19] The Australian Naturopathic Network, 1998–2002, http://www.ann.com.au/MedSci/amino.htm.

[20] C. Polanco González, M. A. Nuño Maganda, M. Arias-Estrada, and G. del Rio, "An FPGA implementation to detect selective cationic antibacterial peptides," *PloS one*, vol. 6, no. 6, article e21399, 2011.

[21] I. H. Lee, C. Zhao, Y. Cho, S. S. L. Harwig, E. L. Cooper, and R. I. Lehrer, "Clavanins, α-helical antimicrobial peptides from tunicate hemocytes," *FEBS Letters*, vol. 400, no. 2, pp. 158–162, 1997.

[22] Y. J. Basir, F. C. Knoop, J. Dulka, and J. M. Conlon, "Multiple antimicrobial peptides and peptides related to bradykinin and neuromedin N isolated from skin secretions of the pickerel frog, Rana palustris," *Biochimica et Biophysica Acta*, vol. 1543, no. 1, pp. 95–105, 2000.

[23] J. M. Conlon, M. Mechkarska, E. Ahmed et al., "Host defense peptides in skin secretions of the Oregon spotted frog Rana pretiosa: implications for species resistance to chytridiomycosis," *Developmental and Comparative Immunology*, vol. 35, no. 6, pp. 644–649, 2011.

[24] J. Goraya, Y. Wang, Z. Li et al., "Peptides with antimicrobial activity from four different families isolated from the skins of the North American frogs Rana luteiventris, Rana berlandieri and Rana pipiens," *European Journal of Biochemistry*, vol. 267, no. 3, pp. 894–900, 2000.

[25] J. M. Conlon, J. Kolodziejek, N. Nowotny et al., "Characterization of antimicrobial peptides from the skin secretions of the Malaysian frogs, Odorrana hosii and Hylarana picturata (Anura:Ranidae)," *Toxicon*, vol. 52, no. 3, pp. 465–473, 2008.

[26] J. Goraya, F. C. Knoop, and J. M. Conlon, "Ranatuerins: antimicrobial peptides isolated from the skin of the American bullfrog, Rana catesbeiana," *Biochemical and Biophysical Research Communications*, vol. 250, no. 3, pp. 589–592, 1998.

[27] A. H. Thompson, A. J. Bjourson, D. F. Orr, C. Shaw, and S. McClean, "A combined mass spectrometric and cDNA sequencing approach to the isolation and characterization of novel antimicrobial peptides from the skin secretions of Phyllomedusa hypochondrialis azurea," *Peptides*, vol. 28, no. 7, pp. 1331–1343, 2007.

[28] K. Conceição, K. Konno, M. Richardson et al., "Isolation and biochemical characterization of peptides presenting antimicrobial activity from the skin of Phyllomedusa hypochondrialis," *Peptides*, vol. 27, no. 12, pp. 3092–3099, 2006.

[29] J. R. S. A. Leite, L. P. Silva, M. I. S. Rodrigues et al., "Phylloseptins: a novel class of anti-bacterial and anti-protozoan peptides from the Phyllomedusa genus," *Peptides*, vol. 26, no. 4, pp. 565–573, 2005.

[30] D. Vanhoye, F. Bruston, S. El Amri, A. Ladram, M. Amiche, and P. Nicolas, "Membrane association, electrostatic sequestration, and cytotoxicity of Gly-Leu-rich peptide orthologs with differing functions," *Biochemistry*, vol. 43, no. 26, pp. 8391–8409, 2004.

[31] R. Galinier, E. Roger, P. E. Sautiere, A. Aumelas, B. Banaigs, and G. Mitta, "Halocyntin and papillosin, two new antimicrobial peptides isolated from hemocytes of the solitary tunicate, Halocynthia papillosa," *Journal of Peptide Science*, vol. 15, no. 1, pp. 48–55, 2009.

[32] R. C. Anderson, *Antimicrobial peptides isolate from ovine blood neutrophils*, thesis, Massey University, Palmerston, North New Zealand, 2005.

[33] M. J. Maclean, C. S. Brinkworth, D. Bilusich et al., "New caerin antibiotic peptides from the skin secretion of the Dainty Green Tree Frog Litoria gracilenta. Identification using positive and negative ion electrospray mass spectrometry," *Toxicon*, vol. 47, no. 6, pp. 664–675, 2006.

[34] R. J. Jackway, J. H. Bowie, D. Bilusich et al., "The fallaxidin peptides from the skin secretion of the Eastern Dwarf Tree Frog Litoria fallax. Sequence determination by positive and negative ion electrospray mass spectrometry: antimicrobial activity and cDNA cloning of the fallaxidins," *Rapid Communications in Mass Spectrometry*, vol. 22, no. 20, pp. 3207–3216, 2008.

[35] K. L. Wegener, C. S. Brinkworth, J. H. Bowie, J. C. Wallace, and M. J. Tyler, "Bioactive dahlein peptides from the skin secretions of the Australian aquatic frog Litoria dahlii: sequence determination by electrospray mass spectrometry," *Rapid Communications in Mass Spectrometry*, vol. 15, no. 18, pp. 1726–1734, 2001.

[36] D. J. M. Stone, R. J. Waugh, J. H. Bowie, J. C. Wallace, and M. J. Tyler, "Peptides from Australian frogs. Structures of

the caeridins from Litoria caerulea," *Journal of the Chemical Society, Perkin Transactions 1*, no. 5, pp. 573–576, 1993.

[37] E. Iwakoshi-Ukena, G. Okada, A. Okimoto, T. Fujii, M. Sumida, and K. Ukena, "Identification and structure-activity relationship of an antimicrobial peptide of the palustrin-2 family isolated from the skin of the endangered frog Odorrana ishikawae," *Peptides*, vol. 32, no. 10, pp. 2052–2057, 2011.

[38] K. S. Moore, C. L. Bevins, M. M. Brasseur et al., "Antimicrobial peptides in the stomach of Xenopus laevis," *Journal of Biological Chemistry*, vol. 266, no. 29, pp. 19851–19857, 1991.

[39] A. M. Cole, P. Weis, and G. Diamond, "Isolation and characterization of pleurocidin, an antimicrobial peptide in the skin secretions of winter flounder," *Journal of Biological Chemistry*, vol. 272, no. 18, pp. 12008–12013, 1997.

[40] N. Boulanger, R. Brun, L. Ehret-Sabatier, C. Kunz, and P. Bulet, "Immunopeptides in the defense reactions of Glossina morsitans to bacterial and Trypanosoma brucei brucei infections," *Insect Biochemistry and Molecular Biology*, vol. 32, no. 4, pp. 369–375, 2002.

[41] V. Čeřovský, J. Slaninová, V. Fučík et al., "New potent antimicrobial peptides from the venom of Polistinae wasps and their analogs," *Peptides*, vol. 29, no. 6, pp. 992–1003, 2008.

[42] S. M. Yoe, C. S. Kang, S. S. Han, and I. S. Bang, "Characterization and cDNA cloning of hinnavin II, a cecropin family antibacterial peptide from the cabbage butterfly, Artogeia rapae," *Comparative Biochemistry and Physiology B*, vol. 144, no. 2, pp. 199–205, 2006.

[43] P. L. Yu, S. D. Choudhury, and K. Ahrens, "Purification and characterization of the antimicrobial peptide, ostricacin," *Biotechnology Letters*, vol. 23, no. 3, pp. 207–210, 2001.

[44] C. Zhao, L. Liaw, I. Hee Lee, and R. I. Lehrer, "cDNA cloning of Clavanins: antimicrobial peptides of tunicate hemocytes," *FEBS Letters*, vol. 410, no. 2-3, pp. 490–492, 1997.

[45] L. M. Cintas, P. Casaus, C. Herranz et al., "Biochemical and genetic evidence that Enterococcus faecium L50 produces enterocins L50A and L50B, the sec-dependent enterocin p, and a novel bacteriocin secreted without an N-terminal extension termed enterocin Q," *Journal of Bacteriology*, vol. 182, no. 23, pp. 6806–6814, 2000.

[46] M. E. Selsted, Y. Q. Tang, W. L. Morris et al., "Purification, primary structures, and antibacterial activities of β-defensins, a new family of antimicrobial peptides from bovine neutrophils," *Journal of Biological Chemistry*, vol. 268, no. 9, pp. 6641–6648, 1993.

[47] H. P. Jia, S. A. Wowk, B. C. Schutte et al., "A novel murine β-defensin expressed in tongue, esophagus, and trachea*," *Journal of Biological Chemistry*, vol. 275, no. 43, pp. 33314–33320, 2000.

[48] O. Shamova, K. A. Brogden, C. Zhao, T. Nguyen, V. N. Kokryakov, and R. I. Lehrer, "Purification and properties of proline-rich antimicrobial peptides from sheep and goat leukocytes," *Infection and Immunity*, vol. 67, no. 8, pp. 4106–4111, 1999.

[49] L. Bagella, M. Scocchi, and M. Zanetti, "cDNA sequences of three sheep myeloid cathelicidins," *FEBS Letters*, vol. 376, no. 3, pp. 225–228, 1995.

[50] D. Romeo, B. Skerlavaj, M. Bolognesi, and R. Gennaro, "Structure and bactericidal activity of an antibiotic dodecapeptide purified from bovine neutrophils," *Journal of Biological Chemistry*, vol. 263, no. 20, pp. 9573–9575, 1988.

[51] I. Zelezetsky, A. Pontillo, L. Puzzi et al., "Evolution of the primate cathelicidin: correlation between structural variations and antimicrobial activity," *Journal of Biological Chemistry*, vol. 281, no. 29, pp. 19861–19871, 2006.

[52] P. Storici, A. Tossi, B. Lenarčič, and D. Romeo, "Purification and structural characterization of bovine cathelicidins, precursors of antimicrobial peptides," *European Journal of Biochemistry*, vol. 238, no. 3, pp. 769–776, 1996.

[53] R. C. Anderson and P. L. Yu, "Isolation and characterisation of proline/arginine-rich cathelicidin peptides from ovine neutrophils," *Biochemical and Biophysical Research Communications*, vol. 312, no. 4, pp. 1139–1146, 2003.

[54] S. S. Park, S. W. Shin, M. K. Kim, D. S. Park, H. W. Oh, and H. Y. Park, "Differences in the skin peptides of the male and female Australian tree frog Litoria splendida: the discovery of the aquatic male sex pheromone splendipherin, together with Phe8 caerulein and a new antibiotic peptide caerin 1.10," *European Journal of Biochemistry*, vol. 267, no. 1, pp. 269–275, 2000.

[55] I. Morishima, S. Suginaka, T. Ueno, and H. Hirano, "Isolation and structure of cecropins, inducible antibacterial peptides, from the silkworm Bombyx mori," *Comparative Biochemistry and Physiology B*, vol. 95, no. 3, pp. 551–554, 1990.

[56] C. Kaletta, K. D. Entian, R. Kellner, G. Jung, M. Reis, and H. G. Sahl, "Pep5, a new lantibiotic: structural gene isolation and prepeptide sequence," *Archives of Microbiology*, vol. 152, no. 1, pp. 16–19, 1989.

[57] C. Samakovlis, P. Kylsten, D. A. Kimbrell, A. Engstrom, and D. Hultmark, "The Andropin gene and its product, a male-specific antibacterial peptide in Drosophila melanogaster," *EMBO Journal*, vol. 10, no. 1, pp. 163–169, 1991.

22

Antigenic Peptides Capable of Inducing Specific Antibodies for Detection of the Major Alterations Found in Type 2B Von Willebrand Disease

Marina de Oliveira Paro,[1] **Cyntia Silva Ferreira,**[1] **Fernanda Silva Vieira,**[1]
Marcos Aurélio de Santana,[1] **William Castro-Borges,**[1]
Maria Sueli Silva Namen-Lopes,[2] **Sophie Yvette Leclercq,**[3]
Cibele Velloso-Rodrigues,[4] **and Milton Hércules Guerra de Andrade**[1]

[1] *Departamento de Ciências Biológicas/DECBI, Instituto de Ciências Exatas e Biológicas/ICEB,*
  Universidade Federal de Ouro Preto, Núcleo de Pesquisas em Ciências Biológicas/NUPEB,
  Campus Universitário Morro do Cruzeiro, 35400-000 Ouro Preto, MG, Brazil
[2] *Fundação Centro de Hematologia e Hemoterapia, Hemominas, Serviço de Pesquisa, Alameda Ezequiel Dias, 321,*
  Bairro Santa Efigênia, 30130-110 Belo Horizonte, MG, Brazil
[3] *Fundação Ezequiel Dias, FUNED, Rua Conde Pereira Carneiro, 80, Bairro Gameleira, 30510-010 Belo Horizonte, MG, Brazil*
[4] *Departamento Básico, Instituto de Ciências Biológicas, Universidade Federal de Juiz de Fora, Campus de Governador Valadares,*
  Área da Saúde, Avenida Doutor Raimundo Monteiro de Resende, 330, Centro, 35010-177 Governador Valadares, MG, Brazil

Correspondence should be addressed to Milton Hércules Guerra de Andrade; miltonguerra00@gmail.com

Academic Editor: Jean-Marie Zajac

Von Willebrand disease (VWD) is an inherited hemorrhagic disorder promoted by either quantitative or qualitative defects of the von Willebrand factor (VWF). The disease represents the most common human coagulopathy afflicting 1.3% of the population. Qualitative defects are subdivided into four subtypes and classified according to the molecular dysfunction of the VWF. The differential diagnosis of the VWD is a difficult task, relying on a panel of tests aimed to assess the plasma levels and function of the VWF. Here, we propose biochemical approaches for the identification of structural variants of the VWF. A bioinformatic analysis was conducted to design seven peptides among which three were representatives of specific amino acid sequences belonging to normal VWF and four encompassed sequences found in the most common VWD subtype 2B. These peptides were used to immunize mice, after which, peptide-specific immunoglobulins were purified. This resulted in four Ig preparations capable of detecting alterations in the subtype 2B VWD plus additional three antibody fractions targeting the normal VWF. The panel of antibodies could serve many applications among them (1) assessment of VWF: antigen interaction, (2) VWF multimer analysis, and (3) production of monoclonal antibodies against VWF for therapeutic purposes as in thrombotic thrombocytopenic purpura.

1. Introduction

Von Willebrand disease (VWD) is an inherited hemorrhagic disturbance related to quantitative and/or qualitative defects of the von Willebrand factor (VWF) [1, 2]. VWD prevalence varies between 0.8 and 2.0%, depending on the investigated population, being considered the most common coagulopathy afflicting humans [3].

The VWF is a multimeric plasma protein, composed of a varying number of 250 kDa monomers, which exhibits an essential role in primary haemostasis. Some of its reported functions are the attachment of platelets to subendothelial collagen at injured sites (platelet plug formation) and protection, binding, and transportation of coagulation factor VIII [4]. Mechanisms leading to the disease present themselves as highly diverse at the molecular level, giving rise to a variety of clinical outcomes. Through different laboratory criteria it is possible to identify three primary types of the disease [5]. Alterations on the plasma levels of VWF are associated with VWD types 1 and 3, whereas structural and functional defects of VWF result in VWD type 2 [3, 6–11]. VWD types 1 and 3 reflect, respectively, partial and complete deficiency of the VWF. VWD type 2 is also classified into four subtypes. In subtype 2A, there is enhanced platelet adhesion caused by the selective deficiency of high molecular mass multimers of VWF (HMW). In subtype 2B VWD, it is observed increased affinity of VWF by platelet glycoprotein Ib (GpIb) associated with loss of HMW-VWF and mild thrombocytopenia. In contrast, in subtype 2M, a pronounced reduction in platelet adhesion is described, even considering the relatively normal size of the VWF multimers. Variants of the VWF in the subtype 2N display reduced affinity by coagulation factor VIII (FVIII). Altogether these six VWD types correlate with diverse clinical outcomes each requiring adequate therapeutic interventions [4, 12]. Whilst some patients with type 2B VWD can be treated with desmopressin, patients who do not show a satisfactory response to this drug should receive VWF/FVIII-containing products [13, 14].

Considering the current difficulties associated with diagnosis of the qualitative defects of the VWF, the development of novel biochemical approaches that would allow detection of such molecular alterations is of great biotechnological interest. The possibility of a precise and direct diagnosis of qualitative defects in the VWF would certainly permit application of a better oriented medical approach to afflicted patients. In this context, it is worth emphasizing that antibodies capable of detecting structural variants of the VWF are not commercially available. The design of peptides for synthesis and further generation of antibodies to be employed in the identification of structural alterations in the VWF might therefore represent a convenient way forward.

It has previously been shown that amino acid substitutions R1306W, R1308C, V1316M, and R1341Q account for 90% prevalence of the subtype 2B VWD [6, 8]. These mutations in the A1 domain of VWF are responsible for a "gain-of-function" defect allowing for an increased affinity of large multimers to platelets in the circulation [15]. In the present investigation we have generated a panel of antipeptide-specific antibodies useful at detecting the aforementioned structural variants of the VWF.

2. Materials and Methods

2.1. Ethics Statement.
All experiments involving mice were conducted according to approved guidelines for animal use and care defined by the Local Ethics Committee on Animal Experimentation (CEUA/UFOP). Healthy individuals were informed previously of the investigatory nature of this study, and after giving their consent, plasma samples were obtained under the procedures approved by the Local Ethics Committee from Hemominas Foundation, MG, Brazil.

2.2. Selection of VWF Peptides for Synthesis.
As stated previously the VWF amino acid substitutions R1306W, R1308C,

V1316M, and R1341Q are collectively found in the majority of subtype 2B VWD patients [6]. In order to design signature peptides that would represent such mutations, *Homo sapiens* VWF sequences were retrieved from the International Society on Thrombosis and Haemostasis database, available at http://www.vwf.group.shef.ac.uk/. Predicted peptide sequences for synthesis obeyed the following criteria: (1) peptide size should be within the 8 to 10 mer range; (2) peptides should contain one aromatic residue for estimation of peptide yield after synthesis and HPLC purification (in the absence of an aromatic residue, this was added at the peptide C-terminal); (3) the location of the peptide sequences in the crystal structure of the VWF (available at http://www.pdbj.org/), revealed by the Swiss-Pdb Viewer 3.7, should demonstrate their exposure to the solvent, meaning that highly hydrophobic peptide sequences were not considered in this study.

The selected peptide sequences were then aligned to *Mus musculus* VWF sequence using ClustalW2 (http://www.ebi.ac.uk/Tools/msa/clustalw2/) to reveal species specific amino acid differences as predictors for successful production of anti-VWF peptide antibodies in mice.

2.3. Synthesis, Characterization, and Production of Polyclonal Antibodies against VWF Peptides.
Peptides selected for synthesis were representatives of the normal and altered versions of the VWF found in the major subtypes of type 2 VWD (Table 1). Peptides were synthesized using the solid-phase protocol essentially as described by Merrifield [16]. Briefly, the support matrix consisted of Rink Amide Resin HL (Merck, Germany) at 0.78 mmol/g for an expected maximum yield of 40 μM of peptides per synthesis, using Fmoc-derivatized amino acids. Synthetic peptides were then purified via reversed-phase chromatography using a C18 column (Shim-pack CLD-ODS Shimadzu, Japan) on a Shimadzu HPLC system. Selected peptide peaks were recovered for analysis through direct injection onto an electrospray-operating mass spectrometer (LCMS-IT-ToF, Shimadzu, Japan) in positive ionization mode and capillary voltage set to 4,300 V. Mass spectrometric data were acquired over 10 ms, after which m/z values obtained were compared with the expected molecular masses for the synthesized peptides.

Synthetic peptides representatives of both normal and altered VWF were individually coupled to the highly immunogen carrier protein keyhole limpet hemocyanin (KLH, Sigma) at 1:1 ratio (1 mg peptide: 1 mg KLH) using glutaraldehyde as the crosslinking agent [17]. Polyclonal anti-KLH peptide antibodies were raised in male *Swiss* mice aged ten weeks. Immunization regimen consisted of three intraperitoneal administrations of 50 μg of the KLH-peptide conjugate prepared in 10% aluminium hydroxide as adjuvant, each at a 15 days interval. Blood was withdrawn after 45 days postimmunization.

2.4. Immobilization of Synthetic Peptides on Sepharose-4B and Purification of Anti-VWF Specific Immunoglobulins.
Affinity columns (0.2 mL) containing immobilized synthetic peptides, at approximately 10 mg/mL, were produced as

Antigenic Peptides Capable of Inducing Specific Antibodies for Detection of the Major Alterations Found in
Type 2B Von Willebrand Disease

177

TABLE 1: Proposed synthetic peptides for detection of the major qualitative alterations found in VWD type 2B.

Major qualitative alterations found in VWD type 2B	Designed 8 mer peptides for synthesis	Theoretical [M + H]$^+$	Observed [M + H]$^+$	Respective peptide from _M. musculus_ VWF
R1306W	MEWLRISY	1097.54	1113.56*	
R1308C	MERLCISY	1014.47	1014.50	MERLHIS
	Respective peptide found in _H. sapiens_ VWF: MERLRISY	1067.56	1082.57*	
V1316M	SQKWVRMA	1005.52	1020.54*	SQKRIRVA
	Respective peptide found in _H. sapiens_ VWF: SQKWVRVA	973.55	972.57 [−1H$^+$]	
R1341Q	RPSELQRY	1048.55	1048.57	RPSELRR
	Respective peptide found in _H. sapiens_ VWF: RPSELRRY	1076.59	1076.62	

*Peptide mass containing an oxidized methionine.

described previously [18]. The antisera from mice immunized with KLH coupled to altered VWF peptides were loaded individually in affinity columns bearing the corresponding normal versions of the VWF peptide. This procedure aimed the subtraction of antibodies targeting the normal VWF. Approximately 100 μL of the unbound fraction was collected for analysis during the first chromatographic step, whilst the remaining were immediately submitted to further six rounds of chromatography to guarantee complete removal of antinormal VWF antibodies. For each chromatographic step bound fractions were eluted in 5 mL 0.1 M glycine pH 2.6 and collected in tubes containing 5 mL of 0.4 M Tris-HCl pH 8.0 for pH neutralization. Alkaline phosphatase labeled anti-mouse IgG, at 1:2000 dilution, was used to detect the presence of IgGs found in both nonretained and retained fractions from each chromatographic step, using the western blotting technique [19].

The next approach involved confirmation that antibodies targeting the normal VWF have been thoroughly subtracted after the six passages in affinity columns containing immobilized altered VWF peptides. For this purpose, 1 mL of pooled plasma from human healthy donors ($n = 5$) was first precipitated in 20% ethanol to enrich for normal VWF. Approximately 10 μg aliquots of the precipitated plasma were resuspended in protein loading buffer and loaded in three different lanes for separation on a 7% SDS-PAGE. The gel was transferred to a PVDF (polyvinylidene fluoride) membrane and each lane western blotted with either a 100 μL fraction obtained from the crude antisera or the two 100 μL nonretained fractions recovered from the first and sixth chromatographic steps.

The combined eluates from each chromatographic step containing immunoglobulins targeting a given altered VWF peptide were lyophilized, resuspended in 1 mL of phosphate buffer pH 7.4, and individually loaded onto the Sepharose-4B affinity column bearing the respective altered VWF peptide. Bound IgGs targeting the altered VWF peptide were eluted exactly as described above.

2.5. Production of Carrier Albumins Bearing Altered VWF Synthetic Peptides and Their Recognition by Anti-VWF Specific Antibodies. Aiming to produce a model sample for detection

of altered VWF through the western blotting technique, purified albumin was conjugated to two representative synthetic peptides (SQKWVRMA and RPSELQRY, Table 1) [20] at a 1:1 ratio, respectively. Briefly, to each 1 mg of the respective peptides it was added 17 μL of N-N′ diisopropylcarbodiimide, diluted 1:25 in dimethylformamide (DMF), and 20 μL of 1 M N-hydroxysuccinimide (also diluted in DMF). Reaction proceeded during 5 min at room temperature, after which 50 μL of 0.4 M sodium acetate (in DMF) was added. After further 2 min incubation, at room temperature, 1 mL of bovine serum albumin (BSA, Sigma), at 1 mg/mL, prepared in 0.05 M ammonium bicarbonate pH 7.5, was combined. Coupling reactions occurred during 1 h followed by dialysis of each preparation in 0.1 M ammonium acetate over 24 h. Peptide-derivatized albumins were lyophilized and resuspended in 100 μL of saline.

Five μg aliquots of peptide-conjugated albumin were loaded into three lanes and separated through 12% SDS-PAGE, followed by Coomassie staining. A replica gel was produced and transferred to a PVDF membrane. Antipeptide-specific IgG purified through affinity chromatography, at approximately 0.15 μg/μL, was used as the primary antibody, at a final dilution of 1:500. The reaction was allowed to proceed for 3 h at room temperature. Development of the reactive band was achieved using alkaline phosphatase labeled anti-mouse IgG (Sigma), at 1:2000 dilution for 2 h, followed by addition of the alkaline phosphatase substrates nitroblue tetrazolium/5-bromo-4-chloro-3-indolyl-phosphate (NBT/BCIP, Sigma) as per the manufacturer's instructions.

3. Results and Discussion

In this study four amino acid sequences, here termed altered VWF peptides (MEWLRISY, MERLCISY, SQKWVRMA, and RPSELQRY), were synthesized for generation of antibodies in mice. These were intended for recognition of the most common qualitative defects of VWF found in subtype 2B VWD. Three additional peptide sequences MERLRISY, SQKWVRVA, and RPSELRRY, representatives of the respective normal versions of the VWF peptides, were also produced and immobilized onto Sepharose 4B, for depleting

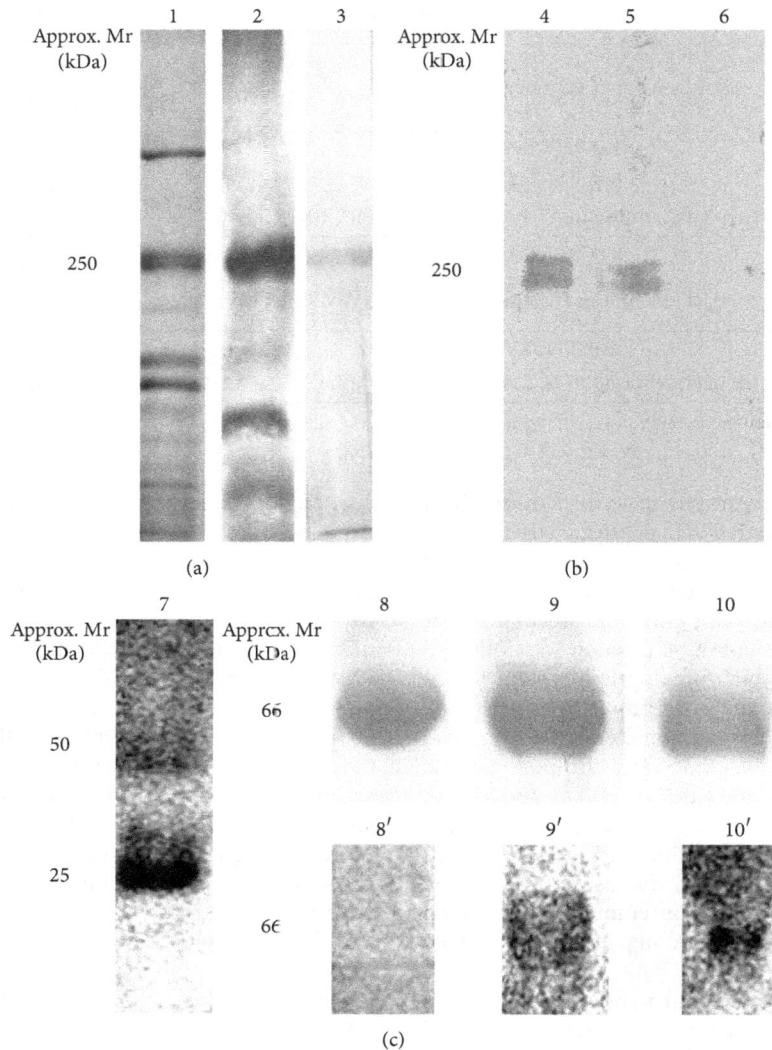

FIGURE 1: SDS-PAGE and western blotting approaches featuring the major steps involved in the production of a panel of anti-(KLH-peptide) antibodies for detection of VWD subtype 2B. (a) *Lane 1*, 10% SDS-PAGE profile of pooled human plasma after enrichment for normal VWF using ethanol precipitation; *Lane 2*, western blotting for detection of normal VWF, at approximately 250 kDa, using the commercially available antibody against VWF/FVIII—note the presence of additional protein bands being recognized, particularly at lower mass; *Lane 3*, a representative Western blotting reaction obtained with the generated panel of anti-(KLH-peptide) antibodies, targeting both normal and altered VWF. (b) Detection of normal VWf by western blotting following the subtractive affinity chromatography; *Lane 4*, detection of VWF prior to depletion of antibodies against the normal factor; *Lanes 5* and *6*, detection of VWF using the nonretained fractions from the first and sixth chromatographic steps, respectively—note that 5 column passages proved sufficient for complete removal of anti-(KLH-peptide) antibodies. (c) *Lane 7*, 10% SDS-PAGE profile representative of the eluates obtained after affinity purification of antibodies targeting specifically altered VWf peptides—note the presence of IgG heavy and light chains at approximately 50 and 25 kDa, respectively; *Lanes 8, 9*, and *10*, 10% SDS-PAGE profile of nonderivatized BSA, BSA-(RPSELRR) and BSA-(SQKRIRVA), respectively; *Lanes 8', 9'*, and *10'*, corresponding western blotting reactions obtained using control sera, anti-(KLH-RPSELRR) and anti-(KLH-SQKRIRVA), respectively.

out immunoglobulins targeting the normal VWF from the antisera raised with the altered VWF peptides. Given that the corresponding *Mus musculus* VWF peptides are MERLHIS, SQKRIRVA, and RPSELRR, we anticipated that the observed amino acid differences between human and mice VWF would justify the peptides capabilities for generating specific antibodies in mice (Table 1).

After immunization, the obtained antisera were tested for their ability to specifically recognize the VWF enriched

from plasma samples of human healthy donors. Figure 1(a), lane 3 is a representative reactivity obtained for all antisera raised with either the altered or normal versions of VWF peptides. A unique band at approximately 250 kDa was observed coinciding exactly with a major band, at the same mass, observed in Figure 1(a), lane 2. This was probed with a commercially available antiserum against VWF-FVIII. In contrast, the latter antibody preparation do recognize other protein bands (particularly of lower masses) in the sample

Antigenic Peptides Capable of Inducing Specific Antibodies for Detection of the Major Alterations Found in
Type 2B Von Willebrand Disease

179

(Figure 1(a), lane 2) revealing that the antisera generated herein are more suitable for specific recognition of the human VWF.

The result obtained in Figure 1(a), lane 3 is a proof that the antisera produced using altered VWF peptides also contain IgGs targeting the normal VWF. Aiming to remove those immunoglobulins from the preparations, a subtraction experiment was conducted. Firstly, each individual antiserum obtained by immunization with a given altered VWF peptide was submitted to affinity chromatography on a Sepharose 4B column, containing the respective immobilized normal VWF peptide. Through collection of both non-retained and retained fractions during 6 chromatographic rounds, it was observed complete removal of IgGs targeting normal VWF. This can be visualized in Figure 1(b) (lanes 4, 5, and 6), which represents the enriched plasma probed with the antiserum prior to affinity chromatography (lane 4) or probed with the retained fractions from the first (lane 5) and sixth (lane 6) column passages. From this result we expected the immunoglobulins targeting specifically the altered VWF peptides to be found in the combined non-retained fractions.

Isolation of anti-altered VWF specific immunoglobulins was achieved through new rounds of affinity chromatographies employing Sepharose 4B bearing immobilized altered VWF peptides. Bound immunoglobulins were eluted from each individual column and the presence of specific IgGs (heavy and light chains at approx. 50 and 25 kDa, resp.) confirmed by western blotting, using alkaline phosphatase labeled anti-mouse IgG (Figure 1(c), lane 7).

Given the experienced difficulty in obtaining a plasma sample from a VWD patient, for whom a qualitative alteration of the subtype 2B has been undoubtedly confirmed, we have engineered a model protein as a manner to test the diagnostic usefulness of our produced antisera. In this approach, nonconjugated BSA, BSA-SQKWVRMA plus BSA-RPSELQRY were separated using 1D 12% SDS-PAGE and the gel lanes western blotted with their respective peptide-specific antisera. As shown in Figure 1(c), nonconjugated albumin, which served as a control, was not detected by any of the anti-altered VWF peptide specific antiserum (Figure 1(c), represented by lane 8/8′). In contrast, the two derivatized albumins were recognized by the respective antisera generated against the two aforementioned altered VWF peptides (Figure 1(c), lanes 9/9′ and 10/10′).

Finally, it is worth mentioning that the protocol used to produce the chimera albumins guarantees minimal coupling of the peptides. This should have resulted in a more realistic model samples being produced. Such information is of relevance considering that high sensitivity for detecting qualitative alterations in the VWF in human plasma is obviously desirable.

4. Conclusions

In this study, a combination of bioinformatic analysis, peptide synthesis, and affinity chromatography allowed the generation of antipeptide specific antibodies capable of recognizing the most common qualitative defects associated with type 2B VWD. These should provide speed and innovation when potentially applied to the diagnosis of human VWD.

Abbreviations

VWF: Von Willebrand factor
VWD: Von Willebrand disease.

Conflict of Interests

Authors do not declare conflict of interests.

Acknowledgments

This work was supported in part by the Conselho Nacional de Desenvolvimento Científico e Tecnológico (MCT/CNPq/CT-SAÚDE no. 57/2010–Clinical Genetics) and Fundação de Amparo a Pesquisa do Estado de Minas Gerais (FAPEMIG-CDS-APQ-03515-10 and CBB-APQ-03096-09).

References

[1] U. Budde and R. Schneppenheim, "von Willebrand factor and von Willebrand disease," *Reviews in Clinical and Experimental Hematology*, vol. 5, no. 4, pp. 335–368, 2001.

[2] A. H. James, "Von Willebrand disease in women: awareness and diagnosis," *Thrombosis Research*, vol. 124, no. 1, pp. S7–S10, 2009.

[3] E. J. Werner, E. H. Broxson, E. L. Tucker, D. S. Giroux, J. Shults, and T. C. Abshire, "Prevalence of von Willebrand disease in children: a multiethnic study," *Journal of Pediatrics*, vol. 123, no. 6, pp. 893–898, 1993.

[4] D. Lillicrap, "Von Willebrand disease-Phenotype versus genotype: deficiency versus disease," *Thrombosis Research*, vol. 120, no. 1, pp. S11–S16, 2007.

[5] S. Keeney and A. M. Cumming, "The molecular biology of von willebrand disease," *Clinical and Laboratory Haematology*, vol. 23, no. 4, pp. 209–230, 2001.

[6] G. Castaman, A. B. Federici, F. Rodeghiero, and P. M. Mannucci, "Von Willebrand's disease in the year 2003: towards the complete identification of gene defects for correct diagnosis and treatment," *Haematologica*, vol. 88, no. 1, pp. 94–108, 2003.

[7] G. Castaman, J. C. J. Eikenboom, R. M. Bertina, and F. Rodeghiero, "Inconsistency of association between type 1 von Willebrand disease phenotype and genotype in families identified in an epidemiological investigation," *Thrombosis and Haemostasis*, vol. 82, no. 3, pp. 1065–1070, 1999.

[8] D. Ginsburg, J. E. Sadler, T. Abe et al., "Von Willebrand disease: a database of point mutations, insertions, and deletions," *Thrombosis and Haemostasis*, vol. 69, no. 2, pp. 177–184, 1993.

[9] F. Rodeghiero, G. Castaman, and A. Tosetto, "Von Willebrand factor antigen is less sensitive than ristocetin cofactor for the diagnosis of type I von Willebrand disease—results based on an epidemiological investigation," *Thrombosis and Haemostasis*, vol. 64, no. 3, pp. 349–352, 1990.

[10] J. E. Sadler, "A revised classification of von Willebrand disease. For the Subcommittee on von Willebrand Factor of the Scientific and Standardization Committee of the International Society on Thrombosis and Haemostasis," *Thrombosis and Haemostasis*, vol. 71, no. 4, pp. 520–525, 1994.

[11] J. E. Sadler, T. Matsushita, Z. Y. Dong, E. A. Tuley, and L. A. Westfield, "Molecular mechanism and classification of von Willebrand disease," *Thrombosis and Haemostasis*, vol. 74, no. 1, pp. 161–166, 1995.

[12] J. E. Sadler, U. Budde, J. C. J. Eikenboom et al., "Update on the pathophysiology and classification of von Willebrand disease: a report of the Subcommittee on von Willebrand factor," *Journal of Thrombosis and Haemostasis*, vol. 4, no. 10, pp. 2103–2114, 2006.

[13] M. S. Enayat, A. M. Guilliatt, W. Lester, J. T. Wilde, M. D. Williams, and F. G. H. Hill, "Distinguishing between type 2B and pseudo-von Willebrand disease and its clinical importance," *British Journal of Haematology*, vol. 133, no. 6, pp. 664–666, 2006.

[14] F. Rodeghiero, G. Castaman, and A. Tosetto, "Optimizing treatment of von Willebrand disease by using phenotypic and molecular data," *Hematology*, vol. 2009, pp. 113–123, 2009.

[15] D. Meyer, E. Fressinaud, L. Hilbert, A.-S. Ribba, J.-M. Lavergne, and C. Mazurier, "Type 2 von Willebrand disease causing defective von Willebrand factor-dependent platelet function," *Best Practice and Research*, vol. 14, no. 2, pp. 349–364, 2001.

[16] R. B. Merrifield, "Solid-phase peptide synthesis," *Advances in Enzymology and Related Areas of Molecular Biology*, vol. 32, pp. 221–296, 1969.

[17] D. Drenckhahn, T. Jöns, and F. Schmitz, "Chapter 2 production of polyclonal antibodies against proteins and peptides," *Methods in Cell Biology*, vol. 37, pp. 7–56, 1993.

[18] I. Matsumoto, Y. Mizuno, and N. Seno, "Activation of sepharose with epichiorohydrin and subsequent immobilization of ligand for affinity adsorbent," *Journal of Biochemistry*, vol. 85, no. 4, pp. 1091–1098, 1979.

[19] H. Towbin, T. Staehelin, and J. Gordon, "Electrophoretic transfer of proteins from polyacrylamide gels to nitrocellulose sheets: procedure and some applications," *Proceedings of the National Academy of Sciences of the United States of America*, vol. 76, no. 9, pp. 4350–4354, 1979.

[20] Bangs Laboratories, *Covalent Coupling Protocols*, Bangs Laboratories, 2008.

23

Modeling the QSAR of ACE-Inhibitory Peptides with ANN and Its Applied Illustration

Ronghai He,[1] Haile Ma,[1, 2] Weirui Zhao,[1] Wenjuan Qu,[1] Jiewen Zhao,[1] Lin Luo,[1] and Wenxue Zhu[2]

[1] *School of Food and Biological Engineering, Jiangsu University, 301 Xuefu Road, Zhenjiang, Jiangsu 212013, China*
[2] *School of Food and Biological Engineering, Henan University of Science and Technology, 48 Xiyuan Road, Luoyang, Henan 471003, China*

Correspondence should be addressed to Haile Ma, mhl@ujs.edu.cn

Academic Editor: Jean-Marie Zajac

A quantitative structure-activity relationship (QSAR) model of angiotensin-converting enzyme- (ACE-) inhibitory peptides was built with an artificial neural network (ANN) approach based on structural or activity data of 58 dipeptides (including peptide activity, hydrophilic amino acids content, three-dimensional shape, size, and electrical parameters), the overall correlation coefficient of the predicted versus actual data points is $R = 0.928$, and the model was applied in ACE-inhibitory peptides preparation from defatted wheat germ protein (DWGP). According to the QSAR model, the C-terminal of the peptide was found to have principal importance on ACE-inhibitory activity, that is, if the C-terminal is hydrophobic amino acid, the peptide's ACE-inhibitory activity will be high, and proteins which contain abundant hydrophobic amino acids are suitable to produce ACE-inhibitory peptides. According to the model, DWGP is a good protein material to produce ACE-inhibitory peptides because it contains 42.84% of hydrophobic amino acids, and structural information analysis from the QSAR model showed that proteases of Alcalase and Neutrase were suitable candidates for ACE-inhibitory peptides preparation from DWGP. Considering higher DH and similar ACE-inhibitory activity of hydrolysate compared with Neutrase, Alcalase was finally selected through experimental study.

1. Introduction

In recent years, some progress have been made in bioinformatics study of functional peptide preparation, such as comparing active peptide sequences in database, hydrolysis enzyme choosing, simulated hydrolysis, activity prediction of hydrolysate, and so forth [1–6]. However, these studies were all based on a known sequence of protein. In fact, bioinformatics application on peptide is still difficult because the majority of proteins have complicated components or unknown sequences.

Besides comparing characterized peptide sequences in databases, peptide quantitative structure-activity relationship (QSAR) models could also be used in peptide bioinformatics study. QSAR models are mathematical functions that describe the relationship between activity and chemical structure expressed by variables. Such models are applied both to predict activity of untested chemical structures and to predict the chemical structure of compounds with specific activity [7]. Several QSAR models have been investigated on ACE-inhibitory peptides. These models were built based on different amino acid descriptors or multivariate statistical regression techniques, such as multiple linear regressions (MLR) or partial least square regression (PLSR), and 3D-QSAR was also used to describe ACE-inhibitory peptide [8–18]. Recently, quantitative sequence-activity model (QSAM) was employed in ACE-inhibitory peptide study [19]. In addition, docking and virtual screening of ACE-inhibitory dipeptides technique was studied, but it also needs experimental verification [20].

An artificial neural network (ANN) is an interdisciplinary technique, involving biology, mathematics, physics, electronics, and computer technology. It is a kind of information processing system based on imitation of the structure and function of brain networks. It is the theoretical model of the human neural network. ANN technique can simulate any nonlinear process; therefore, it can avoid the linear deficiencies [15, 16, 18].

In this study, illustrated by preparation of ACE-inhibitory peptides from defatted wheat germ protein, a QSAR model was built with ANN. The structural characteristics of the ACE-inhibitory peptides were investigated according to the model. Based on the structural characteristics analysis and experimental result of DWGP digestion, appropriate protease was selected to produce high-activity ACE-inhibitory peptides from DWGP isolates.

2. Materials and Methods

2.1. Materials and Chemicals.
Defatted wheat germ protein was purchased from Man Tian Xue Flour Industry (Henan, China). Alcalase 2.4 LFG (2.670 AU/g) and medium temperature amylase 480 L (527.50 KNU/g) were purchased from Novo Co. (Shanghai, China). Angiotensin I-converting enzyme (ACE; EC 3.4.15.1) was purchased from Sigma Chemical Co. (St. Louis, MO, U.S.A.). N-(3[2-Furyl]Acryloyl)-Phe-Gly-Gly (FAPGG) was purchased from Fluka Chemical Corp. (Milwaukee, WI, U.S.A.). All the other reagents were in analytical purity grade.

2.2. Instruments.
The instruments used were as follows: thermostat-controlled water-bath (model HH), Jintan Zhongda Instruments Co., Ltd. (Jintan, Jiangsu, China); pH meter (model PHS-3C), Shanghai Precision & Scientific instrument Co., Ltd. (Shanghai, China); electrothermal blast drying oven, Shanghai Laboratory Instrument Works Co., Ltd. (Shanghai, China); Agilent 1100 HPLC, Agilent Technologies Inc. (Santa Clara, CA, U.S.A.); SPX-250B biochemistry incubator, Changzhou Guohua Electric Co., Ltd. (Changzhou, China); Multiskan Spectrum Microplate Reader, Thermo Scientific Inc. (Hudson, NH, U.S.A.).

2.3. Methods

2.3.1. DWGP Isolates Preparation.
DWGP isolates were prepared according to the method described by XIN Zhi-hong [21] with minor modifications. DWGP was dispersed in 0.2 mol/L NaCl solution at the ratio of 1 : 10 (w/v) and stirred for 30 min at ambient temperature. Then, the suspension's pH was adjusted to 9.5 by using 1 mol/L NaOH. After stirring for 30 min, the suspension was centrifuged at 8000 r/m for 20 min at 4°C. The supernatant was adjusted to pH 7.0 with 1.0 mol/L HCl, then 0.3% (v/v) α-amylase was added in. After stirring for 180 min at 70°C, it was adjusted to pH 4.0 with 1.0 mol/L HCl to precipitate the protein, and the solution was centrifuged at 8000 r/m for 20 min. The precipitate was washed several times with distilled water (pH 4.0), and was then dispersed in a small amount of distilled water, then it was adjusted to pH 7.0 with 0.1 mol/L NaOH. The dispersed precipitate was dried by spraying dryer (model B290, BUCHI Laboratory Equipment Ltd., Switzerland) to get DWGP isolates.

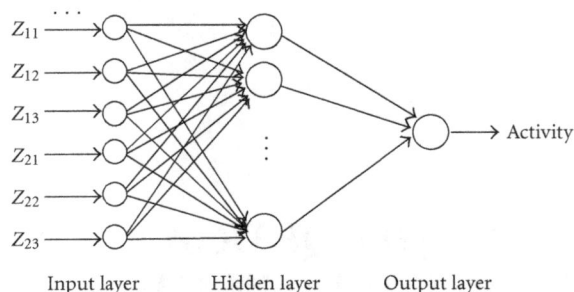

FIGURE 1: Diagram of BP-ANN model of ACE-inhibitory peptides' QSAR.

2.3.2. Hydrolysis of DWGP in a Batch Reactor.
Ten grams of DWGP was dispersed in 1 L distilled water and was digested in batch by Alcalase at pH 9.0, 50°C or by Neutrase at pH 7.0 at 50°C, both at the enzyme/substrate mass ratio of 8% ([E]/[S]). Samples were collected at 0.5, 1, 1.5, 2, 3, 4, and 5 h and were immediately heated in a boiling water bath for 10 min. After cooling, the samples were centrifuged at 10,000 r/m for 15 min, and the supernatants were diluted with distilled water to determine their ACE-inhibitory activities.

2.3.3. Building of QSAR Model on ACE-Inhibitory Peptides.
In this study, Z descriptor was used to predict the ACE-inhibitory activity of peptides, amino acids descriptor selected Z-scales, Z_1, Z_2, and Z_3 means the hydrophilic amino acids, three-dimensional shape, size, and electrical parameters, respectively (Table 1) [22]. Three-layer back propagation (BP) neural network was used to establish a QSAR model to describe relationships between peptide structure and activity.

Fifty-eight kinds of ACE-inhibitory peptides (dipeptides) samples and their activity data (50% inhibitory concentration on ACE, i.e., IC_{50} value) were used in the text and were shown in Table 2. Each dipeptide corresponds to a dependent variable ($\log[1/IC_{50}]$) and six independent variables (Z parameters).

Because of the quite different physical meaning of the input parameters, the following formula was used in this study to make the sample sets data normalized so as to accelerate network convergence and overfitting:

$$Z' = \frac{Z - Z_{min}}{Z_{max} - Z_{min}}, \tag{1}$$

where, Z' is the normalized value of the operator, Z value is the Z operator, Z_{max} and Z_{min} are the maximum and minimum of the Z operator vector before being normalized for each sample.

39 dipeptides were randomly selected as study samples in the neural network model, the rest were test samples. Each of two peptides corresponding to 6 Z operators as a BP neural network input vector. The network output vector is the activity value. Figure 1 is the structure of BP network model.

TABLE 1: Z descriptor scores for amino acids.

Amino acid	Code	Z_1	Z_2	Z_3	Amino acid	Code	Z_1	Z_2	Z_3
Ala	A	0.07	−1.73	0.09	His	H	2.41	1.74	1.11
Val	V	−2.69	−2.53	−1.29	Gly	G	2.23	−5.36	0.30
Leu	L	−4.19	−1.03	−0.98	Ser	S	1.96	−1.63	0.57
Lie	I	−4.44	−1.68	−1.03	Thr	T	0.92	−2.09	−1.40
Pro	P	−1.22	0.88	2.23	Cys	C	0.71	−0.97	4.13
Phe	F	−4.92	1.30	0.45	Tyr	Y	−1.39	2.32	0.01
Trp	W	−4.75	3.65	0.85	Asn	N	3.22	1.45	0.84
Met	M	−2.49	−0.27	−0.41	GIn	Q	2.18	0.53	−1.14
Lys	K	2.84	1.41	−3.14	Asp	D	3.64	1.13	2.36
Arg	R	2.88	2.52	−3.44	Glu	E	3.08	0.39	−0.07

A three-level BP neural network model was built using MATLAB neural network tool (from Matrix Laboratory). Transfer functions of neurons in hidden layer and output layer were Tansig function and Purelin function, respectively. Because the BP neural network is not easily converged or easily falls into local minimum, the following steps were applied to avoid it: (1) network training algorithm using gradient descent momentum Traingdm, (2) network training objectives (mean square error) is set to 10^{-2}, (3) the number of training steps is controlled in 6000. The number of hidden layer neurons was determined through repeated verification.

2.3.4. Determination of Peptides ACE-Inhibitory Activity. N-(3-[2-Furyl]Acryloyl)-Phe-Gly-Gly (FAPGG, purchased from Fluka Chemical Corp., Milwaukee, WI, U.S.A.) was used as substrate in ACE-inhibition assay. The reagents were sequentially added in for test reaction according to Table 3 [23]. The absorbance of each reaction solution was determined by a Multiskan Spectrum Microplate Reader at 340 nm. The initial absorbance of blank ($a1$) and sample ($b1$), and the final absorbance ($a2$ and $b2$, after 30 min reaction at 37°C) were recorded. The absorbance decrease of blank and sample are A (= $a1 - a2$) and B (= $b1 - b2$), respectively. Then, ACE-inhibitory activity (%) was expressed as $I = (A - B)/A$.

2.3.5. Determination of the Degree of Hydrolysis. The degree of hydrolysis (DH) was measured by pH-stat method. The release of amino acids in protein digestion makes pH of the hydrolysate decrease significantly, the alkali solution was added into hydrolysates to maintain pH value. By recording the amount of alkali consumed, the degree hydrolysis of protein and the amount of the rupture protein bonds can be figured out according to the following formula:

$$ DH = \frac{V_{NaOH} \times N_{NaOH}}{\alpha \times M_p \times h_{hot}} \times 100\%, \tag{2} $$

where V_{NaOH} is consumption volume of alkali (mL) in titration; N_{NaOH} is the concentration of alkali (mol/L) in titration; M_p is total protein (g) used; h_{hot} is the total number of peptide bonds per gram of protein (mmol/g, for wheat

germ protein, taking 7.69); α is a-amino acid dissociation degree, it can be calculated according to formula (3):

$$ \alpha = \frac{10^{pH-pK}}{\left(1 + 10^{pH-pK}\right)}, \tag{3} $$

pK is the average pH value of all kinds of amino acids, taking 9.0; pH is response to initial pH.

2.3.6. Analysis of DWGP Amino Acids Composition. Amino acid composition analysis was employed in this study to determine DWGP amino acid composition by o-phthalaldehyde (OPA) precolumn derivatization RP-HPLC determination [24].

3. Results and Discussions

3.1. Building of QSAR Model on ACE-Inhibitory Peptides. In this study, 4–10 hidden layer neurons were selected to build QASR model, each hidden layer neuron was modeled five times in order to identify the optimal number of hidden layer neurons. Network convergence speed rises when the number of neurons increases, but too many or too few of hidden layer neurons will decrease the generalization performance of model. Under the premise of guaranteed network convergence, a fewer number of neurons are preferred. The correlation coefficients R of study samples (the average value of five times of modeling) were shown in Figure 2. It was shown that when the number of hidden layer neurons was 7, the forecast correlation coefficient was the highest. Therefore, seven hidden layer neurons were selected to model the neural network. After repeated modeling, the correlation coefficient R reaches to 0.928, the training set mean square error is 0.0188, and the prediction set mean square error is 0.2091. The predicting results of BP network model to the set of prediction were shown in Figure 3.

3.2. Structural Features Analysis of ACE-Inhibitory Peptides. The back stepping method was used to find out the operator which has the greatest impact on the activity. The steps are as follows: (1) find out which hidden layer neuron has the greatest impact on output (activity), (2) find out which input

TABLE 2: The ACE-inhibitory peptides' sequences with Z descriptor activity.

Peptide	Log(1/IC$_{50}$)	Z_{11}*	Z_{12}	Z_{13}	Z_{21}	Z_{22}	Z_{23}
AA	3.21	0.07	−1.73	0.09	0.07	−1.73	0.09
AW	5	0.07	−1.73	0.09	−4.75	3.65	0.85
DG	1.85	3.64	1.13	2.36	2.23	−5.36	0.3
GF	3.2	2.23	−5.36	0.3	−4.92	1.3	0.45
GP	3.35	2.23	−5.36	0.3	−1.22	0.88	2.23
GR	2.49	2.23	−5.36	0.3	2.88	2.52	−3.44
GW	4.52	2.23	−5.36	0.3	−4.75	3.65	0.85
GY	3.68	2.23	−5.36	0.3	−1.39	2.32	0.01
IF	3.03	−4.44	−1.68	−1.03	−4.92	1.3	0.45
IW	5.7	−4.44	−1.68	−1.03	−4.75	3.65	0.85
IY	5.43	−4.44	−1.68	−1.03	−1.39	2.32	0.01
RF	3.64	2.88	2.52	−3.44	−4.92	1.3	0.45
RP	1.1818	2.88	2.52	−3.44	−1.22	0.88	2.23
VG	2.96	−2.69	−2.53	−1.29	2.23	−5.36	0.3
VW	1.6	−2.69	−2.53	−1.29	−4.75	3.65	0.85
VY	4.66	−2.69	−2.53	−1.29	−1.39	2.32	0.01
YG	2.7	−1.39	2.32	0.01	2.23	−5.36	0.3
RW	4.8	2.88	2.52	−3.44	−4.75	3.65	0.85
AY	4.28	−2.69	−2.53	−1.29	−4.92	1.3	0.45
RP	3.89	−4.44	−1.68	−1.03	−1.22	0.88	2.23
AF	3.72	0.07	−1.73	0.09	−4.92	1.3	0.45
AP	3.64	0.07	−1.73	0.09	−1.22	0.88	2.23
VP	3.38	−2.69	−2.53	−1.29	−1.22	0.88	2.23
IG	2.92	−4.44	−1.68	−1.03	2.23	−5.36	0.3
GI	2.92	2.23	−5.36	0.3	−4.44	−1.68	−1.03
GM	2.85	2.23	−5.36	0.3	−2.49	−0.27	−0.41
GA	2.7	2.23	−5.36	0.3	0.07	−1.73	0.09
GL	2.6	2.23	−5.36	0.3	−4.19	−1.03	−0.98
AG	2.6	0.07	−1.73	0.09	2.23	−5.36	0.3
GH	2.51	2.23	−5.36	0.3	2.41	1.74	1.11
KG	2.49	2.84	1.41	−3.14	2.23	−5.36	0.3
FG	2.43	−4.92	1.3	0.45	2.23	−5.36	0.3
GS	2.42	2.23	−5.36	0.3	1.96	−1.63	0.57
GV	2.34	2.23	−5.36	0.3	−2.69	−2.53	−1.29
MG	2.32	−2.49	−0.27	−0.41	2.23	−5.36	0.3
GK	2.27	2.23	−5.36	0.3	2.84	1.41	−3.14
GE	2.27	2.23	−5.36	0.3	3.08	0.39	−0.07
GT	2.24	2.23	−5.36	0.3	0.92	−2.09	−1.4
WG	2.23	−4.75	3.65	0.85	2.23	−5.36	0.3
HG	2.2	2.41	1.74	1.11	2.23	−5.36	0.3
GQ	2.15	2.23	−5.36	0.3	2.18	0.53	−1.14
GG	2.14	2.23	−5.36	0.3	2.23	−5.36	0.3
QG	2.13	2.18	0.53	−1.14	2.23	−5.36	0.3
SG	2.07	1.96	−1.63	0.57	2.23	−5.36	0.3
LG	2.06	−4.19	−1.03	−0.98	2.23	−5.36	0.3
GD	2.04	2.23	−5.36	0.3	3.64	1.13	2.36
TG	2	0.92	−2.09	−1.4	2.23	−5.36	0.3
EG	2	3.08	0.39	−0.07	2.23	−5.36	0.3
PG	1.77	−1.22	0.88	2.23	2.23	−5.36	0.3
LA	3.51	−4.19	−1.03	−0.98	0.07	−1.73	0.09

TABLE 2: Continued.

Peptide	$\text{Log}(1/IC_{50})$	$Z_{11}{}^*$	Z_{12}	Z_{13}	Z_{21}	Z_{22}	Z_{23}
KA	3.42	2.84	1.41	-3.14	0.07	-1.73	0.09
RA	3.34	2.88	2.52	-3.44	0.07	-1.73	0.09
YA	3.34	-1.39	2.32	0.01	0.07	-1.73	0.09
FR	3.04	-4.92	1.3	0.45	2.88	2.52	-3.44
HL	2.49	2.41	1.74	1.11	-4.19	-1.03	-0.98
DA	2.42	3.64	1.13	2.36	0.07	-1.73	0.09
EA	2	3.08	0.39	-0.07	0.07	-1.73	0.09
DM	2.7782	3.64	1.13	2.36	-2.49	-0.27	-0.41
IP	3.89	2.92	-4.44	-1.68	-1.22	0.88	2.23

$^*Z_{mn}$ the first number (m) behind Z represents the sequence of the amino acid in peptide, and the second number (n, from 1 to 3) represents the hydrophilic amino acids, three-dimensional shape, size, and electrical parameters, respectively.

TABLE 3: Reagents used in determination of ACE inhibiting activity.

	Blank (μL)	Sample (μL)
ACE (0.1 U/mL)	10	10
FAPGG (1 mmol/L)[1]	50	50
HEPES buffer[2]	40	0
Sample	0	40

[1]FAPGG (1.0 mmol/L): prepared with 0.08 M HEPES buffer (pH 8.3) containing 0.3 M NaCl.
[2]HEPES buffer: HEPES 1.910 g, NaCl 1.755 g, dissolved with double-distilled water, pH adjusted with NaOH, and metered volume with double-distilled water to 100 mL, stored at 4°C.

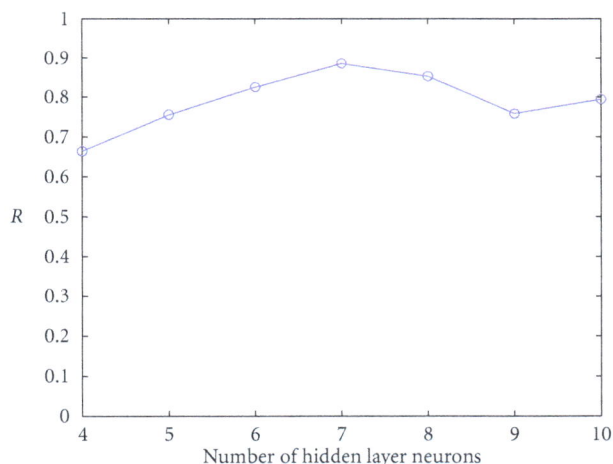

FIGURE 2: The correlation coefficient R values varied with the numbers of hidden layer neurons.

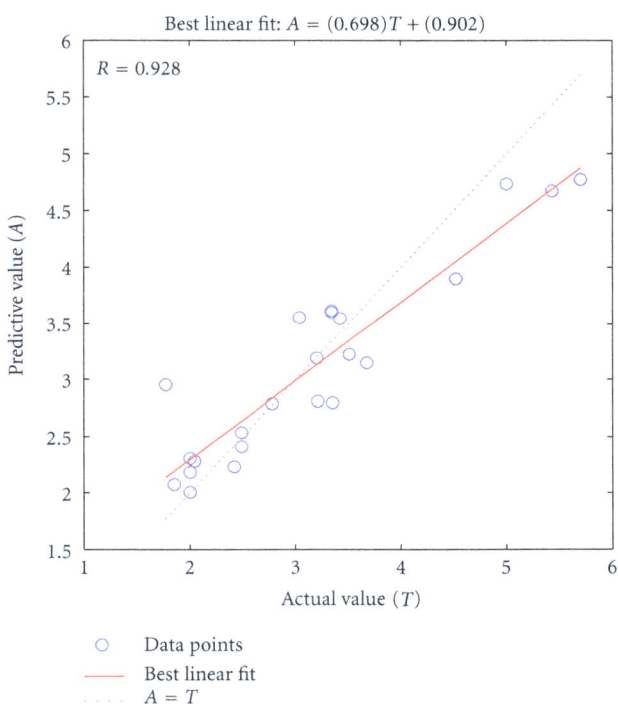

FIGURE 3: The predicting results of the (6-7-1) BP network model on the set of prediction.

neuron (specific Z operator) has impact on the found hidden layer neurons.

In Figure 4, LW(2, 1) refers to the weights when the hidden layer neurons change to the output layer neurons (activity values) through a linear function. If a hidden layer LW(2, 1) is bigger, it means that its corresponding neurons in the hidden layer have a greater impact on the output, on the contrary, if LW(2, 1) becomes small, its corresponding neurons in the hidden layer have little effect on the output.

By searching in the model, the hidden layer neuron with the greatest impact on the output was found. Analysis of the established BP neural network model showed that the weights LW(2, 1) value was (0.70466, 0.74384, -0.63652, -0.37093, 0.49303, -1.3532, 1.1885), the sixth hidden layer neurons value (-1.3532) and seventh value (1.1885) have the greatest impact on output. Then, input neurons (Z operator) which affected the 6 and 7 hidden layer neurons were investigated.

After searching the hidden layer neurons, the input layer neurons with the greatest impact on the hidden layer neurons were subsequently searched. In Figure 4, LW(1, 1) refers to the weights when the output layer neurons change to the

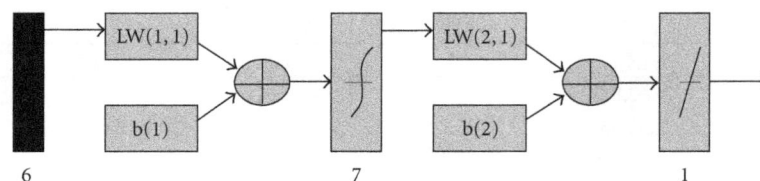

FIGURE 4: The parametric diagram of BP network model.

hidden layer neurons. If LW(1, 1) becomes bigger, it means the input layer neurons have high impact on the hidden layer neurons; when LW(1, 1) becomes smaller, the input layer neurons have little influence on the hidden layer neurons. According to the above searching, we can get the structural features with greater impact on the ACE-inhibitory peptide activity.

Table 4 is weights LW(1, 1) of the input layer to hidden layer neuron. Observing the weight values of the various Z operators on 6 and 7 hidden layer neurons, we found that the Z_{21} parameters (Z_1 operator of the second amino acid, see numbers in Table 4 with $\dagger$ superscript) have the greatest impact on the activity, followed by the Z_{22} (Z_2 operator on No. 2 position, see numbers in Table 4 with $\ddagger$ superscript). As we have defined that the Z_1 operator represents the hydrophobicity of amino acids [23], we could draw a conclusion that hydrophobicity of C-terminal amino acids have the greatest influence on ACE-inhibitory activity; and the greater the hydrophobicity is, the higher the ACE-inhibitory activity is. This result is consistent with some previous studies. Wu et al. [11] used Z descriptors to investigate quantitative structure-activity relationship of ACE-inhibitory dipeptides, and they found that ACE-inhibitory activity was greatly affected by the three-dimensional chemical properties and hydrophobicity of C-terminal amino acids, that is, the higher the volume and the greater hydrophobicity of amino acids were, the nicer the ACE-inhibitory activity was; so some dipeptides with hydrophobic amino acids at the C-terminal, such as phenylalanine, tryptophan, and tyrosine, will have high ACE-inhibitory activity. Cheung et al. [25] have also shown that if C-terminal was aromatic amino acids and proline, N-terminal was branches aliphatic amino acids; the dipeptides could have high ACE-inhibitory activity. Hellberg et al. measured Cheung' peptides samples in the same laboratory, and modeled the QSAR, he found that the dipeptides with positive charge amino acids at the N-terminal and bulky hydrophobic amino acids at C-terminal would have a stronger ACE-inhibitory activity [26]. As for tripeptides, Wu et al. [11] found that strong hydrophobic and small size of N-terminal amino acids, such as valine, leucine, and isoleucine, were more suitable for high-activity tripeptides; for second amino acid from the N-terminal, small electrical bit, large size, and weak hydrophobicity were more suitable. But for C-terminal, a higher electrical, larger volume, and stronger hydrophobic amino acid was more suitable, such as aromatic amino acids. Through the analysis of the three amino acid ACE-inhibitory peptides, Li [27] also reached a conclusion similar to Wu et al. By analyzing ACE-inhibitory peptides from milk sources, Pripp

et al. [7] found that for peptides with less than or equal to 6 amino acids at the C-terminal, the hydrophobicity, the amount of positive charge, and the volume size of amino acids adjacent to the C-terminal greatly affected the ACE-inhibitory peptides activity while the N-terminal amino acid has no direct relationship to the ACE-inhibitory activity. Therefore, the hydrophobicity and size of the C-terminal amino acid have primary effect on ACE-inhibitory activity, and hydrophobic amino acids, aromatic amino acids, or branched-chain amino acids are important components in high-activity peptides. Therefore, protein with high content of hydrophobic amino acid (especially aromatic amino acids) has more potential to produce high activity ACE-inhibitory peptides. By digestion of protein to produce peptides with hydrophobic amino acids at the C-terminal, people will get high ACE-inhibitory activity of hydrolysates.

3.3. Amino Acid Composition and Feature Analysis of Wheat Germ Protein Isolates. The DWGP contains 42.84% hydrophobic amino acids (Table 5), it is similar to rice protein isolate, bovine serum albumin, and casein, and is it significantly higher than mung bean protein isolate and peanut protein isolate [27]. Therefore, DWGP is a good protein resource with abundant hydrophobic amino acid. According to the result of quantitative structure-activity relationship analysis that high content of hydrophobic amino acid protein (especially aromatic amino acids) is suitable as protein material to produce ACE-inhibitory peptides (see Section 2.2 of this paper), wheat germ protein isolate is a good material to produce high-activity ACE-inhibitory peptides.

3.4. Digestion of Defatted Wheat Germ Protein with Different Proteases. Neutrase (a kind of neutral protease) tends to hydrolyze protein to produce peptides whose C-terminals are hydrophobic amino acids, such as Tyr, Try, or Phe. Alcalase (a kind of alkaline protease) tends to hydrolyze protein to obtain peptides whose C-terminals are amino acids with large side-chain and no charge (aromatic and aliphatic amino acids), such as Ile, Leu, Val, Met, Phe, Tyr, or Trp. Moreover, the hydrolysis process will be accelerated when N-terminals of peptides have hydrophobic amino acids [28, 29]. Proteinase K (EC. 3.4.21.14) acts on Phe, Try, Val, Ile, Leu, Trp, Pro, and Met [30]. Chymotrypsin C (EC 3.4.21.2) acts on Try, Leu, Trp, Pro, Met, Glu, Lys, and Pro [28]. The above proteases all tend to hydrolyze protein to generate peptides with hydrophobic amino acids C-terminals, and the QSAR of ACE-inhibitory peptide studies

TABLE 4: The neurons weights from input layer to hidden layer.

Neurons	Z_{11}*	Z_{12}	Z_{13}	Z_{21}	Z_{22}	Z_{23}
(1)	−0.3515	0.28331	0.28286	−0.44043	0.037431	−0.085447
(2)	0.27249	0.041643	−0.9822	−0.087255	0.47412	−0.18318
(3)	0.57416	−0.14092	0.60775	−0.08924	0.11108	−0.5922
(4)	0.24392	0.21597	−0.20115	0.24221	−0.07429	−0.54316
(5)	0.23817	0.07212	0.68217	−0.00285	0.26846	−0.47522
(6)	−0.30588	0.21639	0.036361	−0.41676[†]	−0.2514[‡]	−0.23479
(7)	−0.090315	−0.058403	0.0745	−0.32458[†]	−0.24483[‡]	−0.10795

*Z_{mn} the first number (m) behind Z represents the sequence of the amino acid in peptide, and the second number (from 1 to 3) represents the hydrophilicity of amino acids, three-dimensional shape, size, and electrical parameters, respectively.

TABLE 5: Amino acid composition of wheat germ protein isolates (g/100 g protein).

Amino acid	Content
Asp +Asn	8.40
Glu + Gln	15.28
Ser	4.40
His	3.15
Gly	6.19
Thr	3.94
Arg	9.79
Ala	6.41
Tyr	2.97
Cys	0.39
Val	7.20
Met	1.70
Phe	5.22
Ile	4.94
Leu	8.07
Lys	6.33
Pro	5.63
Trp	0.69
Hydrophobic amino acids	42.84
Aromatic amino acids	8.89

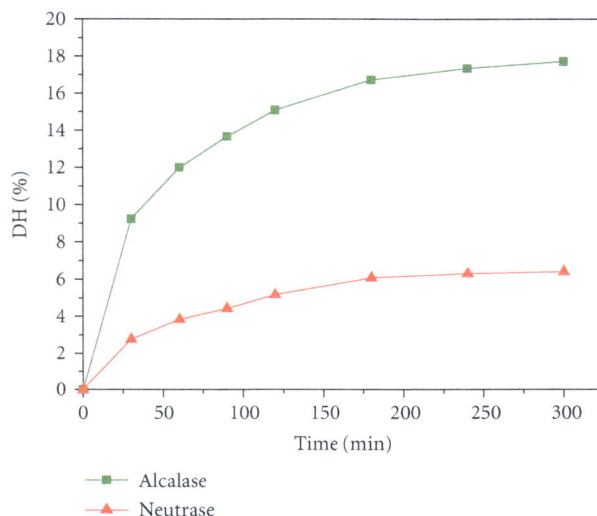

FIGURE 5: The degree of hydrolysis of DWGP hydrolysate treated with Alcalase and Neutrase.

have shown that peptides which have hydrophobic amino acids C-terminals will show potential strong ACE inhibition, so Neutrase, Alcalase, proteinase K, and chymotrypsin C may be the suitable proteases for high-activity ACE-inhibitory peptides preparation. In addition, Alcalase and Neutrase are microbial enzymes which are easily obtained and low cost compared with proteinase K and chymotrypsin C, so they are suitable for industrial application. In this study, Alcalase and Neutrase were investigated to produce ACE-inhibitory hydrolysates by digest DWGP. The degree of hydrolysis (DH) and the ACE-inhibitory activity of DWGP hydrolysates were presented in Figures 5 and 6, respectively.

From Figure 5, we can find DHs of DWGP digested by either alkaline or neutral protease increased significantly before 120 min, and slightly increased during 120∼300 min. Results of Figure 5 imply that the hydrolysis sites of Alcalase and Neutrase are partly similar, but hydrolysis sites of Alcalase exceed Neutrase's; therefore, the former one's hydrolysate has higher DH than the later one's. Figure 6 shows that the ACE-inhibitory rate of hydrolysates digested by Alcalase is remarkably increased during the preceding 120 min, and then it decreases slowly after 120 min. This result indicated that a long-time digestion might cause the excessive degradation of active peptides. Li observed a similar phenomenon in preparations of ACE-inhibitory peptides from Zein, rice protein isolate, mung bean protein isolate, and peanut protein isolate with Alcalase [27]. Pedroche prepared ACE-inhibitory peptides with Alcalase through hydrolysis of chickpea protein also found that the ACE-inhibitory rate reached the maximum at 30 min and then decreased [31]. From Figure 6, we also find that the inhibitory rate of peptides digested by Neutrase rises during the preceding 180 min, and then decreases slowly. The result also indicated that long-time digestion caused the excessive degradation of active peptides. However, during the preceding 120 min, the ACE-inhibitory activity of the Alcalase hydrolysates was significantly higher than the Neutrase hydrolysates at the same time, and both of them reached almost the same activity level after 120 minutes.

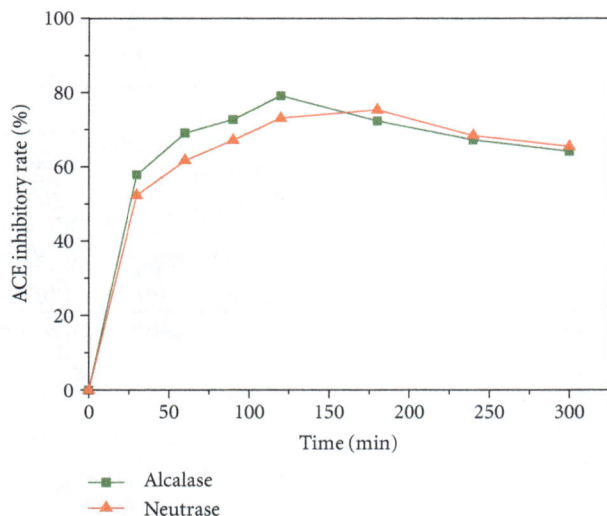

FIGURE 6: The ACE-inhibitory activity of DWGP hydrolysate treated with Alcalase and Neutrase.

According to the average peptide chain length (PLC) formula (PLC = (1/DH) × 100%) of protein digestion [27], higher DH of hydrolysate by Alcalase indicates that more short chain lengths peptides were produced in digestion than by Neutrase. The theoretical conclusion was also proved by the experimental results of Xin et al. [32] and Jia et al. [33], respectively. It has been revealed that the most part of effective ACE-inhibitory peptides after oral administration are small peptides [21], therefore, Alcalase is more suitable for DWGP ACE-inhibitory peptides preparation.

4. Conclusions

Based on data of activity, hydrophilic amino acids, three-dimensional shape, size, and electrical parameters of 58 dipeptides, a quantitative structure-activity relationship (QSAR) of amino acids ACE-inhibitory peptides was built with ANN, the related coefficient is 0.928, and by analyzing the ANN model, it was found that (1) C-terminal is primarily important to ACE-inhibitory activity; (2) proteins containing abundant hydrophobic amino acids are potential good source to produce ACE-inhibitory peptides; (3) as for DWGP, Alcalase was a proper protease for ACE-inhibitory peptides preparation.

Acknowledgments

The authors wish to thank the support from Grants of the China Postdoctoral Science Foundation (20100471386), Jiangsu Postdoctoral Grant (0902029C), and Research Foundation for Talented Scholars of Jiangsu University (08JDG032). The study is a pure academic behavior without financial support of any company.

References

[1] J. Dziuba, P. Minkiewicz, D. Nałecz, and A. Iwaniak, "Database of biologically active peptide sequences," *Nahrung*, vol. 43, no. 7, pp. 190–195, 1999.

[2] J. Dziuba, M. Niklewicz, A. Iwaniak, M. Darewicz, and P. Minkiewicz, "Bioinformatic-aided prediction for release possibilities of bioactive peptides from plant proteins," *Acta Alimentaria*, vol. 33, no. 3, pp. 227–235, 2004.

[3] J. Dziuba, M. Niklewicz, A. Iwaniak, M. Darewicz, and P. Minkiewicz, "Structural properties of proteolytic-accessible bioactive fragments of selected animal proteins," *Polimery/Polymers*, vol. 50, no. 6, pp. 424–428, 2005.

[4] E. Gasteiger, C. Hoogland, A. Gattiker et al., "Protein identification and analysis tools on the ExPASy server," in *The Proteomics Protocols Handbook*, J. M. Walker, Ed., pp. 571–607, Humana Press, Totowa, NJ, USA, 2005.

[5] http://www.expasy.ch.

[6] M. R. Wilkins, I. Lindskog, E. Gasteiger et al., "Detailed peptide characterization using PEPTIDEMASS—a World-Wide-Web-accessible tool," *Electrophoresis*, vol. 18, no. 3-4, pp. 403–408, 1997.

[7] A. H. Pripp, T. Isaksson, L. Stepaniak, T. Sørhaug, and Y. Ardö, "Quantitative structure activity relationship modelling of peptides and proteins as a tool in food science," *Trends in Food Science and Technology*, vol. 16, no. 11, pp. 484–494, 2005.

[8] J. Wu and R. E. Aluko, "Quantitative structure-activity relationship study of bitter di- and tri-peptides including relationship with angiotensin I-converting enzyme inhibitory activity," *Journal of Peptide Science*, vol. 13, no. 1, pp. 63–69, 2007.

[9] G. Z. Liang, P. Zhou, Y. Zhou, Q. X. Zhang, and Z. L. Li, "New descriptors of aminoacids and their applications to peptide quantitative structure-activity relationship," *Acta Chimica Sinica*, vol. 64, no. 5, pp. 393–396, 2006.

[10] A. Givehchi, A. Bender, and R. C. Glen, "Analysis of activity space by fragment fingerprints, 2D descriptors, and multitarget dependent transformation of 2D descriptors," *Journal of Chemical Information and Modeling*, vol. 46, no. 3, pp. 1078–1083, 2006.

[11] J. Wu, R. E. Aluko, and S. Nakai, "Structural requirements of angiotensin I-converting enzyme inhibitory peptides: quantitative structure-activity relationship study of Di- and tripeptides," *Journal of Agricultural and Food Chemistry*, vol. 54, no. 3, pp. 732–738, 2006.

[12] H. U. Mei, Z. H. Liao, Y. Zhou, and S. Z. Li, "A new set of amino acid descriptors and its application in peptide QSARs," *Biopolymers*, vol. 80, no. 6, pp. 775–786, 2005.

[13] HU. Mei, Y. Zhou, LI. L. Sun, and Z. L. Li, "A new descriptor of amino acids and its application in peptide QSAR," *Acta Physico—Chimica Sinica*, vol. 20, no. 8, pp. 821–825, 2004.

[14] F. Tian, P. Zhou, and Z. Li, "T-scale as a novel vector of topological descriptors for amino acids and its application in QSARs of peptides," *Journal of Molecular Structure*, vol. 830, no. 1–3, pp. 106–115, 2007.

[15] H. Wang, B. O. Chen, and S. Yao, "Quantitative structure-activity relationship modeling of Angiotensin Converting enzyme inhibitors by back propagation artificial neural network," *Fenxi Huaxue/ Chinese Journal of Analytical Chemistry*, vol. 34, no. 12, pp. 1674–1678, 2006.

[16] J. Wu, R. E. Aluko, and S. Nakai, "Structural requirements of angiotensin I-converting enzyme inhibitory peptides: quantitative structure-activity relationship modeling of peptides

containing 4-10 amino acid residues," *QSAR and Combinatorial Science*, vol. 25, no. 10, pp. 873–880, 2006.

[17] Z. H. Lin, H. X. Long, Z. Bo, Y. Q. Wang, and YU. Z. Wu, "New descriptors of amino acids and their application to peptide QSAR study," *Peptides*, vol. 29, no. 10, pp. 1798–1805, 2008.

[18] S. Mao, H. Dan-Qun, M. Hu, L. Gui-Zhao, Z. Mei, and L. Zhi-Liang, "New descriptors of amino acids and its applications to peptide quantitative structure-activity relationship," *Chinese Journal of Structural Chemistry*, vol. 27, no. 11, pp. 1375–1383, 2008.

[19] P. Zhou, F. Tian, Y. Wu, Z. Li, and Z. Shang, "Quantitative sequence-activity model (QSAM): applying QSAR strategy to model and predict bioactivity and function of peptides, proteins and nucleic acids," *Current Computer-Aided Drug Design*, vol. 4, no. 4, pp. 311–321, 2008.

[20] A. H. Pripp, "Docking and virtual screening of ACE inhibitory dipeptides," *European Food Research and Technology*, vol. 225, no. 3-4, pp. 589–592, 2007.

[21] G. H. Li, G. W. Le, Y. H. Shi, and S. Shrestha, "Angiotensin I-converting enzyme inhibitory peptides derived from food proteins and their physiological and pharmacological effects," *Nutrition Research*, vol. 24, no. 7, pp. 469–486, 2004.

[22] S. Hellberg, M. Sjöström, B. Skagerberg, and S. Wold, "Peptide quantitative structure-activity relationships, a multivariate approach," *Journal of Medicinal Chemistry*, vol. 30, no. 7, pp. 1126–1135, 1987.

[23] G. Oshima, H. Shimabukuro, and K. Nagasawa, "Peptide inhibitors of angiotensin I-converting enzyme in digests of gelatin by bacterial collagenase," *Biochimica et Biophysica Acta*, vol. 566, no. 1, pp. 128–137, 1979.

[24] D. H. Mou, "Determine amino acid contents with OPA Pre—column derivatization RP—HPLC," *Chinese Journal of Chromatography*, vol. 15, no. 4, pp. 319–321, 1997.

[25] H. S. Cheung, F. L. Wang, M. A. Ondetti, E. F. Sabo, and D. W. Cushman, "Binding of peptide substrates and inhibitors of angiotensin-converting enzyme. Importance of the COOH-terminal dipeptide sequence," *Journal of Biological Chemistry*, vol. 255, no. 2, pp. 401–407, 1980.

[26] S. Hellberg, L. Eriksson, J. Jonsson et al., "Minimum analogue peptide sets (MAPS) for quantitative structure-activity relationships," *International Journal of Peptide and Protein Research*, vol. 37, no. 5, pp. 414–424, 1991.

[27] G. H. Li, "Study on food source angiotensin-converting enzyme inhibitory peptides," Jiangnan University, Wuxi, China, 2005.

[28] http://umbbd.msi.umn.edu/.

[29] A. Pihlanto-Leppälä, T. Rokka, and H. Korhonen, "Angiotensin I converting enzyme inhibitory peptides derived from bovine milk proteins," *International Dairy Journal*, vol. 8, no. 4, pp. 325–331, 1998.

[30] http://www.expasy.org/enzyme.

[31] J. Pedroche, M. M. Yust, J. Girón-Calle, M. Alaiz, F. Millán, and J. Vioque, "Utilisation of chickpea protein isolates for production of peptides with angiotensin I-converting enzyme (ACE)-inhibitory activity," *Journal of the Science of Food and Agriculture*, vol. 82, no. 9, pp. 960–965, 2002.

[32] Z. H. Xin, S. Y. Wu, H. L. Ma, and C. H. Dai, "Study on processing antihypertensive peptides derived from wheat germ protein," *Chinese Journal of Food Science*, vol. 24, no. 10, pp. 101–104, 2003.

[33] J. Jia, H. Ma, W. Zhao et al., "The use of ultrasound for enzymatic preparation of ACE-inhibitory peptides from wheat germ protein," *Food Chemistry*, vol. 119, no. 1, pp. 336–342, 2010.

24

Sequence Determination of a Novel Tripeptide Isolated from the Young Leaves of *Azadirachta indica* A. Juss

M. Rajeswari Prabha and B. Ramachandramurty

Department of Biochemistry, PSG College of Arts & Science, Civil Aerodrome Post, Coimbatore, Tamil Nadu 641014, India

Correspondence should be addressed to M. Rajeswari Prabha; prabhall.bio@gmail.com

Academic Editor: Tzi Bun Ng

The neem tree has long been recognized for its unique properties, both against insects and in improving human health. Every part of the tree has been used as a traditional medicine for household remedy against various human ailments, from antiquity. Although the occurrence of various phytochemicals in neem has been studied, we have identified the presence of a novel tripeptide in the young leaves of neem using a simple and inexpensive paper chromatographic method, detected by Cu(II)-ninhydrin reagent. The peptide nature of the isolated compound is confirmed by spectral studies. The sequence of the peptide is determined using de novo sequencing by tandem MS after purification.

1. Introduction

Small alpha peptides are the most expensive substances, and most of them are not easily available commercially [1]. Pharmacological studies have proved that many peptides, including those isolated from plants, have a potential antitumor effect [2]. These peptides have a number of advantages over other chemical agents including their low molecular weight, relatively simple structure, lower antigenicity, fewer adverse actions, easy absorption, and a variety of routes of administration [3]. Many antibacterial peptide families have been isolated from plants. Pp-thionin, for example, showed activity against *Rhizobium meliloti*, *Xanthomonas campestris*, *Micrococcus luteus*. Circulins A-B and cyclopsychotride A from the cyclotides family showed antibacterial effects against human pathogens such as *Staphylococcus aureus*, *Micrococcus luteus*, *Escherichia coli*, *Pseudomonas aeruginosa*, *Proteus vulgaris*, and *Klebsiella oxytoca* at micromolar concentrations [4]. Various plant extracts are reported to exhibit high antifungal activity due to proteins or peptides [5]. Cardiovascular activity of milk casein-derived tripeptides has also been reported, where bioactive tripeptide-containing milk products attenuated the blood pressure development in spontaneously hypertensive rats [6]. Research on *A indica*

has revealed the occurrence of various compounds such as terpenoids, and flavonoids [7, 8]. But the presence of small alpha peptides has not been reported so far.

Ninhydrin reactions using manual and automated techniques as well as ninhydrin spray reagents are widely used to analyze and characterize amino acids, peptides, and proteins, as well as numerous other ninhydrin-positive compounds in biomedical, clinical, food, forensic, histochemical, microbiological, nutritional, and plant studies [9–11]. Many of the shortcomings of ninhydrin have been met by the synthesis of a variety of ninhydrin analogs. All amino acids and their carboxyl group derivatives like esters and amides, including small peptides, produce a purple color with the classical ninhydrin reagent. This reagent was modified by us by adding cupric ion in order to distinguish qualitatively the carboxyl group derivatives of amino acids from the amino acids on paper after chromatography [12]. Amino acids produce a pink color, and their carboxyl derivatives like esters and amides, including small peptides, produce a yellow color with Cu(II)-ninhydrin reagent. The Cu(II)-ninhydrin method discussed here is a novel one because no other methods presently used can form two different coloured products with a single developing reagent. We have used this method for the detection and purification of amino acid derivatives from different plant products [13, 14]. In this paper, we report the isolation and sequence determination of a small alpha peptide from the young leaves of *A. indica*.

2. Materials and Methods

Ninhydrin was acquired from Pierce (Rockford, IL, USA). Cupric nitrate was of BDH, analytical grade (Mumbai, India). Organic solvents and acids used were of the highest purity available. Whatman No. 1 filter paper discs were obtained from Whatman International Ltd, Maidstone, England. Polyvinylpyrrolidone was purchased from Loba Chemie Pvt. Ltd., Mumbai.

2.1. Preparation of the Crude Extract. The young leaves of *A. indica* were homogenized (1 g/10 mL) with warm 80% ethanol. The extract was filtered through a Whatmann No. 1 filter paper. The extraction was partitioned three times with equal volumes of petroleum ether to remove the pigments. The pigment-free alcohol fraction was evaporated to dryness over a boiling water bath. The resulting residue was treated with 2% polyvinylpyrrolidone (one mL for each gram of leaf used as a starting material) and centrifuged at 3000 g for 10 min at 4°C to remove the phenolic compounds. The clear supernatant obtained was used as the crude source of the small alpha peptide.

2.2. Preparation of Cu(II)-Ninhydrin Reagent. The Cu(II)-ninhydrin reagent was prepared by dissolving cupric nitrate (25 mmol/L) and ninhydrin (1% w/v) in a minimum quantity of a mixture of water and glacial acetic acid (3 : 1 v/v) and diluted with required amount of acetone.

2.3. Circular Paper Chromatography. The crude extract was spotted in the center of a circular Whatman No. 1 filter paper on the arc of a small circle drawn with a pencil. Depending upon the number of samples to be analyzed, the paper may be demarcated. The diameter of the samples spotted was restricted to 0.5 cm by intermittent use of a hot air dryer. The sample spotting may be repeated 15 to 20 times to ensure sufficient concentration of the component to be detected. The chromatography was carried out in an isopropanol : water (4 : 1, v/v) solvent system by connecting a filter paper wick to the solvent through a hole made at the center of the circular paper. After the run which required approximately 20–40 min, the chromatogram was dried at ambient temperature for 30 min. The air-dried chromatogram was developed by spraying uniformly with Cu(II)-ninhydrin reagent followed by drying at 60°C for 10 min.

2.4. Purification of the Alpha Peptide. To subject the compound for various spectral studies, a simple and inexpensive purification procedure was followed. Two circular chromatograms of these compounds were run simultaneously using Whatman No. 1 filter paper discs (12 cm). One was developed with the Cu(II)-ninhydrin reagent. The corresponding region of the paper on the other chromatogram containing the Cu(II)-ninhydrin-positive compounds was

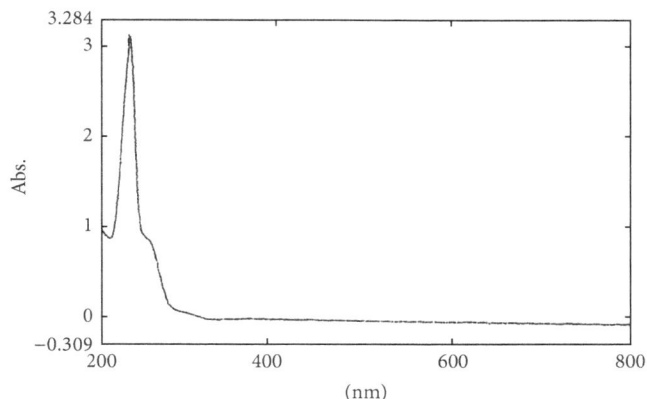

FIGURE 1: UV-Vis spectrum of the small alpha peptide purified from *A. indica*.

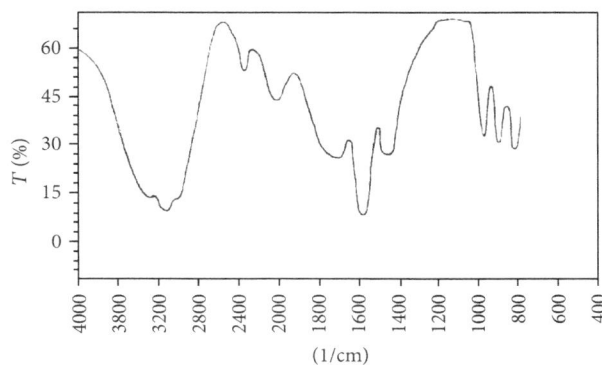

FIGURE 2: FT-IR spectrum of the small alpha peptide purified from *A. indica*.

cut into pieces and eluted in 80% ethanol. The compound, thus obtained, was used for conducting the spectral studies.

2.5. Colorimetric Determination of the Concentration of Cu(II)-Ninhydrin-Positive Compounds. 1 mL of the purified Cu(II)-ninhydrin-positive compound was added with 1 mL of Cu(II)-ninhydrin reagent, and the mixture was incubated at 40°C for 5 min. The yellow color produced was read at 420 nm. The amount of these compounds was determined by using a standard graph constructed with L-glycyl glycine as the standard.

2.6. UV-Vis Spectrophotometry. The purified compound was scanned for its absorption properties, from 200 nm to 900 nm in a Shimadzu, UV-Vis spectrophotometer.

2.7. FT-IR Spectrometry. The purified compound was also subjected to FT-IR analysis using a Shimadzu model FTIR-8300 infrared spectrometer. IR spectra were scanned between 500 and 4,000 wave numbers (per centimeter).

2.8. GC-MS Analysis. The purified Cu(II)-ninhydrin-positive compound was analyzed using the GC/MS

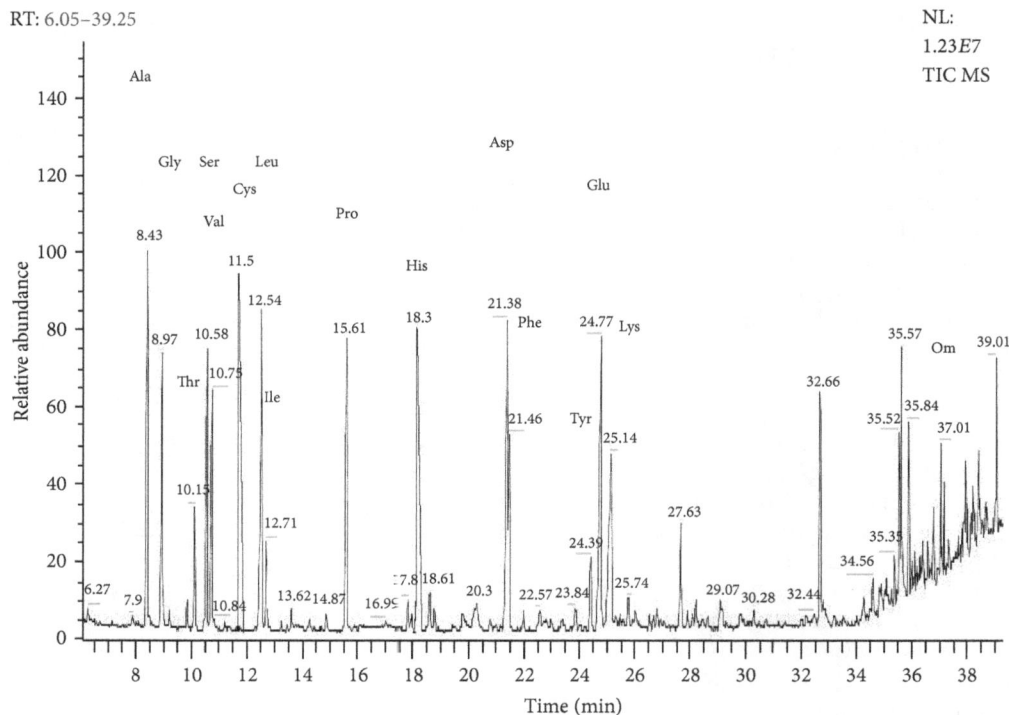

FIGURE 3: Gas chromatography retention spectra of the standard amino acids.

instrument: Trace Ultra version 5.0 produced by Thermo. The separation conditions were as follows: DB-5 Column 30 m × 0.25 mm × 0.25 μm, mobile phase helium at flow rate 1.0 mL/min, injection chamber temperature 220°C, and oven temperature starts at 80°C raised to 250°C at a rate of 8°C per minute. The ionization mode of the mass detector was at 70 eV.

2.9. NMR Studies. The purified Cu(II)-ninhydrin-positive compound in deuterated acetone as a solvent was subjected to ^{1}H and ^{13}C NMR analysis using a Bruker 500 MHz liquid-state NMR spectrometer.

2.10. Acid Hydrolysis of the Purified Compound and Separation by Paper Chromatography. For this experiment, the same sample purified using circular paper chromatography technique was employed. 0.5 mL of this sample aliquot was mixed with 0.5 mL of concentrated HCl in a clean dry test tube (the final concentration of HCl is 6 N). The tube was subsequently sealed with the help of a glass blower. This was placed in an incubator at 110°C for 24 hours after which the sample was reconstituted to 0.5 mL with distilled water. The hydrolyzed sample was spotted on circular Whatman No. 1 filter paper and developed with the isopropanol : water (4 : 1 v/v) system. The chromatograms were uniformly sprayed with ninhydrin reagent and were air dried and heated at 65°C for 10 min.

2.11. De Novo Sequencing Using MS/MS. Peptide sample (purified Cu(II)-ninhydrin-positive compound) was evaporated to dryness at room temperature and resuspended

in 0.1% trifluoroacetic acid buffer and spotted with CHCA (alpha-cyanohydroxycinnamic acid) matrix onto MALDI plate and allowed to dry. A model 4800 Plus MALDI TOF/TOF analyzer (Applied Biosystems Inc., Foster City, CA, USA) was used for direct profiling and MS/MS fragmentation study. Acquisitions were performed in positive ion reflectron mode. MS spectra were accumulated in mass range 400–4000 m/z. Spectra are obtained for the major peptide ions in MS mode, and sequence data are obtained when the spectrometers automatically revert to MS/MS mode. MS/MS was achieved by 2 kV collision-induced dissociation (CID) in positive ion mode. De novo sequencing analysis was used to determine the primary sequence structure for peptides that are not present in currently available databases.

3. Results and Discussion

The plant extract was run on circular paper chromatography using isopropanol : water (4 : 1) solvent system and developed with Cu(II)-ninhydrin reagent. The production of yellow chromaphore indicated the presence of Cu(II)-ninhydrin-positive compound. The detected compound was purified using inexpensive paper chromatographic method, as described. 2 μg concentration of the peptide was obtained from 1 g of the leaf extract. The peptide nature of the purified compound from the young leaves of *A. indica* was confirmed by UV spectrophotometer. Purified alpha peptide from the young leaves of *A. indica* showed maximum absorption at 210 nm (Figure 1) confirming the peptide nature [15]. The peptide nature was further confirmed by identifying functional groups from the FT-IR studies. Figure 2 shows the FT-IR spectrum of the purified compound from the young leaves of *A. indica*. There were sharp peaks at $1600\,cm^{-1}$ and $3380\,cm^{-1}$ indicating the presence of C=O and NH groups respectively, in the compound, confirming the peptide nature [14].

The Cu(II)-ninhydrin-positive compound purified from the young leaves of *A. indica* was analyzed by GC/MS. The retention times of the compound were compared with those of reference standard amino acids under the same conditions. The identification of the amino acids in the sample was based on direct comparison of the retention times and mass spectral data with those for standard compounds, and by computer matching with the Wiley 229, Nist 107, and 21 Library, as well as by comparison of the fragmentation patterns of the mass spectra with those reported in the literature [16]. Figure 3 shows the GC-MS spectra of the standard amino acids. Figure 4 shows the GC-MS spectra of the Cu(II)-ninhydrin-positive compound purified from the young leaves of *A. indica*. From the data obtained, it can be inferred that the peptide isolated from the young leaves of *A. indica* might contain alanine, cysteine, and phenyl alanine.

NMR studies indicated the presence of $-CH_3$, $-CH_2$, and aromatic groups in proton spectra and the presence of $-C=O$, $-CH$, and SH/OH groups in carbon-13 spectra of the peptide purified from *A. indica*. This indicates the presence of aliphatic amino acid, sulphur containing amino acid, and aromatic amino acid in the isolated peptide.

The peptide nature of the compound was also further confirmed from the acid hydrolysis experiment. The isolated compound was hydrolyzed with 6 N HCl. The chromatograms here were developed with the ninhydrin reagent which can detect the amino acids produced by hydrolysis more effectively. Figure 5 shows the acid-hydrolyzed *A. indica* peptide developed in isopropanol : water (4 : 1 v/v) system by circular paper chromatography and sprayed with ninhydrin. The result indicated that there may be presence of three amino acids in the purified peptide.

In the past decade, tandem mass spectrometry (MS/MS) has emerged as a technology of choice for high-throughput proteomics [17]. In spite of the continuously growing sequence databases, de novo sequencing of peptides, that is, sequencing without assistance of a linear sequence database, is still essential in several analytical situations. Figure 6 represents the MS/MS spectra of the peptide isolated from the young leaves of *A. indica*. Table 1 represents the amino acid sequence of the peptide isolated from *A. indica*. The sequence of amino acids was determined by the application of tandem MS. By referring to the mass unit of the respective amino acids, the amino acid sequence of the peptide isolated from the young leaves of *A. indica* is found to be Ala-Phe-Cys (N-alanine-phenylalanine-cysteine-C). The peptide isolated from *A. indica* is a tripeptide, composed of three amino acid residues with alanine at the N-terminus and cysteine at the C-terminus. The determined sequence data was submitted in UniProt Knowledgebase database, and accession number was assigned for the sequence. The sequence data of the peptide

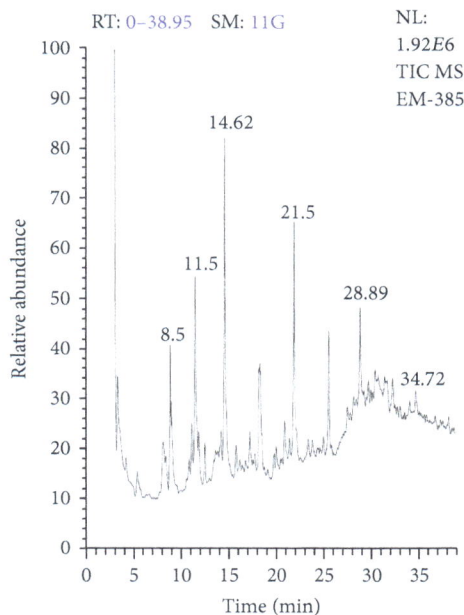

FIGURE 4: Gas chromatography retention spectra of the the Cu(II)-ninhydrin-positive compound purified from *A. indica*.

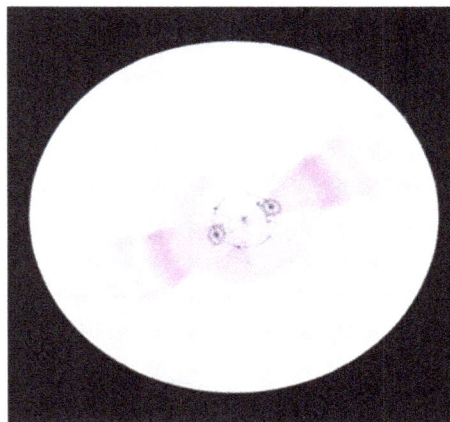

FIGURE 5: The purified small alpha peptide from *A. indica* was subjected to acid hydrolysis, spotted on a circular Whattman paper and developed using isopropanol : water (4 : 1, v/v) system and sprayed with ninhydrin reagent.

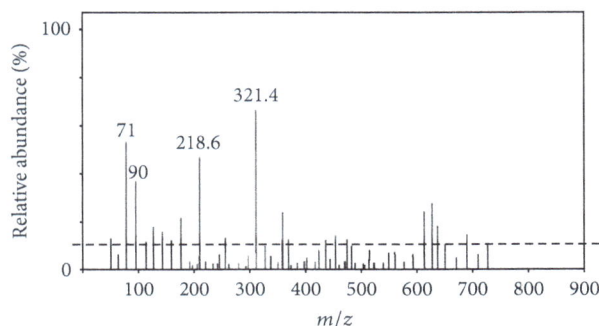

FIGURE 6: MS/MS spectra of peptide isolated from *A. indica*.

TABLE 1: Sequence data summary of the peptide isolated from the young leaves of *Azadirachta indica*.

Protein mass: 321.5 Da		
Fragment ion calculator results		
Sequence: N-Ala-Phe-Cys-C		
Fragment ion table, monoisotopic masses		
Seq	No.	*B*
A	1	71.0203
F	2	218.6125
C	3	321.4521

isolated from the young leaves of *A. indica*, reported in this paper, will appear in the UniProt Knowledgebase under the accession number B3EWR2.

The assessment of the biological role and applications of the purified tripeptide is under study. Compounds with low molecular weight of 500 or less can function as efficient drug molecules [18]. Small peptides containing multifunctional amino acids like L-glutamic acid, L-aspartic acid, L-lysine, L-histidine, L-cysteine, and L-serine can function as potent chelating agents that can be employed in chelation therapy [19]. Novel drugs can also be synthesised by chemical modification of these peptides.

4. Conclusion

From the overall results obtained from this work, it can be inferred that the Cu(II)-inhydrin positive compound purified from the young leaves of *A. indica* is a tripeptide. Unlike amino acids, small peptides are highly expensive, and most of them are not easily available commercially. Chemical synthesis of peptides increases the cost almost exponentially as the length of the peptide increases. If the separation and characterization methods for specific small peptides from inexpensive biological sources are standardized, these peptides can be easily isolated and supplied on demand for research as well as for commercial purposes. The small peptides may serve several purposes in the near future.

Conflict of Interests

The authors have only used the products of the commercial identities referred to in this paper. There is no secondary interest, conflict of interests or financial gain in this paper. They do not have any secondary rights. They are also the sole authors of the paper.

References

[1] B. Ramachandramurty, M. Rajeswari Prabha, and K. C. Raja, "A simple method for the production and detection of small alpha peptides from pulses," *IUP Journal of Life Sciences*, vol. 4, pp. 50–55, 2010.

[2] A. Wélé, Y. Zhang, I. Ndoye, J. P. Brouard, J. L. Pousset, and B. Bodo, "A cytotoxic cyclic heptapeptide from the seeds of Annona cherimola," *Journal of Natural Products*, vol. 67, no. 9, pp. 1577–1579, 2004.

[3] N. H. Tan and J. Zhou, "Plant cyclopeptides," *Chemical Reviews*, vol. 106, pp. 840–895, 2006.

[4] P. Barbosa Pelegrini, R. P. Del Sarto, O. N. Silva, O. L. Franco, and M. F. Grossi-De-Sa, "Antibacterial peptides from plants: what they are and how they probably work," *Biochemistry Research International*, vol. 2011, Article ID 250349, 9 pages, 2011.

[5] A. Jamil, M. Shahid, M. Masud-Ul-Haq Khan, and M. Ashraf, "Screening of some medicinal plants for isolation of antifungal proteins and peptides," *Pakistan Journal of Botany*, vol. 39, no. 1, pp. 211–221, 2007.

[6] P. Jakala, E. Pere, R. Lehtinen, A. Turpeinen, R. Korpela, and H. Vapaatalo, "Cardiovascular activity of milk casein-derived tripeptides and plant sterols in spontaneously hypertensive rats," *Journal of Physiology and Pharmacology*, vol. 60, no. 4, pp. 11–20, 2009.

[7] A. Bose, K. Chakraborty, K. Sarkar et al., "Neem leaf glyco-protein directs T-bet-associated type 1 immune commitment," *Human Immunology*, vol. 70, no. 1, pp. 6–15, 2009.

[8] M. K. Roy, M. Kobori, M. Takenaka et al., "Antiproliferative effect on human cancer cell lines after treatment with nimbolide extracted from an edible part of the neem tree (*Azadirachta indica*)," *Phytotherapy Research*, vol. 21, no. 3, pp. 245–250, 2007.

[9] M. Friedman, J. Pang, and G. A. Smith, "Ninhydrin-reactive lysine in food proteins," *Journal of Food Science*, vol. 49, pp. 10–13, 1984.

[10] K. N. Pearce, D. Karahalios, and M. Friedman, "Ninhydrin assay for proteolysis in ripening cheese," *Journal of Food Science*, vol. 53, pp. 432–438, 1988.

[11] M. Friedman, "Applications of the ninhydrin reaction for analysis of amino acids, peptides, and proteins to agricultural and biomedical sciences," *Journal of Agricultural and Food Chemistry*, vol. 52, no. 3, pp. 385–406, 2004.

[12] V. Ganapathy, B. Ramachandramurty, and A. N. Radhakrishnan, "Distinctive test with copper(II)-ninhydrin reagent for small α-peptides separated by paper chromatography," *Journal of Chromatography*, vol. 213, no. 2, pp. 307–316, 1981.

[13] K. S. Nithya and B. Ramachandramurty, "Screening of some selected spices with medicinal value for Cu (II)-ninhydrin positive compounds," *International Journal of Biological Chemistry*, vol. 1, pp. 62–68, 2007.

[14] B. Ramachandramurty and V. N. Satakopan, "Isolation and partial characterization of a small alpha peptide from Cuminum cyminum L. seeds as detected by Cu(II)—ninhydrin reagent," *International Journal of Chemical Sciences*, vol. 7, no. 4, pp. 2872–2882, 2009.

[15] A. R. Goldfarb, L. J. Saidel, and E. Mosovich, "The ultraviolet absorption spectra of proteins," *The Journal of Biological Chemistry*, vol. 193, no. 1, pp. 397–404, 1951.

[16] M. Culea, "Amino acids quantitation in biological media," *Studia Universitatis Babesbolyai*, vol. 4, pp. 11–15, 2005.

[17] A. M. Frank, M. M. Savitski, M. L. Nielsen, R. A. Zubarev, and P. A. Pevzner, "De novo peptide sequencing and identification with precision mass spectrometry," *Journal of Proteome Research*, vol. 6, no. 1, pp. 114–123, 2007.

[18] A. R. Fersht, J. P. Shi, J. Knoll-Jones, D. M. Lowe, A. J. Wilkinson, and D. M. Blow, "Hydrogen bonding and biological specificity analysed by protein engineering," *Nature*, vol. 314, pp. 235–238, 1985.

[19] T. Storr, M. Markel, G. X. Song-Zhao et al., "Synthesis, characterisation and metal coordinating ability to multifunctional carbohydrate containing compounds for Alzheimer's therapy," *Journal of the American Chemical Society*, vol. 129, no. 23, pp. 4753–7463, 2007.

25

Proline Rich Motifs as Drug Targets in Immune Mediated Disorders

Mythily Srinivasan[1] and A. Keith Dunker[2]

[1] Department of Oral Pathology, Medicine and Radiology, Indiana University School of Dentistry,
   Indiana University Purdue University at Indianapolis 1121 West Michigan Street, DS290, Indianapolis, IN 46268, USA
[2] Department of Biochemistry and Molecular Biology and School of Informatics, Indiana University School of Medicine,
   Indiana University Purdue University at Indianapolis, Indianapolis, IN, USA

Correspondence should be addressed to Mythily Srinivasan, mysriniv@iupui.edu

Academic Editor: Jean-Marie Zajac

The current version of the human immunome network consists of nearly 1400 interactions involving approximately 600 proteins. Intermolecular interactions mediated by proline-rich motifs (PRMs) are observed in many facets of the immune response. The proline-rich regions are known to preferentially adopt a polyproline type II helical conformation, an extended structure that facilitates transient intermolecular interactions such as signal transduction, antigen recognition, cell-cell communication and cytoskeletal organization. The propensity of both the side chain and the backbone carbonyls of the polyproline type II helix to participate in the interface interaction makes it an excellent recognition motif. An advantage of such distinct chemical features is that the interactions can be discriminatory even in the absence of high affinities. Indeed, the immune response is mediated by well-orchestrated low-affinity short-duration intermolecular interactions. The proline-rich regions are predominantly localized in the solvent-exposed regions such as the loops, intrinsically disordered regions, or between domains that constitute the intermolecular interface. Peptide mimics of the PRM have been suggested as potential antagonists of intermolecular interactions. In this paper, we discuss novel PRM-mediated interactions in the human immunome that potentially serve as attractive targets for immunomodulation and drug development for inflammatory and autoimmune pathologies.

1. Protein-Protein Interactions

Protein-protein interactions (PPIs) are critical for most biological functions and cellular processes [1, 2]. Under appropriate environmental conditions, the PPIs take place through an interface governed by shape, chemical complementarity, and flexibility of the interacting molecules. Different types of PPIs have been described. Homo- or heterologous oligomeric PPI complexes represent isclogous or heterologous association of identical protein units. PPI complexes of interdependent protomer units are referred to as obligate complexes as opposed to nonobligate complexes that occur independently [3, 4]. The strength of PPI is represented by the dissociation constant (K_D) expressed in molar concentration and derived from the ratio between the dissociation and association rate constants. Based on duration and affinity, PPIs can be classified as strong interactions that exhibit K_D values with μM concentrations and weak or transient interactions with values in the mM or higher concentrations. Transient PPIs are further divided into strong and weak transient interactions. While strong transient PPIs require a molecular trigger such as ligand binding to shift the oligomeric equilibrium, weak transient interactions are mediated by binding between a few critical residues [3, 5].

Traditionally, PPIs are thought to be mediated by "lock and key" or "induced-fit" interaction between large structured domains [6]. However, characterization of increasing numbers of protein sequences and structures has suggested that the interacting modules of multidomain proteins can be distinguished as globular domains, as short peptide functional sites, and as long peptides that interact with their partners over extensive regions. Thus, three distinct protein-protein interfaces are recognized [7]. While the evolutionarily conserved domain-domain interfaces are large and relatively stable, the evolutionarily plastic domain peptide interfaces are smaller and transient [8–10]. Protein interaction domains can be classified based on sequence homology, ligand-binding properties, or structural similarity [11]. Thus, a typical class of ligand binding proteins may contain a variety of protein interaction domains that recognize a common ligand, whereas a family classified based on sequence homology contains a single fold that may recognize a variety of ligands. Some families function in a narrow cellular context, while others participate in a diverse range of processes [11, 12].

2. Protein:Peptide Interactions: Linear Motifs and Molecular Recognition Features (MoRFs)

Interactions between globular proteins or domains and peptide have been investigated in parallel using biological data and/or computational methods [7, 13–19]. Short segments of structured binding sites within long disordered regions of proteins are referred to as linear motifs (LMs) or as molecular recognition features (MoRFs) [13–15, 18]. While LMs are identified/predicted by sequence patterns, MoRFs are identified by sequence features associated with disorder prediction [13, 14, 17, 18].

A systematic survey of LM-mediated protein interactions estimated that 15–40% of all interactions in a typical eukaryotic cell are mediated through protein-peptide interactions [20]. Predictions over nine proteomes of MoRFs that often form α-helices upon binding indicate that about $44 \pm 4\%$ of eukaryotic proteins contain potential helix-forming MoRFs [21]. These LM and MoRF frequency estimates concur in the suggestion that a large fraction of macromolecular complexes is affected either directly or indirectly by peptide-binding events. The protein interface is predefined and ready to accommodate the binding peptide. For efficient interaction, the peptide "scans" the protein surface for a large enough pocket into which it anchors through a small number of residues or core motif that contribute maximally to the free energy of binding [22]. The hot spot residues show a tendency to be localized in the center, and the number of hot spots is partially dependent on the length of the interacting motif [14, 23]. In general, the average solvent-accessible surface area that is buried upon peptide binding is less than half the area buried in protein-protein complexes. Furthermore, the peptides tend to bind in a more planar fashion, optimize hydrogen bonds, and display better packing than proteins at the interface [10, 20].

3. Amino Acid Propensies in LMs

In terms of amino acid composition, the sequence of LMs can be distinguished into typical patterns. Each LM possesses a set of conserved residues having restricted identities that serve as specificity determinants and a second set of fully variable residues that likely act as spacers [14, 24]. The spacers and flanking regions exhibit preferential presence of disorder promoting charged residues. The restricted identities of LM are enriched in proline as well as in hydrophobic residues including phenylalanine, leucine, tryptophan, tyrosine, and isoleucine [14, 25]. The relative paucity of glycine and alanine in the LMs may represent strategies to simultaneously curb excessive flexibility and restrict the tendency to form strong secondary structural elements [25–27]. Systematic analyses suggested that, as compared to the general disordered regions, MoRFs are also enriched in hydrophobic residues, in particular the aromatic residues [15, 18, 25]. In many cases, LMs and MoRFs identify the same region of sequence. The function of LMs is essentially embodied in the primary amino acid sequence independent of tertiary structure and is strongly context dependent which defines the natural constraints that act on these motifs. Within protein structures, LMs are predominantly observed in the solvent-exposed regions such as the loops, in intrinsically disordered regions or between defined domains [28].

4. Significance of Proline in LMs

Of special significance is the preponderance of proline both in the conserved identities and in the flanking regions of LMs [24, 25]. Among the naturally occurring amino acids, proline is unique in several features. It is the only residue with substituted amide nitrogen. Proteins that recognize the δ carbon on the substituted amide nitrogen within the context of the otherwise standard peptide backbone can select precisely for proline at a given position without making extended contacts with the rest of the side chain [29]. This facilitates sequence-specific recognition without requiring a particularly high-affinity interaction [30]. Such specific and weak bindings are important for cellular communication and signaling functions that require rapidly reversible interactions [31].

Proline is the only naturally occurring amino acid in which the side chain atoms form a pyrrolidine ring with the backbone atoms. This cyclic structure mediates the slow isomerization between cis/trans conformations [32]. The polyproline stretches can adopt two unique helical conformations, I and II [33]. Polyproline type I (PP_I) is a right-handed helix consisting of cis-prolines. While poly-L proline in apolar solvents can adopt the PP_I conformation, there is paucity of PP_I helical segments in proteins [33, 34]. PP_{II} helix is a left-handed helix, consists of proline in trans-conformation, but also accommodates frequently other amino acids such as glutamine, serine, and arginine [35, 36]. With three residues per turn, the PP_{II} helix is an extended structure and has an overall shape resembling a triangular prism. PP_{II} helices are widely distributed in the eukaryotic proteome and hence are of greater biological significance [37, 38].

The unusual shape of the proline side chain imposes structural constraints on adjacent residues such that the proline rich motif (PRM) preferentially adopts the left-handed PP_{II} helical conformation [37, 39]. In PP_{II} helix, both the side chains and the backbone carbonyls point out from the helical axis into solution at regular intervals [40]. Furthermore, the lack of intramolecular hydrogen bonds primarily due to the absence of a backbone hydrogen-bond donor on proline leaves these carbonyls free to participate in intermolecular hydrogen bonds. Thus, both side chains and carbonyls can easily be "read" by interacting proteins making PP_{II} helix an excellent recognition motif [37]. In addition, since the backbone conformation is already restricted, the entropic cost of binding is reduced [41]. In contrast to the enthalpy-induced associations such as the lock and key model, PP_{II} helices are entropy driven and behave as "adaptable gloves" in order to obtain the correct recognition. Indeed, in a recent study that reported significantly lower configurational entropy for known peptide inhibitors, polyproline peptides were among those with lowest entropy values [42]. While the intrinsic properties of the proline facilitate the PP_{II} helix formation, the conformation is potentially stabilized by the surrounding water molecules supporting the preponderance of PRM in solvent exposed loops/disordered regions of proteins [37–39]. Furthermore, it has been observed that, in addition to the enrichment of proline and hydrophobic residues, the LMs are also rich in charged residues including arginine and aspartic acid [24, 25]. Positively charged residues both local and nonlocal to the PP_{II} helices satisfy the H-bond donor potential of the main-chain carbonyls and stabilize the PP_{II} conformation [43]. An advantage of focusing on such distinct chemical features is that such interactions can be discriminatory without resorting to extremely high affinities [37]. Indeed, PRM-mediated interactions exhibit fast on and off rates of binding adopted for effective control and regulatory functions [31, 38].

5. Nature of PRM-Binding Domain (PRD)

The binding between the PRM and the protein domains relies on interactions with core-flanking epitopes of the motif and the PRD interface residues to achieve the necessary specificities [14, 25, 44]. The amino and carboxyl termini of the PRD are generally located relatively close together allowing the domains to slot into their respective host proteins with minimal disruption of the overall protein structure [45]. Structurally, the PRD themselves are found in exposed and accessible regions to recruit target proteins. The PRDs are enriched in aromatic residues that often selectively interact with the critical proline of the PRM in the binding interface. The planar structure of the aromatic side chains appears to be highly complementary to the ridges and grooves presented on the PP_{II} helix formed by the PRM [45, 46].

Summarizing, if a recognition event involves the distinctive property of proline among the 20 natural amino acids, the interaction does not have to be of particularly high affinity to be selective [47]. The benefits of weak, but specific, interactions in intracellular signaling pathways may help explain the preponderance of proline-based recognition motifs in the eukaryotic proteome [48]. Indeed, a recent survey revealed an abundance of the polyproline motif "PXXP" in various gene ontology groups of proteins including enzymes, cytoskeletal proteins, nucleic acid-binding proteins, transport proteins, splicing factors, metal-binding proteins, and ribosomal proteins suggesting an evolutionary conservation of protein-protein networks centered on PRMs [49].

6. PRD:PRM Interactions in the Immunie Responses

The human immunome network elicited so far consists of nearly 1400 interactions involving approximately 600 proteins [50]. Several protein:peptide interactions have been shown to be pivotal for the formation of molecular assemblies of functional complexes that include membrane bound receptors, cytoplasmic signaling molecules, and transcriptional regulatory proteins [51–53]. The life time of the protein complexes as well as the regulatory processes are tightly controlled for proper functioning [29, 44].

PRM-mediated intermolecular interactions are observed in many facets of the immune response including antigen recognition, cell-cell communication, and signaling [46]. Stimulation of lymphocytes with a specific antigen initiates a cascade of signal transduction events that are integrated by numerous adapter proteins which function to establish larger protein complexes and promote complete activation. Many of these adapter proteins possess specific protein domains such as the Src homology 3 (SH3) domains and the WW (named for two tryptophans (W)) domains that selectively recognize proline rich regions in their interacting partners [12, 46]. In addition, many cell surface and intracellular proteins in the immunome exhibit one or two proline-rich regions that interact with highly conserved hydrophobic residues in their binding partners and mediate transient protein-protein interactions [14, 44, 48]. The advantages of transient protein-peptide interactions for functioning of the immune system can be enumerated as follows.

(1) The small interface between the peptides and their protein domain partners facilitates low-affinity weak interactions that are easily formed and disrupted to regulate cellular responses. Indeed cell surface receptors that mediate immune responses are often coupled to intracellular signaling pathways by recognition of modular protein interaction domains that bind a short LM for example, CD2:CD2BP interaction ($K_D = \mu M$) [54].

(2) One protein can bind multiple peptides providing an elegant mechanism that uses transient interactions for bringing together different combinations of complexes each with different functions leading to a different signal and response, for example, CD80:CD28/CD152 interactions [55].

(3) The low-affinity binding allows for large number of interactions of short durations ranging between 10 s and 100 s decreasing the possibility of sustained adherence and facilitating fleeting contacts critical for cell-cell communications such as the interactions between the antigen presenting cells and the T cells mediated by integrins [56, 57].

In this paper, we discuss novel PRD:PRM interactions that could potentially serve as attractive targets for immunomodulation and drug development for inflammatory and autoimmune pathologies. The paper does not include the widely recognized SH3 domain:peptide interactions. Readers interested in these interactions are referred to the excellent reviews [45, 58].

7. PRD:PRM Interactions in Cell Surface Immunome

Proteins located at the surface of immune cells are of particular significance in migration, in specific antigen recognition, in modulating the function of receptors for immune response mediators such as the cytokines as well as in highly focused fine control of intercellular interactions between proteins on opposing cells. Many such interactions have been characterized using monoclonal antibodies [51, 59]. Differential expression and/or function of cell surface proteome in health and disease have been reported in several immune-mediated pathologies substantiating the potential role of select immune cell surface proteins as excellent targets for diagnostic and therapeutic interventions [60].

8. T-Cell Costimulatory Receptor: Ligand Interactions

A manually curated database suggested that over 20% of the human cell surface immunome consists of members of the immunoglobulin superfamily which includes the T-cell costimulatory molecules [61, 62]. In addition to the antigenic stimulation, complete activation of lymphocytes requires costimulatory receptor-ligand interactions that modulate the strength, course, and duration of the immune response [63, 64]. One of the better characterized complexes include the interactions between the costimulatory receptors CD28/CD152 or ICOS expressed on T cells and the CD80/CD86 and ICOSL ligands on the antigen presenting cells. Structurally, the CD28, CD152, and ICOS are composed of a single extracellular IgV domain linked to a stalk region and a transmembrane segment followed by a relatively short cytoplasmic tail, which contains at least one tyrosine-based signaling motif [65]. The receptors exhibit significant homology in primary sequence and share a consensus sequence consisting of three consecutive prolines in the complementarity determining region-3 like region/FG loop [56]. In CD28 and CD152, these prolines are embedded within the MYPPPY sequence, while, in ICOS, they are embedded in the FDPPPF sequence. Mutagenesis experiments indicate that the polyproline motif is essential for the binding of CD152, CD28, and ICOS to their respective ligands [66, 67]. The crystal structures of the murine CD152 bound to CD80 or CD86 reveal that the binding interface is formed predominately by contacts between the MYPPPY sequence of the CD152 FG loop and a concave surface on the front sheet of the CD80 ligand (Figure 1). The three proline residues in the polyproline motif adopt a unique open cis-trans-cis main chain configuration that exhibits geometric complementarity to the binding pocket of the ligand [62, 68]. Secondary structure prediction by PROSS suggests that the second proline of CD152 in the bound complex with CD80 adopts PP_{II} helical conformation with the ϕ and Ψ angles of -76 and 164.6, respectively [69, 70]. In the molecular model of ICOS built using the solution structure of CD152 as template, the critical proline that interacts with the ICOSL binding interface exhibits ϕ and Ψ angles of -53.4 and 167.4, respectively [71] (Tables 1 and 2; Figure 1). The sequence and structural homology together with functional similarity of the FG loops suggest that these receptors share a common mode of recognition for their ligands. The ectodomains of the CD80, CD86, and ICOSL possess a membrane proximal IgC and a distal IgV domain that makes substantial contact with the solvent exposed polyproline motif in the FG loop of the receptors through a surface with considerable hydrophobic character [71–73]. The extended PP_{II} helical conformation facilitates the backbone atoms of the PRM to form hydrogen bonds with the conserved tyrosine at the CD80/ICOSL interface [30, 37, 68]. Intriguingly, it has been suggested that the presence of phenylalanine, a more hydrophobic and poor hydrogen bond acceptor/donor, in the binding interface of CD86 perhaps contributes to the absence of PP_{II} helical conformation and lower affinity for the interactions between CD152 and CD86 [73]. Thus, the local environment of the CD80 binding pocket and the orientation of the functionally important proline in the PRM of CD152 may account for the difference in the strength of interaction between the CD152 and the CD80/CD86. Knowledge derived from the contact preferences of amino acids at PPI interface and the residue propensity to form PP_{II} helix has been adopted in the design of a small peptide, the CD80-competitive antagonist peptide (CD80-CAP) that mimics the ligand binding conformation of the receptor and inhibits CD28/CD152:CD80 interactions [69] (Figure 1). Treatment with CD80-CAP suppressed T-cell-mediated inflammatory responses in mouse models of rheumatoid arthritis, multiple sclerosis, and inflammatory bowel disease [74–76].

9. PRM:PRD Interactions in the Cytoskeleton

Efficient accomplishment of immune responses in inflammation and infection requires a finely regulated cytoskeleton to enable cellular membrane reorganization, receptor localization, and recruitment of signaling molecules, all of which are crucial for immune cell activation, proliferation, secretion, migration, and survival [77]. The cytoskeleton consists of filamentous structures composed mainly of actin, vimentin, and tubulin. Signaling from cell surface receptors

TABLE 1: The dihedral angles of the proline-rich motifs in selected immune-response-related proteins.

Class			Motif	Critical proline	
	CD152			Phi	Psi
	Mouse	1DQT	MYPPPY	−84	167
	Human	1I8L	MYPPPY	−76	164.6
	CD28				
Cell surface proteins	Mouse	Model	MYPPPY	−53.4	166.5
	Human		MYPPPY	−55.4	165.3
	ICOS				
	Mouse	Model	FDPPP	−53.4	167.4
	Human		FDPPP	−83.9	124.2
Cytoskeletal proteins	**WIP**	2IFS	LPPP	−65.9	157.3
		1MKE	LPPP	−75	158.1
	GILZ				
	Human	Molecular Model	PEAP	−67.5	142.5
Transcriptional factors	Mouse		PEAP	−72.5	162.3
	SMRT				
	Human	2ODD	PPP	−70	158.4

TABLE 2: The proline-rich motif (PRM) and the proline-rich motif binding (PRB) domain in immune-related proteins.

	Protein receptor	PRM	PRM interactant	PRD critical residue	Evidence
	CD28	MYPPPY	CD80	Y71	Mutagenesis
Membrane associated	CD152	MYPPPY	CD86	F	Mutagenesis, structural analysis
	ICOS	FDPPP	ICOSL	Y53	Mutagenesis, molecular model
Cytoplasmic	WIP	LPPP	WASP	WHI1 domain, W54	Structural analysis
Signal transduction	GILZ	PXX	p65	TAD (F534. F542)	Immunoprecipitation
	p53	PXXP	p300	(SPC1 192–337) SPC-2 (1737–1913)	ChIP
Transcriptional cofactors	SMRT	PXLXP	p65	TAD	GST pull-down assays
	SMRT		MCTF	MYND domain	Mutagenesis
			EA1	MYND domain	Solution structure
			ETO	MYND domain	Solution structure

Myc-related cellular transcription factor; viral oncoproteins EA1; ETO (a nuclear corepressor protein) chromatin immunoprecipitation.

and migration are mediated by rapid assembly of actin filaments predominantly at the plasma membrane or cortical cytoskeleton. The assembly of cytoskeletal network is regulated by multiple classes of actin binding proteins that initiate, polymerize, sever, depolymerize, and terminate filament formation [78]. The interaction between cytoskeletal binding proteins and fibers is often transient and of low affinity. Recently, alterations in regulators of cortical actin cytoskeletal proteins have been implicated in immune deficiency and autoimmune diseases [79].

10. Wiskott-Aldrich Syndrome Protein (WASP): WASP-Interacting Protein (WIP) Binding

The importance of actin-mediated cytoskeletal regulation in humans is exemplified by Wiskott-Aldrich syndrome (WAS), an immune deficiency disease characterized by recurrent infections, eczema, thrombocytopenia, and an increased risk

of autoimmunity and malignancy as a result of abnormal lymphocyte activation. It is an X-linked disease caused by mutations in WASP, a member of actin regulators that function as scaffolds transducing a wide range of signals between proteins or from proteins to membranes to mediate dynamic changes in the actin cytoskeleton [80].

WASP functions in multiple cellular processes in immune responses. It links the T-cell receptor (TCR)/CD3 complex to the actin cytoskeleton, enhancing the efficiency of the immunological synapse formation and cytokine secretion [81]. It also promotes homeostasis of regulatory T-cells and controls T-cell activation and effector functions [80, 82]. Most WASP molecules in the cytoplasm of T cells are associated with WASP-interacting protein/WIP, which acts as a chaperone to localize WASP to areas of active actin polymerization including the immunological synapse [83, 84]. Absence of either WASP or WIP induces impaired T cell proliferation in response to TCR/CD3 ligation as well as defective T-cell homing. Importantly, WASP levels can be restored to normal by expressing WIP in WIP-deficient cells suggesting that the WIP stabilizes and regulates the absolute cellular levels of WASP [85, 86].

The WIP has also been shown to affect T-cell activation independent of WASP [86]. In resting cells, WIP remains in a complex with WASP and an adapter protein, CrkL [83]. In activated T cells, the CrkL interacts with phosphorylated ZAP-70, a critical adapter molecule near the cell membrane and recruits the WASP-WIP-CrKL complex to the immunological synapse [87]. At the synapse, WIP is phosphorylated by PKCθ, resulting in the release of WASP which upon activation by the membrane-bound Cdc42 kinase initiates actin polymerization. Free WIP binds to newly formed actin filaments and helps stabilize the immunological synapse [82, 86]. Furthermore, WIP regulates the activity of the NF-AT/AP-1 transcription factor complex. AP-1 is a heterodimer of Fos and Jun proteins that both regulate the transcription of multiple biological mediators including the cell surface receptors and directly facilitate the entry of T cells into cell cycle. WIP overexpression is associated with increased actin stabilization and enhanced AP-1 in activated T cells [88].

Structurally, WASP proteins are multidomain proteins consisting of a conserved enabled/VASP homology-1 (EVH1) domain, also referred as the WASP homology 1 (WH1) domain, a GTPase binding domain, a proline-rich region, and a basic motif connected through a central region to a WH-2 motif that binds the actin nucleating complex [80, 82]. Most missense mutations in WAS involve the residues in the EVH1 domain of WASP [80]. EVH1 domains have been found in ~630 human genes and are classified into four distinct protein families based on amino acid sequence analysis that includes WASP; enabled/vasodilator-stimulated phosphoprotein (Ena/VASP); Homer/Vesl; sprout-related proteins with an EVH1 domain (SPRED). Each EVH1 subclass recognizes a distinct pattern of amino acids, but all of them bind proline-rich sequence in the left-handed PP_{II} conformation. Residues flanking the PRM contribute to the binding specificity of the EVH1 complexes [12].

The primary structure of WIP consists of a highly conserved verprolin homology (VH) domain that binds actin filaments, multiple putative SH3-binding domains for interacting with adapter/signaling molecules and a WASP-binding domain (WBD) [86, 88]. The WBD of WIP is approximately 30 residues long with a highly conserved central proline-rich motif and two short epitopes on either side of the motif. Nuclear magnetic resonance studies and glutathione S-transferase pull-down assays have demonstrated that the conserved polyproline motif of WIP occupies the canonical binding site in the WH1/EVH1 domain of WASP. The WIP polyproline motif forms a PP_{II} helical turn and straddles the highly conserved tryptophan side chain of WASP at the WH1 domain interface, the binding contributing nearly 40% of the total buried surface of the WIP-EVH1 complex [82, 83, 86, 88]. Secondary structure analysis of the WBD of WIP by PROSS showed that the residues exhibited ϕ and Ψ angles of −65.9 and 157.3, respectively, consistent with PP_{II} helical conformation [70] (Tables 1 and 2). Interestingly, the WIP polyproline motif has been shown to bind WASP in the opposite direction through an elongated WBD as compared with other EVH1:peptide complexes [92] (Figure 2). The interactions between the conserved phenylalanine residues in the epitope preceding the polyproline motif and the hydrophobic surface of the WH1 domain of WASP (Val, Ala) as well as the formation of a salt bridge between the conserved glutamine in the binding pocket of the WH1 domain of WASP and the acidic residue (K/R) localized in an epitope following the polyproline motif of WIP have been shown to facilitate the reverse orientation [83, 92]. Mutation of WASP residues involved in interaction with any of the three WIP epitopes reduces the WASP binding. These structural features support the observations that different missense mutations disrupt the intermolecular interactions and accelerate degradation of WASP similarly in WAS patients with different genotypes. Thus, the WIP/WASP structure exhibits semi-independent composite linear motifs that are recognized in an extended conformation with enhanced specificity [12, 80, 86]. A similar example of a scaffolding interaction with bidirectional binding of composite linear motifs has been reported for complexes that regulate mitogen-activated protein kinase pathways [12, 92].

Recently, it has been reported that treatment with a peptide derived from the proline-rich WBD of WIP restored WASP to physiological levels in lymphocytes from patients with mutations in the WBD of WASP. Furthermore, treatment with the WIP peptide ameliorated the defects in the reorganization of actin cytoskeleton in T cells from these patients [93] (Figure 2).

11. PRM:PRD Interactions in Transcriptional Regulation

Eukaryotic gene expression is a dynamic process regulated by multiple signaling networks mediated by rapid and reversible PPI complexes. Formation of ternary complexes of transcriptional regulatory proteins is critical for gene transcription. These complexes include cross-talk between different families of transcription factors and the interactions of transcription factors with coactivators or corepressors. Transcription initiation requires the formation of an initiation complex that consists of RNA polymerase II, the basal transcription machinery (made up of TFIIA, TFIIB, TFIID, TFIIE, TFIIF, and TFIIH), and the sequence-specific promoter binding transcription factors [94]. Many transcription factors mediate transcriptional activation by interacting through their transactivation domain with one or more components of the basal transcription machinery. Such direct interactions are thought to bring the activation domain over large distances into close proximity with the initiation complex close to the transcription start site [95]. In this context, it is interesting to note that PRMs have been frequently observed in many transcription factors suggesting that the flexibility offered by such segments potentially contribute to the interactions involved in the formation of functional multiprotein transcriptional complex [96].

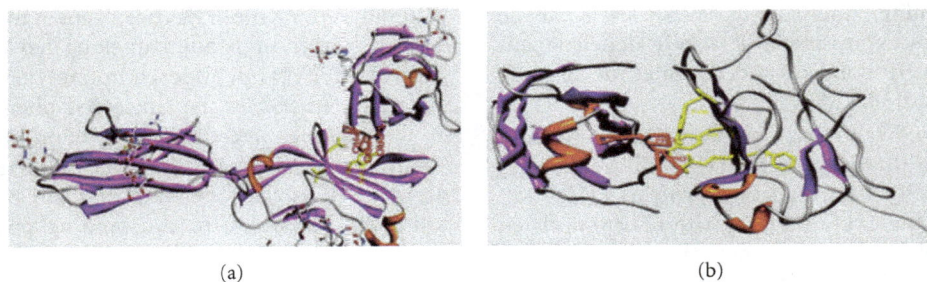

(a)   (b)

FIGURE 1: (a) represents the solution structure of the CD152:CD80 complex (PDB1I8 L), and (b) represents the complex of ICOS:ICOS L. Homology modeling of ICOS and ICOSL was predicted using the structures of CD152 (PDB 1AH1 and PDB 1DQT), CD80 (PDB 1DR9) respectively, as templates by Geno 3D [89] and SWISS MODEL [90]. Prediction of the structural complex was performed by ClusPro [91], the complex with least energy is shown.

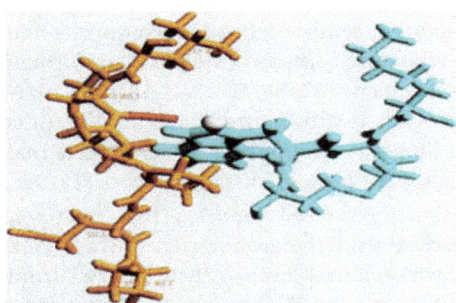

FIGURE 2: PDB 1MKE depicting the solution structure of WASP:WIP complex, the PRM is labeled.

12. PRM in p300:p53 Interactions

The p300 protein is a versatile coactivator with several conserved domains including the bromodomain which recognizes acetylated residues; cysteine-histidine-(CH-) rich domains; a KIX domain; an ADA2 homology domain. While the amino and carboxy termini of p300 activate transcription, the histone acetyltransferase activity is mediated by the central region. The modular organization of p300 provides a scaffold for assembly of multicomponent coactivator complexes that regulate transcription through multiple mechanisms. These include providing a scaffold for recruiting many transcription factors, acting as a bridge to connect sequence specific transcription factors to the basal transcription apparatus, mediating complete activation of select transcription factors via an intrinsic histone acetylase activity, as well as influencing chromatin activity by modulating nucleosomal histones [97]. Interestingly, phage-peptide display analysis suggested that the p300 protein exhibits a strong affinity to bind proline rich peptides [98].

The p53 is one of the most well-studied eukaryotic transcription factors that functions as a homotetramer. It is upregulated in response to cellular stress and induces up- or downregulation of genes involved in cell cycle arrest, DNA repair, apoptosis, antiangiogenesis, and senescence pathways [99]. Structurally, it has a modular domain architecture consisting of independently folded DNA-binding domain and tetramerization domains flanked by natively unfolded regions in the amino and carboxy termini. Activation of p53 is associated with various posttranslational modifications of multiple lysine residues at the carboxyl terminus. The amino terminus consists of an acidic transactivation domain (TAD) including a proline-rich region. The TAD binds many components of the transcription machinery including the coactivators p300/cAMP-response element binding protein (CREB), binding protein (CBP), as well as the negative regulators MDM-2 [99, 100]. The binding with p300 is essential for the transcriptional function of p53 [98].

The p300 binds the carboxy terminal regulatory domain of p53 predominantly via its bromodomain [99]. This docking releases the p53 of its intrinsic conformational constraints, allowing phosphorylation of critical threonine and serine residues in the activation domain, thus facilitating stabilization. Additionally, the TAD at the amino terminus of p300 interacts with the proline-rich "PXXP" motifs of the p53 activation domain [99, 100]. The binding of the amino and carboxy termini of the p300 with the transactivation regulatory regions of p53 induces a conformational alteration that promotes the sequence-specific DNA binding of p53. The acetylation of critical lysine residues of p53 by the histone acetyltransferase activity of p300 then promotes the transcriptional activity. This concomitant binding of p300 to a relatively ubiquitous proline repeat motif and the classic hydrophobic LXXLL motif of the p53 terminal regions highlights an additional layer of combinatorial regulation of the core p300 protein-protein interactions at a promoter region [98, 101]. Chromatin immunoprecipitation studies showed that the deletion of the proline repeat motif of p53 prevents DNA-dependent acetylation of p53 by occluding p300 from the p53-DNA complex [98]. Although the pathological role of p53 in many neoplasms has been well characterized, the mechanisms of DNA damage and the contribution of abnormal p53 in autoimmune inflammatory diseases such as ulcerative colitis and rheumatoid arthritis are recently recognized. Intriguingly, peptides derived from the "PXXP" containing proline repeat domain of p53 have been shown to bind p300 and inhibit sequence-specific DNA-dependent acetylation of p53 [98].

13. Glucocorticoid-Induced Leucine Zipper: p65 Interaction

The mammalian NF-κB family of inducible transcription factors is responsible for regulating specific sets of genes in many cell types and participates in many cellular processes, including inflammation, proliferation, and cell survival. The most common form of NF-κB is a p65:p50 heterodimer, which in resting T cells remains in the cytoplasm as an inactive complex bound to the IκB inhibitor proteins. Following T-cell activation, degradation of IκB releases NF-κB, allowing the subunits to translocate to the nucleus [102]. The ability of p65 to recruit the histone acetyltransferase activity-associated complex consisting of p300 and other coactivators within the nucleus governs the transcriptional regulation. The p300 induces acetylation of a critical lysine residue in the rel homology domain of p65 which then binds to specific sites in the promoter regions of target DNA elements and transiently activates transcription of proteins involved in immune or inflammatory responses and cell growth control [103]. Misregulation of NF-κB is linked to a wide variety of human diseases including infections, inflammatory autoimmune disorders, and various cancers. Hence, specific inhibitors of this nuclear factor are being sought and tested as treatments [104].

Glucocorticoids are well characterized anti-inflammatory and immunosuppressive agents. Glucocorticoids act by binding the glucocorticoid receptor in the cytoplasm, thus activating the receptor, which then translocates to the nucleus where it suppresses p65 acetylation by competing with the p300 histone acetyltransferase activity [105]. In addition, the intranuclear glucocorticoid receptor binds specific negative and positive glucocorticoid response elements in target DNA to directly suppress immune response and to activate transcription of anti-inflammatory genes [106].

The glucocorticoid-induced leucine zipper (GILZ) was recently identified during a systematic study of genes transcriptionally induced by glucocorticoids [107]. Expression of GILZ is downregulated following T-cell activation [108]. Blockade of T-cell activation either by interfering with T-cell costimulatory molecules or by blocking intracellular signaling pathway has been shown to upregulate GILZ [74, 108]. In addition, treatment with exogenous GILZ has been shown to suppress inflammatory responses [109–111]. Mechanistically, GILZ-mediated effects on immune and inflammatory responses have been attributed to its ability to inhibit NF-κB activation [112].

GILZ has been shown to physically bind the p65 subunit of NF-κB through a protein-protein interaction [113]. Since the interaction is independent of the rel-homology domain and the phosphorylation of inhibitory proteins, it has been suggested that the GILZ binds the transactivation domain of the p65 molecule. Analyses of structural complexes of interactions wherein the binding depends on the presence of one or more prolines have shown that the functionally critical proline/s in the interface of one protein often are in contact with aromatic residues from the other component [114]. In this context, it is interesting to observe that

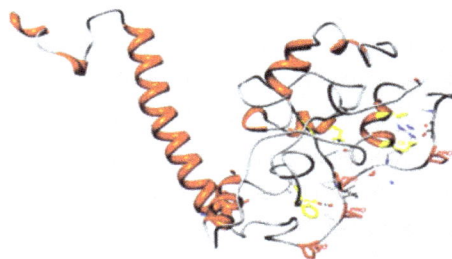

FIGURE 3: The GILZ:p65 complex: homology modeling of GILZ and p65 was predicted using the structures of delta sleep inducing peptide (PDB:1DIP) and (PDB: 2IW3), respectively, as templates by Geno 3D [89] and SWISS model [90]. Prediction of the structural complex was performed by ClusPro [91]; the complex with least energy is shown.

p65-transactivation domain that potentially interacts with the GILZ-COOH presents two highly conserved aromatic residues, F^{534} and F^{542}, which are critical residues for p65 transactivation [115]. Structure prediction of p65 and GILZ were generated by homology modeling using Geno3D and Swiss model protein structure prediction servers [89, 90] (data not shown). Docking of p65 with GILZ suggested that the Pro-120 of GILZ was within 5 Å distance of F^{534} of p65 TAD [91] (Figure 3).

The primary sequence of GILZ consists of an amino terminal leucine-zipper motif and a proline-rich carboxy terminus. Mutational analysis localized the site of interaction with the p65 to the proline-rich carboxy terminus of GILZ (GILZ-COOH). The GILZ-COOH consists of three consecutive (PXX) motifs with a proline as every third residue [110]. Secondary structure assignment based on backbone dihedral angles by PROSS showed that the Pro^{120} of GILZ exhibited a ϕ angle of $-67° \pm 5°$ and a Ψ angle of $142.5° \pm 15°$ (Tables 1 and 2), thus adopting a PP$_{II}$ helical conformation [70]. Additionally, the presence of multiple glutamic acid residues in the region increases the net charge further promoting the extended conformation by electrostatic repulsion [116]. Recently, a small peptide mimic of GILZ, GILZ-P, has been developed by conceptually integrating the mechanism of action glucocorticoids and the knowledge derived from the structural analysis of GILZ and its interaction with the p65 subunit of NF-κB. Treatment with GILZ-P suppressed T-cell activation and inflammation in a mouse model for multiple sclerosis [117]. It is speculated that the low-molecular-weight GILZ-P can provide promising leads for developing small molecule NF-κB inhibitors.

14. Silencing Mediator for Retinoic and Thyroid Hormone Receptors (SMRT) and p65 Interaction

As stated above, the regulated activation and repression of transcription are critical in many biological processes. Controlled repression of transcription is observed in cell-fate decisions during development and cellular differentiation, as well as in the maintenance of homeostasis [94]. SMRT and nuclear receptor corepressor (NCoR) are large homologous corepressor proteins that mediate transcriptional repression by many different nuclear receptors [118]. SMRT and NCoR are also recruited by many other DNA-binding transcription factors, such as BCL6, Kaiso, ETO, MEF2C, CNOT2, and CBF1. Mechanistically, SMRT and NcOR complexes associate with histone deacetylase (HDAC) enzymes and mediate transcriptional repression through deacetylation and condensation of chromatin [119].

Consistent with observation that the acetylation of p65 governs transcriptional activation, the components of a corepressor complex for NF-κB include HDAC, SMRT, and NCoR [120]. SMRT and NCoR do not exhibit an enzymatic activity but trigger the catalytic activity of HDAC. The SMRT-HDAC complex is responsible for basal repression of classical NF-κB-regulated gene targets in the unstimulated state [121]. In resting cells, NF-κB remains in the cytoplasm complexed with the IKBα and IKBβ heterodimer. Following stimulation, the IκB kinase-α (IKK) and IKKβ mediate phosphorylation of the IκB proteins and release the NF-κB subunits for nuclear translocation [102]. The IKKα phosphorylates SMRT and initiates derepression, thus preventing HDAC chromatin association. This allows the active p50-RelA/p65 of NF-κB to bind DNA and potentiate transcription [118, 119, 122]. Thus, the SMRT-dependent transcriptional regulation of NF-κB plays a critical role in controlling cellular proliferation.

Structurally mammalian SMRT has two amino terminal SNT (Swi/Ada/N-CoR/TFIID)/DNA-binding domains and two receptor interaction domains that present corepressor nuclear receptor (CoRNR) motif near the carboxy terminus [119, 121]. In addition, SMRT also contains three repression domains (RDs) that recruit diverse proteins. The third RD domain of SMRT includes proline-rich regions and has been shown to mediate transcriptional regulation by interacting with the ligand-activated glucocorticoid receptor [120]. Yeast two hybrid system and glutathione S transferase pull-down assays showed that the residues encompassing the proline rich SMRT RD3 region specifically and selectively bind the residues of the transactivation domain of p65. Significantly treatment with SMRT peptide derived from this proline-rich region has been shown to physically interact with the p65 and inhibit transactivation of inflammatory proteins via recruitment of HDACs [123] (Figure 4).

15. MYND:PRM Interactions

The MYND domain is a zinc-binding domain present in a large number of proteins that participate in many protein-protein interactions involved in transcriptional regulation. Some of the proteins with MYND domain include BS69, a transcriptional corepressor; the chimeric fusion protein of acute myelogenous leukemia (AML) and ETO (a nuclear protein that interacts with corepressor molecules) (AML-1-ETO); the bone morphogenesis protein receptor-associated molecule 1 (BRAM1), deformed epidermal autoregulatory factor-1 (DEAF-1), and SET and MYND domain-containing proteins (SMYD) [124–126]. Functionally, many MYND

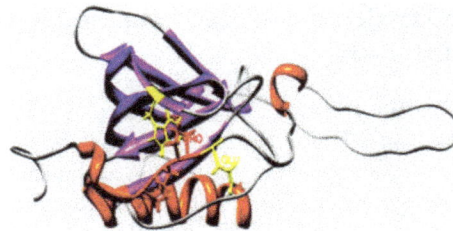

FIGURE 4: SMRT: p65 complex: the SMRT structure was derived from the Chain A of the PDB 2ODD. Homology model of p65 was predicted using the structures of PDB:2IW3 as templates by Geno 3D by Geno 3D [89] and SWISS MODEL [90]. Prediction of the structural complex was performed by ClusPro [91], the complex with least energy is shown.

domain containing proteins have been involved in diverse cellular processes including proliferation, apoptosis, adhesion, and migration. Although the role of most MYND domains has been investigated with respect to tumorigenesis, their role in hematopoietic development suggests a potential role in normal immune response as well as in immunopathology [127, 128].

The MYND domain typically recognizes proline-rich motifs in partner proteins. For example, molecular studies have shown that the BS69 MYND binds the viral oncoproteins EA1 and EBNA1 as well as the Myc-related cellular transcription factor (MGA) through "PXLXP" motif conserved in all three interacting partners [124]. The chimeric AML1-ETO protein contains the DNA-binding domain of AML1 and nearly all of ETO [129]. The ETO hosts a MYND domain at the carboxy terminus and has been shown to physically associate with N-CoR/SMRT and their associated HDACs to aberrantly repress transcription [130]. Mutational studies suggested that the proline-rich motif "PPPLI" in the SMRT-RD3 specifically interacts with the ETO MYND domain [131]. Interestingly, a peptide derived from the SMRT-RD3 has been shown to specifically bind the MYND domain of AML/ETO. Solution structure of the MYND-SMRT peptide complex suggested that the SMRT PRM binds in an extended conformation to a hydrophobic pocket in MYND [130] (Figure 5). The ϕ and Ψ angles of the critical proline residues are consistent with PP$_{II}$ helical conformation (Tables 1 and 2) [70]. The side chain of the critical proline in the SMRT PRM is packed on top of a highly conserved tryptophan in the ligand-binding region of the MYND domain. The carbonyls of the second and third prolines of the SMRT PRM form hydrogen bonds with the conserved glutamine and serine residues of the MYND domain at the binding interface. Although hydrogen bonds to the backbone generally cannot provide specificity, the relative geometrical positions of the highly conserved tryptophan, glutamine, and serine in MYND-binding domain are thought to favor interaction with the elongated conformation of the SMRT/NcOR PRM [130].

Since transcriptional regulation involves direct interactions between the transactivation domains of a transcription

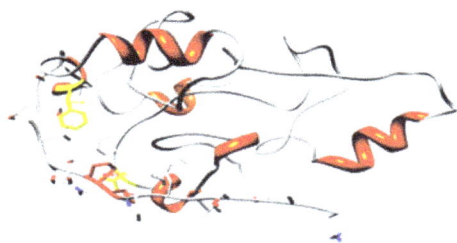

FIGURE 5: The SMRT:MYND complex (PDB 2ODD) with proline in the interface highlighted.

factor with either coactivators/corepressors in the transcriptional machinery to initiate or suppress transcription, molecules that directly block the formation of these complexes would then function as transcriptional modulators [132]. Peptides or small molecule mimics of the transactivation domain of transcription factors should be able to competitively interfere with its natural counterpart. However, translation of this concept has been highly challenging as evidenced by the few synthetic transactivation domain inhibitors reported in recent years.

16. Conclusion

Increasing knowledge of the interactome in the physiological and pathological immune responses provides an unprecedented opportunity for identification and characterization of potential diagnostic and therapeutic targets. Although, it is recognized that protein-protein interaction interfaces may be dissected into much smaller contact points, and only a small number of amino acids are critical to the specificity of the interactions, comprehensive rules are still difficult to derive [133]. Despite this, an often used strategy in the discovery of peptide drugs is an exploitation of the complementary surfaces of naturally occurring binding partners. It is expected that these peptides function as competitive inhibitors, masking an interaction site and making it inaccessible for the binding of the protein from which it has been derived. The inhibitory peptides could serve as potential drugs by themselves, and also more importantly, knowledge about the structure of the critical amino acids at the interface could be used as a basis to design a collection of potential mimetics [134].

References

[1] T. Pawson and P. Nash, "Assembly of cell regulatory systems through protein interaction domains," *Science*, vol. 300, no. 5618, pp. 445–452, 2003.

[2] L. O. Sillerud and R. S. Larson, "Design and structure of peptide and peptidomimetic antagonists of protein-protein interaction," *Current Protein and Peptide Science*, vol. 6, no. 2, pp. 151–169, 2005.

[3] O. Keskin, A. Gursoy, B. Ma, and R. Nussinov, "Towards drugs targeting multiple proteins in a systems biology approach," *Current Topics in Medicinal Chemistry*, vol. 7, no. 10, pp. 943–951, 2007.

[4] I. M. A. Nooren and J. M. Thornton, "Diversity of protein-protein interactions," *EMBO Journal*, vol. 22, no. 14, pp. 3486–3492, 2003.

[5] S. E. Ozbabacan, H. B. Engin, A. Gursoy, and O. Keskin, "Transient protein-protein interactions," *Protein Engineering Design Selection*, vol. 24, pp. 635–648, 2011.

[6] S. Ren, V. N. Uversky, Z. Chen, A. K. Dunker, and Z. Obradovic, "Short Linear Motifs recognized by SH2, SH3 and Ser/Thr Kinase domains are conserved in disordered protein regions," *BMC Genomics*, vol. 9, no. 2, article S26, 2008.

[7] P. Tompa, M. Fuxreiter, C. J. Oldfield, I. Simon, A. K. Dunker, and V. N. Uversky, "Close encounters of the third kind: disordered domains and the interactions of proteins," *BioEssays*, vol. 31, no. 3, pp. 328–335, 2009.

[8] R. Jothi, P. F. Cherukuri, A. Tasneem, and T. M. Przytycka, "Co-evolutionary analysis of domains in interacting proteins reveals insights into domain-domain interactions mediating protein-protein interactions," *Journal of Molecular Biology*, vol. 362, no. 4, pp. 861–875, 2006.

[9] A. Stein, R. B. Russell, and P. Aloy, "3did: interacting protein domains of known three-dimensional structure," *Nucleic Acids Research*, vol. 33, pp. D413–D417, 2005.

[10] A. Zucconi, S. Panni, S. Paoluzi, L. Castagnoli, L. Dente, and G. Cesareni, "Domain repertoires as a tool to derive protein recognition rules," *FEBS Letters*, vol. 480, no. 1, pp. 49–54, 2000.

[11] R. R. Copley, T. Doerks, I. Letunic, and P. Bork, "Protein domain analysis in the era of complete genomes," *FEBS Letters*, vol. 513, no. 1, pp. 129–134, 2002.

[12] F. C. Peterson and B. F. Volkman, "Diversity of polyproline recognition by EVH1 domains," *Frontiers in Bioscience*, vol. 14, pp. 833–846, 2009.

[13] N. E. Davey, N. J. Haslam, D. C. Shields, and R. J. Edwards, "SLiMSearch 2.0: biological context for short linear motifs in proteins," *Nucleic Acids Research*, vol. 39, supplement 2, pp. W56–W60, 2011.

[14] N. E. Davey, K. Van Roey, R. J. Weatheritt et al., "Attributes of short linear motifs," *Molecular BioSystems*, vol. 8, no. 1, pp. 268–281, 2012.

[15] A. Mohan, C. J. Oldfield, P. Radivojac et al., "Analysis of molecular recognition features (MoRFs)," *Journal of Molecular Biology*, vol. 362, no. 5, pp. 1043–1059, 2006.

[16] V. Neduva and R. B. Russell, "DILIMOT: discovery of linear motifs in proteins," *Nucleic Acids Research*, vol. 34, pp. W350–W355, 2006.

[17] J. C. Obenauer, L. C. Cantley, and M. B. Yaffe, "Scansite 2.0: proteome-wide prediction of cell signalling interactions using short sequence motifs," *Nucleic Acids Research*, vol. 31, no. 13, pp. 3635–3641, 2003.

[18] V. Vacic, C. J. Oldfield, A. Mohan et al., "Characterization of molecular recognition features, MoRFs, and their binding partners," *Journal of Proteome Research*, vol. 6, no. 6, pp. 2351–2366, 2007.

[19] M. B. Yaffe and L. C. Cantley, "Mapping specificity determinants for protein-protein association using protein fusions and random peptide libraries," *Methods in Enzymology*, vol. 328, pp. 157–170, 2000.

[20] V. Neduva and R. B. Russell, "Peptides mediating interaction networks: new leads at last," *Current Opinion in Biotechnology*, vol. 17, no. 5, pp. 465–471, 2006.

[21] Y. Cheng, C. J. Oldfield, J. Meng, P. Romero, V. N. Uversky, and A. K. Dunker, "Mining α-helix-forming molecular recognition features with cross species sequence alignments," *Biochemistry*, vol. 46, no. 47, pp. 13468–13477, 2007.

[22] A. Stein and P. Aloy, "Contextual specificity in peptide-mediated protein interactions," *PLoS ONE*, vol. 3, no. 7, Article ID e2524, 2008.

[23] E. Petsalaki, A. Stark, E. García-Urdiales, and R. B. Russell, "Accurate prediction of peptide binding sites on protein surfaces," *PLoS Computational Biology*, vol. 5, no. 3, Article ID e1000335, 2009.

[24] C. Chica, F. Diella, and T. J. Gibson, "Evidence for the concerted evolution between short linear protein motifs and their flanking regions," *PloS ONE*, vol. 4, no. 7, Article ID e6052, 2009.

[25] M. Fuxreiter, P. Tompa, and I. Simon, "Local structural disorder imparts plasticity on linear motifs," *Bioinformatics*, vol. 23, no. 8, pp. 950–956, 2007.

[26] A. K. Dunker, M. S. Cortese, P. Romero, L. M. Iakoucheva, and V. N. Uversky, "Flexible nets: the roles of intrinsic disorder in protein interaction networks," *FEBS Journal*, vol. 272, no. 20, pp. 5129–5148, 2005.

[27] P. Romero, Z. Obradovic, X. Li, E. C. Garner, C. J. Brown, and A. K. Dunker, "Sequence complexity of disordered protein," *Proteins*, vol. 42, no. 1, pp. 38–48, 2001.

[28] A. Via, C. M. Gould, C. Gemünd, T. J. Gibson, and M. Helmer-Citterich, "A structure filter for the Eukaryotic Linear Motif Resource," *BMC Bioinformatics*, vol. 10, article 351, 2009.

[29] L. J. Ball, R. Kühne, J. Schneider-Mergener, and H. Oschkinat, "Recognition of proline-rich motifs by protein-protein-interaction domains," *Angewandte Chemie*, vol. 44, no. 19, pp. 2852–2869, 2005.

[30] M. P. Williamson, "The structure and function of proline-rich regions in proteins," *Biochemical Journal*, vol. 297, no. 2, pp. 249–260, 1994.

[31] P. Van der Merwe and A. N. Barclay, "Transient intercellular adhesion: the importance of weak protein-protein interactions," *Trends in Biochemical Sciences*, vol. 19, no. 9, pp. 354–358, 1994.

[32] W. J. Wedemeyer, E. Welker, and H. A. Scheraga, "Proline cis-trans isomerization and protein folding," *Biochemistry*, vol. 41, no. 50, pp. 14637–14644, 2002.

[33] A. A. Adzhubei and M. J. E. Sternberg, "Left-handed polyproline II helices commonly occur in globular proteins," *Journal of Molecular Biology*, vol. 229, no. 2, pp. 472–493, 1993.

[34] A. A. Adzhubei and M. J. E. Sternberg, "Conservation of polyproline II helices in homologous proteins: implications for structure prediction by model building," *Protein Science*, vol. 3, no. 12, pp. 2395–2410, 1994.

[35] M. A. Kelly, B. W. Chellgren, A. L. Rucker et al., "Host-Guest study of left-handed polyproline II helix formation," *Biochemistry*, vol. 40, no. 48, pp. 14376–14383, 2001.

[36] B. J. Stapley and T. P. Creamer, "A survey of left-handed polyproline II helices," *Protein Science*, vol. 8, no. 3, pp. 587–595, 1999.

[37] M. V. Cubellis, F. Caillez, T. L. Blundell, and S. C. Lovell, "Properties of polyproline II, a secondary structure element implicated in protein-protein interactions," *Proteins*, vol. 58, no. 4, pp. 880–892, 2005.

[38] B. K. Kay, M. P. Williamson, and M. Sudol, "The importance of being proline: the interaction of proline-rich motifs in signaling proteins with their cognate domains," *FASEB Journal*, vol. 14, no. 2, pp. 231–241, 2000.

[39] J. F. Gibrat, B. Robson, and J. Garnier, "Influence of the local amino acid sequence upon the zones of the torsional angles φ and ψ adopted by residues in proteins," *Biochemistry*, vol. 30, no. 6, pp. 1578–1586, 1991.

[40] A. Rath, A. R. Davidson, and C. M. Deber, "The structure of "unstructured" regions in peptides and proteins: role of the polyproline II helix in protein folding and recognition," *Biopolymers*, vol. 80, no. 2-3, pp. 179–185, 2005.

[41] B. W. Chellgren and T. P. Creamer, "Side-chain entropy effects on protein secondary structure formation," *Proteins*, vol. 62, no. 2, pp. 411–420, 2006.

[42] E. B. Unal, A. Gursoy, and B. Erman, "Conformational energies and entropies of peptides, and the peptide-protein binding problem," *Physical Biology*, vol. 6, no. 3, Article ID 036014, 2009.

[43] N. Eswar, C. Ramakrishnan, and N. Srinivasan, "Stranded in isolation: structural role of isolated extended strands in proteins," *Protein Engineering*, vol. 16, no. 5, pp. 331–339, 2003.

[44] Y. Mansiaux, A. P. Joseph, J.-C. Gelly, and A. G. de Brevern, "Assignment of polyproline ii conformation and analysis of sequence—structure relationship," *PLoS ONE*, vol. 6, no. 3, Article ID e18401, 2011.

[45] A. Zarrinpar, R. P. Bhattacharyya, and W. A. Lim, "The structure and function of proline recognition domains," *Science's STKE*, vol. 2003, no. 179, p. RE8, 2003.

[46] C. Freund, H. G. Schmalz, J. Sticht, and R. Kühne, "Proline-rich sequence recognition domain (PRD): ligands, function and inhibition," in *Protein–Protein Interactions as New Drug Targets*, E. Klussmann and J. Scott, Eds., pp. 407–428, Springer, Berlin, Germany, 2008.

[47] W. Gu and V. Helms, "Dynamical binding of proline-rich peptides to their recognition domains," *Biochimica et Biophysica Acta*, vol. 1754, no. 1-2, pp. 232–238, 2005.

[48] V. Neduva and R. B. Russell, "Proline-rich regions in transcriptional complexes: heading in many directions," *Science's STKE*, vol. 2007, no. 369, article pe1, 2007.

[49] B. R. Chandra, R. Gowthaman, R. R. Akhouri, D. Gupta, and A. Sharma, "Distribution of proline-rich (PxxP) motifs in distinct proteomes: functional and therapeutic implications for malaria and tuberculosis," *Protein Engineering, Design and Selection*, vol. 17, no. 2, pp. 175–182, 2004.

[50] R. Montañez, I. Navas-Delgado, M. A. Medina, J. F. Aldana-Montes, and F. Sánchez-Jiménez, "Information integration of protein-protein interactions as essential tools for immunomics," *Cellular Immunology*, vol. 244, no. 2, pp. 84–86, 2006.

[51] M. C. Díaz-Ramos, P. Engel, and R. Bastos, "Towards a comprehensive human cell-surface immunome database," *Immunology Letters*, vol. 134, no. 2, pp. 183–187, 2011.

[52] C. Ortutay and M. Vihinen, "Efficiency of the immunome protein interaction network increases during evolution," *Immunome Research*, vol. 4, no. 1, article 4, 2008.

[53] C. Ortutay and M. Vihinen, "Immunome Knowledge Base (IKB): an integrated service for immunome research," *BMC Immunology*, vol. 10, article 3, 2009.

[54] C. Freund, R. Kühne, H. Yang, S. Park, E. L. Reinherz, and G. Wagner, "Dynamic interaction of CD2 with the GYF and the SH3 domain of compartmentalized effector molecules," *EMBO Journal*, vol. 21, no. 22, pp. 5985–5995, 2002.

[55] A. V. Collins, D. W. Brodie, R. J. C. Gilbert et al., "The interaction properties of costimulatory molecules revisited," *Immunity*, vol. 17, no. 2, pp. 201–210, 2002.

[56] D. R. Fooksman, S. Vardhana, G. Vasiliver-Shamis et al., "Functional anatomy of T cell activation and synapse formation," *Annual Review of Immunology*, vol. 28, pp. 79–105, 2010.

[57] M. Gunzer, C. Weishaupt, A. Hillmer et al., "A spectrum of biophysical interaction modes between T cells and different antigen-presenting cells during priming in 3-D collagen and in vivo," *Blood*, vol. 104, no. 9, pp. 2801–2809, 2004.

[58] S. Hong, T. Chung, and D. Kim, "SH3 domain-peptide binding energy calculations based on structural ensemble and multiple peptide templates," *PLoS ONE*, vol. 5, no. 9, Article ID e12654, pp. 1–8, 2010.

[59] H. Zola, "Medical applications of leukocyte surface molecules—the CD molecules," *Molecular Medicine*, vol. 12, no. 11-12, pp. 312–316, 2006.

[60] M. Srinivasan and R. W. Roeske, "Immunomodulatory peptides from IgSF proteins: a review," *Current Protein and Peptide Science*, vol. 6, no. 2, pp. 185–196, 2005.

[61] J. P. C. Da Cunha, P. A. F. Galante, J. E. De Souza et al., "Bioinformatics construction of the human cell surfaceome," *Proceedings of the National Academy of Sciences of the United States of America*, vol. 106, no. 39, pp. 16752–16757, 2009.

[62] P. Anton Van Der Merwe, "Modeling costimulation," *Nature Immunology*, vol. 1, no. 3, pp. 194–195, 2000.

[63] M. Srinivasan, I. E. Gienapp, S. S. Stuckman et al., "Suppression of experimental autoimmune encephalomyelitis using peptide mimics of CD28," *Journal of Immunology*, vol. 169, no. 4, pp. 2180–2188, 2002.

[64] M. Srinivasan, R. M. Wardrop, I. E. Gienapp, S. S. Stuckman, C. C. Whitacre, and P. T. P. Kaumaya, "A retro-inverso peptide mimic of CD28 encompassing the MYPPPY motif adopts a polyproline type II helix and inhibits encephalitogenic T cells in vitro," *Journal of Immunology*, vol. 167, no. 1, pp. 578–585, 2001.

[65] P. Loke and J. P. Allison, "Emerging mechanisms of immune regulation: the extended B7 family and regulatory T cells," *Arthritis Research and Therapy*, vol. 6, no. 5, pp. 208–214, 2004.

[66] P. A. Van Der Merwe and S. J. Davis, "Molecular interactions mediating T cell antigen recognition," *Annual Review of Immunology*, vol. 21, pp. 659–684, 2003.

[67] S. Wang, G. Zhu, K. Tamada, L. Chen, and J. Bajorath, "Ligand binding sites of inducible costimulator and high avidity mutants with improved function," *Journal of Experimental Medicine*, vol. 195, no. 8, pp. 1033–1041, 2002.

[68] C. C. Stamper, Y. Zhang, J. F. Tobin et al., "Crystal structure of the B7-1/CTLA-4 complex that inhibits human immune responses," *Nature*, vol. 410, pp. 608–611, 2001.

[69] M. Srinivasan, D. Lu, R. Eri, D. D. Brand, A. Haque, and J. S. Blum, "CD80 binding polyproline helical peptide inhibits T cell activation," *Journal of Biological Chemistry*, vol. 280, no. 11, pp. 10149–10155, 2005.

[70] R. Srinivasan and G. D. Rose, "A physical basis for protein secondary structure," *Proceedings of the National Academy of Sciences of the United States of America*, vol. 96, no. 25, pp. 14258–14263, 1999.

[71] K. Chattopadhyay, S. Bhatia, A. Fiser, S. C. Almo, and S. G. Nathenson, "Structural basis of inducible costimulator ligand costimulatory function: determination of the cell surface oligomeric state and functional mapping of the receptor binding site of the protein," *Journal of Immunology*, vol. 177, no. 6, pp. 3920–3929, 2006.

[72] S. Ikemizu, R. J. C. Gilbert, J. A. Fennelly et al., "Structure and dimerization of a soluble form of B7-1," *Immunity*, vol. 12, no. 1, pp. 51–60, 2000.

[73] X. Zhang, J. C. D. Schwartz, S. C. Almo, and S. G. Nathenson, "Crystal structure of the receptor-binding domain of human B7-2: insights into organization and signaling," *Proceedings of the National Academy of Sciences of the United States of America*, vol. 100, no. 5, pp. 2586–2591, 2003.

[74] S. Dudhgaonkar, A. Thyagarajan, and D. Sliva, "Suppression of the inflammatory response by triterpenes isolated from the mushroom Ganoderma lucidum," *International Immunopharmacology*, vol. 9, no. 11, pp. 1272–1280, 2009.

[75] R. Eri, K. N. Kodumudi, D. J. Summerlin, and M. Srinivasan, "Suppression of colon inflammation by CD80 blockade: evaluation in two murine models of inflammatory bowel disease," *Inflammatory Bowel Diseases*, vol. 14, no. 4, pp. 458–470, 2008.

[76] M. Srinivasan, R. Eri, S. L. Zunt, D. J. Summerlin, D. D. Brand, and J. S. Blum, "Suppression of immune responses in collagen-induced arthritis by a rationally designed CD80-binding peptide agent," *Arthritis and Rheumatism*, vol. 56, no. 2, pp. 498–508, 2007.

[77] M. Radulovic and J. Godovac-Zimmermann, "Proteomic approaches to understanding the role of the cytoskeleton in host-defense mechanisms," *Expert Review of Proteomics*, vol. 8, no. 1, pp. 117–126, 2011.

[78] A. Disanza, A. Steffen, M. Hertzog, E. Frittoli, K. Rottner, and G. Scita, "Actin polymerization machinery: the finish line of signaling networks, the starting point of cellular movement," *Cellular and Molecular Life Sciences*, vol. 62, no. 9, pp. 955–970, 2005.

[79] D. C. Wickramarachchi, A. N. Theofilopoulos, and D. H. Kono, "Immune pathology associated with altered actin cytoskeleton regulation," *Autoimmunity*, vol. 43, no. 1, pp. 64–75, 2010.

[80] M. P. Blundell, A. Worth, G. Bouma, and A. J. Thrasher, "The Wiskott-Aldrich syndrome: the actin cytoskeleton and immune cell function," *Disease Markers*, vol. 29, no. 3-4, pp. 157–175, 2010.

[81] Y. Sasahara, R. Rachid, M. J. Byrne et al., "Mechanism of recruitment of WASP to the immunological synapse and of its activation following TCR ligation," *Molecular Cell*, vol. 10, no. 6, pp. 1269–1281, 2002.

[82] A. J. Thrasher and S. O. Burns, "WASP: a key immunological multitasker," *Nature Reviews Immunology*, vol. 10, no. 3, pp. 182–192, 2010.

[83] F. C. Peterson, Q. Deng, M. Zettl et al., "Multiple WASP-interacting protein recognition motifs are required for a functional interaction with N-WASP," *Journal of Biological Chemistry*, vol. 282, no. 11, pp. 8446–8453, 2007.

[84] N. Ramesh, I. M. Antón, J. H. Hartwig, and R. S. Geha, "WIP, a protein associated with Wiskott-Aldrich syndrome protein, induces actin polymerization and redistribution in lymphoid cells," *Proceedings of the National Academy of Sciences of the United States of America*, vol. 94, no. 26, pp. 14671–14676, 1997.

[85] S. L. Bras, M. Massaad, S. Koduru et al., "WIP is critical for T cell responsiveness to IL-2," *Proceedings of the National Academy of Sciences of the United States of America*, vol. 106, no. 18, pp. 7519–7524, 2009.

[86] N. Ramesh and R. Geha, "Recent advances in the biology of WASP and WIP," *Immunologic Research*, vol. 44, no. 1–3, pp. 99–111, 2009.

[87] J. L. Cannon, C. M. Labno, G. Bosco et al., "WASP recruitment to the T cell: APC contact site occurs independently of Cdc42 activation," *Immunity*, vol. 15, no. 2, pp. 249–259, 2001.

[88] X. Dong, G. Patino-Lopez, F. Candotti, and S. Shaw, "Structure-function analysis of the WIP role in T cell receptor-stimulated NFAT activation: evidence that WIP-WASP dissociation is not required and that the WIP NH2 terminus is inhibitory," *Journal of Biological Chemistry*, vol. 282, no. 41, pp. 30303–30310, 2007.

[89] C. Combet, M. Jambon, G. Deléage, and C. Geourjon, "Geno3D: automatic comparative molecular modelling of protein," *Bioinformatics*, vol. 18, no. 1, pp. 213–214, 2002.

[90] F. Kiefer, K. Arnold, M. Künzli, L. Bordoli, and T. Schwede, "The SWISS-MODEL repository and associated resources," *Nucleic Acids Research*, vol. 37, no. 1, pp. D387–D392, 2009.

[91] N. Petrovsky, "Immunome research—five years on," *Immunome Research*, vol. 7, no. 1, pp. 1–4, 2011.

[92] B. F. Volkman, K. E. Prehoda, J. A. Scott, F. C. Peterson, and W. A. Lim, "Structure of the N-WASP EVH1 domain-WIP complex: insight into the molecular basis of Wiskott-Aldrich Syndrome," *Cell*, vol. 111, no. 4, pp. 565–576, 2002.

[93] M. J. Massaad, N. Ramesh, S. Le Bras et al., "A peptide derived from the Wiskott-Aldrich syndrome (WAS) protein-interacting protein (WIP) restores WAS protein level and actin cytoskeleton reorganization in lymphocytes from patients with WAS mutations that disrupt WIP binding," *Journal of Allergy and Clinical Immunology*, vol. 127, no. 4, pp. 998–1005.e2, 2011.

[94] R. Tjian and T. Maniatis, "Transcriptional activation: a complex puzzle with few easy pieces," *Cell*, vol. 77, no. 1, pp. 5–8, 1994.

[95] S. Triezenberg, "Structure and function of transcriptional activation domains," *Current Opinion in Genetics and Development*, vol. 5, no. 2, pp. 190–196, 1995.

[96] H. P. Gerber, K. Seipel, O. Georgiev et al., "Transcriptional activation modulated by homopolymeric glutamine and proline stretches," *Science*, vol. 263, no. 5148, pp. 808–811, 1994.

[97] H. M. Chan and N. B. La Thangue, "p300/CBP proteins: HATs for transcriptional bridges and scaffolds," *Journal of Cell Science*, vol. 114, no. 13, pp. 2363–2373, 2001.

[98] D. Dornan, H. Shimizu, L. Burch, A. J. Smith, and T. R. Hupp, "The proline repeat domain of p53 binds directly to the transcriptional coactivator p300 and allosterically controls DNA-dependent acetylation of p53," *Molecular and Cellular Biology*, vol. 23, no. 23, pp. 8846–8861, 2003.

[99] A. C. Joerger and A. R. Fersht, "Structural biology of the tumor suppressor p53," *Annual Review of Biochemistry*, vol. 77, pp. 557–582, 2008.

[100] A. C. Joerger and A. R. Fersht, "The tumor suppressor p53: from structures to drug discovery," *Cold Spring Harbor Perspectives in Biology*, vol. 2, no. 6, p. a000919, 2010.

[101] N. Dumaz, D. M. Milne, L. J. Jardine, and D. W. Meek, "Critical roles for the serine 20, but not the serine 15, phosphorylation site and for the polyproline domain in regulating p53 turnover," *Biochemical Journal*, vol. 359, no. 2, pp. 459–464, 2001.

[102] P. A. Baeuerle and T. Henkel, "Function and activation of NF-κB in the immune system," *Annual Review of Immunology*, vol. 12, pp. 141–179, 1994.

[103] M. E. Gerritsen, A. J. Williams, A. S. Neish, S. Moore, Y. Shi, and T. Collins, "CREB-binding protein/p300 are transcriptional coactivators of p65," *Proceedings of the National Academy of Sciences of the United States of America*, vol. 94, no. 7, pp. 2927–2932, 1997.

[104] M. A. Calzado, S. Bacher, and M. L. Schmitz, "NF-κB inhibitors for the treatment of inflammatory diseases and cancer," *Current Medicinal Chemistry*, vol. 14, no. 3, pp. 367–376, 2007.

[105] L. I. McKay and J. A. Cidlowski, "CBP (CREB binding protein) integrates NF-κB (nuclear factor-κB) and glucocorticoid receptor physical interactions and antagonism," *Molecular Endocrinology*, vol. 14, no. 8, pp. 1222–1234, 2000.

[106] N. C. Nicolaides, Z. Galata, T. Kino, G. P. Chrousos, and E. Charmandari, "The human glucocorticoid receptor: molecular basis of biologic function," *Steroids*, vol. 75, no. 1, pp. 1–12, 2010.

[107] F. D'Adamio, O. Zollo, R. Moraca et al., "A new dexamethasone-induced gene of the leucine zipper family protects T lymphocytes from TCR/CD3-activated cell death," *Immunity*, vol. 7, no. 6, pp. 803–812, 1997.

[108] E. Ayroldi, O. Zollo, A. Bastianelli et al., "GILZ mediates the antiproliferative activity of glucocorticoids by negative regulation of Ras signaling," *Journal of Clinical Investigation*, vol. 117, no. 6, pp. 1605–1615, 2007.

[109] E. Ayroldi and C. Riccardi, "Glucocorticoid-induced leucine zipper (GILZ): a new important mediator of glucocorticoid action," *FASEB Journal*, vol. 23, no. 11, pp. 3649–3658, 2009.

[110] C. Riccardi, "GILZ (glucocorticoid-induced leucine zipper), a mediator of the anti-inflammatory and immunosuppressive activity of glucocorticoids," *Annali di Igiene*, vol. 22, no. 1, supplement 1, pp. 53–59, 2010.

[111] L. Cannarile, S. Cuzzocrea, L. Santucci et al., "Glucocorticoid-induced leucine zipper is protective in Th1-mediated models of colitis," *Gastroenterology*, vol. 136, no. 2, pp. 530–541, 2009.

[112] E. Ayroldi, G. Migliorati, S. Bruscoli et al., "Modulation of T-cell activation by the glucocorticoid-induced leucine zipper factor via inhibition of nuclear factor κB," *Blood*, vol. 98, no. 3, pp. 743–753, 2001.

[113] B. Di Marco, M. Massetti, S. Bruscoli et al., "Glucocorticoid-induced leucine zipper (GILZ)/NF-κB interaction: role of GILZ homo-dimerization and C-terminal domain," *Nucleic Acids Research*, vol. 35, no. 2, pp. 517–528, 2007.

[114] R. Bhattacharyya and P. Chakrabarti, "Stereospecific interactions of proline residues in protein structures and complexes," *Journal of Molecular Biology*, vol. 331, no. 4, pp. 925–940, 2003.

[115] M. L. Schmitz, M. A. Dos Santos Silva, H. Altmann, M. Czisch, T. A. Holak, and P. A. Baeuerle, "Structural and functional analysis of the NF-κB p65 C terminus. An acidic and modular transactivation domain with the potential to adopt an α-helical conformation," *Journal of Biological Chemistry*, vol. 269, no. 41, pp. 25613–25620, 1994.

[116] A. K. Dunker, J. D. Lawson, C. J. Brown et al., "Intrinsically disordered protein," *Journal of Molecular Graphics and Modelling*, vol. 19, no. 1, pp. 26–59, 2001.

[117] M. Srinivasan and S. Janardhanam, "Novel p65 binding glucocorticoid-induced leucine zipper peptide suppresses experimental autoimmune encephalomyelitis," *Journal of Biological Chemistry*, vol. 286, no. 52, pp. 44799–44810, 2011.

[118] J. Oberoi, M. W. Richards, S. Crumpler, N. Brown, J. Blagg, and R. Bayliss, "Structural basis of poly(ADP-ribose) recognition by the multizinc binding domain of Checkpoint with Forkhead-associated and RING domains (CHFR)," *Journal of Biological Chemistry*, vol. 285, no. 50, pp. 39348–39358, 2010.

[119] M. L. Privalsky, "The role of corepressors in transcriptional regulation by nuclear hormone receptors," *Annual Review of Physiology*, vol. 66, pp. 315–360, 2004.

[120] K. De Bosscher, W. Vanden Berghe, and G. Haegeman, "Cross-talk between nuclear receptors and nuclear factor κB," *Oncogene*, vol. 25, no. 51, pp. 6868–6886, 2006.

[121] J. Li, J. Wang, J. Wang et al., "Both corepressor proteins SMRT and N-CoR exist in large protein complexes containing HDAC3," *EMBO Journal*, vol. 19, no. 16, pp. 4342–4350, 2000.

[122] J. E. Hoberg, A. E. Popko, C. S. Ramsey, and M. W. Mayo, "IκB kinase α-mediated derepression of SMRT potentiates acetylation of RelA/p65 by p300," *Molecular and Cellular Biology*, vol. 26, no. 2, pp. 457–471, 2006.

[123] S. K. Lee, J. H. Kim, Y. C. Lee, J. Cheong, and J. W. Lee, "Silencing mediator of retinoic acid and thyroid hormone receptors, as a novel transcriptional corepressor molecule of activating protein-1, nuclear factor-κB, and serum response factor," *Journal of Biological Chemistry*, vol. 275, no. 17, pp. 12470–12474, 2000.

[124] S. Ansieau and A. Leutz, "The conserved Mynd domain of BS69 binds cellular and oncoviral proteins through a common PXLXP motif," *Journal of Biological Chemistry*, vol. 277, no. 7, pp. 4906–4910, 2002.

[125] P. D. Gottlieb, S. A. Pierce, R. J. Sims et al., "Bop encodes a muscle-restricted protein containing MYND and SET domains and is essential for cardiac differentiation and morphogenesis," *Nature Genetics*, vol. 31, no. 1, pp. 25–32, 2002.

[126] J. Lausen, S. Cho, S. Liu, and M. H. Werner, "The nuclear receptor co-repressor (N-CoR) utilizes repression domains I and III for interaction and co-repression with ETO," *Journal of Biological Chemistry*, vol. 279, no. 47, pp. 49281–49288, 2004.

[127] J. M. Matthews, M. Bhati, E. Lehtomaki, R. E. Mansfield, L. Cubeddu, and J. P. Mackay, "It takes two to tango: the structure and function of LIM, RING, PHD and MYND domains," *Current Pharmaceutical Design*, vol. 15, no. 31, pp. 3681–3696, 2009.

[128] H. P. Liu, P. J. Chung, C. L. Liang, and Y. S. Chang, "The MYND domain-containing protein BRAM1 inhibits lymphotoxin beta receptor-mediated signaling through affecting receptor oligomerization," *Cellular Signalling*, vol. 23, no. 1, pp. 80–88, 2011.

[129] B. A. Hug and M. A. Lazar, "ETO interacting proteins," *Oncogene*, vol. 23, no. 24, pp. 4270–4274, 2004.

[130] Y. Liu, W. Chen, J. Gaudet et al., "Structural basis for recognition of SMRT/N-CoR by the MYND domain and its contribution to AML1/ETO's activity," *Cancer Cell*, vol. 11, no. 6, pp. 483–497, 2007.

[131] M. Yan, S. A. Burel, L. F. Peterson et al., "Deletion of an AML1-ETO C-terminal NcoR/SMRT-interacting region strongly induces leukemia development," *Proceedings of the National Academy of Sciences of the United States of America*, vol. 101, no. 49, pp. 17186–17191, 2004.

[132] A. K. Dunker and V. N. Uversky, "Drugs for 'protein clouds': targeting intrinsically disordered transcription factors," *Current Opinion in Pharmacology*, vol. 10, no. 6, pp. 782–788, 2010.

[133] L. L. Conte, C. Chothia, and J. Janin, "The atomic structure of protein-protein recognition sites," *Journal of Molecular Biology*, vol. 285, no. 5, pp. 2177–2198, 1999.

[134] B. Groner, "Peptides as drugs, discovery and development," in *Peptides as Drugs, Discovery and Development*, B. Groner, Ed., pp. 1–8, Wiley-VCH, Weinheim, Germany, 2009.

Permissions

The contributors of this book come from diverse backgrounds, making this book a truly international effort. This book will bring forth new frontiers with its revolutionizing research information and detailed analysis of the nascent developments around the world.

We would like to thank all the contributing authors for lending their expertise to make the book truly unique. They have played a crucial role in the development of this book. Without their invaluable contributions this book wouldn't have been possible. They have made vital efforts to compile up to date information on the varied aspects of this subject to make this book a valuable addition to the collection of many professionals and students.

This book was conceptualized with the vision of imparting up-to-date information and advanced data in this field. To ensure the same, a matchless editorial board was set up. Every individual on the board went through rigorous rounds of assessment to prove their worth. After which they invested a large part of their time researching and compiling the most relevant data for our readers. Conferences and sessions were held from time to time between the editorial board and the contributing authors to present the data in the most comprehensible form. The editorial team has worked tirelessly to provide valuable and valid information to help people across the globe.

Every chapter published in this book has been scrutinized by our experts. Their significance has been extensively debated. The topics covered herein carry significant findings which will fuel the growth of the discipline. They may even be implemented as practical applications or may be referred to as a beginning point for another development. Chapters in this book were first published by Hindawi Publishing Corporation; hereby published with permission under the Creative Commons Attribution License or equivalent.

The editorial board has been involved in producing this book since its inception. They have spent rigorous hours researching and exploring the diverse topics which have resulted in the successful publishing of this book. They have passed on their knowledge of decades through this book. To expedite this challenging task, the publisher supported the team at every step. A small team of assistant editors was also appointed to further simplify the editing procedure and attain best results for the readers.

Our editorial team has been hand-picked from every corner of the world. Their multi-ethnicity adds dynamic inputs to the discussions which result in innovative outcomes. These outcomes are then further discussed with the researchers and contributors who give their valuable feedback and opinion regarding the same. The feedback is then collaborated with the researches and they are edited in a comprehensive manner to aid the understanding of the subject.

Apart from the editorial board, the designing team has also invested a significant amount of their time in understanding the subject and creating the most relevant covers. They scrutinized every image to scout for the most suitable representation of the subject and create an appropriate cover for the book.

The publishing team has been involved in this book since its early stages. They were actively engaged in every process, be it collecting the data, connecting with the contributors or procuring relevant information. The team has been an ardent support to the editorial, designing and production team. Their endless efforts to recruit the best for this project, has resulted in the accomplishment of this book. They are a veteran in the field of academics and their pool of knowledge is as vast as their experience in printing. Their expertise and guidance has proved useful at every step. Their uncompromising quality standards have made this book an exceptional effort. Their encouragement from time to time has been an inspiration for everyone.

The publisher and the editorial board hope that this book will prove to be a valuable piece of knowledge for researchers, students, practitioners and scholars across the globe.

List of Contributors

Ceslava Kairane, Riina Mahlapuu, Kersti Ehrlich, Kalle Kilk, Mihkel Zilmer and Ursel Soomets
The Centre of Excellence of Translational Medicine, Department of Biochemistry, Faculty of Medicine, University of Tartu, Ravila Street 19, 50411 Tartu, Estonia

P. Anantha Reddy, Sean T. Jones, Anita H. Lewin and F. Ivy Carroll
Center for Organic and Medicinal Chemistry, Discovery Sciences Research Triangle Institute, Research Triangle Park, NC 27709-2194, USA

Ekaterina F. Kolesanova, Maxim A. Sanzhakov and Oleg N. Kharybin
Orekhovich Institute of Biomedical Chemistry, Russian Academy of Medical Sciences, 10 Pogodinskaya Ulica, Moscow 119121, Russia

Ramon Bernal-Pedraza
Departamento de Neurobiologıa del Desarrollo y Neurofisiologıa, Instituto de Neurobiologıa, Universidad Nacional Autonoma de Mexico, UNAM-Campus Juriquilla, 76230 Juriquilla, QRO, Mexico
Departamento de Farmacobiolog´ıa, Cinvestav-IPN, Mexico City, DF, Mexico

Fernando Pena-Ortega
Departamento de Neurobiologıa del Desarrollo y Neurofisiologıa, Instituto de Neurobiologıa, Universidad Nacional Autonoma de Mexico, UNAM-Campus Juriquilla, 76230 Juriquilla, QRO, Mexico

M. Khazaei and Z. Tahergorabi
Department of Physiology, Isfahan University of Medical Sciences, Isfahan 81743638, Iran

Mohammed Munaf, Pierpaolo Pellicori, Victoria Allgar and Kenneth Wong
Department of Cardiovascular and Respiratory Studies, Hull and East Yorkshire Medical Research and Teaching Centre, Daisy Building, Castle Hill Hospital, Castle Road, Kingston upon Hull HU16 5JQ, UK

Chris Tikellis and M. C. Thomas
Division of Diabetic Complications, Baker IDI Heart and Diabetes Institute, P.O. Box 6492 Melbourne, VIC 8008, Australia

Yosuke Suzuki, Hiroki Itoh, Kohei Amada, Ryota Yamamura, Yuhki Sato and Masaharu Takeyama
Department of Clinical Pharmacy, Oita University Hospital, Hasama-machi, Oita 879-5593, Japan

Diana R. Engineer
Division of Diabetes, Endocrinology and Metabolism, Michael E DeBakey Veterans Affairs Medical Center, Houston, TX 77030, USA
Baylor College of Medicine, 2002 Holcombe Boulevored, Building 109, Room 210, Houston, TX 77030, USA
Division of Diabetes, Department of Medicine, Endocrinology and Metabolism, St Luke's Episcopal Hospital, Houston, TX 77030, USA

Jose M. Garcia
Division of Diabetes, Endocrinology and Metabolism, Michael E DeBakey Veterans Affairs Medical Center, Houston, TX 77030, USA
Baylor College of Medicine, 2002 Holcombe Boulevored, Building 109, Room 210, Houston, TX 77030, USA
Huffington Center of Aging, Baylor College of Medicine, Houston, TX 77030, USA

Masayuki Komatsu, Sunita Ghimire Gautam and Koichi Nishigaki
Department of Functional Materials Science, Graduate School of Science and Engineering, Saitama University, 255 Shimo-okubo, Sakura-ku, Saitama-shi, Saitama 338-8570, Japan
Rational Evolutionary Design of Advanced Biomolecules, Saitama (REDS), Saitama Small Enterprise Promotion Corporation, No. 552, Saitama Industrial Technology Center, 3-12-18 Kami-Aoki, Kawaguchi, Saitama 333-0844, Japan

Sunita Ghimire Gautam and Koichi Nishigaki
Department of Functional Materials Science, Graduate School of Science and Engineering, Saitama University, 255 Shimo-okubo, Sakura-ku, Saitama-shi, Saitama 338-8570, Japan

K. M. Bhaskara Reddy, Dokka Mallikharjunasarma, Kamana Bulliraju, Vanjivaka Sreelatha and Kuppanna Ananda
Chemical Research Division, Mylan Laboratories Ltd., Anrich Industrial Estate, Bollaram, Hyderabad 502325, India

Y. Bharathi Kumari
Department of Chemistry, College of Engineering, Jawaharlal Nehru Technological University Hyderabad, Kukatpally, Hyderabad 500085, India

Armando I. Gutiérrez-Lerma
Departamento de Neurobiologia del Desarrollo y Neurofisiologia, Instituto de Neurobiologia, UNAM, Boulevard Juriquilla 3001, 16230 Queretaro, Mexico
Departamento de Farmacobiologia, Cinvestav-IPN, Calzada de los Tenorios 235, Col. Granjas Coapa, 14330 Mexico, DF, Mexico

Benito Ordaz and Fernando Peña-Ortega
Departamento de Neurobiologia del Desarrollo y Neurofisiologia, Instituto de Neurobiologia, UNAM, Boulevard Juriquilla 3001, 16230 Queretaro, Mexico

Coralie Sclavons, Carmen Burtea, Sophie Laurent and Luce Vander Elst
Department of General, Organic and Biomedical Chemistry, NMR and Molecular Imaging Laboratory, University of Mons, Mendeleiev Building, 19 Avenue Maistriau, 7000 Mons, Belgium

Robert N. Muller
Department of General, Organic and Biomedical Chemistry, NMR and Molecular Imaging Laboratory, University of Mons, Mendeleiev Building, 19 Avenue Maistriau, 7000 Mons, Belgium
Center for Microscopy and Molecular Imaging (CMMI), 8 Rue Adrienne Bolland, 6041 Gosselies, Belgium

Sébastien Boutry
Center for Microscopy and Molecular Imaging (CMMI), 8 Rue Adrienne Bolland, 6041 Gosselies, Belgium

Shawn Keogan, Shendra Passic and Fred C. Krebs
Department of Microbiology and Immunology, Center for Molecular Virology and Translational Neuroscience and Center for Sexually Transmitted Disease, Institute for Molecular Medicine and Infectious Disease, Drexel University College of Medicine, Philadelphia, PA 19102, USA

Mau Sinha, Sanket Kaushik, Punit Kaur, Sujata Sharma and Tej P. Singh
Department of Biophysics, All India Institute of Medical Sciences, Ansari Nagar, New Delhi 110029, India

Laura H. Vahatalo
Department of Pharmacology, Drug Development and Therapeutics and Turku Center for Disease Modeling, University of Turku, Itainen Pitkakatu 4B, 20520 Turku, Finland
Fin Pharma Doctorate Program Drug Discovery Section, University of Turku, Itainen Pitkakatu 4B, 20520 Turku, Finland

Suvi T. Ruohonen
Department of Pharmacology, Drug Development and Therapeutics and Turku Center for Disease Modeling, University of Turku, Itainen Pitkakatu 4B, 20520 Turku, Finland

Eriika Savontaus
Department of Pharmacology, Drug Development and Therapeutics and Turku Center for Disease Modeling, University of Turku, Itainen Pitkakatu 4B, 20520 Turku, Finland
Unit of Clinical Pharmacology, Turku University Hospital, Itainen Pitkakatu 4B, 20520 Turku, Finland

Brian Hall, Carley Squires and Keith K. Parker
Department of Biomedical and Pharmaceutical Sciences (MPH I02), Center for Structural and Functional Neuroscience, Skaggs School of Pharmacy, The University of Montana, 32 Campus Drive No. 1552, Missoula, MT 59812-1552, USA

Dolores Javier Sanchez-Gonzalez
Subseccion de Biologıa Celular y Tisular, Escuela Medico Militar, Universidad del Ejercito y Fuerza Aerea, 11200 Mexico City, MEX, Mexico
Sociedad Internacional para la Terapia Celular con Celulas Madre, Medicina Regenerativa y Antienvejecimiento S.C. (SITECEM), 53840 Naucalpan, MEX, Mexico

Enrique Mendez-Bolaina
Facultad de Ciencias Quımicas, Universidad Veracruzana, 94340 Orizaba, VER, Mexico
Centro de Investigaciones Biomedicas-Doctorado en Ciencias Biom´edicas, Universidad Veracruzana, 91000 Xalapa, VER, Mexico

Nayeli Isabel Trejo-Bahena
Sociedad Internacional para la Terapia Celular con Celulas Madre, Medicina Regenerativa y Antienvejecimiento S.C. (SITECEM), 53840 Naucalpan, MEX, Mexico
Area de Medicina Fısica y Rehabilitacion, Hospital Central Militar, 11200 Mexico City, MEX, Mexico

Maria Cristina Vianna Braga
CAT/CEPID, Instituto Butantan, Avenida Vital Brasil 1500, 05503-900 Sao Paulo, SP, Brazil
Ministerio da Ciencia, Tecnologia e Inovacao, Esplanada dos Ministerios, Bloco E, 70067-900 Brasılia, DF, Brazil

Arthur Andrade Nery and Henning Ulrich
Departamento de Bioquımica, Instituto de Quımica, Universidade de Sao Paulo, Av. Lineu Prestes 748, 05508-900 Sao Paulo, SP, Brazil

Katsuhiro Konno
Institute of Natural Medicine, University of Toyama, 2630 Sugitani, Toyama 930-0194, Japan

Juliana Mozer Sciani and Daniel Carvalho Pimenta
Laboratorio de Bioquımica e Biofısica, Instituto Butantan, Avenida Vital Brasil 1500, 05503-900 Sao Paulo, SP, Brazil

Thomas Lampert
Department of Biological Sciences, State University of New York at Buffalo, 109 Cooke Hall, Buffalo, NY 14260, USA

Cheryl Nugent, John Weston, Nathanael Braun and Heather Kuruvilla
Department of Science and Mathematics, Cedarville University, 251 North Main Street, Cedarville, OH 45314, USA

C. Polanco
Centro de Investigaciones Quımicas, Universidad Autonoma del Estado de Morelos, Avenida Universidad 1001, 62209 Cuernavaca, MOR, Mexico
Subdireccion de Epidemiologia Hospitalaria y Control de Calidad de la Atencion Medica, Instituto Nacional de Ciencias Medicas y Nutricion Salvador Zubiran, Vasco de Quiroga 15, Piso 4, Colonia Seccion XVI, 14000 Mexico City, DF, Mexico

J. L. Samaniego
Departamento de Matematicas, Facultad de Ciencias, Universidad Nacional Autonoma de Mexico, Ciudad Universitaria, 04510 Mexico City, DF, Mexico

T. Buhse
Centro de Investigaciones Quımicas, Universidad Autonoma del Estado de Morelos, Avenida Universidad 1001, 62209 Cuernavaca, MOR, Mexico

J. A. Castanon-Gonzalez
Subdireccion de Epidemiologia Hospitalaria y Control de Calidad de la Atencion Medica, Instituto Nacional de Ciencias Medicas y Nutricion Salvador Zubiran, Vasco de Quiroga 15, Piso 4, Colonia Seccion XVI, 14000 Mexico City, DF, Mexico

F. G. Mosqueira
Direccion General de Divulgacion de la Ciencia, Universidad Nacional Autonoma de Mexico, Ciudad Universitaria, P.O. Box 70487, 04510 Mexico City, DF, Mexico

A. Negron-Mendoza and S. Ramos-Bernal
Instituto de Ciencias Nucleares, Universidad Nacional Autonoma de Mexico, Ciudad Universitaria, P.O. Box 70543, 04510 Mexico City, DF, Mexico

Marina de Oliveira Paro, Cyntia Silva Ferreira, Fernanda Silva Vieira, Marcos Aurélio de Santana, William Castro-Borges and Milton Hércules Guerra de Andrade
Departamento de Ciencias Biologicas/DECBI, Instituto de Ciencias Exatas e Biologicas/ICEB, Universidade Federal de Ouro Preto, Nucleo de Pesquisas em Ciencias Biologicas/NUPEB, Campus Universitario Morro do Cruzeiro, 35400-000Ouro Preto, MG, Brazil

Maria Sueli Silva Namen-Lopes
Fundacao Centro de Hematologia e Hemoterapia, Hemominas, Servico de Pesquisa, Alameda Ezequiel Dias, 321, Bairro Santa Efigenia, 30130-110 Belo Horizonte, MG, Brazil

Sophie Yvette Leclercq
Fundacao Ezequiel Dias, FUNED, Rua Conde Pereira Carneiro, 80, Bairro Gameleira, 30510-010 Belo Horizonte, MG, Brazil

Cibele Velloso-Rodrigues
Departamento Basico, Instituto de Ciencias Biologicas, Universidade Federal de Juiz de Fora, Campus de Governador Valadares, Area da Saude, Avenida Doutor Raimundo Monteiro de Resende, 330, Centro, 35010-177 Governador Valadares, MG, Brazil

Ronghai He, Weirui Zhao, Wenjuan Qu, Jiewen Zhao and Lin Luo
School of Food and Biological Engineering, Jiangsu University, 301 Xuefu Road, Zhenjiang, Jiangsu 212013, China

Wenxue Zhu
School of Food and Biological Engineering, Henan University of Science and Technology, 48 Xiyuan Road, Luoyang, Henan 471003, China

Haile Ma
School of Food and Biological Engineering, Jiangsu University, 301 Xuefu Road, Zhenjiang, Jiangsu 212013, China
School of Food and Biological Engineering, Henan University of Science and Technology, 48 Xiyuan Road, Luoyang, Henan 471003, China

M. Rajeswari Prabha and B. Ramachandramurty
Department of Biochemistry, PSG College of Arts & Science, Civil Aerodrome Post, Coimbatore, Tamil Nadu 641014, India

Mythily Srinivasan
Department of Oral Pathology, Medicine and Radiology, Indiana University School of Dentistry, Indiana University Purdue University at Indianapolis 1121 West Michigan Street, DS290, Indianapolis, IN 46268, USA

A. Keith Dunker
Department of Biochemistry and Molecular Biology and School of Informatics, Indiana University School of Medicine, Indiana University Purdue University at Indianapolis, Indianapolis, IN, USA

www.ingramcontent.com/pod-product-compliance
Lightning Source LLC
Chambersburg PA
CBHW080633200326
41458CB00013B/4611